Introductic [illegible] ural Mechanics

P. S. Smith

School of Engineering
University of East London

palgrave
macmillan

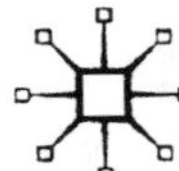

First published 2001 by
PALGRAVE MACMILLAN

Palgrave Macmillan in the UK is an imprint of Macmillan Publishers Limited, registered in England, company number 785998, of Houndmills, Basingstoke, Hampshire RG21 6XS.

Palgrave Macmillan in the US is a division of St Martin's Press LLC, 175 Fifth Avenue, New York, NY 10010.

Palgrave Macmillan is the global academic imprint of the above companies and has companies and representatives throughout the world.

Palgrave® and Macmillan® are registered trademarks in the United States, the United Kingdom, Europe and other countries.

ISBN - 13: 978-0-3396-255-8 paperback
ISBN - 10: 0-333-96255-9 hardback

This book is printed on paper suitable for recycling and made from fully managed and sustained forest sources. Logging, pulping and manufacturing processes are expected to conform to the environmental regulations of the country of origin.

A catalogue record for this book is available from the British Library.

A catalog record for this book is available from the Library of Congress.

Printed in Great Britain by Cpod, Trowbridge, Wiltshire

Contents

Preface

In writing this book I was mindful of the comments of both students and practitioners gathered over more than twenty years in civil engineering, latterly in education. Most centred on the fact that, having learned the background and basic theory, you are left trying to find suitable situations and examples of application of the acquired knowledge. Tracking a problem through often helps to gel the theory, and with practice can focus the mind on what is trying to be achieved.

In this text you will find both theory and worked examples side-by-side. The examples start from a fundamental level and progress to more difficult, intricate and taxing problems. Early examples assume little or no experience in the subject matter but soon propel you forward to solving complex, practical-based structural systems. The emphasis is on problem-based learning, and the procedures adopted are reinforced by their application to a plentiful supply of worked examples.

You are invited to reinforce you understanding of the structural theories and applications, or to use the examples as a reference to test your ability to apply your knowledge to the various problems presented here. Each chapter commences by providing details of the basic theory and procedures to be adopted in the solutions, supported by worked examples to demonstrate the application, and concluded with further problems to test your understanding.

To both aspiring and practising engineers this book will provide a powerful learning or revision text which will enable you to test and develop your understanding of structural analysis.

1. Basic Concepts and Load Path

In this chapter we will learn about:
- ***Newton's Laws of Motion***
- ***Equilibrium***
- ***Force vectors***
- ***Moments***
- ***Loading and member types***
- ***Load transfer and load path***

1.1 Introduction

The analysis of structures considered here will be based on a number of fundamental concepts which follow from simple Newtonian mechanics; it is necessary that we first review Newton's Laws of Motion. The word 'Laws' is often replaced with 'Axioms', as they cannot be proved in the normal experimental sense but are self-evident truths which are believed to be correct because all results obtained assuming them to be true agree with experimental observations.

In 1687 Sir Isaac Newton published a work that clearly set out the Laws of Mechanics. He proposed the following three laws to govern motion:

Newton's Laws of Motion

Law 1: Every body will continue in a state of rest or uniform motion in a straight line unless acted on by a resultant force.

Law 2: The change in momentum per unit time is proportional to the impressed force, and takes place in the direction of the straight line along the axis in which the force acts.

Law 3: Action and reaction are equal and opposite.

Based on these Laws, we are able to define some basic concepts which will assist us in our analysis of structures.

1.2 Equilibrium

Equilibrium is an unchanging state; it is a state of 'balance'. In the analysis of structures this will be achieved when the total of all applied forces, reactions and moments equate to zero. In this condition, the structure will be in balance and no motion will occur. We will now consider three types of static equilibrium. Figure 1.1 shows an object, in this case a ball, placed on three differently shaped surfaces.

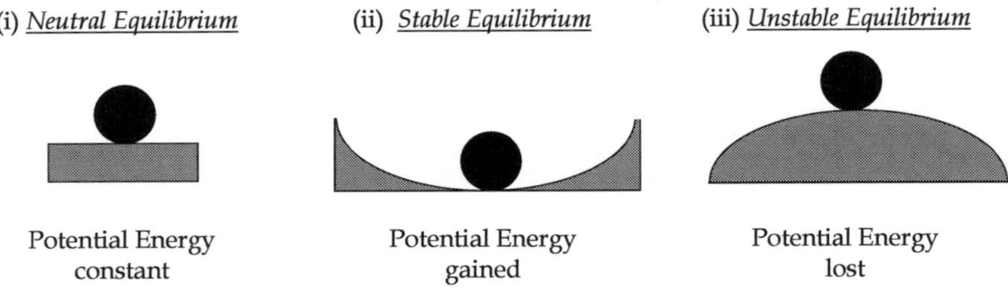

Figure 1.1: Equilibrium

(Note: Potential Energy = change in energy if the body is displaced. Potential energy is the energy the body possesses by virtue of its position above a known datum, in this case the apex of the surface on which it rests.)

In (i) we have a neutral equilibrium position. The ball will remain at rest unless acted on by a force. The potential energy of the system is constant. In system (ii) any movement of the ball will require a gain in potential energy. When released, the ball will try to achieve equilibrium by returning to its original position. In (iii) any motion will cause the ball to move. The shape of the surface on which it rests will further promote this movement. Relative to its original position, potential energy will be lost.

Static equilibrium is achieved by having a zero force resultant. It is perhaps worth noting here that our early analysis of structures will be based wholly on the principles of statics alone. That is, forces will be constant with respect to time. Hence we will consider the 'static analysis' of structures. The study of structures subject to forces that vary with time is known as 'dynamic analysis'.

1.2.1 Force

From Newton's Second Law, and since: momentum = mass × velocity [1.1]

we can derive an expression for force such that:

Force = change in momentum per unit time = mass × acceleration [1.2]

or

$F = m \times a$ where F = force
m = mass

$$\text{a} = \text{acceleration} = \frac{\text{change in velocity}}{\text{time taken}}$$

Acceleration is a vector quantity since it has both direction and magnitude. It is a measure of change of speed (velocity) over time taken. We commonly look at the acceleration rates of cars as the time it takes to go from 0 to 60 miles per hour (mph) or 0 to 96 kilometres per hour (kph), around 5 seconds for a Ferrari! This can be represented differentially as:

$$\text{a} = \frac{\text{d}v}{\text{d}t} \qquad [1.3]$$

where a = acceleration.
dv = change in velocity (a vector quantity, i.e. one that has both direction and magnitude)
dt = change in time.

We also know that velocity is a measure of distance covered over time, e.g. at a velocity (or more familiarly a speed) of 96 kph a car would cover 96 km in one hour or 0.027 m/s. This can also be represented differentially as:

$$a = \frac{\mathrm{d}s/\mathrm{d}t}{\mathrm{d}t} \qquad \text{or} \qquad a = \frac{\mathrm{d}^2 s}{\mathrm{d}t^2} \qquad [1.4]$$

where $\mathrm{d}s$ = change in distance
and $\mathrm{d}t$ = change in time.

Acceleration (and thus velocity) may be rectilinear (in a straight line) or rotational.

In order to determine forces on a structure we first need to consider the differences between *weight* and *mass*.

The *weight* of an object is defined as the force acting on it due to the influence of gravitational attraction, or gravity. Thus, attaching an object to a spring balance and noting the extension will enable us to determine its weight. From our knowledge of physics we know that within the elastic range, extension is proportional to force (Hooke's Law), and most spring balances are calibrated to read weight directly. Consider an object of mass m and weight W. If the object is held at a certain height above the Earth's surface and released it will fall to the ground. Its acceleration, in this case, will be the acceleration due to gravitational force; this is normally denoted by g and taken to be 9.81 m/s^2. Since, from equation [1.2]:

$$\text{Force} = \text{mass} \times \text{acceleration}$$

the force acting on the object, that is its weight, will be:

$$W = m\,g$$

where W = weight
m = mass
g = acceleration due to gravity.

Hence we can derive the force (weight) of various objects by multiplying its mass by its acceleration due to gravity, e.g. an object of mass 1 kg will have a force of 1 kg $\times$ 9.81 m/s^2 = 9.81 kg.m/s^2 or 9.81 Newton. The units of force are the Newton and are normally denoted as N. (Note: 100g is approximately 1 N – the weight of an average apple!)

It is important to note that the acceleration due to gravity is not actually constant over the whole Earth's surface. This is due to the Earth being ellipsoidal in shape. It is also interesting to note that the weight of a body on the Moon will be approximately one-sixth of that on the Earth. This is because the intensity of the gravity on the Moon is one-sixth of that on the Earth. The mass of an object is therefore constant, whereas its weight will vary in magnitude with variations in gravitational intensity g.

Having determined the force exerted by an object, some basic geometric properties may be defined. All forces are vector quantities, which means that they have both magnitude and direction. They may therefore be the subject of vector addition. Consider the situation shown in Figure 1.2.

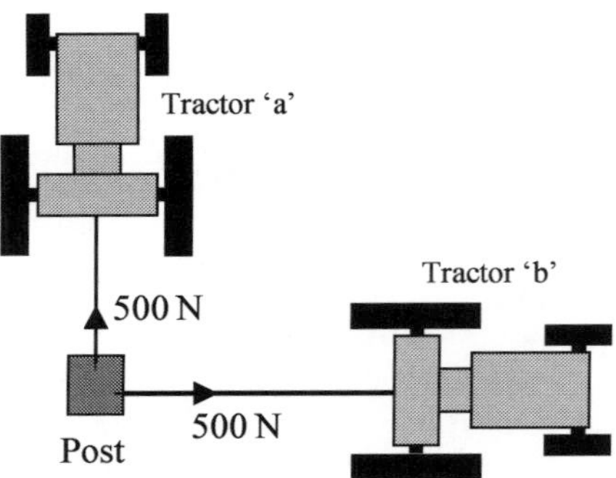

Figure 1.2: Force vectors

Two tractors (seen here in plan) are used to remove (pull out) a post from the ground by exerting horizontal forces as shown. Both tractors are attached by cables to the post and exert forces of 500 N in the directions indicated. In force vector terms we can represent our system as shown in Figure 1.3(i). The forces are represented by straight lines, which can be drawn to scale, denoting both the direction and magnitude of the force. Any suitable scale can be used to construct the diagram, however in most cases such a diagram will not be necessary.

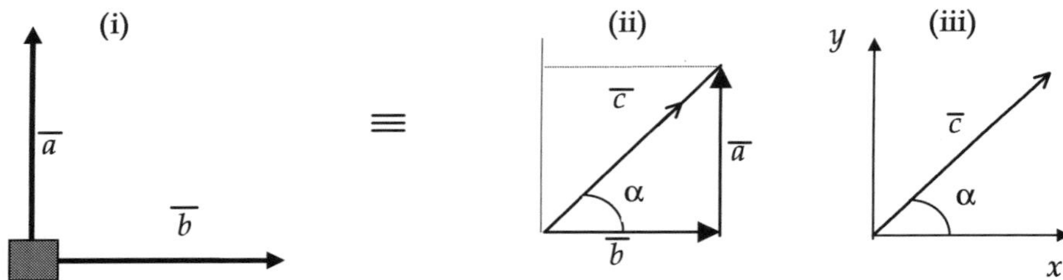

Figure 1.3: Graphical representation of forces

It is possible to achieve the same overall result by replacing the forces exerted by tractors 'a' and 'b' with a single tractor 'c' and therefore replacing the system represented in Figure 1.3(i). In order to determine the direction and magnitude of force required by the single tractor we analyse our initial system. Using elementary geometrical relationships we can determine the magnitude and direction of the vector $\overline{c}$. This can be calculated using the Pythagoras Theorem and standard Sine, Cosine and Tangent relationships. The analysis will be completed using normal Cartesian co-ordinates as shown graphically in Figure 1.3(ii) and (iii). (Note the bar on top of the letters, for example $\overline{a}$, indicates a vector quantity.)

Therefore, to calculate the direction and magnitude of the new vector c we can use simple vector algebra:

Direction: $\quad \text{Tan}\,\alpha = \dfrac{\overline{\text{a}}}{\overline{\text{b}}} = \dfrac{500}{500} = 1.00 = 45^{\circ}$

Magnitude $\quad \overline{c}^2 = \overline{a}^2 + \overline{b}^2 \qquad c = \sqrt{500^2 + 500^2} = 707.1 \text{ N}$

Thus we can replace our original system with that shown in Figure 1.4.

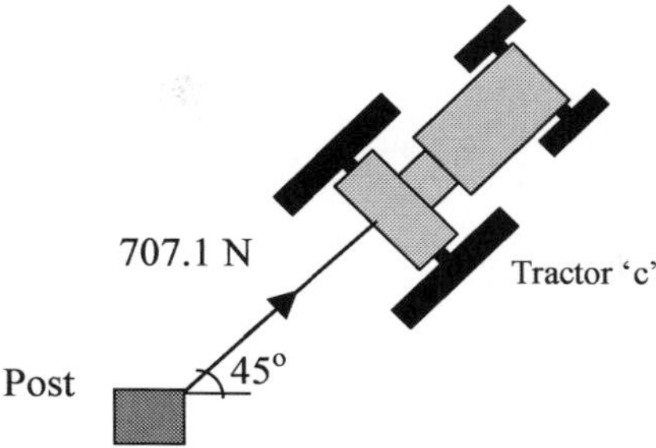

Figure 1.4: Single force system

Similarly we can break down a single force *c* into its mutually orthogonal components, *a* and *b*. From Figure 1.3(iii) we find the vector:

$$\overline{c}\cos\alpha = \overline{b} \text{ in the '}x\text{' direction}$$

$$\overline{c}\sin\alpha = \overline{a} \text{ in the '}y\text{' direction}$$

Force vectors can therefore be resolved into a resultant, or broken down into horizontal and vertical components.

In this book we will only consider two-dimensional (2D) structures; that is, structures with both breadth and height only. The proposed co-ordinate system is shown in Figure 1.5.

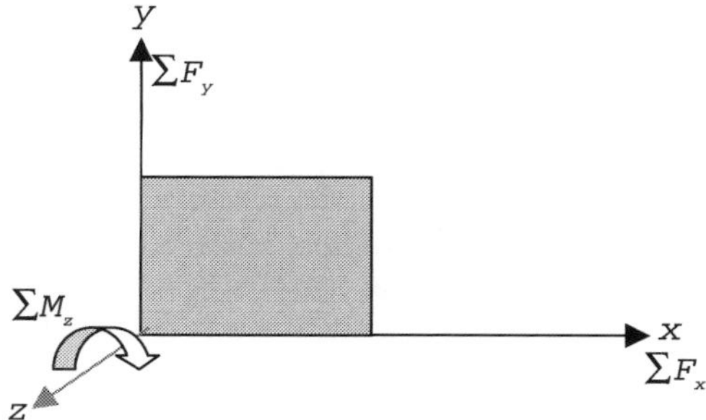

Figure 1.5: Two-dimensional co-ordinate system

We will apply this co-ordinate system across the entire structure. The co-ordinate system will therefore be considered as 'global'. In some methods of analysis the co-ordinate system may be orientated to the individual member axis and will then be considered as a 'local' system. For equilibrium in a two dimensional system we must ensure that the summation of the forces in the *x* direction, the summation of the forces in the *y* direction and the moments about the *z* axis equate to zero. Or:

$$\Sigma F_x = 0 \qquad [1.5a]$$

$$\Sigma F_y = 0 \qquad [1.5b]$$

$$\Sigma M_z = 0 \qquad [1.5c]$$

However, real structures exist in three-dimensional (3D) space. In three dimensions our co-ordinate system will be that shown in Figure 1.6. For stable equilibrium of a rigid 3D body at the origin, we must now consider six equations.

$$\Sigma F_x = 0 \qquad [1.6a]$$
$$\Sigma M_x = 0 \qquad [1.6b]$$
$$\Sigma F_y = 0 \qquad [1.6c]$$
$$\Sigma M_y = 0 \qquad [1.6d]$$
$$\Sigma F_z = 0 \qquad [1.6e]$$
$$\Sigma M_z = 0 \qquad [1.6f]$$

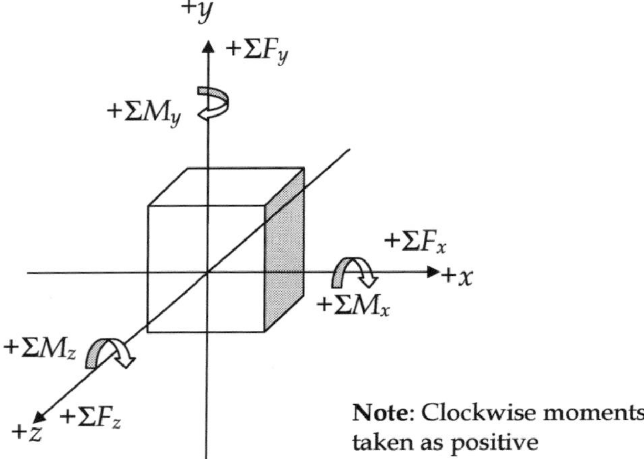

Figure 1.6: Three-dimensional co-ordinate system

Therefore, in two-dimensional analysis we are only required to solve for three equations in order to ensure static equilibrium of a rigid body. For three-dimensional structures the analysis is much more complex, requiring the solution of six equations, and will not form part of these early studies.

Practice has shown that, in the formative years of study, it is easier to analyse structures using the global Cartesian co-ordinates concept and resolving forces into horizontal and vertical components in order to check for equilibrium. This may require the resolution of a number of concurrent forces in order to determine the total horizontal and vertical forces applied at a particular position on the structure. Consider the two forces applied at point T as shown in Figure 1.7.

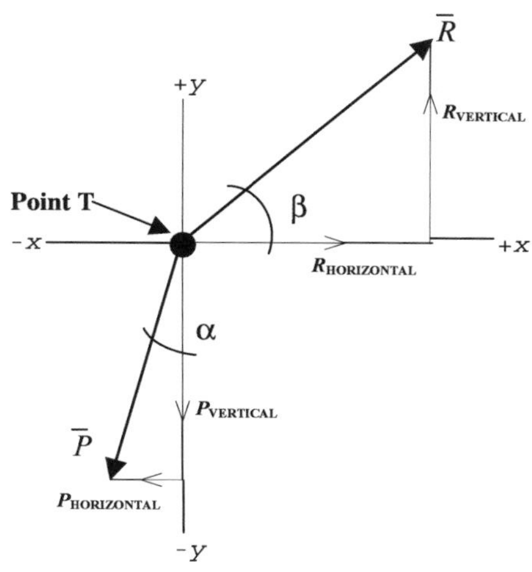

Figure 1.7: Concurrent forces applied to a joint

In order to simplify our analysis we will resolve each force into its horizontal and vertical components and then sum the results to find the resultant horizontal and vertical forces. Hence:

For vector (force) *R*:

$R_{\text{VERTICAL}} = R \sin \beta$

$R_{\text{HORIZONTAL}} = R \cos \beta$

For vector (force) *P*:

$P_{\text{VERTICAL}} = P \cos \alpha$

$P_{\text{HORIZONTAL}} = P \sin \alpha$

Hence resultant forces are:

Resultant vertical force $= +R \sin \beta - P \cos \alpha$

Resultant horizontal force $= +R \sin \beta - P \sin \alpha$

Note that the positive and negative signs are generated in normal Cartesian co-ordinates by the force directions (as shown in Figure 1.7). Also note that we have dropped the 'bar' convention on the vectors to simplify the equations. We can therefore resolve any number of forces into a single horizontal and a single vertical component by adding all horizontal and vertical forces respectively acting at the point under consideration.

Example 1.1: Resolution of four forces at a point

Consider the point J shown in Figure 1.8 and the forces applied to it. Determine the magnitude and direction of the horizontal and vertical forces *H* and *V* required to ensure equilibrium.

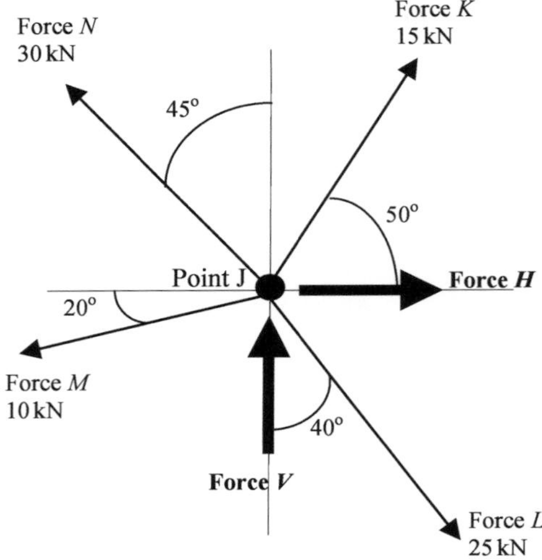

Figure 1.8: Example 1.1 – Forces applied at point J

In order to calculate the magnitude and direction of the forces *H* and *V* we must first determine the components of each of the forces (Figure 1.9) and the resultant out-of-balance force.

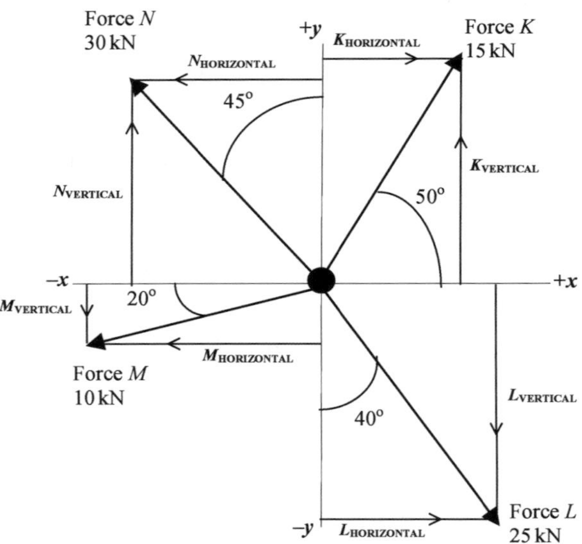

Figure 1.9: Example 1.1 – Components of applied forces

Note that the directions of the horizontal and vertical components of forces are determined to produce the same action as the original force. Thus, for instance, the force *K* is upwards and to the right of our co-ordinate system, hence the resultant horizontal is to the right and the vertical component is upwards. Likewise, force *M* is downwards and to the left, hence are derived the direction of its components. We can now calculate the components of each force, however we must remember to indicate its direction with a '+' or '–' according to its orientation. The results are shown in Figure 1.10.

Force	Horizontal component	Vertical component
K	+15 cos 50° = 9.642	+15 sin 50° = 11.49
L	+25 sin 40° = 16.07	–25 cos 40° = –19.15
M	–10 cos 20° = –9.397	–10 sin 20° = –3.42
N	–30 sin 45° = –21.213	+30 cos 45° = 21.213
TOTAL	– 4.898 kN	10.133 kN

Figure 1.10: Example 1.1 – Table of force components and resultant

The cosine and sine values are determined with respect to the position of the known angle, therefore the relationship is that normally applied to Sine, Cosine and Tangent functions. The resultant values indicate that the joint is not in equilibrium with a resultant force of 4.898 kN in the horizontal (negative x) direction and 10.133 kN in the upward y direction. Therefore, in order to place the joint in equilibrium we would need to apply a vertical downward force of 10.133 kN and a horizontal force of 4.898 kN (from left to right as shown in Figure 1.11) in order to counteract these out-of-balance forces.

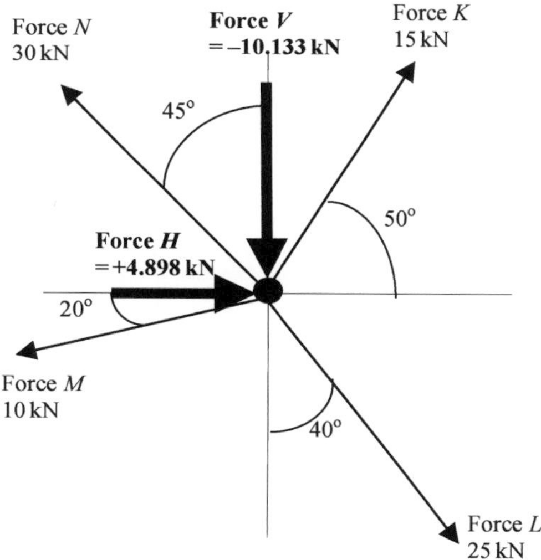

Figure 1.11: Example 1.1 – Forces required for equilibrium at point J

We can therefore determine ΣF_x and ΣF_y for any system and, if these equate to zero, the system will be in equilibrium.

◇

1.2.2 Moments

Summing all moments to zero is the basis for deriving the third equilibrium equation. This requires us to resolve forces applied at some distance from an origin. This may be completed as follows.

Consider the forces applied in the form of a couple as shown in Figure 1.12. The figure shows what is termed as a 'pure couple', that is equal force applied in opposite directions to a member about an origin. Therefore force F_1 equals force F_2 (in this case). The couple will, if allowed to move freely, produce a rotation about the support shown at the origin 'O'. The forces induce a 'twisting' action at the origin as a result of the applied moments. The magnitude of a moment is the product of the force and the distance, from the point of origin, about which it acts. In the SI[1] system of units commonly used in structural analysis, forces are measured in Newtons 'N' or kilo-Newtons 'kN'. Distance is measured in metres or millimetres hence the units of moment are normally N mm or kN m.

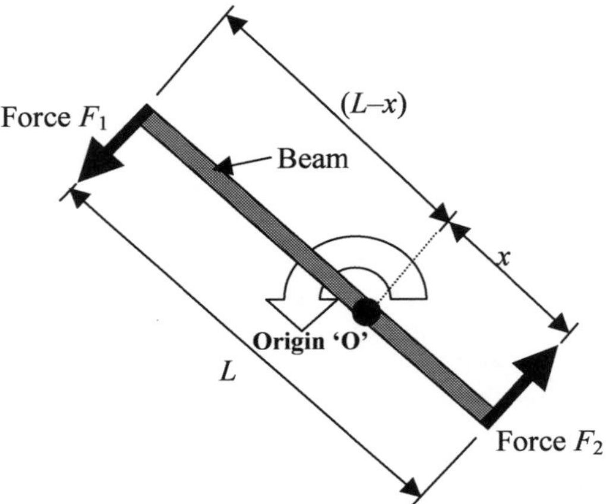

Figure 1.12: Diagram showing a 'pure' couple system

We can replace the system shown in Figure 1.12 with a moment and a force applied at the origin. The value of the moment will be the sum of the applied moments from forces F_1 and F_2. The sum of the forces will be F_1 minus F_2 (forces are applied in opposing directions). Hence:

$$\text{Total force} = F_1 - F_2 \text{ (in opposite directions)}$$

but since $F_1 = F_2$

$$\text{Total force} = F_1 - F_1 = 0$$

Moment of force F_1

$$\text{Moment } (F_1) = \text{force} \times \text{distance from origin} = F_1 \times (L–x)$$
$$\text{Moment } (F_1) = F_1\,(L–x)$$

Moment of force F_2

$$\text{Moment } (F_2) = \text{force} \times \text{distance from origin} = F_2 \times (x)$$
$$\text{Moment } (F_2) = F_2\,(x)$$

Summing the moments and noting that clockwise moments are considered positive (+) and anti-clockwise moments are considered negative (-) we have:

Summation of moments $\quad \Sigma M = -F_1\,(L–x) - F_2\,(x) \quad$ (anticlockwise moments)

But, since $F_1 = F_2$ and expanding:

$$\Sigma M = -F_1\,(L) + F_1\,(x) - F_1\,(x)$$

Hence:

$$\Sigma M = -F_1\,(L)$$

This is always the case for a pure couple.

1. Système International d'Unités; Comité International des Poids et Measure, France, 1960.

We can therefore replace our pure couple with a system consisting of a moment and a force at the origin, such as that shown in Figure 1.13.

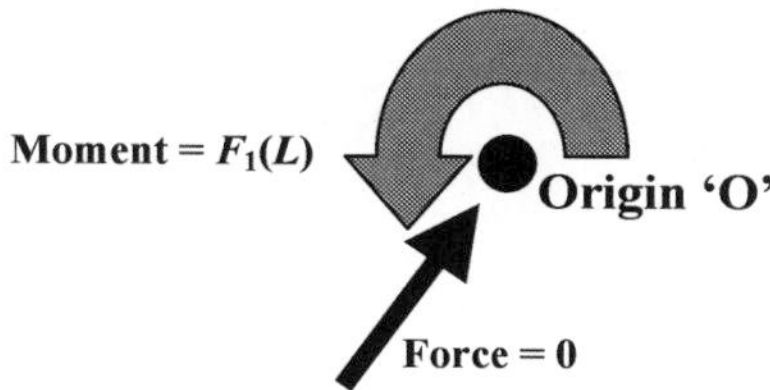

Figure 1.13: Equivalent moment and force system for a pure couple

Any force acting at a known distance from an origin will cause a rotation about that origin. This rotation is due to the moment induced by the force. Any system can be simplified to a force + moment system, as shown in Figure 1.14.

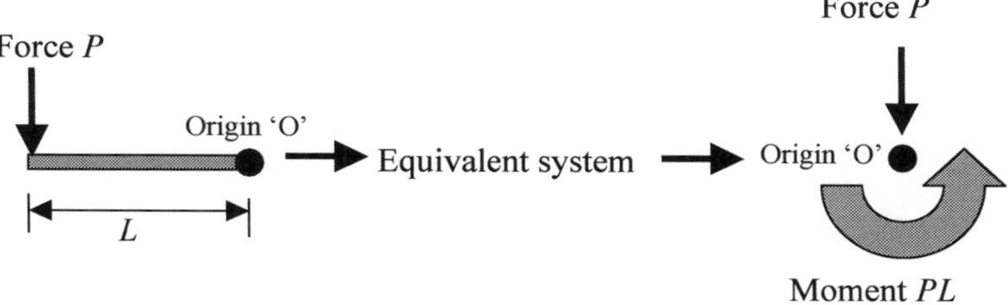

Figure 1.14: Equivalent force and moment system for simple beam member

Note here that the distance L is always the distance from the line of action of the force (produced as necessary) of a normal line to the point, O, of interest (i.e. perpendicular, 90°, orthogonal).

It will be necessary to calculate the resultant for a combination of concurrent forces. Two or more such forces may be acting at a point and these can be combined vectorially as before (using algebra, parallelograms, triangle of forces or graphically). If required they can be transferred to act at a point by use of the (force + moment) method previously described and illustrated in Figure 1.15 which shows the addition of the moments from two co-planar forces.

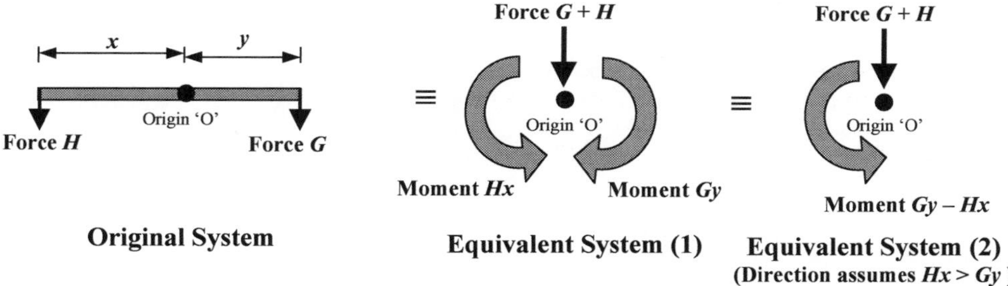

Figure 1.15: Addition of two co-planar forces

We can therefore find the resultant moment about an origin for two or more forces by adding the calculated value of the moment, accounting for its rotational direction using a clockwise positive system. It is possible, using this method, to calculate the moment at any point on a structure. It does **not** matter where in the plane these moments are applied, because the summation will be the same.

We will study further the derivation of forces and bending moments for various structures in Chapters 3 and 4.

1.3 Load Paths – Determination of Load

In the design of structures it is essential that we make an accurate assessment of the loading to which the structural members will be subjected in order that the most accurate response may be predicted. A structural member will be subjected to both direct and indirect loading, that is loads applied directly to the member and loads transferred through other structural elements to it. It is also essential that we include all loads when designing a structural member and, in order to do this, we need to be able to trace the 'load path' of elements to the point under consideration. Therefore, not only must we accurately calculate the area supported by each element but also the loading that it carries.

1.3.1 Loading Types

Loading may be categorised into two different types:

1. Those forces which are due to gravity and may include the weight of the structure or gravity forces distributed throughout the structure.

2. Forces applied at the boundary; these may include reactions at supports.

Forces applied to the structure may be further subdivided into *dead* and *imposed* (or *live*) loads.

Dead loads are due to self-weight and the permanent weight of elements supported by the structure. Examples of this type of loading are the roof tiles, battens and felt, roof joist and other building elements shown in the typical roof structure of Figure 1.16a or the plywood flooring, floor joist and plasterboard shown in the typical floor detail of Figure 1.16b

The masses of materials are listed in various national standards; for example, in the United Kingdom, we would normally refer to British Standard BS 648: 1969[2] or masses can be determined from manufacturers' brochures. From this can be calculated the load that the element will exert on the members supporting it. Figure 1.17 shows details of such a calculation for the total dead load for the roof of Figure 1.16a, calculated from the mass of each of the building elements that it supports. This is calculated per metre depth of the building. Note that in order to calculate the force per metre depth for rafters and joist members, we will need to calculate the number of members for each 1-metre depth of building. Assuming the rafters and joists are spaced at 450 mm centres, we can calculate the number as (1/0.45) or 2.22 joists per metre. The battens will also be spaced along the roof joists, in this case at 200 mm centres; we therefore will have (1/0.2) or 5 battens per metre.

2. British Standards Institution, BS 648: 1969: Schedule of weights of building materials.

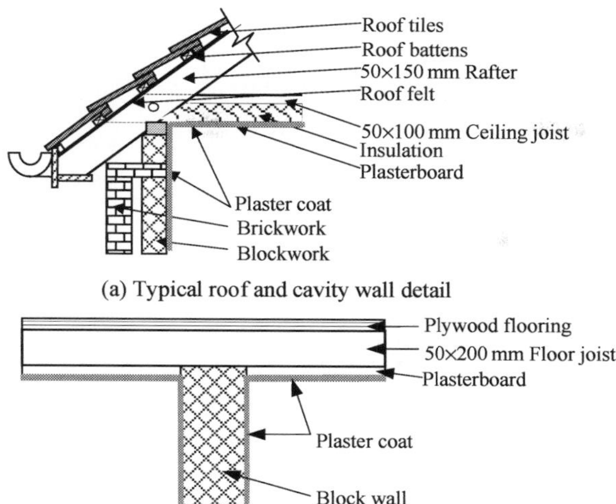

(a) Typical roof and cavity wall detail

(b) Typical floor and internal wall detail

Figure 1.16: Section showing typical construction details for roof, floor and walls

	Element	Mass	Area/m	*g*	Total (N/m^2)
1	Roof tiles	70 kg/m^2	—	9.81 m/s^2	686.70
2	Battens	590 kg/m^3	25×50×(1/0.2)	9.81 m/s^2	36.88
3	Felt	3.5 kg/m^2	—	9.81 m/s^2	34.34
4	Rafter	590 kg/m^3	50×150×(1/0.45)	9.81 m/s^2	96.50
5	Ceiling joist	590 kg/m^3	50×100×(1/0.45)	9.81 m/s^2	64.31
6	Insulation	4.0 kg/m^2	—	9.81 m/s^2	39.24
7	Plasterboard	12.5 kg/m^2	—	9.81 m/s^2	122.63
8	Plaster coat	22 kg/m^2	—	9.81 m/s^2	215.82
				Total	1296.42

Figure 1.17: Calculation of force from dead load of typical roof construction

In most cases the mass is calculated for a specified material thickness and is therefore given in kg/m^2. For timber members and materials such as brick, block and plaster, the mass of the element is specified in kg/m^3. In order to calculate its weight per m^2 we must calculate the volume of each element per m^2 of the structure.

If we require to calculate the force exerted on a structural member, such as a lintel over a window, from the dead load of the roof, we can use the value calculated in Figure 1.17 of approximately 1.3 kN/m^2. If, however, we were only designing the rafter to the roof the dead load required would be the total of the forces exerted by the tiles (1), battens (2) and felt (3), which may be calculated as 0.758 kN/m^2, since these are the only elements supported by that member.

We can calculate the forces exerted by the materials of the floor and cavity wall structure shown in Figure 1.16a and b in a similar manner (Figure 1.18), noting that the thickness of material must be included in the calculation of materials with specified mass in m^3.

	Element	Mass	Thickness (mm)	g	Total (N/m^2)
CAVITY WALL					
1	Brickwork	2200 kg/m^3	100	9.81 m/s^2	2158.20
2	Blockwork	1176 kg/m^3	102	9.81 m/s^2	1176.72
3	Plaster coat	1800 kg/m^3	12	9.81 m/s^2	211.90
				TOTAL	3546.82
FLOOR					
1	Floor joist	590 kg/m^3	50×200×(1/0.45)	9.81 m/s^2	128.50
2	Plywood floor	35 kg/m^2	—	9.81 m/s^2	343.35
3	Plasterboard	12.5 kg/m^2	—	9.81 m/s^2	122.63
4	Plaster coat	1800 kg/m^3	12	9.81 m/s^2	211.90
				TOTAL	806.38

Figure 1.18: Calculation of force from dead load of typical floor and wall construction

The dead loads for structures can therefore be calculated by simple conversion to force units and the addition of the resultant for each of the elements used to form it. Note that the internal wall of Figure 1.16b can be calculated in a similar manner to be approximately 2.2 kN/m^2.

Imposed loads are those loads that may or may not be permanent but are the loads that are imposed when the structure is in use. These might include the load applied by a crowd at a football match or by traffic over a bridge. These loads therefore vary with time. For instance, an empty grandstand will have little imposed load applied, whereas maximum load might be considered when full; alternatively, a roof structure in winter may well have a covering of snow which is not present in summer. Note that though time is a factor with these types of load, the periods are considered to be long and therefore they are taken to be static loads. The movement of the crowd however would constitute a dynamic load!

Imposed loads are specified in national codes and standards. In the UK British Standard 6399: Part 1: 1984[3] for dead and imposed loads is commonly used, and some typical values for imposed loads from this code are shown in Figure 1.19.

Class of building	Distributed load (kN/m^2)	Concentrated Load (kN)
Residential – Roof Imposed Load	0.75	—
Residential – Ceiling Imposed Load	0.25	—
Residential – Floor Imposed Load	1.5	1.4
Industrial – School Classroom Imposed Floor Load	3.0	2.7
Industrial – Boiler Room Imposed Floor Load	7.5	4.5

Figure 1.19: Typical imposed load values for different classes of structure

When analysing structural members we will generally calculate the total load to be, that which will cause the worst-case situation. The final design of structural elements may require the use of factored load values. Most modern codes of practice base design on a load factor method. This requires the designer to apply factors to the calculated Dead, Imposed and Wind loads in order to provide a factor of safety in the design. Typical values taken from British Standard 8110: Part 1: 1985[4] for the design of concrete members would be 1.4 and 1.6 respectively when the design is based on dead and imposed loads alone, or 1.2:1.2:1.2 when the design includes dead, imposed and wind loads.

3. British Standards Institution, BS 6399: Part 1: 1984: Loading on Building: Code of practice on dead and imposed loads, 1984.
4. British Standards Institution, BS 8110: Structural use of concrete: Part 1: 1985: Code of practice for the design and construction, 1985.

A third class of loading that should also be considered is that due to wind. **Wind load** is the load applied to a structure due to thermal movements in the airflow over the surface of the Earth. It is influenced by a number of factors including environmental conditions such as temperature and pressure gradients, and physical conditions such as local topography, location and height of structure. Codes of Practice and national standards[5] are available to the designer which specify the parameters to be considered in the design of structural elements subject to wind loading. It is important to note that wind load can be applied in both a sideways / lateral load as well as an upward / suction load.

Each load, and its combinations, will have a different effect on our structural elements and will impose different design problems. Structural deformations will vary with the load conditions applied. In our studies we will consider two important types of static load, that is loads applied at a specific point (point loads) and those applied over a finite length (distributed load). The latter can vary in shape as well as intensity. These will be discussed further in Chapter 3.

We will first consider the effects of external forces on a structure.

1.3.2 Member Loads

Axial Loads

An external load applied to a structural member in a direction along its longitudinal axis is called an axial load. There are two types of axial load:

(1) Tensile axial load;

(2) Compressive axial load.

Tensile Axial Load

If a load is applied to a member that pulls it away from its support, possibly causing an extension in the member, this is called a *tensile force*. A member that supports a tensile load is called a ***tie***. Consider the analysis of the member shown in Figure 1.20. The hinge is required only to demonstrate the action of the system under tensile load. In our analysis we will require the system to be in equilibrium. We therefore require that the support remains in its position in space. The support is considered to be fixed in position, an unyielding boundary. The effect of a pulling force at the free end is to induce a downward movement in the member in the direction of the force. In order that the end remains in an equilibrium position, the member must exert an equal and opposite force through its length, which will be transferred to the support and be balanced by a support reaction.

Pure tension members are only stressed in tension, and they generally do not require any resistance to lateral force. We can therefore insert a hinge (pinned) joint in the member without affecting its structural stability. Clearly such a member will have little lateral (sideways) resistance, so that if it were subjected to a lateral load the member would be able to rotate. Under the action of a tensile load however, the effect would be to bring the member back into line, stretching it back into shape. This means that members can comprise steel cables or ropes which have no compression or bending resistance and are extremely flexible. Supports must be capable of supporting the load from such members, which in some cases will mean that the load may be inclined, such as on suspension bridges. Very thin members can act as tensile membranes.

5. British Standards Institution, BS 6399: Part 2: 1997: Loading for Building: Code of practice for wind loads,1997.

Examples include temporary sports halls were the structure is inflated in order to form a covered arena. Further examples of pure tension members would be the guy ropes or stays which support very tall masts and towers.

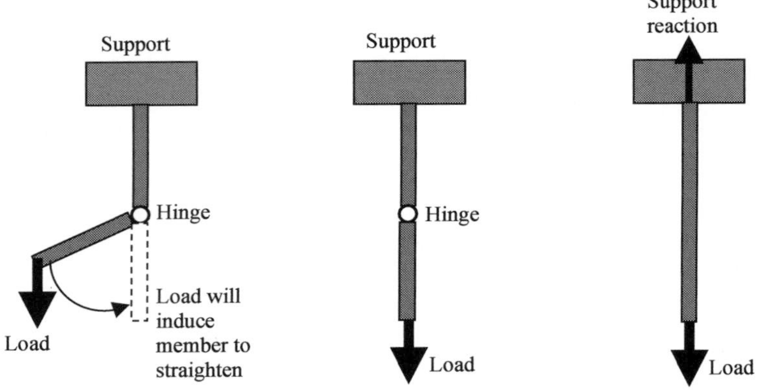

Figure 1.20: Tensile axial load applied to structural member

Compressive Axial Load

If a load is applied to a member which pushes it towards its support and possibly causes a shortening of the member, this is called a *compressive force*. A member which supports a compressive load is called a ***strut*** or in larger structural frames a column. Consider the analysis of the member shown in Figure 1.21. We again require equilibrium in the member at its support reaction. We therefore consider that the support remains in its position in space. The effect of a pushing force at the free end is to try and move the member in the direction of the force. In order that the end remains in its position, the member must resist by exerting an equal and opposite force through its length, which is transferred to the support and resisted by the support reaction.

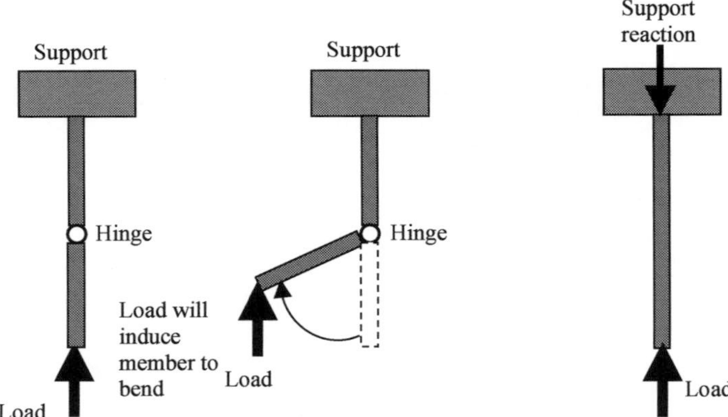

Figure 1.21: Compressive axial load applied to structural member

The types of structural elements that might carry such loads are columns in large frames, piers and foundations. Concrete and masonry (brickwork) work extremely well in compression but are very weak in tension, whilst steel is a very good tensile material but, depending on the section's geometric properties, may be very weak in compression. (i.e. steel reinforcing bars)

Bending Moments

If a member is subjected to a force applied at some distance from its support, an applied moment will be induced which will cause the member to rotate or bend. Two examples are shown in Figure 1.22a and b. Figure 1.22a shows a vertical column, fixed at its base and subjected to a horizontal force. The force will cause the column to bend and a horizontal displacement may occur. This will induce shortening in the length of the column member on the side opposite to the force and stretching on the side on which the force is applied. Since the member is being extended on one side, this face is in tension. The side which is being shortened is in compression.

The beam member of Figure 1.22b illustrates a similar problem. The beam deforms (deflects) in a downward direction due to the load, which is downward. The top face of the beam shortens and therefore is in compression, whilst the bottom extends and is in tension. The actual amount of deformation in a real structure will be small in comparison to its length. If the members were constructed in plain concrete deformation might cause cracking in the tension (lengthening) face. If the column were masonry, the force might be large enough to cause a tensile failure in the member by breaking the bond between the mortar and brickwork or by causing a tensile failure in the brick. It is worth noting that structural engineers tend to design mortar in masonry structures to be weaker than the brick, so that failure always occurs in the mortar joints. If the force is large enough, cracks would extend through the member until they reached such a depth that failure will occur.

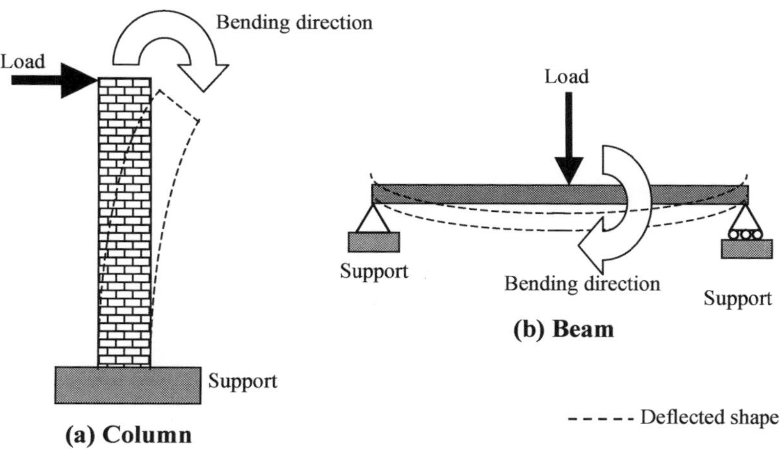

Figure 1.22: Bending moment applied to structural member

In both cases, concrete and masonry, the characteristics of the member can be modified by placing steel reinforcement on the tension side. The steel, which is good in tension, will increase the tensile resistance and, in effect, hold the member together.

Shear Force

If we again consider the column and beam of Figure 1.22a and 1.22b, we note that the applied force will induce a bending moment. As discussed in section 1.1.3, this moment will be accompanied by a force. For equilibrium to exist, the force must be resisted at the supports by an equal and opposite reaction. If we now consider the forces at the reaction of the column and beam, we find that we have an applied force and a resistance acting in opposite directions at the interface between the member and its support, as shown in Figure 1.23.

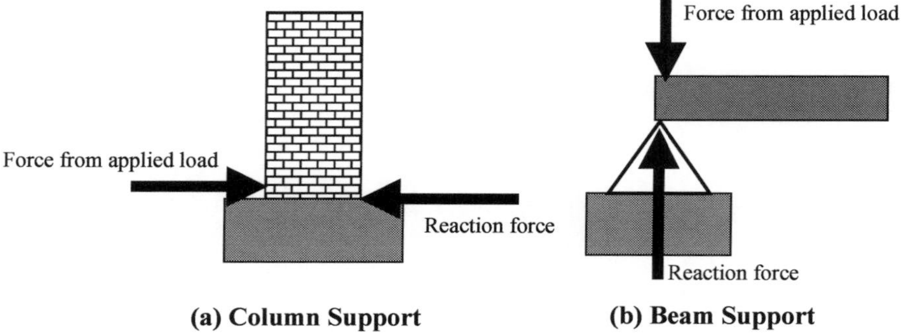

Figure 1.23: Shear force at column and beam supports

The loads are applied at right angles to the longitudinal axis of the member. Their effect is to try and *tear* the structure apart along the line of the interface between the member and the support. This tearing action is known as *shear* and can occur in combination with bending loads. Shear force will not only take place at the support but along the members' entire length.

Torsion

These forces will cause a rotation about the axis of the member. They may therefore be considered as a twisting moment. They are caused by bending moments applied along the longitudinal axis of the member, that is the out-of-plane z axis, of our two-dimensional system shown in Figure 1.5. They will cause a rotation (twisting) of the member AB shown in Figure 1.24.

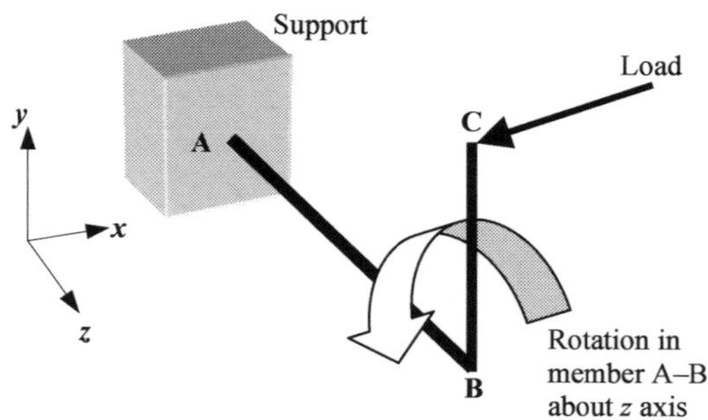

Figure 1.24: Torsion moment

The figure shows a three-dimensional structure ABC consisting of a horizontal beam AB parallel to the z axis and a member BC parallel to the y axis. The load is applied in a direction parallel to the x axis. This will induce a moment in the member BC. This moment will induce twisting in the member AB. The amount of this twisting moment will be equivalent to the maximum applied moment in BC.

Torsional effects create very complex problems. Elementary examples will be considered later in this introductory text. It is however worth noting that, in very large structures, torsional resistance is often provided by constructing very rigid concrete lift shafts or stairwells at the core of the building

Having introduced various types of load and their effects on different structural members, we will now consider the determination of the amount of load to which the various elements will be subjected.

1.3.3 Load Transfer

Load will be transferred through a structure and to its foundations. Thus the load imposed from a roof structure must, somehow, be transferred through other structural elements until it is finally resisted by the earth supporting the foundations. In order to design the roof members, we will first need to determine how much load they are to carry. Having designed the roof, we might then consider the design of the walls supporting the roof structure and, as such, we must determine how much of the roof load will be transferred to the wall. We might then consider the floor or other structural member. Whatever member we decide to design, we must first determine how much load it will be required to carry. This will be based on its position in the structure and on how the loads are transferred from part to part. To demonstrate the problem, consider the simple structure of Example 1.2, shown in Figure 1.25.

Example 1.2: Calculation of loads for a simple structure

For the structure shown in Figure 1.25, determine the unfactored total load, per metre depth of the building, at the base of the foundations across the section A–A using the loads indicated in the table of Figure 1.26.

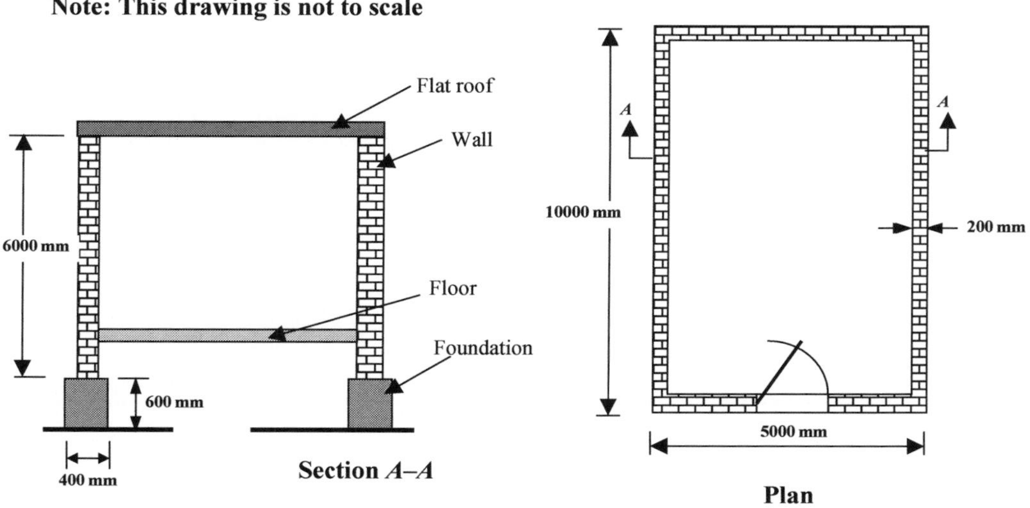

Figure 1.25: Simple structure – Example 1.2

	Dead Load	Imposed Load
Flat roof	0.75 kN/m^2	0.75 kN/m^2
Wall	3.5 kN/m^2	—
Floor	0.85 kN/m^2	1.5 kN/m^2
Foundation	5.6 kN/m	—

Figure 1.26: Table of applied dead and imposed loads for the structure of Example 1.2.

First it is important to note that we will be calculating the load per metre depth of the building. Secondly, the dead load induced by the foundation is given for the structure in Figure 1.26. Often, only the density of the concrete is known, in which case we will need to calculate the force per metre from the volume of the foundation. Finally, there is no imposed load on the wall. This is because, in this case, the imposed loads have been included in the load-carrying floor and roof elements and, as we are only calculating vertical imposed loads, the wind effects need not be considered here. This will simplify the analysis of the structure.

In order to complete the load analysis of the structure, we must determine how much of each load is transferred to each of the foundations.

By observation, and assuming roof and floor span in the shortest direction, as indicated in Figure 1.25, we see that half of the roof and floor loads (2.5 m), all the wall load (6 m) and the foundation load are transferred to the ground on either side of the building. It only requires us to calculate each value. This is best completed in table form as shown in Figure 1.27.

	Dead Load (kN/m)		Imposed Load (kN/m)	
Flat roof	$0.75\ \text{kN/m}^2 \times (5/2)\,\text{m}$	$= 1.875$	$0.75\ \text{kN/m}^2 \times (5/2)\,\text{m}$	$= 1.875$
Wall	$3.5\ \text{kN/m}^2 \times 6\,\text{m}$	$= 21.0$		
Floor	$0.85\ \text{kN/m}^2 \times (5/2)\,\text{m}$	$= 2.125$	$1.5\ \text{kN/m}^2 \times (5/2)\,\text{m}$	$= 3.75$
Foundation		$= 5.6$		
	TOTAL DEAD	**30.6**	**TOTAL IMPOSED**	**5.625**

Figure 1.27: Total loads at base of foundation for the structure of Example 1.2

It can be seen from Figure 1.27 that the total dead load at the underside of the foundation is 30.6 kN/m run of building and the total imposed load is 5.625 kN/m run of building. The total unfactored load is therefore:

$$\begin{aligned}\text{Total Load} &= 30.6 + 5.625 \\ &= 36.225\ \text{kN/m run}\end{aligned}$$

We can now follow this simple procedure to calculate the loads at various points on any structure. However, it is important to note that not all distributed loads will be of a simple rectangular or uniform nature. Also, certain simplifications are made in our analysis, some of which will be considered in the garage structure shown in Figure 1.28 of Example 1.3.

Example 1.3: Load path calculations for a garage structure

Consider the simple garage structure shown in Figure 1.28.

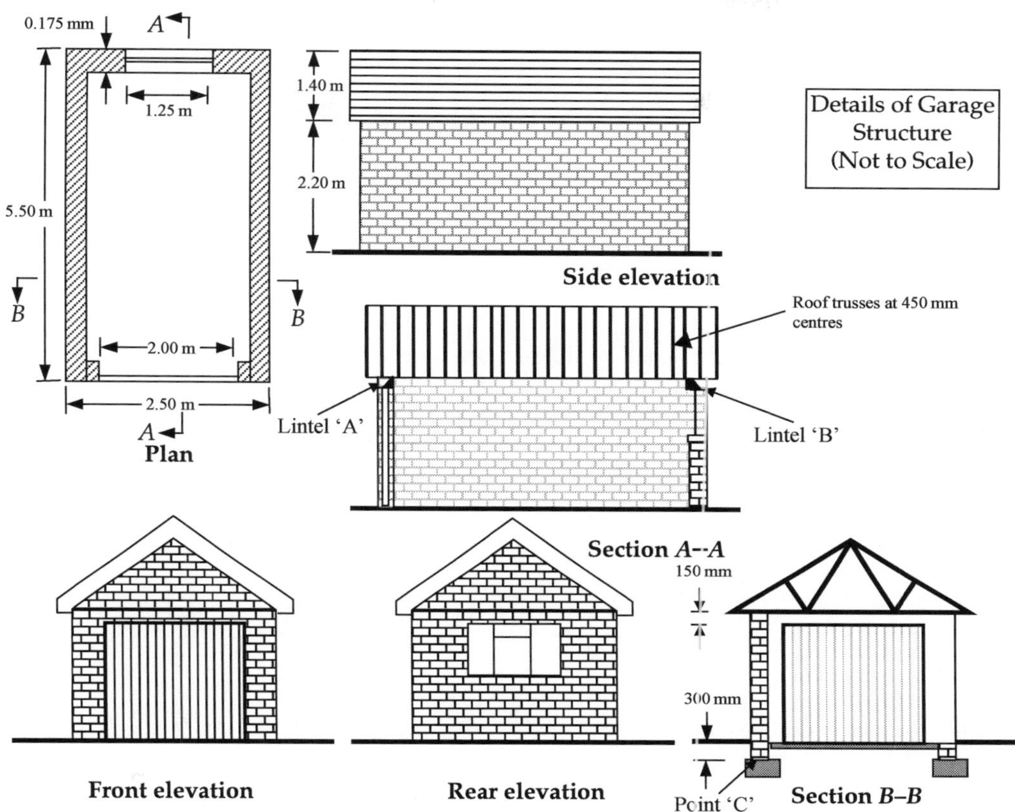

Figure 1.28: Details of garage– Example 1.3

The structural loads have already been calculated and are tabulated in the table of Figure 1.29.

	Dead Load (kN/m^2)	Imposed Load (kN/m^2)
Pitched roof	0.85	1.00
Walls	2.30	—
Floor	1.50	1.50

Figure 1.29: Structural loads for garage structure – Example 1.3

For the given loading, calculate the total unfactored load carried by the lintels A and B and the load per metre depth of the building just above the foundation at point C.

Before commencing our analysis of loading for the lintels A and B, let us consider the pitched roof and its effect on the openings. Although the roof spans from side to side on the garage and is supported on the walls, in practice some roof load will be transmitted to the lintels. This is because, in this case, when constructing the roof, an overhanging section will be formed. This section can transfer load to the end (or gable) walls. Also, some load will be transmitted because the trusses do not tightly abut the end walls. In making this allowance some 'engineering judgement' must be used. In this example we will consider the end walls support one span (i.e. 450 mm) of roof. A detail of the lintel loads is shown in Figure 1.30.

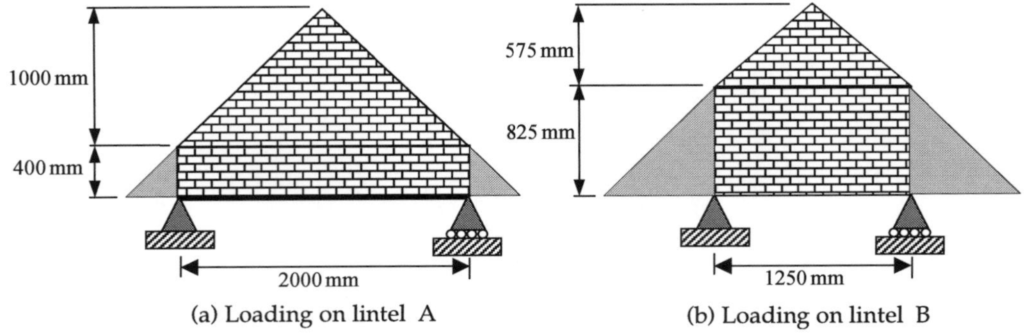

(a) Loading on lintel A (b) Loading on lintel B

Figure 1.30: Load applied by brickwork to lintels A and B

Using basic trigonometry we can calculate the area of wall supported by each lintel. To calculate the total load we multiply the area by the value for dead and imposed load given in Figure 1.30 plus an allowance for the roof load hence:

(Brickwork)

Total load on lintel A $= (0.4\,\text{m} \times 2.0\,\text{m} \times (2.3\,\text{kN/m}^2)) + (0.5 \times 1.0\,\text{m} \times 2.0\,\text{m} \times (2.3\,\text{kN/m}^2))$

(Pitched roof)

$+ \quad (0.45\,\text{m} \times 2\,\text{m} \times (0.85{+}1.00)\,\text{kN/m}^2)$

$= \quad 1.84 \quad + \quad 2.3 \quad + \quad 1.66$

$= \quad 5.8\,\text{kN}$

(Brickwork)

Total load on lintel B $= (0.825\,\text{m} \times 1.25\,\text{m} \times (2.3\,\text{kN/m}^2)) + (0.5 \times 0.575\,\text{m} \times 1.25\,\text{m} \times (2.3\,\text{kN/m}^2))$

(Pitched roof)

$+ \quad (0.45\,\text{m} \times 1.25\,\text{m} \times (0.85{+}1.00)\,\text{kN/m}^2)$

$= \quad 2.37 \quad + \quad 0.83 \quad + \quad 1.00$

$= \quad 4.2\,\text{kN}$

We assume that the lintels support an area equivalent to one truss.

Next we calculate the load transferred to the foundations. We assume that the floor is supported on the walls and also that the roof load can be considered as being uniformly distributed. This second assumption may seem a bit of an anomaly since the trusses are spaced at 450 mm centres, however the load from each truss will be distributed throughout the brickwork. At the foundation level the distribution will be complete. Figure 1.31 shows the calculation of loads at foundation level.

	Dead Load (kN/m)		Imposed Load (kN/m)	
Pitched roof	$0.85\,\text{kN/m}^2 \times (2.5/2)\,\text{m}$	= 1.063	$1.00\,\text{kN/m}^2 \times (2.5/2)\,\text{m}$	= 1.250
Wall	$2.3\,\text{kN/m}^2 \times (2.2\,\text{m}{+}0.3\,\text{m})$	= 5.750		
Floor	$1.50\,\text{kN/m}^2 \times ((2.5{-}2 \times (0.175))/2)\,\text{m}$	= 1.613	$1.50\,\text{kN/m}^2 \times 2.15\,\text{m}/2$	= 1.613
	TOTAL DEAD	**8.426**	**TOTAL IMPOSED**	**2.863**

Figure 1.31: Total loads at the top of the foundation (point C) for garage – Example 1.3

Therefore the total load at point C, per metre depth of the building, will be:

Total load per metre depth of building at point C $= 8.426 + 2.863$

$= 11.289$ kN/m run (of side wall)

Example 1.4: Load path calculations for a building structure

Consider the section through the building shown in Figure 1.32.

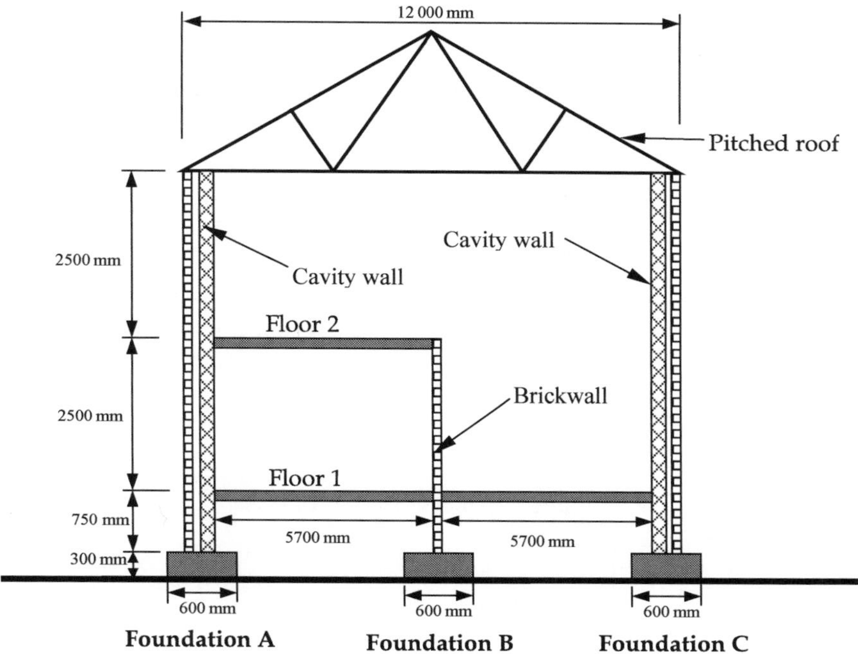

Figure 1.32: Section through building – Example 1.4

Given the dead and imposed loads in the table of Figure 1.33 and assuming the unit weight of concrete is 24 kN/m^3 calculate the load, per metre depth of building, at the underside of foundations A, B and C.

	Dead Load (kN/m^2)	Imposed Load (kN/m^2)
Pitched roof	0.90	1.00
Cavity wall	4.00	—
Brick wall	2.00	—
Floor	1.00	1.50

Figure 1.33: Structural loads for building – Example 1.4

Solution

	Dead Load (kN/m^2)	Imposed Load (kN/m^2)	Total Load (kN/m^2)
Pitched roof	0.90	1.00	1.90
Cavity wall	4.00	—	4.00
Brick wall	2.00	—	2.00
Floor	1.00	1.50	2.50

Figure 1.34: Total loads for building – Example 1.4

	Foundation A	Foundation B	Foundation C
Pitched roof	$1.9\,kN/m^2 \times (12/2)\,m$ = 11.4 kN/m		$1.9\,kN/m^2 \times (12\,m/2)$ = 11.4 kN/m
Cavity wall	$4.00\,kN/m^2 \times (2.5+2.5+0.75)m$ = 23.0 kN/m		$4.00\,kN/m^2 \times (2.5+2.5+0.75)\,m$ = 23.0 kN/m
Floor 2	$2.5\,kN/m^2 \times (5.7/2)\,m$ = 7.13 kN/m	$2.5\,kN/m^2 \times (5.7/2)\,m$ = 7.13 kN/m	
Floor 1	$2.5\,kN/m^2 \times (5.7/2)\,m$ = 7.13 kN/m	$2.5\,kN/m^2 \times (2 \times (5.7/2))\,m$ = 14.26 kN/m	$2.5\,kN/m^2 \times (5.7/2)\,m$ = 7.13 kN/m
Brick wall		$2.00\,kN/m^2 \times (2.5 + 0.75)\,m$ = 6.5 kN/m	
Foundation	$24\,kN/m^3 \times (0.6 \times 0.3)\,m$ = 4.32 kN/m	$24\,kN/m^3 \times (0.6 \times 0.3)\,m$ = 4.32 kN/m	$24\,kN/m^3 \times (0.6 \times 0.3)\,m$ = 4.32 kN/m
TOTAL	52.98 kN/m run	32.21 kN/m run	45.85 kN/m run

Figure 1.35: Total loads at the base of the foundations A, B and C for building – Example 1.4

Example 1.5: Load path calculations for a beam and foundation

Given the dead and imposed loads in the table of Figure 1.36 calculate the factored loads, per metre depth of building, at beam A and at the top of the points B and C, as shown in Figure 1.37. Assume the dead loads are to be factored by 1.4 and all imposed loads by 1.6.

	Dead Load (kN/m^2)	Imposed Load (kN/m^2)
Flat Roof	0.85	0.75
Cavity wall	3.50	—
Block wall	2.50	—
Floor 1	1.50	1.50
Floor 2	0.85	1.50

Figure 1.36: Structural loads for building – Example 1.5

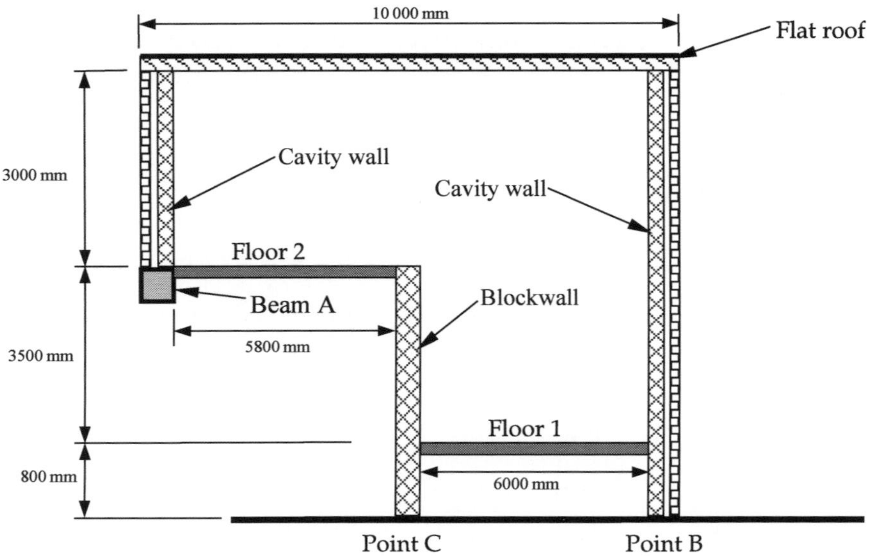

Figure 1.37: Section through building – Example 1.5

Solution

	Dead Load (kN/m^2) (1)	Imposed Load (kN/m^2) (2)	Factored Dead Load ((1)×1.4)	Factored Imposed Load ((2)×1.6)	Total Load (Factored) (kN/m^2)
Flat roof	0.85	0.75	1.19	1.20	2.39
Cavity wall	3.50	—	4.90	—	4.90
Block wall	2.50	—	3.50	—	3.50
Floor 1	1.50	1.50	2.10	2.40	4.50
Floor 2	0.85	1.50	1.19	2.40	3.59

Figure 1.38: Total loads for building – Example 1.5

	Beam A	Point B	Point C
Flat roof	$2.39\,kN/m^2 \times (10/2)\,m = 11.95\,kN/m$	$2.39\,kN/m^2 \times (10/2)\,m = 11.95\,kN/m$	
Cavity wall	$4.90\,kN/m^2 \times 3.0\,m = 14.70\,kN/m$	$4.90\,kN/m^2 \times (3.0+3.5+0.8)\,m = 35.77\,kN/m$	
Block wall			$3.5\,kN/m^2 \times (3.5+0.8)\,m = 15.05\,kN/m$
Floor 1		$4.5\,kN/m^2 \times (6.0/2)\,m = 13.50\,kN/m$	$4.5\,kN/m^2 \times (6.0/2)\,m = 13.50\,kN/m$
Floor 2	$3.59\,kN/m^2 \times (5.8/2)\,m = 10.41\,kN/m$		$3.59\,kN/m^2 \times (5.8/2)\,m = 10.41\,kN/m$
TOTAL	37.06 kN/m run	61.22 kN/m run	38.96 kN/m run

Figure 1.39: Total loads on beam A and at points B and C for building – Example 1.5

◇

We are not able to calculate the bearing pressure beneath foundations B and C unless we include the weight of the foundations themselves. In a like manner, we can design beam A to carry a factored distributed load of 37.06 kN/m, however it is sometimes necessary to include the self-weight of the beam in our calculations. This may be particularly important in the design of large concrete members. This is easily achieved by calculating the weight of the beam, per metre length, and, after factorising, adding it to the previously calculated load of 37.06 kN/m.

Hence, if beam A is a concrete section of density 2400 kg/m^3 and cross-section 300 mm × 450 mm the factored self-weight would be:

$$\text{Self-weight} = ((2400\ kg/m^3 \times (9.81\ m/s^2)) \times 0.3\ m \times 0.45\ m) \times 1.4 = 4449.8\ N/m\ run = 4.45\ kN/m\ run$$

Therefore, the load on beam A including its self-weight will be:

$$\text{Total factored load (including self-weight of beam)} = 37.06 + 4.45 = 41.51\ kN/m\ run$$

A similar procedure can be adopted to calculate loads transmitted to various elements of any structure.

Having completed this chapter you should now be able to:

- ***Calculate force vectors***
- ***Calculate moments***
- ***Define dead, imposed and wind loads***
- ***Describe member reactions and loads in terms of***
  - ***Axial tension and compression***
  - ***Bending***
  - ***Shear***
  - ***Torsion***
- ***Determine the load paths for simple structures***
- ***Calculate the distribution of loading on various structural elements***

Further Problems 1

1. For the system of forces shown in Figure Question 1:

   (a) Calculate the horizontal and vertical components for each of the forces;
   (b) Determine if the given system of forces are in static equilibrium.

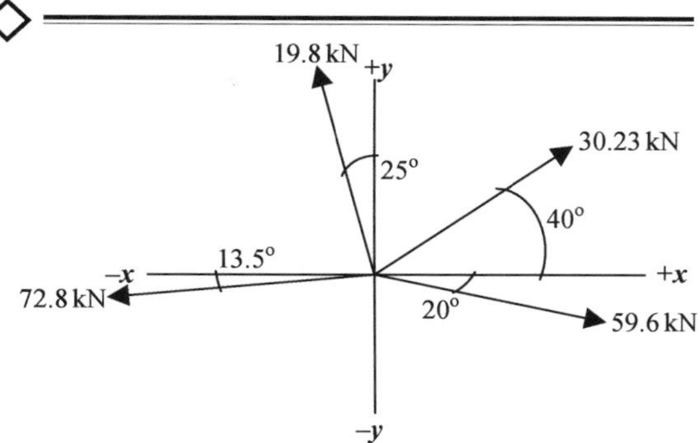

Figure Question 1

2. Figure Question 2 shows a system of forces applied to a beam which has a hinged support at joint P.

   Determine a single force and moment system that could be used to produce a similar effect at joint P.

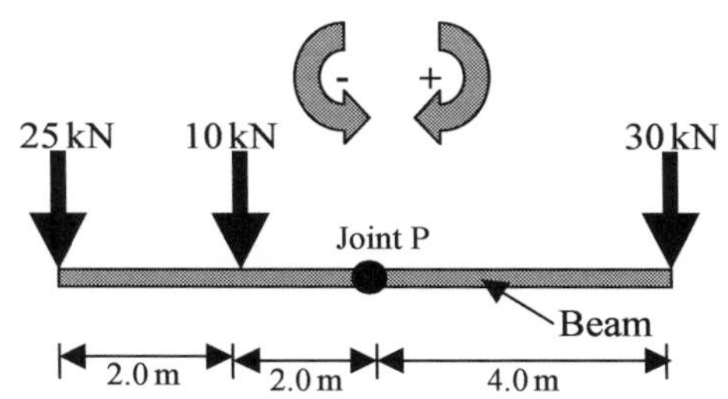

Figure Question 2

3. Figure Question 3 shows a section through a building. Below is a table indicating the loads imposed by each of the structural elements.

   Assuming a load factor for dead loads of 1.4 and for imposed loads of 1.6, calculate the total factored loads at beam B and foundations A and B. The self-weight of beam B can be ignored in this case.

	Dead Load (kN/m^2)	Imposed Load (kN/m^2)
Pitched roof 1	0.95	1.0
Pitched roof 2	0.75	1.0
Cavity wall	3.00	—
Brick wall	1.80	—
Floor 1	1.00	1.5
Floor 2	0.85	1.5
Foundation	4.00	—

Table of loads – Question 3

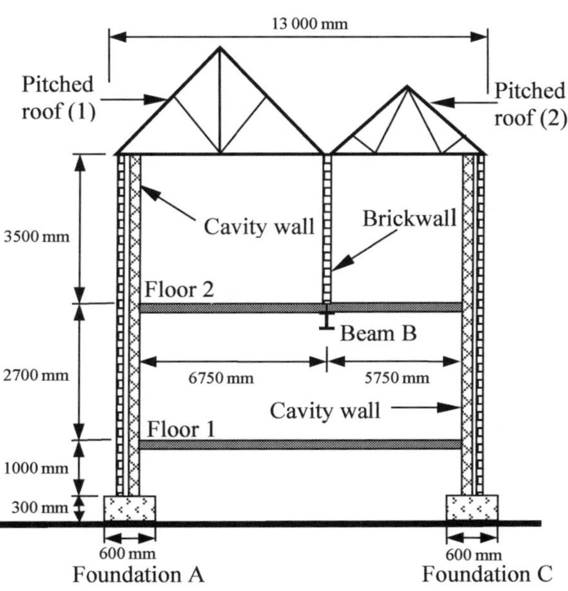

Figure Question 3

4. Figure Question 4 shows a section through a building. Below is a table indicating the dead and imposed loads contributed by each of the structural elements.

Calculate the total factored loads at beam A and foundations B, C and D. Self-weight of the beam should be included. Concrete density may be assumed to be 2400 kg/m^3.

	Total Load (kN/m^2)
Flat roof	2.00
Cavity wall	4.00
Brick wall	6.00
Block wall	2.50
Floor 1	2.80
Floor 2	2.00
Floor 3	2.00

Table of total loads – Question 4

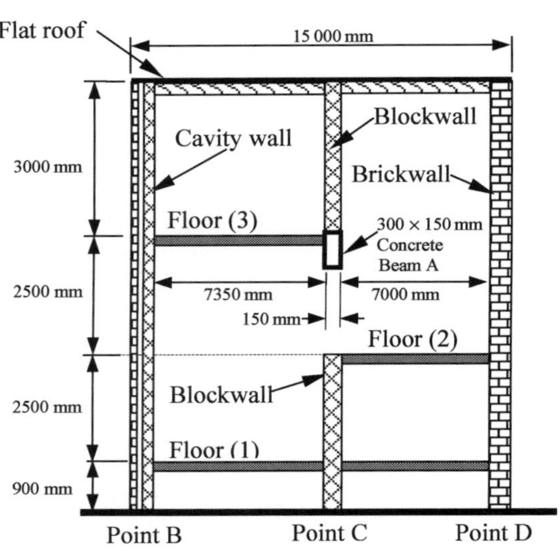

Figure Question 4

5. A concrete beam is to be used as a support, as shown in Figure Question 5. The total load imposed is shown in the table below. Calculate the total load, per unit length, to which the beam will be subjected including its self-weight.

	Dead Load (kN/m^2)	Imposed Load (kN/m^2)
Flat roof	0.83	0.75
Cavity Wall	3.33	—
Floor	1.00	1.5

Table of applied loads – Question 5

Density of concrete, 2400 kg/m^3

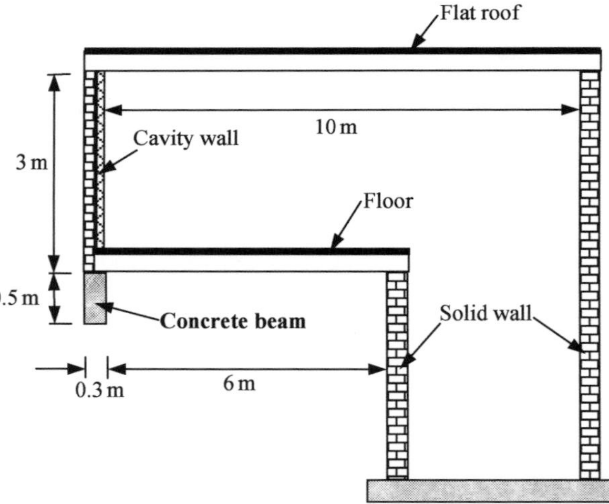

Figure Question 5

Solutions to Further Problems 1

1. Resolution of forces

	Horizontal	Vertical
Force A	30.23 cos 40° = 23.16 kN	30.23 sin 40° =19.43 kN
Force B	59.6 cos 20° = 56.00 kN	–59.6 sin 40° = –20.38 kN
Force C	–72.8 cos 13.5° = –70.79 kN	–72.8 sin 13.5° = –16.99 kN
Force D	–19.8 sin 25° = –8.37 kN	19.8 cos 25° =17.94 kN
TOTAL	**0.00 kN**	**0.00 kN**

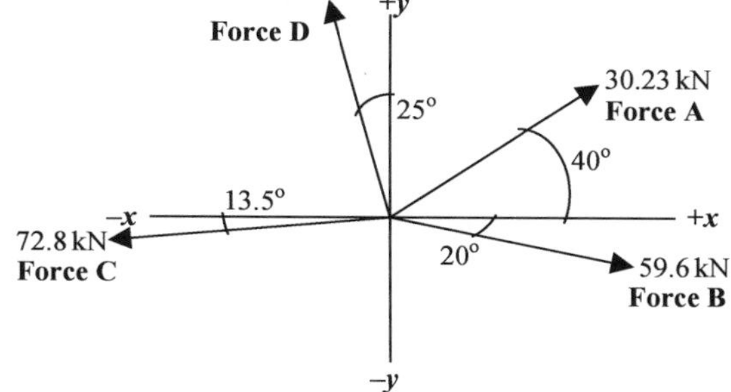

Σ Horizontal forces = 0 and
Σ Vertical forces = 0
∴ System is in equilibrium.

2. Resolution of moments and forces

Moments of loads:

Load A = –25 kN × (2.0 m + 2.0 m)
= –100 kN m

Load B = –10 kN × 2.0 m
= –20 kN m

Load C = 30 kN × 4.0 m
= 120 kN m

Sum of all moment at joint P:
= –100 – 20 + 120 kN m = **0.0 kN m**

Sum of all forces at joint P:
= –25 kN – 10 kN – 30 kN = **–65 kN**

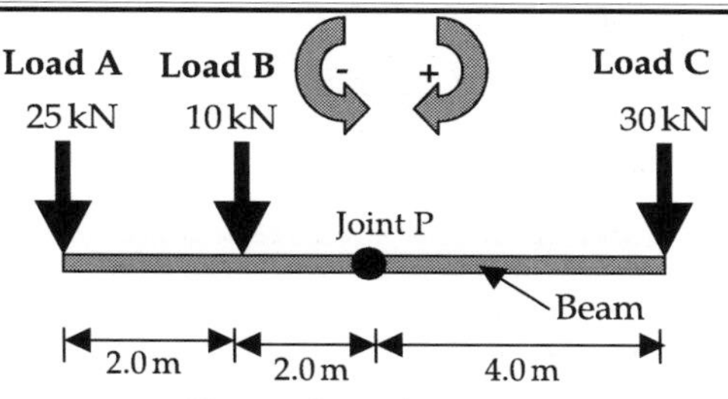

Figure Question 2

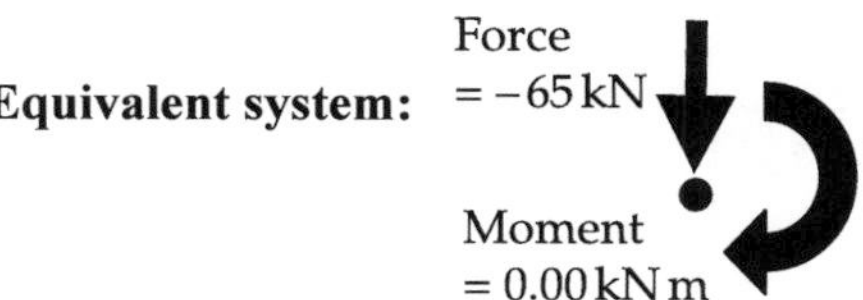

3. Loads on foundations and beam

	Dead Load × 1.4 (kN/m²)	Imposed Load × 1.4 (kN/m²)	Total Factored Loads (kN/m²)
Pitched roof 1	1.33	1.6	2.93
Pitched roof 2	1.05	1.6	2.65
Cavity wall	4.20	—	4.20
Brick wall	2.52	—	2.52
Floor 1	1.40	1.5	3.80
Floor 2	1.19	1.5	3.59
Foundation	5.60	—	5.60

Table of factored loads

	Found. A (kN/m)	Beam B (kN/m)	Found. C (kN/m)
Pitched roof (1)	9.89	9.89	—
Pitched roof (2)	—	7.62	7.62
Cavity wall	30.24	—	30.24
Brick wall	—	8.82	—
Floor 1	23.75	—	23.75
Floor 2	12.12	22.44	10.32
Foundation	5.60	—	5.6
TOTAL	**81.60**	**48.77**	**77.53**

Table of total loads

4. Load path to beam and foundations

Self-weight of concrete beam
$= (2400\,\text{kg/m}^3 \times 0.3 \times 0.15 \times 9.81)/1000$
$= 1.06\,\text{kN/m}$ run

	Beam A (kN/m)
Flat roof	14.35
Cavity wall	—
Brickwall	—
Blockwall	7.50
Floor 1	—
Floor 2	—
Floor 3	7.35
Self-weight	1.06
TOTAL	**30.26**

Total loads for beam A

	Point B (kN/m)	Point C (kN/m)	Point D (kN/m)
Flat roof	7.35	—	7.00
Cavity wall	35.60	—	—
Brickwall	—	—	53.4
Blockwall	—	8.50	—
Floor 1	10.29	20.09	9.80
Floor 2	—	7.00	7.00
Floor 3	7.35	—	—
Self-weight	—	—	—
TOTAL	**60.59**	**35.59**	**77.2**

Total loads at points B, C and D

5. Loads applied to concrete beam

	Dead Load (kN/m^2)	Imposed Load (kN/m^2)	Total Load (kN/m^2)
Flat roof	0.83	0.75	1.58
Cavity wall	3.33	—	3.33
Floor	1.00	1.5	2.5

Table of total loads

	Load on concrete beam (kN/m)	
Flat roof	$1.58\,\text{kN/m}^2 \times 10\,\text{m}/2 =$	7.90
Cavity wall	$3.33\,\text{kN/m}^2 \times 3.0\,\text{m} =$	9.99
Floor 1	$2.5\,\text{kN/m}^2 \times 6\,\text{m}/2 =$	7.50
Self-weight		3.53
TOTAL		**28.92**

Load on concrete beam

Self-weight of concrete beam
$= (2400\,\text{kg/m}^3 \times 0.5 \times 0.3 \times 9.81)/1000$
$= 3.53\,\text{kN/m}$ run

2. Section Properties and Moment of Resistance

In this chapter we will learn about:

- ***Direct stress***
- ***The 'Engineers' bending equation'***
- ***Position of a sections centroid***
- ***Second Moment of Area***
- ***Bending stress***
- ***Moment of resistance***

2.1 Introduction

As engineers we are required to design structural members that are safe and adequately meet the serviceability requirements of the client. Safety must be our prime consideration and whilst some concessions may be made in serviceability, NONE can be made with safety.

Structural elements are required to carry various amounts of load over varying distances and in varying positions. Serviceability requirements will also vary. Whilst a vertical displacement to span ratio of 1:360[1] might be acceptable in a traditionally constructed dwelling, in a structure clad with glass a ratio of 1:3000 or higher might be more appropriate. This is because in traditional construction a small amount of flexibility to take up movement in mortar joints, the elasticity of materials or other finishes is acceptable; whilst glass, which is very brittle, may well shatter when stressed by small movements of the restraining structure.

The serviceability requirements will also be governed by the degree of comfort required by the building user and the acceptable amount of damage/movement specified in the design standards. In design we must ensure that structures meet both safety and serviceability requirements. This demands that we design members which are adequate to support the worst-case load to which the structure will be subjected. The amount of load a structural member can carry depends on many factors including load position, type, pattern, span and support conditions. We could specify members by tabulating each of these factors but, since there are an infinite number of such combinations to be considered, we generally design for each individual case.

To ensure a member is adequate for its purpose, we design it such that the extreme fibre stress does not exceed a specified maximum in either tension or compression. This specified value, in many instances, has been determined following numerous tests on the material to be used. The specified value may also have a further factor of safety applied through factoring the loads as discussed in Chapter 1.

1. British Standards Institution, BS 5950: 1985: Structural use of steelwork in buildings: Part 1: Code of practice for the design of simple and continuous construction: hot rolled sections.

2.2 Section Properties

2.2.1 Stress

We have already found direct stress for a simple tensile or compressive model is given by:

$$\text{Direct stress}\,(\sigma) = \frac{\text{Force}}{\text{Area}} = \frac{F}{A} \qquad [2.1]$$

For the tensile sample shown in Figure 2.1 the applied forces F pulls the member in the direction of the force. In order to maintain its equilibrium position, and preserve the position of the ends of the member, the member must respond with an internal resistance to the force. Since the forces are pulling in an outward direction, in relation to the member, the internal resistance must be inward and hence tensile stresses are induced.

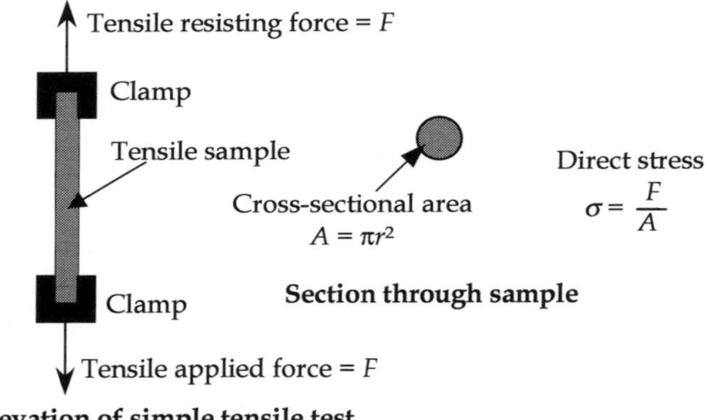

Figure 2.1: Tension-loaded member

The problem when designing beams and other similar members is that they are often subjected to bending stresses rather than direct tensile or compressive stress. A relationship exists, commonly referred to as the Engineers' bending equation, which can be used to calculate stress distribution across a member and is defined as:

$$\frac{M}{I} = \frac{\sigma}{y} = \frac{E}{R} \qquad [2.2]$$

This is a very important equation which will prove to be extremely useful in structural design and analysis. You should therefore endeavour to remember it. It is extremely important to ensure that when using this equation a consistent set of units is adopted. The most common units for stress are N/mm^2 and it may be necessary to convert values to ensure consistency.

We will now consider each part of the equation in turn.

M* = *Moment (N mm)

This is the bending moment, normally taken as the maximum bending moment calculated, as this will produce the maximum extreme fibre stress.

***I* = *Second Moment of Area* (mm^4)**

This refers to a geometric property for the section and is related to its shape and dimensions. We will discuss the determination of the *I*-value for various shapes of sections later in this chapter.

σ = *Stress* (N/mm^2)

This is the stress calculated at the position of the applied moment, at any distance from the centroid of the cross-section. It is important to note that the specified maximum allowable stress for certain materials may be different in tension and compression. For instance, concrete is a very good compressive material but very weak in tension.

y* = *Distance in vertical direction (mm)

This is the distance from the centroid to the point at which we wish to calculate the stress. Maximum stress will normally occur in the extreme fibres of the member, in which case *y* is the distance from the centroid to the top or bottom of the beam. However, we can calculate the stress at any point through the section by varying the distance *y*. For a rectangular section loaded in the vertical downward plane, the stress distribution will generally be of the form shown in Figure 2.2. Note that if stress is required horizontally across the section, *x* distances will be substituted for *y* distances.

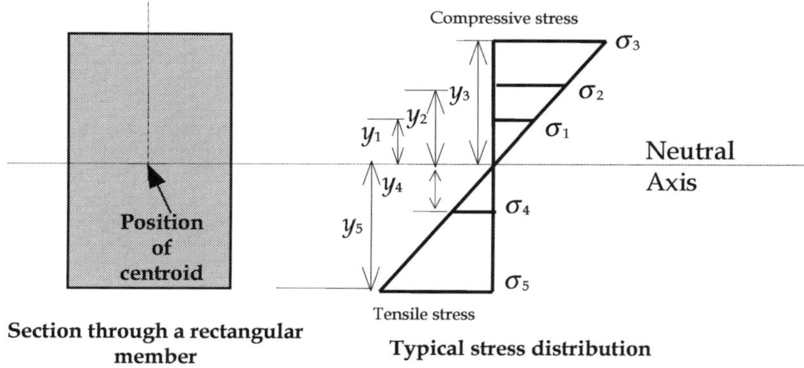

Figure 2.2: Typical stress distribution for a rectangular section

Along the neutral axis of the object (which passes through the centroid) the stress is zero. This is the position at which the stress changes from positive (compressive) to negative (tensile) for a beam that deflects in the form shown in Figure 2.3. In this case the beam is loaded on the top face and this will result in a downward displacement causing a shortening in the top of the beam and lengthening in the bottom. The amount is usually very small in relation to the member's overall length and is shown in an exaggerated form in the figure.

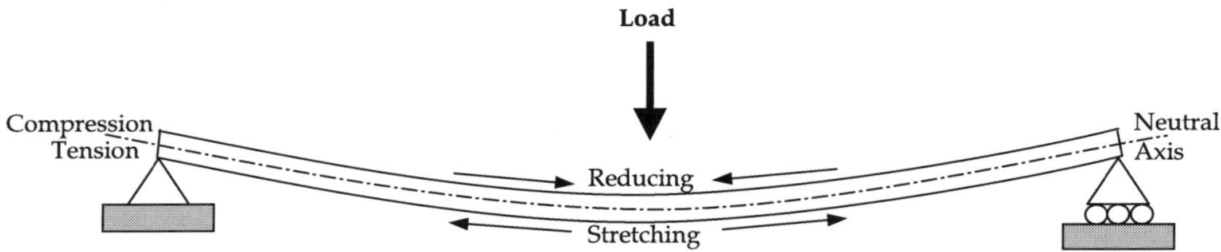

Figure 2.3: Deflected form of beam supporting a vertical downward load

The tension induced at the bottom of the beam tends to stretch the material. If the beam is constructed of steel or other ductile material, this will not normally be a problem as long as maximum tensile stresses are not exceeded. If, however, it is constructed of concrete, which is very weak in tension, we find that cracks will appear in the tension zone. These cracks can enlarge and in extreme cases cause failure. To improve the bending characteristics of concrete we often insert a ductile material, such as steel reinforcement, in the tension zone.

E = Young's Modulus or Elastic Modulus (N/mm^2)

This is a property of the material and relates its stress to strain characteristics. If we applied a load to a test steel sample of known area, we can calculate the direct stress (σ) from:

$$\sigma = \frac{\text{Force}}{\text{Area}}$$

Strain (ε) is a measure of the deformation of the material, that is how much its length varies under load and is given by the formula:

$$\varepsilon = \frac{\text{Change in length}}{\text{Original length}} \qquad [2.3]$$

Note that strain is measured as 'change in length' or 'extension' over original length and **not** extended length over original length. By measuring these properties we will be able to plot a graph to show how they vary, for a specified material under load, as shown in Figure 2.4.

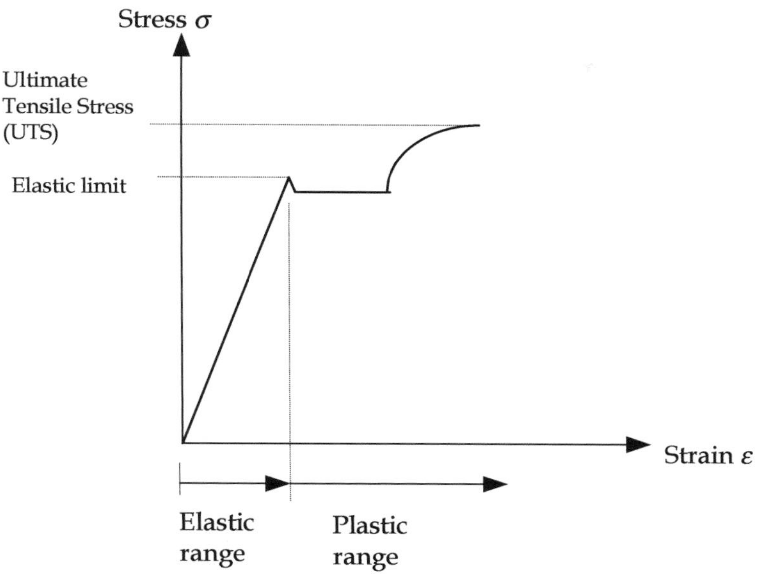

Figure 2.4: Typical stress versus strain graph for steel

Within the elastic range, the sample may be loaded and the strain recorded. If the sample is then unloaded it will return to its original size and shape. In this region, stress is proportional to strain. This is known as Hooke's Law. Loading beyond the elastic limit will cause a permanent deformation such that the sample will not return to its original length. The stress–strain path back to the unloaded state will however generally be parallel to the slope of the original line in the elastic region, as shown in the Figure 2.5. The sample will fail at its Ultimate Tensile Stress (indicated in Figure 2.4).

We normally design structural members within the elastic range so that, on unloading, they will recover their original shape. Later in your studies you may well consider designs completed in the plastic range but this is for a very specialised range of structures.

Within the elastic range we can define a property specific to the material, known as its Young's or Elastic Modulus. This is given by calculating the gradient (slope of the line) within the elastic range:

$$E = \frac{\text{Stress}}{\text{Strain}} = \frac{\sigma}{\varepsilon} = \frac{\text{Force/Area}}{\text{Extension/Original length}} \qquad [2.4]$$

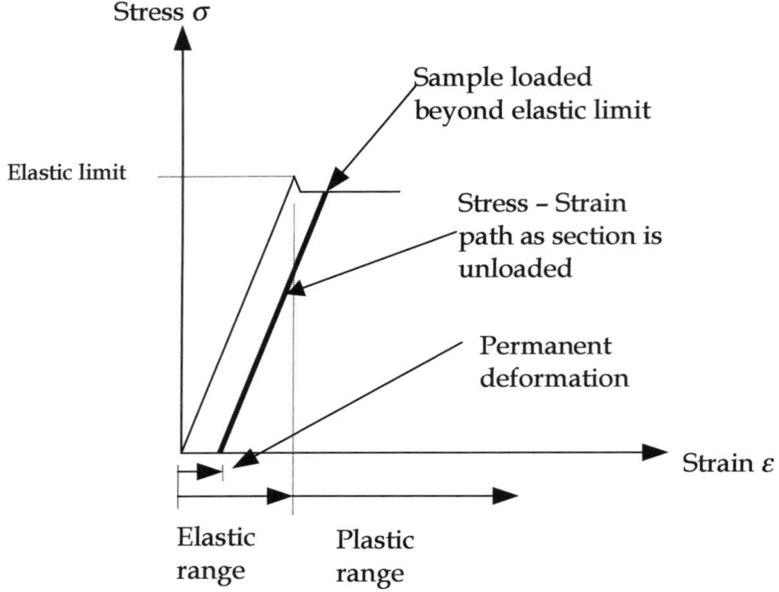

Figure 2.5: Graph for a sample loaded beyond the elastic range

Many tests have been completed on samples in order to determine its E value. These tests have also been used to determine limits of stress for the safe design of structures. These values are usually specified as the 'maximum permissible stress'. The limits are often given as a percentage of the actual elastic limit in order to account for variations in material properties and quality of construction, and add a further factor of safety to that provided by the load factors mentioned in Chapter 1.

R = Radius of curvature (mm)

This is the radius of curvature of the member under the specified loading. It is useful to note that this property may be used to help calculate the overall displacement of a beam, and it will be discussed in detail in Chapter 5.

In order to calculate the stress for a simple member we need only determine the applied moment (M), the Second Moment of Area for the section (I) and the distance from the centroid of the section to the point at which the stress is to be calculated (y), since:

$$\frac{M}{I} = \frac{\sigma}{y} = \frac{E}{R}$$

and as

$$\frac{M}{I} = \frac{\sigma}{y} \qquad [2.5]$$

so

$$\sigma = \frac{My}{I} \qquad [2.6]$$

2.2.2 Fundamental Section Properties

For any section we will be able to determine a number of properties:

Area (A mm^2)

Calculation of area depends on the size and the shape of the object being considered. Areas of complex sections can be calculated by reducing them to simple basic shapes and then adding the parts together to determine the area of the compound section. Typical areas of some common shapes are shown in Figure 2.6.

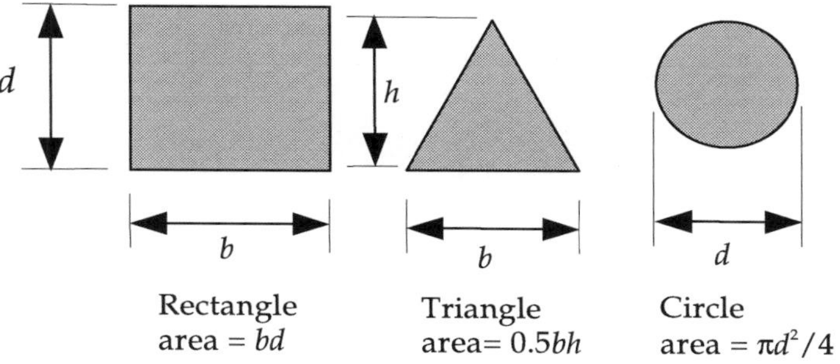

Figure 2.6: Areas of common sections

Centroid ($\bar{x}$ or $\bar{y}$ mm)

This is the centre of the area for the section and is a very important point since any bending will take place about the longitudinal axis through this point. Its distance from a known datum in normal Cartesian co-ordinates defines the position of the centroid. The datum can be taken at any point within the section but for ease the Horizontal Datum is normally considered at the bottom and the Vertical Datum to the left of the section, so that distances are always measured in a positive direction as shown for the rectangular section in Figure 2.7.

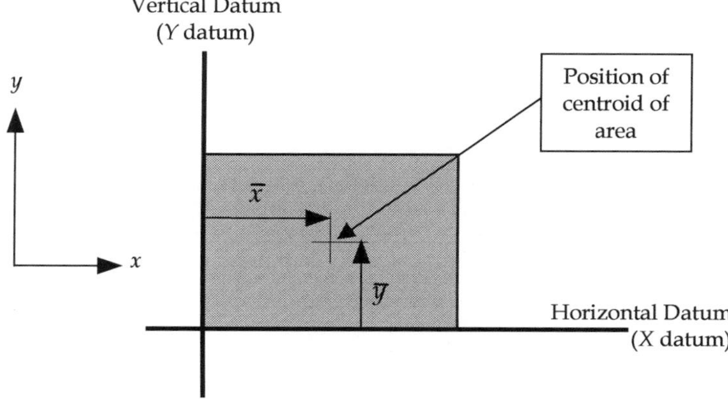

Figure 2.7: Co-ordinate system for calculation of centroid position

The position of the centroid for various common sections is shown in Figure 2.8:

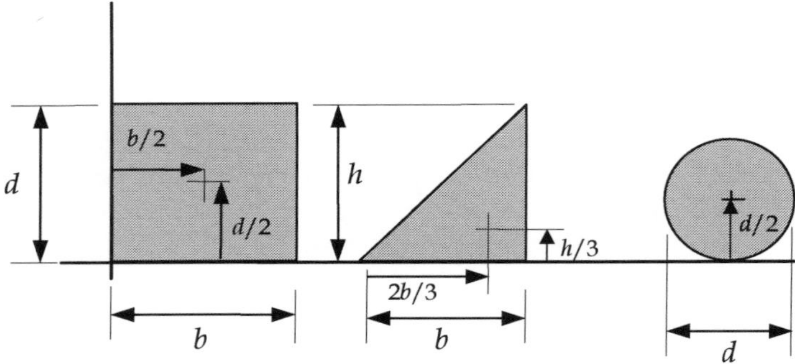

Figure 2.8: Centroid positions for some common shapes

The centroid of compound sections, that is a section made up of these standard shapes, may be calculated using the following formula (Figure 2.7):
For the distance from the *Y* datum to the centroid in the horizontal direction:

$$\bar{x} = \frac{\Sigma Ax}{\Sigma A} \qquad [2.7]$$

For the distance from the *X* datum to the centroid in the vertical direction:

$$\bar{y} = \frac{\Sigma Ay}{\Sigma A} \qquad [2.8]$$

◇

Example 2.1: Centroid of a 'tee' section

Consider the compound section of Example 2.1 shown in Figure 2.9 below. Calculate the position of the centroid for the compound section:

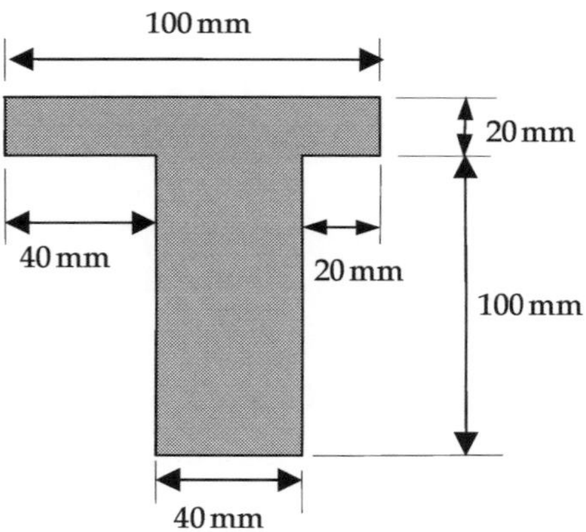

Figure 2.9: Example 2.1 – details of 'tee' section

First, we locate our datum for the x and y axes to the bottom and left of the shape, as shown in Figure 2.10, and we split the complex shape into two simple rectangles.

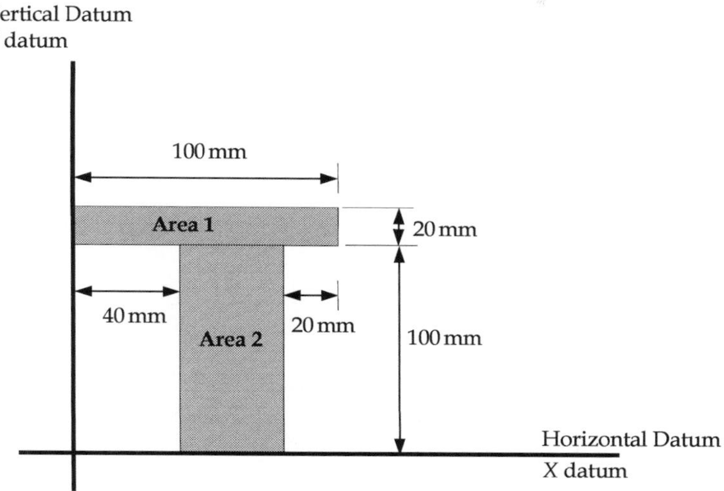

Figure 2.10: Example 2.1 – location of *X* and *Y* datums and details of parts

We can now calculate the area of each part as follows:

Area 1 = 100 mm × 20 mm = 2000 mm
Area 2 = 100 mm × 40 mm = 4000 mm

To calculate the y distance for each part, we must determine the distance from the X datum to the centroid of the part. For a rectangle, the centroid is located as shown in Figure 2.8 at half its depth in the vertical direction. However, for each part we must calculate the distance to the centroid from the datum such that:

Distance to the centroid of Part 1 from X datum: $y_1 = 100\,\text{mm} + 20/2\,\text{mm} = 110\,\text{mm}$
Distance to the centroid of Part 2 from X datum: $y_2 = 100/2\,\text{mm} = 50\,\text{mm}$

x distances are calculated in a similar way. However, in this case we are calculating the distance from the Y datum to the centroid of each part such that:

Distance to the centroid of Part 1 from Y datum: $x_1 = 100/2\,\text{mm} = 50\,\text{mm}$
Distance to the centroid of Part 2 from Y datum: $x_2 = 40\,\text{mm} + 40/2\,\text{mm} = 60\,\text{mm}$

To calculate the distance to the centroid for the compound section, measured from the X and Y datums, we use the formulae given in equations [2.7] and [2.8], hence:

$$\bar{x} = \frac{\Sigma Ax}{\Sigma A} = \frac{(2000 \times 50) + (4000 \times 60)}{(2000 + 4000)} = 56.67\text{ mm}$$

$$\bar{y} = \frac{\Sigma Ay}{\Sigma A} = \frac{(2000 \times 110) + (4000 \times 50)}{(2000 + 4000)} = 70\text{ mm}$$

The centroid is therefore located 56.67 mm measured from the Y datum and 70 mm measured from the X datum as shown in Figure 2.11.

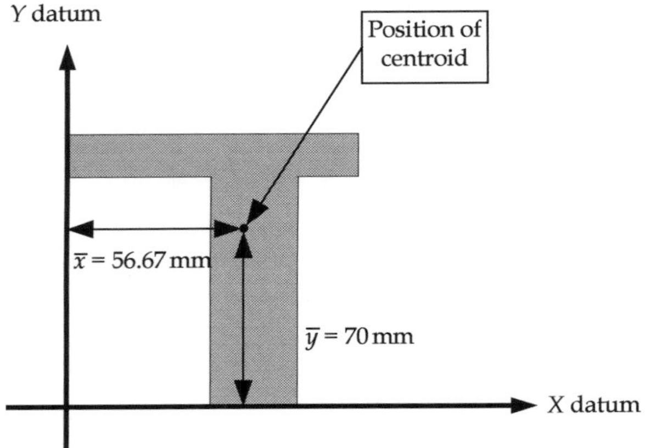

Figure 2.11: Position of centroid for compound 'tee' section of Example 2.1

The general formulae can be applied to all sections which can be broken down into standard shapes, and the centroid position found.

Second Moment of Area (I mm^4)

This section property is always defined about a specified axis passing through the centroid of the section. This is because the types of section used by engineers are specifically designed to exploit the strength of a material when oriented in a specified direction. For instance, timber floor beams are normally placed with their narrow axis horizontal and the longer axis vertical, as shown in Figure 2.12. This is because bending is designed to take place about the vertical axis whilst the beam will be subjected to little or no horizontal load.

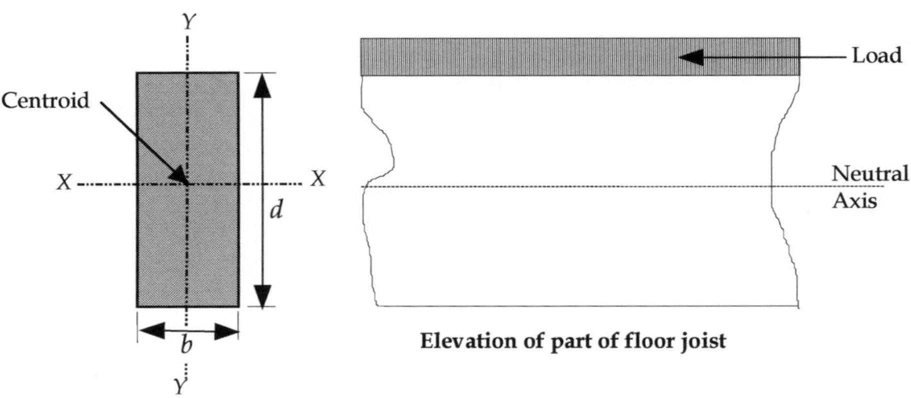

Figure 2.12: Typical orientation of timber floor joist (beam)

The *I*-value is therefore usually specified as I_{xx} or I_{yy} depending on which axis the value is calculated about. In calculations you must be careful to select the correct value, depending on the axis about which bending will take place. In the case of our floor beam, bending will occur about the *X–X* axis and so I_{xx} is the appropriate value to use in design. Note that both the *X–X* and *Y–Y* axes pass through the centroid of the compound section, and should not be confused with the *X* and *Y* datums which are used to determine the centroid position.

Mathematically, the Second Moment of Area is defined as the moment of the area about a known datum. Consider a small area *A*, a known distance *h* from a specified *X* datum, as shown in Figure 2.13. (Note that a similar relationship can be derived about the *Y* datum.)

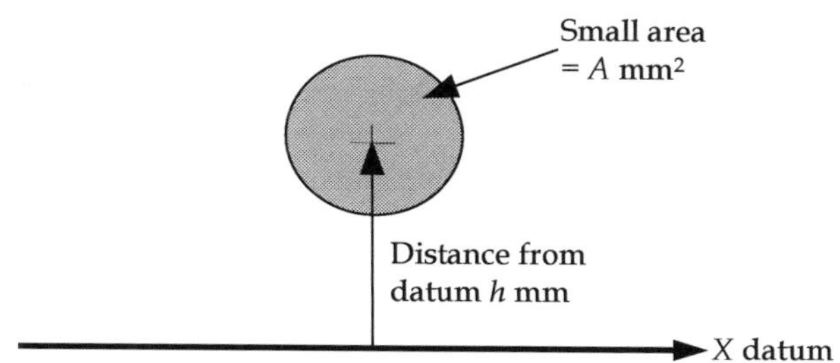

Figure 2.13: Small area with centroid located at *h* from *X* datum

The following properties can be calculated for the section about the datum:

Moment of Area about *X* datum = Ah [2.9]

Second Moment of Area about *X* datum = Ah^2 [2.10]

The Second Moment of Area for a number of common shapes is shown in Figure 2.14:

Section	Area *A*	I_{xx}	I_{yy}
Y, X, X, d, b, Y	bd	$\frac{bd^3}{12}$	$\frac{db^3}{12}$
Y, X, X, h, b, Y	$0.5bh$	$\frac{bh^3}{36}$	$\frac{hb^3}{36}$
d, X, X	$\frac{\pi d^2}{4}$	$\frac{\pi d^4}{64}$	$\frac{\pi d^4}{64}$

Figure 2.14: Second Moment of Area for some common shapes

An area has a Second Moment of Area about an axis through its centroid I_{cg}. All the I-values in Figure 2.14 are given relative to the centroid. It is possible using, for instance, the Parallel Axis Theorem to calculate the Second Moment of Area about a datum some distance from the centroid.

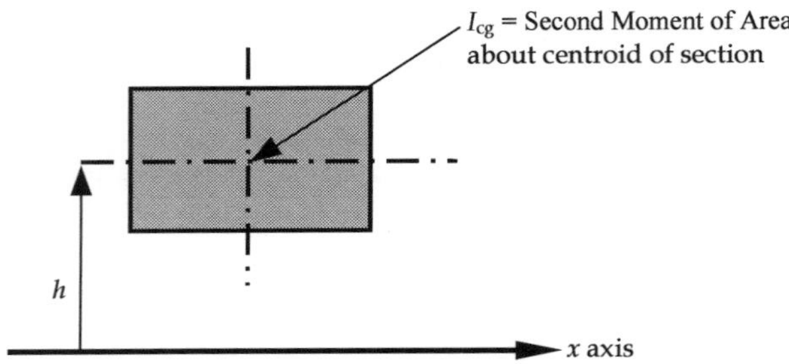

Figure 2.15: Second Moment of Area for a section about a given axis

The I-value about the datum is given by the Parallel Axis Theorem as shown in Figure 2.15 by:

$$I_{\text{about datum}} = I_{cg} + Ah^2 \qquad [2.11]$$

where I_{cg} = Second Moment of Area for the section about its centroid
A = Area of the section
h = Distance from x axis about which Second Moment of Area is to be calculated.

In this case, since the distances are measured from the X datum, the Second Moment of Area would be given for the x axis and described as an I_{xx} value. We can modify the Parallel Axis Theorem in order to help us calculate the Second Moment of Area for a compound section through its centroid about either the X–X or Y–Y axes. In this case the above formula may be rewritten as:

$$I_{xx}\text{ (compound section)} = \Sigma\, I_{xx} + \Sigma\, A(\bar{y} - y)^2 \qquad [2.12a]$$

Also

$$I_{yy}\text{ (compound section)} = \Sigma\, I_{yy} + \Sigma\, A(\bar{x} - x)^2 \qquad [2.12b]$$

It is often easier to complete calculations using a table, as this will help to reduce and identify any errors. A simple table is proposed in Figure 2.16:

Part No.	Area A (mm^2)	Distance Y or X (mm)	$A \times Y$ or $A \times X$ (mm^3)	I_{xx} or I_{yy} (mm^4)	h or $(\bar{y} - y)$ or $(\bar{x} - x)$ (mm)	Ay^2 or $A(\bar{y} - y)^2$ or $A(\bar{x} - x)^2$ (mm^4)
Column (1)	Column (2)	Column (3)	Column (4)	Column (5)	Column (6)	Column (7)
	ΣA		ΣAy or ΣAx	ΣI_{xx} or ΣI_{yy}		$\Sigma A(\bar{y} - y)^2$ or $\Sigma A(\bar{x} - x)^2$

Figure 2.16: Table for calculation of Second Moment of Area

The terms required in the table have been previously defined but it is worth reiterating a number of points. In column 1 (Part No.), it is often helpful to draw a simple sketch of the part and mark its dimensions. This will save constant reference to the original drawings and hence assist in reducing the possibility of errors. The part should be the simplest shape possible. Column 2 (Area) is for the area of the section. Column 3 is for the distance from either the *X* or *Y* datum. The table can be used for either or extra columns can be added if both are required, however, it is often easier to use separate tables for each axis. Again, columns 4 and 5 may be modified for calculations about the particular axis required. Column 5 requires the calculation of the Second Moment of Area for the **individual part about its centroid** using the formula given in Figure 2.14. In column 6 we need to calculate the distance from the centroid for the compound section to the centroid of the individual part. Since we have already calculated the distance from our datum to the centroid of the individual part and placed the result in column 2 – and since we can calculate the distance to the centroid of the compound section as previously described, again measured from the *X* and *Y* datums – the distance from the centroid of the individual part to the centroid of the compound section must be $(\bar{y}-y)$ or $(\bar{x}-x)$ as shown for Part 1 in Figure 2.17.

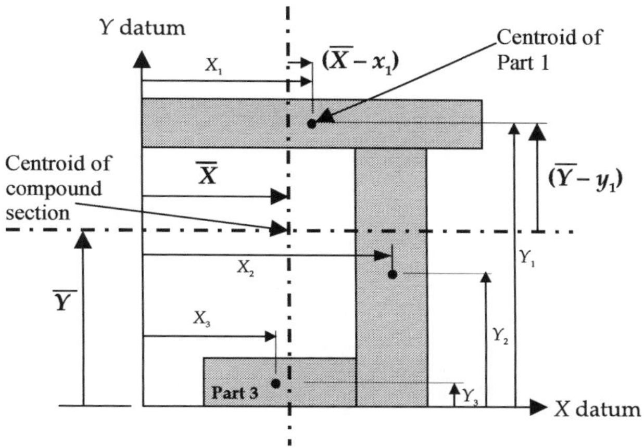

Figure 2.17: Relationship of centroid of compound section and centroid of parts

The table can be used to calculate either *I*-value by substituting the correct *X* or *Y* distance and the correct Second Moment of Area for each part. Distance from the centroid to the part is designated as *h*.

Example 2.2: Position of centroid and Second Moment of Area calculations for a 'tee' section

Calculate the Second Moment of Area about the *X–X* and *Y–Y* axis for the section shown in Figure 2.18.

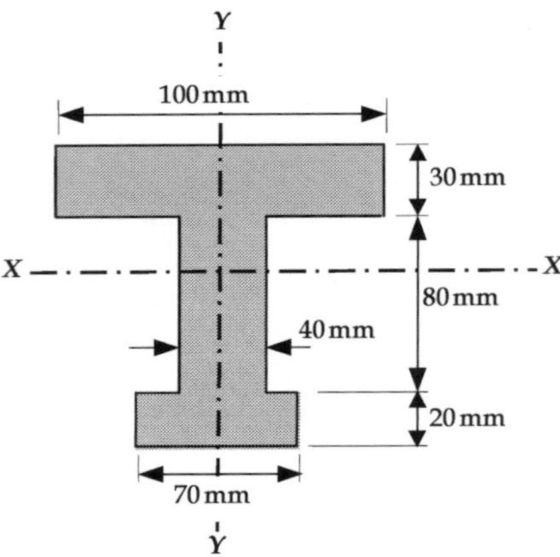

Figure 2.18: Compound 'tee' section – Example 2.2

First, we note that the section is symmetrical about the *Y–Y* axis. It is therefore possible to locate the position of the centroid; by inspection, $\bar{x} = 50\,\text{mm}$, as shown in Figure 2.19. We must now split the compound section into parts, as shown, and locate the datums to the bottom and left of the section from which all distances will be measured.

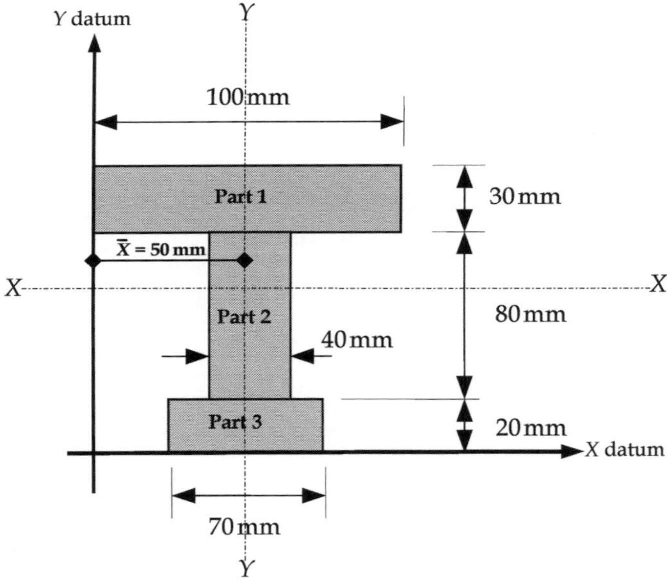

Figure 2.19: Breakdown of compound section – Example 2.2

We then set up a table, as shown in Figure 2.20, in this case to calculate the Second Moment of Area about the *X–X* axis (I_{xx}).

Part No.	Area A (mm^2)	Distance Y (mm)	$A \times Y$ (mm^3)	I_{xx} (mm^4)	$(\bar{y}-y)$ (mm)	$A(\bar{y}-y)^2$ (mm^4)
PART 1 (100 mm × 30 mm)	3000	$80+20+30/2=115$	345 000			
PART 2 (40 mm × 80 mm)	3200	$20+80/2=60$	192 000			
PART 3 (70 mm × 20 mm)	1400	$20/2=10$	14 000			
	$\Sigma A = 7600$		$\Sigma Ay =$ 551 000			

Figure 2.20: Part of table to calculate I_{xx} for compound section – Example 2.2

The distance y is calculated as the distance from the X datum to the centroid of each individual part.

After calculating the area for each section and determining the y distance to the centroid of each part from the X datum, we can sum the total area (ΣA) and the 'area $\times$ y distance' (ΣAy) for each part, and hence calculate the centroid position in the y direction. We use the totals of column (2) and (4) as follows:

$$\bar{y}=\frac{551000}{7600}=72.5 \text{ mm (measured from the } X \text{ datum located at the bottom of Part 3)}$$

In order to calculate the Second Moment of Area we must first complete the table as shown in Figure 2.21.

Part No.	Area A (mm^2)	Distance Y (mm)	$A \times Y$ (mm^3)	I_{xx} (mm^4)	$(\bar{y}-y)$ (mm)	$A(\bar{y}-y)^2$ (mm^4)
PART 1 (100 mm × 30 mm)	3000	$80+20+30/2=115$	345 000	$bd^3/12 = 100 \times 30^3/12 = 225\,000$	$72.5-115=-42.5$	$3000 \times (-42.5)^2 = 5\,418\,750$
PART 2 (40 mm × 80 mm)	3200	$20+80/2=60$	192 000	$bd^3/12 = 40 \times 80^3/12 = 1\,706\,666.67$	$72.5-60=12.5$	$3200 \times (12.5)^2 = 50\,000$
PART 3 (70 mm × 20 mm)	1400	$20/2=10$	14 000	$bd^3/12 = 70 \times 20^3/12 = 46\,666.67$	$72.5-10=62.5$	$1400 \times (62.5)^2 = 5\,468\,750$
	$\Sigma A =$ 7600		$\Sigma Ay =$ 551000	$\Sigma I_{xx} =$ 197 833.34		$\Sigma A(\bar{y}-y)^2 =$ 11 387 500

Figure 2.21: Completed table to calculate I_{xx} for compound section – Example 2.2

It should be noted that the negative values in the column containing the $(\bar{y} - y)$ can be expected as some of the centroid positions for the parts will be located above the centroid of the compound section and others will be below it. Since the values will be squared in the following column the negative sign does not affect the calculation, the negative sign merely indicating that the centroid of the part is above the centroid of the compound section. To calculate I_{xx} we then use the Parallel Axis Theorem equation [2.12a] and add the total of columns (5) and (7) giving:

$$I_{xx} \text{ (compound section)} = \Sigma I_{xx} + \Sigma A(\bar{y} - y)^2$$
$$I_{xx} = 197\,833.34 + 11\,387\,500$$
$$I_{xx} = 11\,585\,333.34 \text{ mm}^4$$

The Second Moment of Area calculated is that about the X–X axis which passes through the centroid of the compound section and is therefore about the neutral axis for the section. It is also important to note that the units used here are mm^4. It is often the case in reference tables that other units are used, such as cm^4. We must ensure that the units we use are consistent in order to provide a meaningful solution.

We can calculate the Second Moment of Area about the Y–Y axis (I_{yy}) in a similar manner and the completed table is shown Figure 2.22. Note that, as previously stated, it is not necessary to calculate the x distance to the centroid for a symmetrical figure such as this.

Part No.	Area A (mm^2)	Distance X (mm)	$A \times X$ (mm^3)	I_{yy} (mm^4)	$(\bar{x}-x)$ (mm)	$A(\bar{x}-x)^2$ (mm^4)
PART 1 30mm 100mm	3000	100/2 =50	150 000	$db^3/12 =$ $30\times100^3/12$ $= 2\,500\,000$	50–50 =0	$3000\times(0)^2$ =0
PART 2 80mm 40mm	3200	50	160 000	$Db^3/12 =$ $80\times40^3/12$ $= 426\,666.67$	0	0
PART 3 20mm 70mm	1400	50	70 000	$db^3/12 =$ $20\times70^3/12$ $= 571\,666.67$	0	0
	$\Sigma A =$ 7600		$\Sigma Ax =$ 380 000	$\Sigma I_{yy} =$ 3 498 333.34		$\Sigma A(\bar{x}\times x)^2$ =0

Figure 2.22: Completed table to calculate I_{yy} for compound section – Example 2.2

It should also be noted that since the centroid of each of the parts and the centroid of the compound section are in vertical alignment the final two columns of the table are not required, so that using the Parallel Axis Theorem equation [2.12b]:

$$I_{yy} \text{ (compound section)} = \Sigma I_{yy}$$
$$I_{yy} = 3\,498\,333.34 \text{ mm}^4$$

◇

Example 2.3: Centroid and Second Moment of Area for a compound section

Calculate the Second Moment of Area about the $X-X$ and $Y-Y$ axis for the section shown in Figure 2.23.

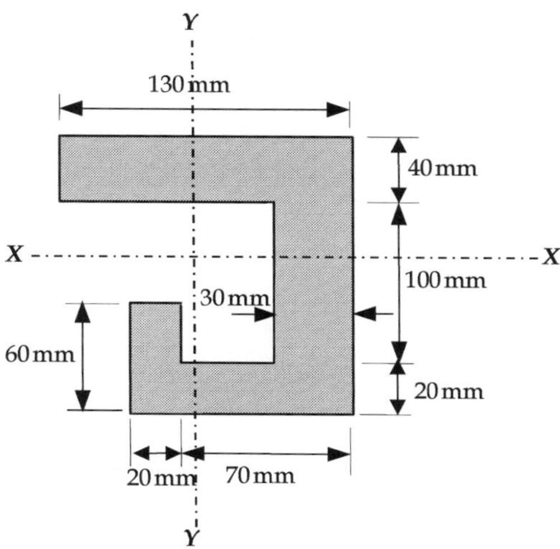

Figure 2.23: Compound section – Example 2.3

Again we must first split the compound section into parts, locate our X and Y datums and set up the table. Initially we will calculate the Second Moment of Area about the $X–X$ axis through the centroid (I_{xx}) as shown in Figure 2.24.

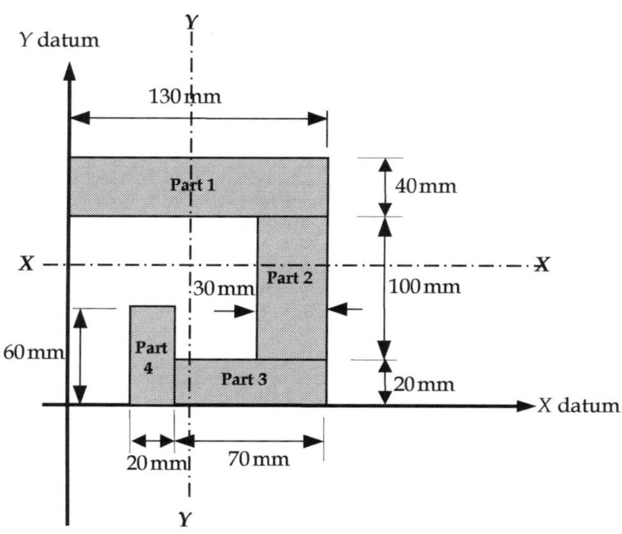

Figure 2.24: Breakdown of compound section – Example 2.3

We can then set up and calculate values from our table as before (Figure 2.25).

Part No.	Area A (mm^2)	Distance Y (mm)	$A \times Y$ (mm^3)	I_{xx} (mm^4)	$(\bar{y}-y)$ (mm)	$A(\bar{y}-y)^2$ (mm^4)
PART 1 40mm 130mm	5200	$100+20$ $+40/2$ $=140$	728 000	$bd^3/12 =$ $130\times40^3/12$ $=$ 693 333.33	91.48 -140 $=-48.52$	$5200\times$ $(-48.52)^2$ $=$ 12 241 790.1
PART 2 100mm 30mm	3000	$20+100/2$ $=70$	210 000	$bd^3/12 =$ $30\times100^3/12$ $=$ 2 500 000	91.48 -70 $=21.48$	$3000\times(21.48)^2$ $=$ 1 384 171.2
PART 3 20mm 70mm	1400	$20/2$ $=10$	14 000	$bd^3/12 =$ $70\times20^3/12$ $=46\,666.67$	91.48 -10 $=81.48$	$1400\times(81.48)^2$ $=$ 9 294 586.6
PART 4 60mm 20mm	1200	$60/2 = 30$	36 000	$bd^3/12 =$ $20\times60^3/12$ $=$ 360 000	91.48 -30 $=61.48$	$1200\times(61.48)^2$ $=$ 4 535 748.5
	$\Sigma A =$ 10 800		$\Sigma Ay =$ 988 000	$\Sigma I_{xx} =$ 3 600 000		$\Sigma A(\bar{y}-y)^2 =$ 27 456 296.4

Figure 2.25: Completed table to calculate I_{xx} for compound section – Example 2.3

To calculate the centroid position in the y direction we use the totals of columns (2) and (4):

$$\bar{y} = \frac{988000}{10800} = 91.48 \text{ mm (from the } X \text{ datum at the bottom of Part 3)}$$

Note that it is necessary to calculate this value at an early stage, normally after completion of column (4), as it is required for later calculations within the table $(\bar{y}-y)$.

To calculate I_{xx} using the Parallel Axis Theorem, we add the total of columns (5) and (7) giving:

$$I_{xx} \text{ (compound section)} = \Sigma I_{xx} + \Sigma A(\bar{y}-y)^2$$
$$I_{xx} = 3\,600\,000 + 27\,456\,296.4$$
$$I_{xx} = 31\,056\,296.4 \text{ mm}^4$$

We can calculate I_{yy} in a similar manner. The completed table is shown in Figure 2.26:

Part No.	Area A (mm^2)	Distance X (mm)	$A \times X$ (mm^3)	I_{yy} (mm^4)	$(\bar{x}-x)$ (mm)	$A(\bar{x}-x)^2$ (mm^4)
PART 1 40mm 130mm	5200	130/2 = 65	338 000	$db^3/12 =$ $40 \times 130^3/12$ $= 7\,323\,333.33$	81.1 −65 = 16.1	$5200(16.1)^2$ = 1 347 892
PART 2 100mm 30mm	3000	100 + 30/2 = 115	345 000	$db^3/12 =$ $100 \times 30^3/12$ $= 225\,000$	81.1 −115 = −33.9	$3000(-33.9)^2$ = 3 447 630
PART 3 20mm 70mm	1400	60 + 70/2 = 95	133 000	$db^3/12 =$ $20 \times 70^3/12$ $= 571\,666.67$	81.1 −95 = −13.9	$1400(-13.9)^2$ = 270 494
PART 4 60mm 20mm	1200	40 + 20/2 = 50	60 000	$db^3/12 =$ $60 \times 20^3/12$ $= 40\,000$	81.1 −50 = 31.1	$1200(31.1)^2$ = 1 160 652
	$\Sigma A =$ 10 800		$\Sigma Ax =$ 876 000	$\Sigma I_{yy} =$ 8 160 000		$\Sigma A(\bar{x}-x)^2 =$ 6 226 668

Figure 2.26: Completed table to calculate I_{xx} for compound section

$$\bar{x} = \frac{876\,000}{10\,800} = 81.1 \text{ mm (from the } Y \text{ datum at left of compound section)}$$

I_{yy} (compound section) $= \Sigma I_{yy} + \Sigma A(\bar{x}-x)^2$

$I_{yy} = 8\,160\,000 + 6\,226\,668$

$I_{yy} = 14\,386\,668 \text{ mm}^4$

Note, that the centroid does not have to be located on a physical part of the compound section, as it actually relates to a point at which the areas above and below the *X–X* axis and to the left and right of the *Y–Y* axis are in balance.

◇

Example 2.4: Centroid and Second Moment of Area for a compound section

Calculate the Second Moment of Area about the *X–X* and *Y–Y* axis for the compound section shown in Figure 2.27 (all dimensions are in millimetres).

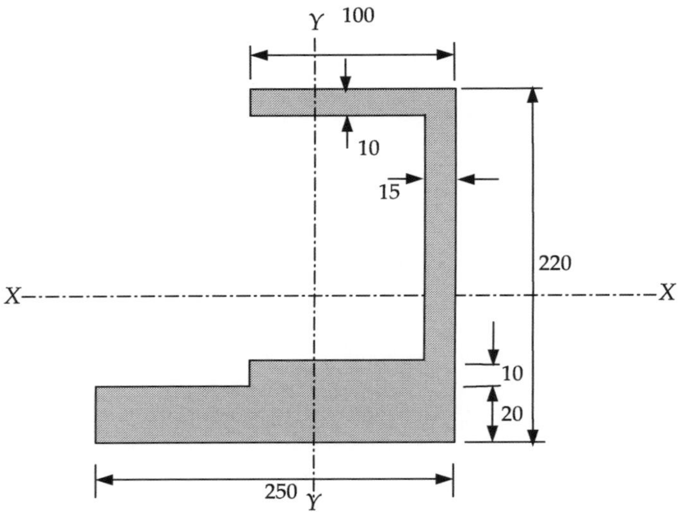

Figure 2.27: Compound section – Example 2.4

First split the compound section into parts and locate the X and Y datums as shown in Figure 2.28.

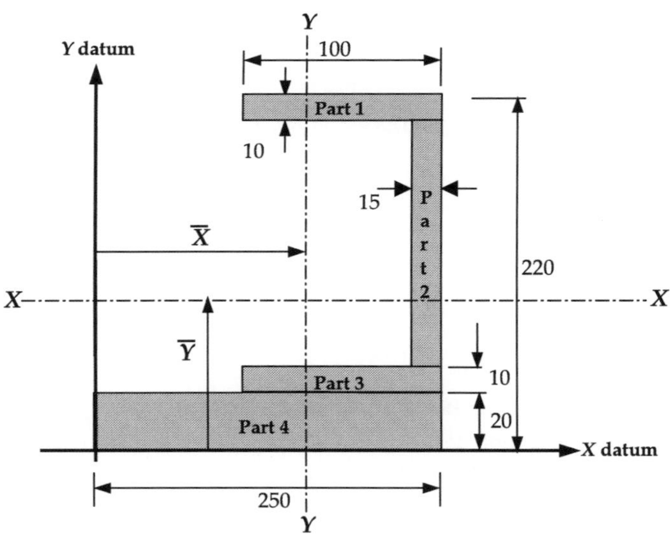

Figure 2.28: Breakdown of compound section – Example 2.4

We can now set up tables to calculate the Second Moment of Area about each axis, noting that for a rectangular section I_{xx} is given by $bd^3/12$ and I_{yy} by $db^3/12$ (I_{xx} is another useful formula to remember).

Part No.	Area A (mm^2)	Distance Y (mm)	$A \times Y$ (mm^3)	I_{xx} (mm^4)	$(\bar{y}-y)$ (mm)	$A(\bar{y}-y)^2$ (mm^4)
PART 1 (100 × 10)	1000	$220-5=215$	215 000	8333.333	−151.7	23 012 890
PART 2 (15 × 180)	2700	$20+10+180/2$ $=120$	324 000	7 290 000	−56.7	8 680 203
PART 3 (100 × 10)	1000	$20+10/2=25$	25 000	8333.333	38.3	1 466 890
PART 4 (250 × 20)	5000	$20/2=10$	50 000	166 666.667	53.3	14 204 450
	$\Sigma A=$ 9700		$\Sigma Ay=$ 614 000	$\Sigma I_{xx}=$ 7 473 333.33		$\Sigma A(\bar{y}-y)^2=$ 47 364 433

Figure 2.29: Completed table to calculate I_{xx} for compound section – Example 2.4

To calculate the centroid position in the y direction we use the totals of columns (2) and (4):

$$\bar{y}=\frac{614\,000}{9700}=63.3 \text{ mm (from the datum at the bottom of Part 4)}$$

To calculate I_{xx} using the Parallel Axis Theorem we add the totals of columns (5) and (7) giving:

$$I_{xx} \text{ (compound section)} = \Sigma I_{xx} + \Sigma A(\bar{y}-y)^2$$
$$I_{xx} = 7\,473\,333.33 + 47\,364\,433$$
$$I_{xx} = 54\,837\,766.33 \text{ mm}^4$$

We can calculate I_{yy} in a similar manner (as shown in Figure 2.30).

Part No.	Area A (mm^2)	Distance X (mm)	$A \times X$ (mm^3)	I_{yy} (mm^4)	$(\bar{x}-x)$ (mm)	$A(\bar{x}-x)^2$ (mm^4)
PART 1 (100 × 10)	1000	$250-100/2$ $=200$	200 000	833 333.333	−26.83	719 848.9
PART 2 (15 × 180)	2700	$250-15/2$ $=242.5$	654 750	50 625	−69.33	12 977 952.03
PART 3 (100 × 10)	1000	$250-100/2$ $=200$	200 000	833 333.333	−26.83	719 848.9
PART 4 (250 × 20)	5000	$250/2=125$	625 000	26 041 666.67	48.17	11 601 744.5
	$\Sigma A=$ 9700		$\Sigma Ax=$ 1 679 750	$\Sigma I_{yy}=$ 27 758 958.34		$\Sigma A(\bar{x}-x)^2=$ 26 019 394.33

Figure 2.30: Completed table to calculate I_{yy} for compound section – Example 2.4

$\overline{x}=\dfrac{1679750}{9700}=173.17$ mm (from the Y datum at the left of the compound section)

I_{yy} (compound section) $= \Sigma\, I_{yy} + \Sigma\, A(\overline{x}-x)^2$
$I_{yy} = 27\,758\,958.34 + 26\,019\,394.33$
$I_{yy} = 53\,778\,352.67\ \text{mm}^4$

Note that we could have simplified the table to calculate the I_{yy} value by combining the results for Parts 1 and 3 which have the same values throughout. We cannot do this for the I_{xx} values as the distance to the centroid for each part, measured from the X datum, varies. The table of Figure 2.29 could therefore have been modified as shown in Figure 2.31.

Part No.	Area A (mm^2)	Distance X (mm)	$A \times X$ (mm^3)	I_{yy} (mm^4)	$(\overline{x}-x)$ (mm)	$A(\overline{x}-x)^2$ (mm^4)
PARTS 1 & 3 (100 × 10)	1000 ×2 = 2000	250 – 100/2 = 200	200 000 ×2 = 400 000	833 333.333 ×2 = 1 666 666.67	−26.83	719 848.9 ×2 = 1 439 697.8
PART 2 (15 × 180)	2700	250 – 15/2 = 242.5	654 750	50 625	−69.33	12 977 952.03
PART 4 (250 × 20)	5000	250/2 = 125	625 000	26 041 666.67	48.17	11 601 744.5
	ΣA = 9700		ΣAx = 1 679 750	ΣI_{yy} = 27 758 958.34		$\Sigma\, A(\overline{x}-x)^2$ = 26 019 394.33

Figure 2.31: Simplified table to calculate I_{yy} for compound section – Example 2.4

◇

Alternatively, the problem might have been calculated by considering the section as a solid rectangle and then deducting the parts that were not solid, as shown in Example 2.5.

◇

Example 2.5: Centroid and Second Moment of Area for a compound section

Consider once again the compound section of Example 2.4 shown in Figure 2.27, and calculate the Second Moment of Area about the X–X and Y–Y axes as before.

In this case we will supplement our compound section by adding sections to form a rectangle. We locate the X and Y datums as before. We can add two rectangular blocks, Parts 2 and 3, as shown in Figure 2.32. Part 1 is then a rectangle with dimensions 250 × 220 mm. I_{xx} and I_{yy} are calculated in a similar manner to that previously used except, in this case, we deduct the missing Parts 2 and 3 from the totals.

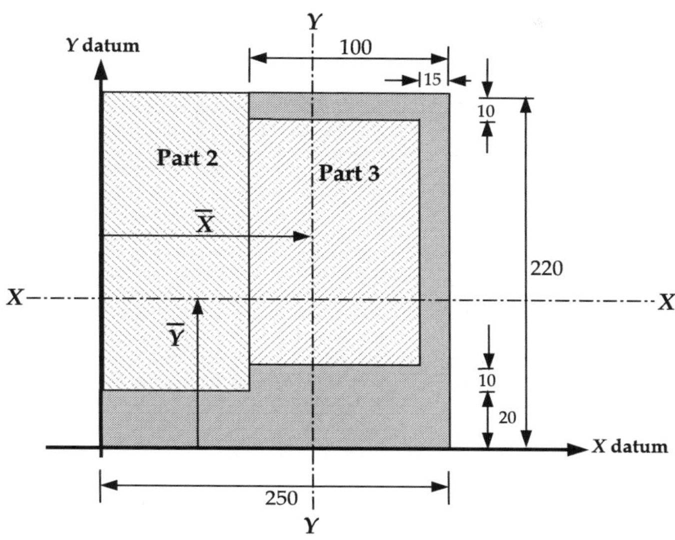

Figure 2.32: Breakdown of compound section – Example 2.5

The tables are shown in Figures 2.33 and 2.34.

Part No.	Area A (mm^2)	Distance Y (mm)	$A \times Y$ (mm^3)	I_{xx} (mm^4)	$(\bar{y}-y)$ (mm)	$A(\bar{y}-y)^2$ (mm^4)
PART 1 (250 × 220)	55 000	220/2 = 110	6 050 000	221 833 333.3	−46.7	119 948 950
PART 2 (150 × 200)	−30 000	20 + 200/2 = 120	−3 600 000	−100 000 000	−56.7	−96 446 700
PART 3 (85 × 180)	−15 300	20 + 10 + 180/2 = 120	−1 836 000	−41 310 000	−56.7	−49 187 817
	ΣA = 9700		ΣAy = 614 000	ΣI_{xx} = 80 523 333.3		$\Sigma A(\bar{y}-y)^2$ = −25 685 567

Figure 2.33: Completed table to calculate I_{xx} for compound section – Example 2.5

To calculate the centroid position in the y direction we use the totals of column (2) and (4):

$$\bar{y} = \frac{614\,000}{9700} = 63.3 \text{ mm} \quad \text{(from the } X \text{ datum at the bottom of the rectangle)}$$

To calculate I_{xx} using the Parallel Axis Theorem we add the total of columns (5) and (7) giving:

$$I_{xx}\text{ (compound section)} = \Sigma I_{xx} + \Sigma A(\bar{y}-y)^2$$
$$I_{xx} = 80\ 523\ 333.3 + (-25\,685\ 567)$$
$$I_{xx} = 54\ 837\ 766.33\ \text{mm}^4$$

Similarly I_{yy} can be calculated as before.

Part No.	Area A (mm^2)	Distance X (mm)	$A \times X$ (mm^3)	I_{yy} (mm^4)	$(\bar{x}-x)$ (mm)	$A(\bar{x}-x)^2$ (mm^4)
PART 1 220 250	55 000	250/2 = 125	6 875 000	286 458 333.3	48.17	127 619 189.5
PART 2 200 150	−30 000	150/2 = 75	−2 250 000	−56 250 000	98.17	−289 120 467
PART 3 180 85	−15 300	150 + 85/2 = 192.5	−2 945 250	−9 211 875	−19.33	−5 716 828.17
	ΣA = 9700		ΣAx = 1 679 750	ΣI_{yy} = 220 996 458.3		$\Sigma A(\bar{x}-x)^2$ = −167 218 105.7

Figure 2.34: Completed table to calculate I_{yy} for compound section – Example 2.5

To calculate the centroid position in the x direction we use the totals of columns (2) and (4):

$$\bar{x} = \frac{1679\,750}{9700} = 173.17\ \text{mm}$$ (from the Y datum at the left-hand side of the rectangle)

To calculate I_{yy} using the Parallel Axis Theorem we add the total of columns (5) and (7) giving:

$$I_{yy}\text{ (compound section)} = \Sigma I_{yy} + \Sigma A(\bar{x}-x)^2$$
$$I_{yy} = 220\ 996\ 458.3 + (-167\ 218\ 105.7)$$
$$I_{yy} = 53\ 778\ 352.67\ \text{mm}^4$$

It can be seen that the results obtained from either method will yield the same final values for the position of the centroid and I_{xx} and I_{yy} for the compound section.

Holes in sections can be dealt with in a similar manner by deducting the values entered in the table from the totals.

In Example 2.6 we will consider a compound section which contains a triangular element. These can be analysed using the methods previously described, the only difference being the formula required to calculate the area of the part and Second Moment of Area. Also the position of the centroid of the triangular part will be required and is located as shown in Figure 2.6.

Example 2.6: Centroid and Second Moment of Area for a compound section

Consider the compound section shown in Figure 2.35 and calculate the Second Moment of Area about the *X–X* and *Y–Y* axes. (All dimensions in millimetres.)

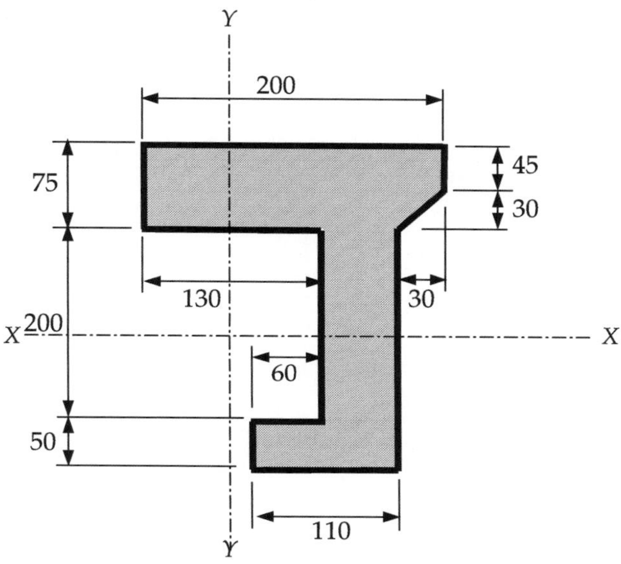

Figure 2.35: Compound section – Example 2.6

First break down the section into basic shapes and locate the *X* and *Y* datums.

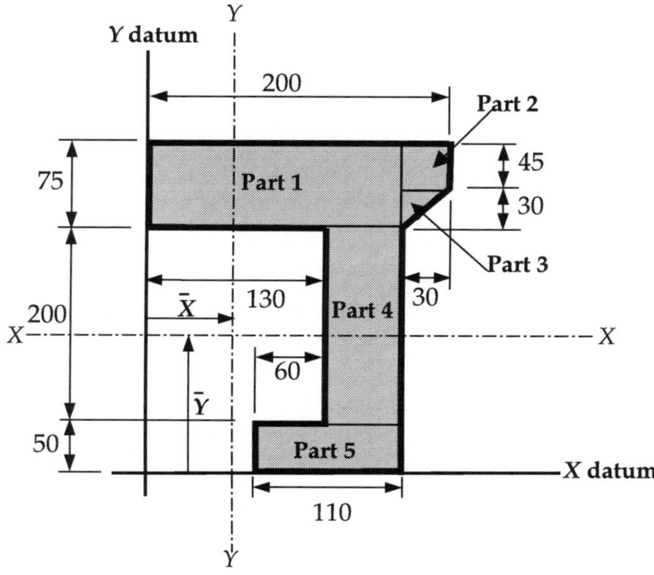

Figure 2.36: Breakdown of compound section – Example 2.6

Tables for the calculation of the position of the centroid and for the Second Moment of Area about each axis can be created as before.

For I_{xx}:

Part No.	Area A (mm^2)	Distance Y (mm)	$A \times Y$ (mm^3)	I_{xx} (mm^4)	$(\bar{y}-y)$ (mm)	$A(\bar{y}-y)^2$ (mm^4)
PART 1 75 170	17 500	50 + 200 + 75/2 = 287.5	3 665 625	5 976 562.5	−93.84	112 275 806.4
PART 2 45 30	1350	50 + 200 + 30 + 45/2 = 302.5	408 375	227 812.5	−108.84	15 992 296.56
PART 3 30 30	450	50 + 200 + (2/3 × 30) = 240	108 000	$bh^3/36$ = 22 500	−46.34	966 328.02
PART 4 200 50	10 000	50 + 200/2 = 150	1 500 000	33 333 333.33	43.66	19 061 956
PART 5 50 110	5500	50/2 = 25	137 500	10 526.667	168.86	15 825 347.8
	ΣA = 30 050		ΣAy = 5 819 500	ΣI_{xx} = 40 706 041.67		$\Sigma A(\bar{y}-y)^2$ = 305 121 734.8

Figure 2.37: Completed table to calculate I_{xx} for compound section – Example 2.6

$$\bar{y} = \frac{5\,819500}{30050} = 193.66 \text{ mm (from the } X \text{ datum at the bottom of Part 5)}$$

I_{xx} (compound section) $= \Sigma I_{xx} + \Sigma A(\bar{y}-y)^2$

$I_{xx} = 40\ 706\ 041.67 + 305\ 121\ 734.8$

$I_{xx} = 345\ 827\ 776.5$ mm^4

For I_{yy}:

Part No.	Area A (mm^2)	Distance X (mm)	$A \times X$ (mm^3)	I_{yy} (mm^4)	$(\bar{x}-x)$ (mm)	$A(\bar{x}-x)^2$ (mm^4)
PART 1 75 170	15 000	170/2 = 85	1 275 000	30 706 250	42.9	27 606 022.1
PART 2 45 30	1350	200 – 30/2 = 185	249 750	101 250	−57.1	4 401 553.5
PART 3 30 30	450	200 – (2/3×30) = 180	81 000	$hb^3/36$ = 22 500	−52.1	1 221 484.5
PART 4 200 50	10 000	130 + 50/2 = 155	1 550 000	2 083 333.3	−27.1	7 344 100
PART 5 50 110	5500	70 + 110/2 = 125	687 500	5 545 833.3	2.9	46 255
	ΣA = 30 050		ΣAx = 3 843 250	ΣI_{yy} = 38 459 166.67		$\Sigma A(\bar{x}-x)^2$ = 40 619 543

Figure 2.38: Completed table to calculate I_{yy} for compound section – Example 2.6

$$\bar{x} = \frac{3843250}{30050} = 127.9 \text{ mm (from the } Y \text{ datum at the left of section)}$$

I_{yy} (compound section) $= \Sigma I_{yy} + \Sigma A(\bar{x} - x)^2$

$I_{yy} = 38\,459\,166.67 + 40\,619\,543$

$I_{yy} = 79\,078\,709.67 \text{ mm}^4$

◇

Whilst it is often necessary to calculate the section properties of a unique section, in practice it is sometimes necessary to calculate the properties of a modified section, for instance when a plate has been added to a Universal Beam or Channel. In the United Kingdom, steel sections are normally produced in 'serial' sizes, that is, manufactured to specific dimensions. This enables rolling mills to produce large quantities of specific sections. Section properties within serial sizes can also be adjusted by increasing the thickness of the web and flanges. (Note: for an I-shaped section the top and bottom horizontal elements are termed the flanges, the web being the vertical element.) Serial sizes are usually specified by depth d, breadth b and the mass of the section per metre, hence a $457 \times 191 \times 98$ kg beam would specify a beam of serial depth 457 mm, breadth 191 mm and a mass per metre of 98 kg. A steel channel $229 \times 89 \times 32.76$ kg would be a channel of serial depth 229 mm, breadth 89 mm and mass of 32.76 kg per metre. The actual size and dimensions of the section may differ slightly from the overall serial size and the actual dimensions and section properties are listed in tables supplied by manufacturers.[2]

In certain instances it may be necessary to modify a serial sized section. Sometimes this is necessary in order to increase the capacity of the section, but it is more often required to provide adequate bearing for the structure supported by the beam. Consider the case shown in Figure 2.39a. The designer has found that a beam $178 \times 102 \times 19$ kg is adequate to support a cavity wall over an opening of specified width. The beam width of 102 mm is not adequate to support the cavity wall above it as this will cause an overhang and increase the local compressive stresses in the base of the wall. Therefore, a plate is added to the top. It is assumed that the method employed to fix the plate will result in a monolithic (single) section. The addition of the plate will alter the section properties of the beam. Quite often, engineers are required to utilise recycled material, such as beam sections, and it may be necessary to improve their load-carrying characteristics with the addition of extra plates. Figure 2.39b shows a similar situation using a steel channel. We will now consider the determination of properties for these modified sections.

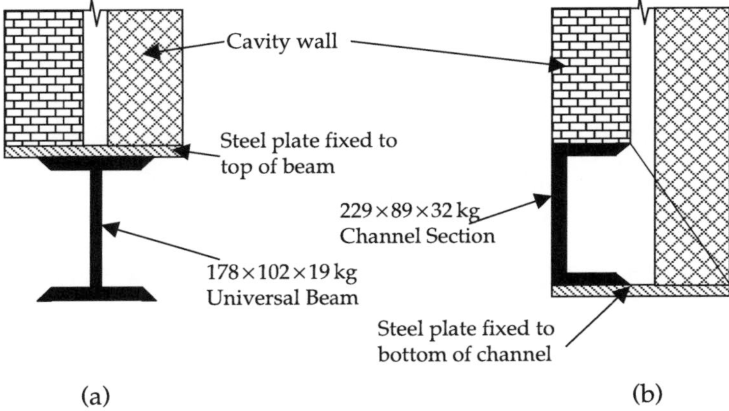

Figure 2.39a and b: Typical modified beam and channel sections

2. *Steelwork: Design guide to BS 5950: Part 1: 1985: Volume 1, Section properties and member capacities*, second edn, The Steel Construction Institute, 1987.

Example 2.7: Position of centroid and Second Moment of Area for a modified Universal Beam Section

Figure 2.40 shows an abstract from part of a steel table for a serial size Universal Beam (UB). It is proposed to modify the beam by adding a 250 mm × 25 mm thick mild steel plate to the top. The method of fixing the plate will ensure a monolithic compound section and the plate centroid will be aligned along the *Y–Y* axis of the beam as shown in Figure 2.41. Calculate the modified Second Moment of Area about the *X–X* axis.

Serial Size	Mass per metre	Depth (mm)	Area (cm^2)	Second Moment of Area (cm^4)		Elastic Modulus (cm^3)	
				X–X axis	*Y–Y* axis	*X–X* axis	*Y–Y* axis
457 × 191	98	467.4	125.3	45700	2340	1960	243

Figure 2.40: Abstract of steel table showing properties of section – Example 2.7

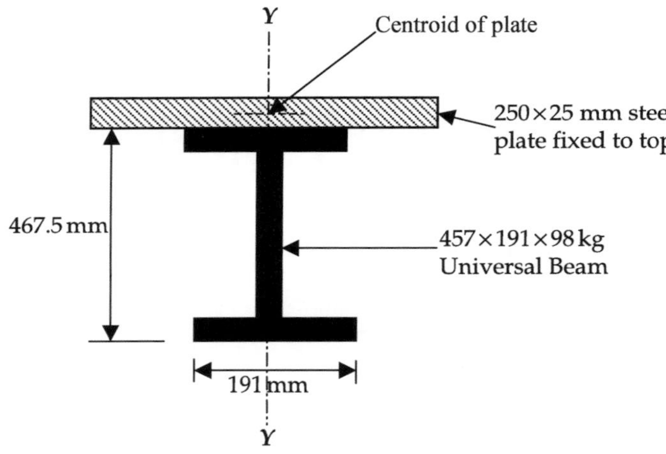

Figure 2.41: General arrangement of section – Example 2.7

It is important to note that the modified section is symmetrical about the *Y–Y* axis, therefore the position of the centroid about the *Y–Y* axis will remain unchanged. Also note that in this case, we will only calculate the I_{xx} value for the section. The process for determining the I_{yy} value is similar to the manner adopted for I_{xx} substituting the appropriate horizontal values.

The details contained in Figure 2.40 are typical of those to be found in many steel tables published by manufacturers. Note that the area is given in cm^2 whilst the Second Moment of Area for the beam is in cm^4. In order to determine the properties of the compound section a consistent set of units will be required, normally millimetres.

We must first calculate the centroid position for the compound section in the vertical direction. This will be located above that for the beam alone. The reason that we know it will be above that for the beam is, as previously discussed, the centroid is the balance point for the section. The addition of the plate will shift its balance toward the top of the compound section. This fact will help us to verify our calculations.

We set up the table, as before, however, since the properties of the Universal Beam have previously been determined the table is very much simpler than those we have used before. The X datum will be located at the bottom of the universal beam section and all y distances determined from this datum.

Part No.	Area A (mm^2)	Distance Y (mm)	$A \times Y$ (mm^3)	I_{xx} (mm^4)	$(\bar{y}-y)$ (mm)	$A(\bar{y}-y)^2$ (mm^4)
Steel Plate 250 × 25	6250	467.4 + 25/2 = 479.9	2 999 375	325 520.833	−164.26	168 633 422.5
457×191×98kg Universal Beam	12 530	467.4/2 = 233.7	2 928 261	457 000 000	−81.94	84 128 469.9
	ΣA = 18 780		ΣAy = 5 927 636	ΣI_{xx} = 457 325 520.8		$\Sigma A(\bar{y}-y)^2$ = 252 761 892.4

Figure 2.42: Completed table to calculate I_{xx} for compound section – Example 2.7

$$\bar{y} = \frac{5927636}{18780} = 315.4 \text{ mm (from the } X \text{ datum at the bottom of beam)}$$

Note this is located above the original centroid of the beam alone (233.7 mm).

$$I_{xx} \text{ (compound section)} = \Sigma I_{xx} + \Sigma A(\bar{y}-y)^2$$
$$I_{xx} = 457\,325\,520.8 + 252\,761\,892.4$$
$$I_{xx} = 710\,087\,413.2 \text{ mm}^4$$

The addition of the plate has therefore increased the I_{xx} of the section by nearly 56% and will have obvious implications on its load-carrying characteristics.

Example 2.8: Position of centroid and Second Moment of Area for a modified Rolled Steel Channel section

The table of Figure 2.43 below contains an abstract from part of a steel table for a serial sized Rolled Steel Channel (RSC) section. It is proposed to modify the channel by adding a 300 mm × 12 mm thick mild steel plate to the bottom. The method of fixing used will ensure a homogeneous (single) compound section and the compound beam is shown in Figure 2.44. Calculate the modified Second Moment of Area about the X–X axis.

Serial Size	Mass per metre	Depth (mm)	Area (cm^2)	Second Moment of Area (cm^4)	
				X–X axis	Y–Y axis
229 × 89	32.76	228.6	41.7	3390	285

Figure 2.43: Abstract of steel table showing properties of channel section – Example 2.8

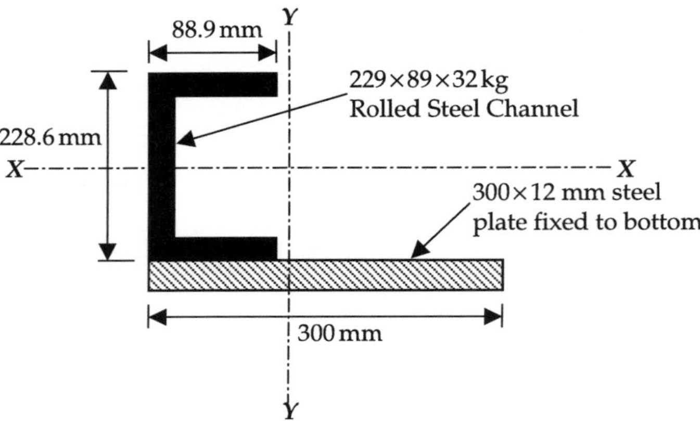

Figure 2.44: General arrangement of channel section and plate – Example 2.8

In this case the section is not symmetrical about the *Y–Y* axis, therefore the position of the centroid about the *Y–Y* axis will need to be calculated if the I_{yy} value is required. This can be determined using the correct values from the steel manufacturer's tables in a similar manner to the I_{xx} value calculated here. Also note that we will again only calculate the I_{xx} value for the compound section.

Remember also that in order to determine the properties of the compound section a consistent set of units must be employed.

We set up our table, as before the *X* datum will be located at the bottom of the steel plate and all *y* distances determined from that datum.

Part No.	Area A (mm^2)	Distance Y (mm)	$A \times Y$ (mm^3)	I_{xx} (mm^4)	$(\bar{y}-y)$ (mm)	$A(\bar{y}-y)^2$ (mm^4)
229×89×32kg Rolled Steel Channel	4170	12 + 228.6/2 = 126.3	526 671	33 900 000	-55.74	12 955 971.5
Steel Plate 300 × 12	3600	12/2 = 6	21 600	43 200	64.463	15 006 171.5
	ΣA = 7770		ΣAy = 548 271	ΣI_{xx} = 33 943 200		$\Sigma A(\bar{y}-y)^2$ = 27 962 143

Figure 2.45: Completed table to calculate I_{xx} for compound section – Example 2.8

$$\bar{y} = \frac{548271}{7770} = 70.563 \text{ mm (from the } X \text{ datum at the bottom of plate)}$$

Note that this is located below the position of the centroid of the channel (126.3 mm).

I_{xx} (compound section) $= \Sigma I_{xx} + \Sigma A(\bar{y}-y)^2$
$I_{xx} = 33\ 943\ 200 + 27\ 962\ 143$
$I_{xx} = 61\ 905\ 343 \text{ mm}^4$

The addition of the plate will again increase the load-carrying capacity of the section.

Having calculated the section properties for a particular member it is then necessary to calculate its load-carrying characteristics. As discussed in section 2.2.1 a relationship exists which allows us to calculate the stress in a section using equation [2.6] such that (for bending about the *X–X* axis):

$$\sigma = \frac{My}{I_{xx}} \qquad [2.6]$$

Therefore, if the vertical applied moment (*M*), the Second Moment of Area (I_{xx}), about the axis perpendicular to the applied moment, and the depth (*y* in the vertical direction) to the point at which the stress is to be calculated are known, we can calculate the stress at that point. A similar relationship exists for calculating stress induced by horizontal applied moments by substituting I_{yy} and *x* distances.

In design, however, it is often the case that the allowable maximum stress that the material can sustain is known. This has usually been determined from physical testing of the material and is often modified to include a factor of safety. Since this value expresses the maximum permissible stress for the material, we can use it to calculate the maximum allowable moment or **Moment of Resistance** that a member can safely sustain without being overstressed.

2.3 Moment of Resistance

2.3.1 Introduction

We have previously investigated the determination of various section properties including area, position of centroid and Second Moment of Area. We can now use these properties to determine the bending stresses induced in a section due to a specified loading.

2.3.2 Stress Distribution across a Section

As discussed in Section 2.2.1, it is possible to determine stresses across a particular section for bending about the vertical and horizontal axis from the transposed part of the Engineers' bending equation [2.6a and b]:

$$\sigma_{\text{in vertical direction}} = \frac{My}{I_{xx}} \qquad \sigma_{\text{in horizontal direction}} = \frac{Mx}{I_{yy}} \qquad [2.6\text{a and b}]$$

Thus if we know the applied bending moment (*M*) , the Second Moment of Area about the bending axis (*I*) and the distance to the extreme fibre (*x* or y_{top} and *x* or y_{bottom}), we can calculate the extreme fibre stresses.

Example 2.9: Stress distribution across a rectangular beam

Calculate the maximum stresses in the timber beam shown in Figure 2.46.

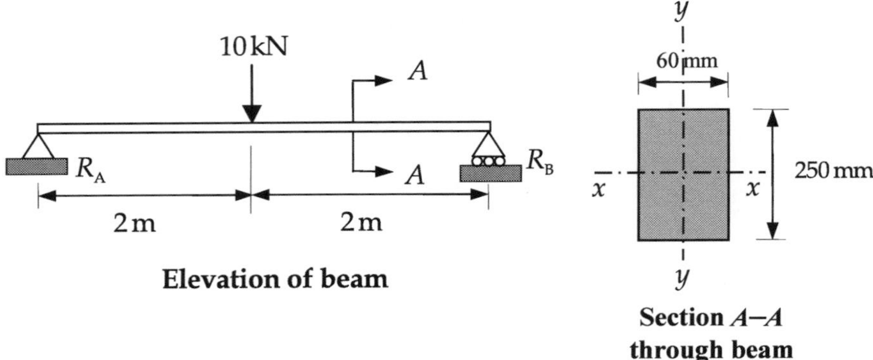

Figure 2.46: General arrangement and section of beam – Example 2.9

First we need to calculate the applied maximum bending moment. We will later learn how to calculate bending moments for any part of a beam, however the beam of Figure 2.46 can be solved using a standard formula. This is only possible because the beam is symmetrically loaded with a point load in the centre of the span. The forces at the reactions, required to ensure equilibrium, and the maximum applied bending moment may be calculated from:

Reactions: $R_A = R_B = \dfrac{P}{2}$ where P = value of point load [2.13]

and

Maximum bending moment (at mid-span) $= \dfrac{PL}{4}$ where L = total span of beam [2.14]

Therefore, for the beam of Figure 2.46, R_A and R_B = 5 kN each and the maximum bending moment, which occurs at 2 m from R_A will be:

$$R_A = \frac{10 \times 4}{4} = 10 \text{ kN m}$$

We must now calculate the Second Moment of Area about the bending axis, in this case I_{xx}. As the section is rectangular we know:

$\bar{y} = d/2 = 250/2 = 125$ mm and

$$I_{xx} = \frac{bd^3}{12} = \frac{60 \times 250^3}{12} = 78\,125\,000 \text{ mm}^4$$

Maximum stress will occur at the maximum distance from the centroid at the top or bottom of the beam. In this case the distance y is the same, 125 mm, as the section is rectangular.

Therefore, for the rectangular section of Example 2.9, the stresses will vary from a maximum at the top and bottom to zero at the Neutral Axis through the centroid. The distribution of stress will vary according to the dimensions and shape of the section under consideration and it may, in some cases, be necessary to calculate the stress at both the top and bottom of a section in order to determine the maximum value. It may also be necessary to calculate both values for sections composed of materials which have differing maximum allowable tensile and compressive stresses, such as unreinforced concrete.

However, for the section shown in Example 2.9:

$$\text{Maximum stress} = \sigma = \frac{My}{I} = \frac{10 \times 10^6 \times 125}{78\ 125\ 000} = 16\ \text{N/mm}^2$$

We can also calculate the stress distribution across the section by varying the y distance from the centroid from 0 to 125 mm as shown in Figure 2.47.

Depth across section from bottom (mm)	Distance from centroid (y) (mm)	$\sigma = \frac{My}{I}$ (N/mm^2)
0	125	16
25	100	12.8
50	75	9.6
75	50	6.4
100	25	3.2
125	0	0
150	–25	–3.2
175	–50	–6.4
200	–75	–9.6
225	–100	–12.8
250	–125	–16

Figure 2.47: Stress distribution across section – Example 2.9

Stress across the section can now be plotted.

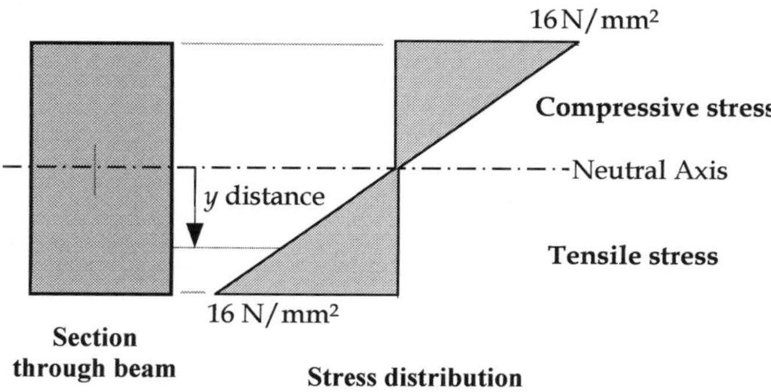

Figure 2.48: Stress distribution across section – Example 2.9

It can be seen that the bending stress varies linearly across the section with a maximum value at the extreme fibres. Positive and negative values in the table indicate tensile and compressive stresses which are determined by the direction of the applied load. In Example 2.9, bending will take place about the X–X axis with the load applied at the top, and the beam will bend downwards stretching the bottom of the section and shortening the top. Thus the bottom of the beam will be in tension and the top in compression (see Figure 2.3).

In I-section beams, bending stresses are predominantly carried by the flanges whilst shear stresses are carried by the web. It is therefore essential to ensure that the flanges are as far apart as possible as this increase in the distance between the flanges will increase the moment-carrying capacity of the section by reducing stresses.

In Example 2.10 we will consider another rectangular beam similar to that in Example 2.9.

Example 2.10: Stress distribution across a rectangular beam

In this example we will increase the depth of the beam of Example 2.9 and reduce its width, whilst retaining the same overall area for the section at 15 000 mm².

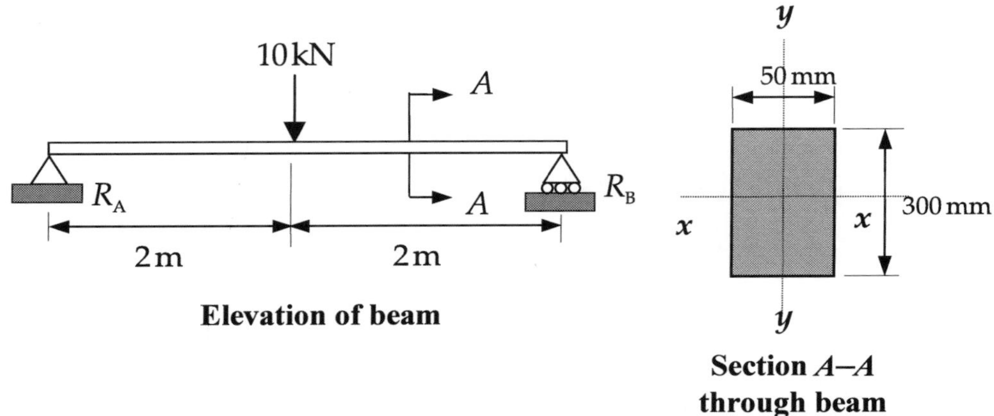

Figure 2.49: General arrangement and section of beam – Example 2.10

Maximum bending moment is again 10 kNm

$\bar{y} = d/2 = 300/2 = 150\,\text{mm}$ and

$$I_{xx} = \frac{bd^3}{12} = \frac{50 \times 300^3}{12} = 112\,500\,000\ \text{mm}^4$$

$$\text{Maximum stress} = \sigma = \frac{My}{I} = \frac{10 \times 10^6 \times 150}{112\,500\,000} = 13.33\ \text{N/mm}^2$$

We can again calculate the stress distribution across the section by varying the y distance from the centroid from 0 to 150 mm as shown in Figure 2.50.

Depth across section from bottom (mm)	Distance from centroid (y) (mm)	$\sigma = \frac{My}{I}$ (N/mm^2)
0	150	13.33
25	125	11.1
50	100	8.89
75	75	6.67
100	50	4.45
125	25	2.22
150	0	0
175	– 25	– 2.22
200	– 50	– 4.45
225	– 75	– 6.67
250	– 100	– 8.89
275	– 125	– 11.1
300	– 150	– 13.33

Figure 2.50: Stress distribution across section – Example 2.10

Stress distribution across the section can also be plotted.

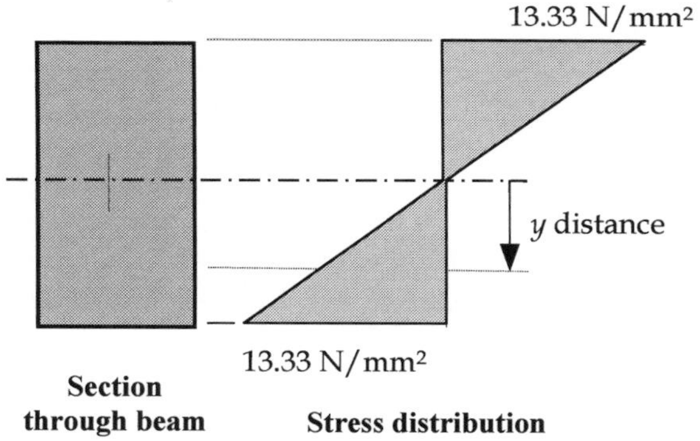

Figure 2.51: Stress distribution across section – Example 2.10

Note that by increasing the depth of the section whilst still using the same amount of material, a reduction of 17% in the extreme fibre stresses has been achieved.

Example 2.11: Stress distribution across a tee-shaped beam

The section shown in Figure 2.52 is to be used to form a beam supporting an evenly distributed load of 25 kN/m applied to the top flange over a clear span of 4 m. Calculate the stress distribution across the section for maximum bending moment and sketch the stress distribution at 25 mm intervals of depth through the section.

As with the previous example of a beam supporting a point load at mid-span, there exists a standard formula for calculating the reactions and maximum bending moment for a beam supporting an evenly (or uniformly) distributed load over its entire length. For a constant uniformly distributed load, across the whole length of the beam, the reactions and maximum bending moment are given by:

Reactions: $R_A = R_B = \dfrac{wl}{2}$ where w = load per metre [2.15]

and l = total span of the beam

and

Maximum bending moment (at mid-span) $= \dfrac{wl^2}{8}$ [2.16]

Therefore, R_A and R_B = $(25 \times 4)/2$ = 50 kN each

Maximum bending moment (M) = $\dfrac{wl^2}{8} = \dfrac{25 \times 4^2}{8}$ = 50 kN m

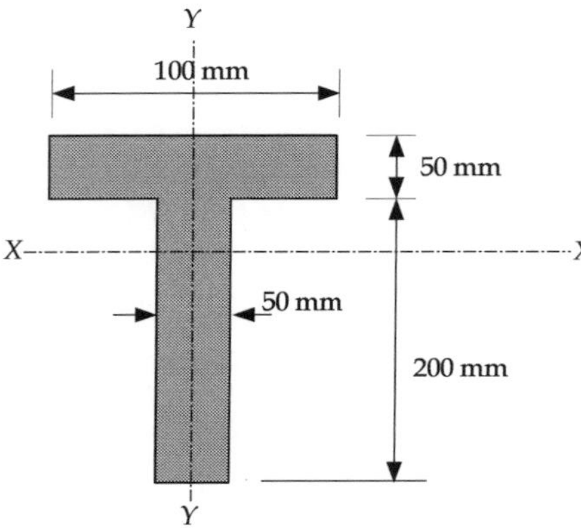

Figure 2.52: Cross-section of beam – Example 2.11

First, we calculate the properties of the section, the position of the centroid and the Second Moment of Area. We start by noting that the 'tee' section is symmetrical about the Y–Y axis and hence the position of the centroid in the x direction can be determined by observation as being 50 mm from the Y datum. We then split the section into parts and calculate the required values as before.

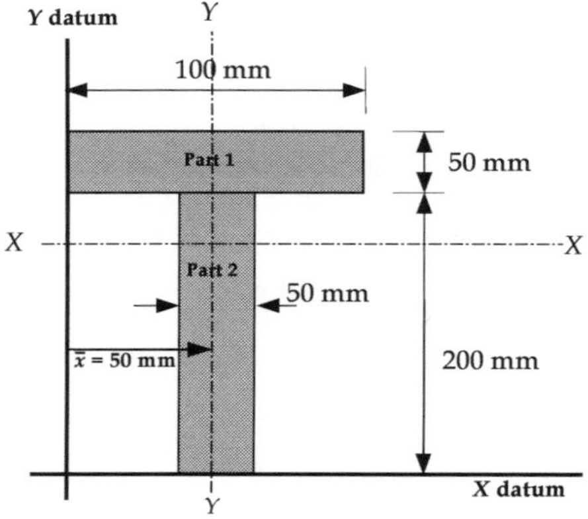

Figure 2.53: Breakdown of section – Example 2.11

In order to calculate the position of the centroid and I_{xx} we set up a table as before (Figure 2.54):

Part No.	Area A (mm^2)	Distance Y (mm)	$A \times Y$ (mm^3)	I_{xx} (mm^4)	$(\bar{y}-y)$ (mm)	$A(\bar{y}-y)^2$ (mm^4)
PART 1 50 100	5000	225	11 250 000	1 041 666.67	–83.33	34 719 444.5
PART 2 200 50	10 000	100	1 000 000	33 333 333.3	41.67	17 363 889
	$\Sigma A =$ 15 000		$\Sigma Ay =$ 21 250 000	$\Sigma I_{xx} =$ 34 375 000		$\Sigma A(\bar{y}-y)^2 =$ 52 083 333.5

Figure 2.54: Table to calculate section properties – Example 2.11

$\bar{y} = 141.67\,\text{mm}$ $\qquad$ $I_{xx} = 86\,458\,333.5\ \text{mm}^4$

We can calculate bending stress distribution, as before, and the results are shown in Figure 2.55.

Depth across section from bottom (mm)	Distance from centroid (y) (mm)	$\sigma = \dfrac{My}{I}$ N/mm^2
0	141.67	81.93
25	116.67	67.47
50	91.67	53.01
75	66.67	38.56
100	41.67	24.1
125	16.67	9.64
150	–8.33	–4.82
175	–33.33	–19.28
200	–58.33	–33.73
225	–83.33	–48.19
250	–108.33	–62.64

Figure 2.55: Stress distribution across section – Example 2.11

Hence bending stress distribution across the section can be plotted as shown in Figure 2.56.

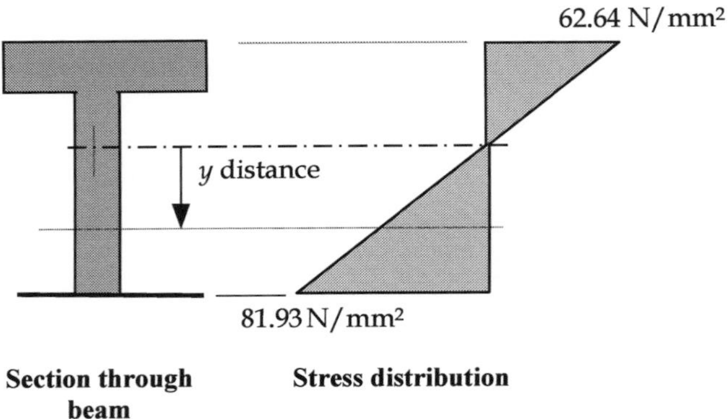

Figure 2.56: Stress distribution across section – Example 2.11

For design it is often only necessary to determine the extreme fibre stresses in order to check the adequacy of the required member size, however when designing in reinforced concrete it is sometimes necessary to calculate stress at specific points in order to determine positions requiring reinforcement.

In most cases the detailed design of steel sections will be based on the use of serial sizes; that is to say pre-fabricated sections of fixed cross-sectional dimensions as previously discussed. Having determined the applied moment we can use an iterative technique (or more commonly trial and error!) to determine the adequacy of a section. This will be limited by the allowable maximum extreme fibre stress.

If the allowable maximum extreme fibre stress is known for the material, it is sometimes easier to calculate the Moment of Resistance (*MoR*) of a given section and compare this with the calculated applied maximum moment (*M*).

2.3.3 Determination of Moment of Resistance

Again, from the Engineers' bending equation we have:

$$\frac{M}{I} = \frac{\sigma}{y} = \frac{E}{R}$$

From this we have:

$$\frac{M}{I} = \frac{\sigma}{y}$$

and

$$M = \frac{\sigma \, I}{y} \qquad [2.17]$$

Thus if the maximum allowable extreme fibre stress, σ, is known, and the distance from the centroid to the extreme fibre, y, can be calculated along with the Second Moment of Area, I, for the section, we can calculate the Moment of Resistance for a given member. Consider Example 2.9 again.

Example 2.12: Moment of Resistance for a rectangular beam

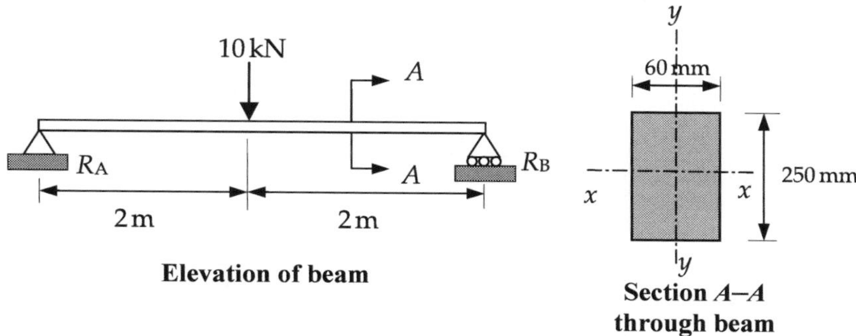

Figure 2.57: General arrangement and section of beam – Example 2.12

Calculate the Moment of Resistance for the timber beam shown in Figure 2.57 if the maximum allowable tensile and compressive stress is 16 N/mm^2. Is the section adequate to support the applied point load of 10 kN as shown?

To calculate the Moment of Resistance of the section we must first calculate the Second Moment of Area for the section about the bending axis, in this case I_{xx}. As the section is rectangular we know:

$\bar{y} = d/2 = 250/2 = 125$mm and

$$I_{xx} = \frac{bd^3}{12} = \frac{60 \times 250^3}{12} = 78\ 125\ 000\ \text{mm}^4$$

Maximum stress will occur at maximum distance from the centroid at the top or bottom of the beam and, in this case, the distance is the same, 125 mm, as the section is rectangular.

$$\text{Moment of Resistance} = \frac{\sigma\, I}{y} = \frac{16 \times 78\ 125\ 000}{125 \times 10^6} = 10\ \text{kN m}$$

(Note: result converted from N mm to kN m by dividing by 10^6.)

The applied maximum bending moment is again given by the standard formula;

$$M = \frac{PL}{4} = \frac{10 \times 4}{4} = 10\,\text{kN m (as before)}$$

Moment of Resistance must be greater than or equal to Applied Moment (M)

10 kN m = 10 kN m

The beam is adequate!

The calculation can be modified in order to accommodate the different factors involved in the design of a beam, such as the allowable applied load for a given span as shown in Example 2.13.

Example 2.13: Determination of the maximum allowable point load applied to a rectangular beam

For the beam shown in Figure 2.58, determine the maximum value for the point load P if the maximum allowable bending stress (top and bottom) must not exceed 16 N/mm^2.

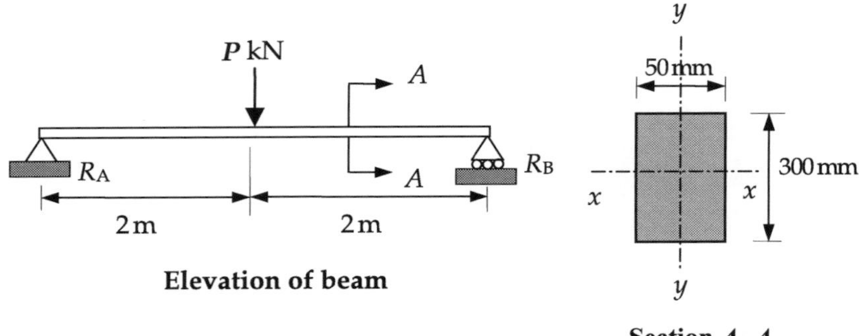

Figure 2.58: General arrangement and section of beam – Example 2.13

$\bar{y} = d/2 = 300/2 = 150\,\text{mm}$ and

$$I_{xx} = \frac{bd^3}{12} = \frac{50 \times 300^3}{12} = 112\ 500\ 000\ \text{mm}^4$$

$$\text{Moment of Resistance} = \frac{\sigma\ I}{y} = \frac{16 \times 112\ 500\ 000}{150 \times 10^6} = 12\ \text{kN m}$$

The applied maximum bending moment is again given by:

$$M = \frac{PL}{4}$$ and since at the limit of stress $MoR = M$ then

$$12\ \text{kN m} = \frac{P \times 4}{4}$$

which gives $P = 12$ kN.

Moment of Resistance is therefore a very useful property since from it we can compare our calculated applied moment to the resistance moment of the beam to determine its adequacy.

Example 2.14: Moment of Resistance for a tee-section beam

Calculate the Moment of Resistance for the beam shown in Figure 2.59 if the maximum allowable tensile and compressive stress is 30 N/mm^2. Is the beam adequate to support a uniformly distributed load of 25 kN/m applied to the top flange over a span of 4 m?

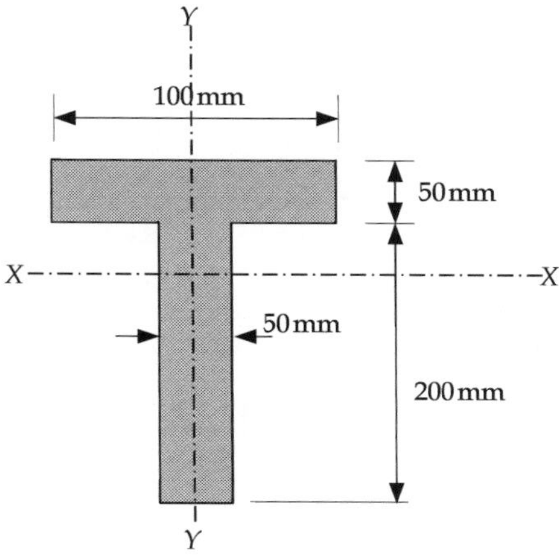

Figure 2.59: Cross-section of beam – Example 2.14

First we determine the applied moment from the standard formula of equation [2.16] such that:

$$M = \frac{wl^2}{8} = \frac{25 \times 4^2}{8} = 50 \text{ kN m}$$

From our previous calculations (Example 2.11) we know that:

$$\overline{y} = 141.67 \text{mm} \qquad I_{xx} = 86\ 458\ 333.5 \text{mm}^4$$

In this case we must calculate the minimum Moment of Resistance. This is because the distance from the centroid to the top of the section and that to the bottom are different and the beam will fail just after the minimum Moment of Resistance is exceeded by the applied moment.

$$\text{Moment of Resistance (bottom)} = \frac{\sigma\ I}{y} = \frac{30 \times 86\ 458\ 333.5}{141.67} = 18.3 \text{ kN m}$$

$$\text{Moment of Resistance (top)} = \frac{\sigma\ I}{y} = \frac{30 \times 86\ 458\ 333.5}{(250 - 141.67)} = 23.94 \text{ kN m}$$

We now have two Moments of Resistance, however the beam will become unserviceable at the lowest value as the extreme fibres at the top will then be at the limit of the allowable stress. The top of the beam will become overstressed at an applied moment just above 18.3 kN m, much less that the 50 kN m specified here. The beam is therefore not adequate for this particular situation.

◇

We are now able to determine both the section properties and Moments of Resistance for various structural members and thus determine their adequacy in bending for their designed purpose.

◇

Example 2.15: Moment of Resistance for a I-section beam

For the beam shown in Figure 2.60 determine its adequacy to support the applied load if the maximum allowable extreme fibre stress is 50 N/mm^2.

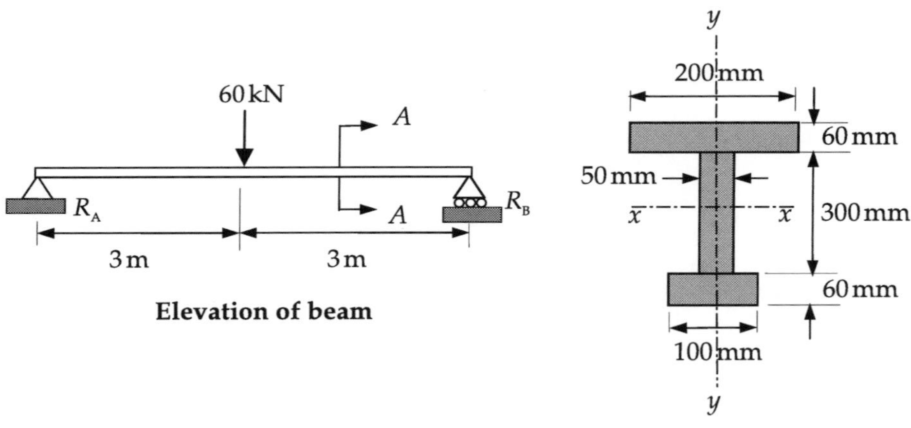

Figure 2.60: General arrangement and section of beam – Example 2.15

First we must determine the section properties. In this case we will need to calculate the position of the centroid from an X datum and the Second Moment of Area about the X–X axis through the centroid. We split the section into parts and locate our X and Y datums at the bottom and to the left of the section, as before, to calculate the centroid position.

Part No.	Area A (mm^2)	Distance Y (mm)	$A \times Y$ (mm^3)	I_{xx} (mm^4)	$(\bar{y}-y)$ (mm)	$A(\bar{y}-y)^2$ (mm^4)
PART 1 (200 × 60)	12 000	390	4 680 000	3 600 000	−147.27	260 261 434.8
PART 2 (50 × 300)	15 000	210	3 150 000	112 500 000	32.73	16 068 793.5
PART 3 (100 × 60)	6000	30	180 000	1 800 000	212.73	271 524 317.4
	$\Sigma A =$ 33 000		$\Sigma Ay =$ 8 010 000	$\Sigma I_{xx} =$ 166 500 000		$\Sigma A(\bar{y}-y)^2 =$ 547 854 545.7

Figure 2.61: Table to calculate section properties – Example 2.15

$\bar{y} = 242.73$ mm $\qquad I_{xx} = 714\ 354\ 545.7$ mm^4

$$\text{Moment of Resistance (top)} = \frac{\sigma\ I}{y} = \frac{50 \times 714\ 354\ 545.7}{(420 - 242.73)} = 201.9 \text{ kN m}$$

$$\text{Moment of Resistance (bottom)} = \frac{\sigma\ I}{y} = \frac{50 \times 714\ 354\ 545.7}{242.73} = 147.15 \text{ kN m}$$

$$\text{Applied bending moment from point load} = M = \frac{PL}{4} = \frac{60 \times 6}{4} = 90 \text{ kN m}$$

BEAM IS ADEQUATE SINCE minimum MoR = 147.15 kN m > 90 kN m = Applied Moment

◇

Example 2.16: Moment of Resistance for a haunched I-section beam

The beam section shown in Figure 2.62 has allowable maximum stress in both compression and tension of 155 N/mm^2.

(a) Calculate the beam's Moment of Resistance about the X–X axis.
(b) The beam is to be loaded at the top with a maximum moment of 760 kN m. Comment on the adequacy of the section.

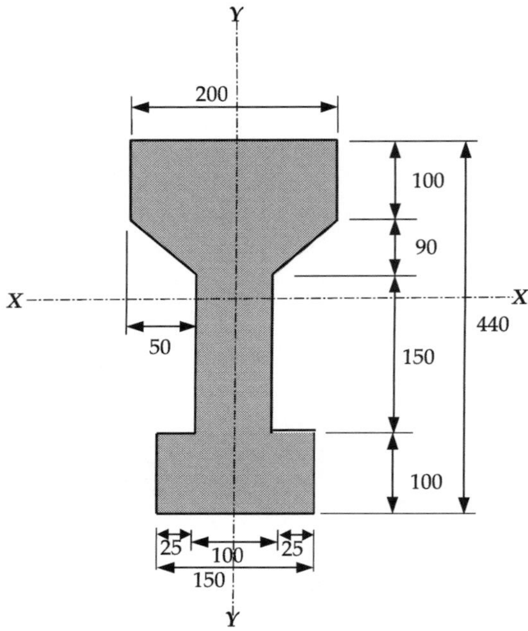

Figure 2.62: Section through beam – Example 2.16

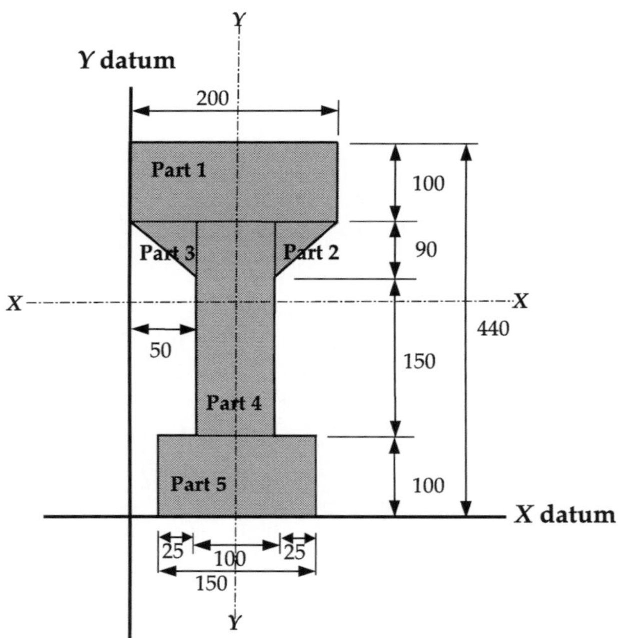

Figure 2.63: Breakdown of section – Example 2.16

For I_{xx}:

Part No.	Area A (mm^2)	Distance Y (mm)	$A \times Y$ (mm^3)	I_{xx} (mm^4)	$(\bar{y}-y)$ (mm)	$A(\bar{y}-y)^2$ (mm^4)
PART 1 (200 × 100)	20 000	390	7 800 000	16 666 666.67	−150.24	451 441 152
PART 2 (50 × 90)	2250	100+150+ (2/3×90) = 310	697 500	$bh^3/36$ = 1 012 500	−70.24	11 100 729.6
PART 3 (50 × 90)	2250	310	697 500	$bh^3/36$ = 1 012 500	−70.24	11 100 729.6
PART 4 (100 × 240)	24 000	220	5 280 000	115 200 000	19.76	9 370 982.4
PART 5 (150 × 100)	15 000	50	750 000	12 500 000	189.76	540 132 864
	ΣA = 63 500		ΣAy = 15 225 000	ΣI_{xx} = 146 391 666.7		$\Sigma A(\bar{y}-y)^2$ = 1 023 146 458

Figure 2.64: Completed table to calculate I_{xx} for compound section – Example 2.16

$$\bar{y}=\frac{15\,225\,000}{63\,500}=239.76 \text{ mm}$$

$$I_{xx} = 146\,391\,666.7 + 1\,023\,146\,458 \text{ mm}^4$$

$$I_{xx} = 1\,169\,538\,124 \text{ mm}^4$$

$$\text{Moment of Resistance (top)} = \frac{\sigma\, I}{y} = \frac{155\times 1\,169\,538\,124}{(440-239.76)\times 10^6} = 905.3 \text{ kN m}$$

$$\text{Moment of Resistance (bottom)} = \frac{\sigma\, I}{y} = \frac{155\times 1\,169\,538\,124}{239.76\times 10^6} = 756.1 \text{ kN m}$$

Applied bending moment = M = 760 kNm

BEAM IS INADEQUATE SINCE minimum MoR = 756.1 kN m < 760 kN m = Applied Moment

---

Example 2.17: Moment of Resistance for a Universal Beam section

It is also possible to calculate the Moment of Resistance for a modified 'serial' section. Figure 2.65 shows an abstract from part of a steel table for a serial sized universal beam (UB). It is proposed to modify the beam by adding a 300 mm × 30 mm thick mild steel plate to the top. The method of fixing of the plate will ensure a monolithic compound section and the plate centroid will be aligned along the $Y–Y$ axis of the beam as shown in Figure 2.66. Calculate the Moment of Resistance for the section about the $X–X$ and $Y–Y$ axes if the allowable extreme fibre stress is 165 N/mm^2.

Serial Size	Mass per metre (kg)	Depth (mm)	Width (mm)	Area (cm^2)	Second Moment of Area (cm^4)	
					X–X axis	Y–Y axis
533 × 210	82	528.3	208.7	104	47 500	2010

Figure 2.65: Abstract of steel table showing properties of section – Example 2.17

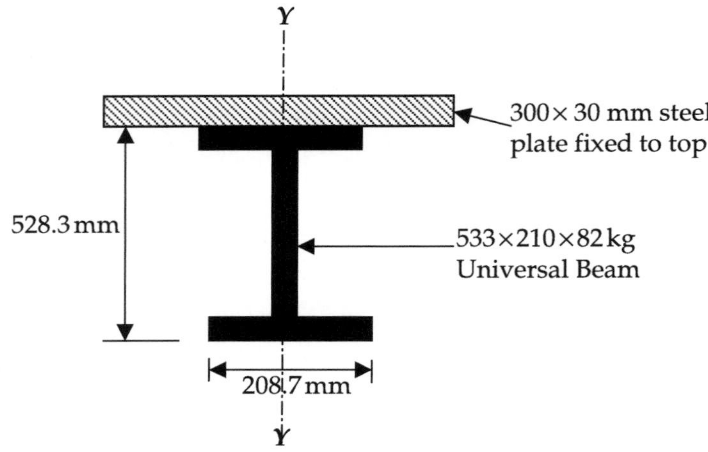

Figure 2.66: General arrangement of compound section – Example 2.17

For I_{xx}:

Part No.	Area A (mm^2)	Distance Y (mm)	$A \times Y$ (mm^3)	I_{xx} (mm^4)	$(\bar{y}-y)$ (mm)	$A(\bar{y}-y)^2$ (mm^4)
Steel Plate 300 × 30	9000	528.3 + 30/2 = 543.3	4 889 700	675 000	−149.7	201 690 810
533×210×82 kg Universal Beam	10 400	528.3/2 = 264.15	2 747 160	475 000 000	129.45	174 275 946
	$\Sigma A =$ 19 400		$\Sigma Ay =$ 7 636 860	$\Sigma I_{xx} =$ 475 675 000		$\Sigma A(\bar{y}-y)^2 =$ 375 966 756

Figure 2.67: Completed table to calculate I_{xx} for compound section – Example 2.17

$$\bar{y} = \frac{7\,636\,860}{19\,400} = 393.65 \text{ mm (from the } X \text{ datum at the bottom of beam)}$$

$$I_{xx} = 475\,675\,000 + 375\,966\,756$$

$$I_{xx} = 851\,641\,756 \text{ mm}^4$$

Moment of Resistance (top) about $X–X$ axis $== \dfrac{\sigma\ I}{y} = \dfrac{165 \times 851\ 641\ 756}{(558.3 - 393.65) \times 10^6} = 853.45$ kN m

(without plate MoR = 296.7 kN m)

Moment of Resistance (bottom) about $X–X$ axis $= \dfrac{\sigma\ I}{y} = \dfrac{165 \times 851\ 641\ 756}{393.65 \times 10^6} = 356.7$ kN m

(without plate MoR = 296.7 kN m)

For I_{yy}:

Part No.	Area A (mm^2)	Distance X (mm)	$A \times X$ (mm^3)	I_{yy} (mm^4)	$(\bar{x}-x)$ (mm)	$A(\bar{x}-x)^2$ (mm^4)
Steel Plate 300 × 30	9000	150	1 350 000	$db^3/12$ = 675 000	0	0
533×210×82kg Universal Beam	10 400	(300−208.7)/2 + 208.7/2 = 150	1 560 000	20 100 000	0	0
	ΣA = 19 400		ΣAx = 2 910 000	ΣI_{yy} = 20 775 000		$\Sigma\ A(\bar{x}-x)^2$ = 0

Figure 2.68: Completed table to calculate I_{yy} for compound section – Example 2.17

$\bar{x} = \dfrac{2\ 910\ 000}{19400} = 150$ mm (from the Y datum at the extreme left of plate)

(Note: this could have been determined by observation here – symmetrical figure)

$I_{yy} = 20\ 775\ 000\ \text{mm}^4$

Moment of Resistance about $Y–Y$ axis $= \dfrac{\sigma\ I}{y} = \dfrac{165 \times 20\ 775\ 000}{150 \times 10^6} = 22.85$ kN m

(without plate MoR = 22.11 kN m)

Note that I-section beams are primarily designed to resist bending about the $X–X$ axis and so the Moment of Resistance in this direction is very high. The addition of the plate increases the Moment of Resistance by 60 kN m (approximately 20%) about the $X–X$ axis but has little effect on its resistance to bending about the $Y–Y$ axis.

◇

Having completed this chapter you should now be able to:

- ***Calculate the position of the centroid for various sections***
- ***Calculate the distribution of stress across a section***
- ***Calculate the Second Moment of Area about a specific axis***
- ***Determine the maximum tensile and compressive stresses for a member***
- ***Calculate the Moment of Resistance for a given section***
- ***Calculate the section properties and Moment of Resistance for a modified 'standard' steel section***

Further Problems 2

(all dimensions in millimetres)

1. The section shown in Figure Question 1 is to be formed in steel plate.

   (a) Calculate the position of the centroid in the x and y directions from a defined datum.

   (b) Calculate the Second Moment of Area for the plate about the X–X and Y–Y axes through the centroid.

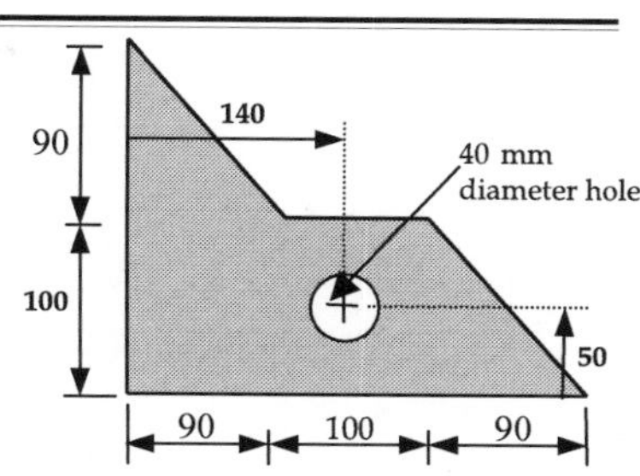

Figure Question 1

2. For the section shown in Figure Question 2 determine:

   (a) the position of the centroid from a defined datum;

   (b) the Second Moment of Area about the X–X and Y–Y axes through the centroid.

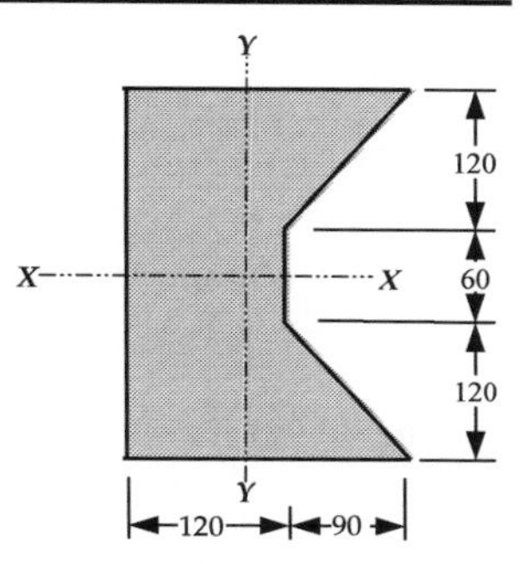

Figure Question 2

3. The section shown in Figure Question 3 is to form a beam which will support an ultimate moment of 1100 kN m applied in a horizontal direction. The maximum allowable compressive and tensile stress is 190 N/mm^2.

   Determine:

   (a) the position of the centroid of the section;

   (b) the Second Moment of Area about the Y–Y axis through the centroid;

   (c) the Moment of Resistance about the Y–Y axis.

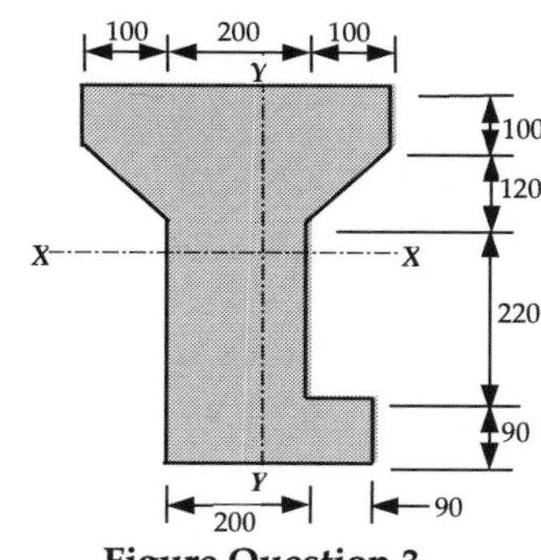

Figure Question 3

4. A steel plate is to be fixed to the bottom of a Rolled Steel Channel (RSC) section to form a beam (Figure Question 4). The fixing will ensure both channel and plate act as a single member. Details of the channel are given below. Determine the Moment of Resistance for the section about the X–X axis if the allowable stress is limited to 165 N/mm^2.

Serial Size	Mass per metre	Depth (mm)	Area (cm^2)	Second Moment of Area (cm^4)	
				X–X axis	Y–Y axis
254×76	28.29	254.0	36.0	3370	163

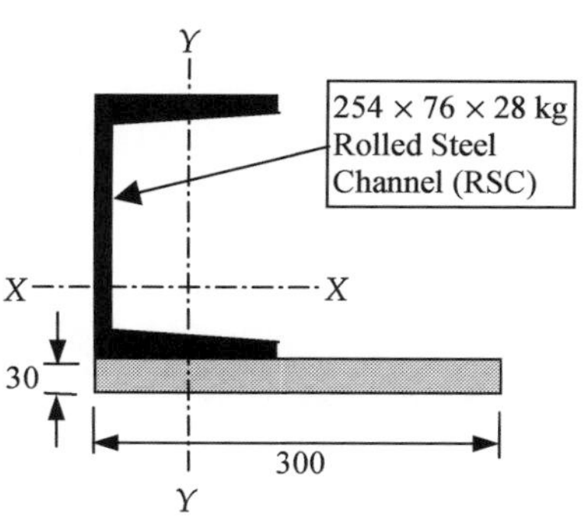

Figure Question 4

Solutions to Further Problems 2

1. **For I_{xx}**

Part No.	Area (mm^2)	Y (mm)	$A \times Y$ (mm^3)	I_{xx} (mm^4)	$(\bar{y}-y)$ (mm)	$A(\bar{y}-y)^2$ (mm^4)
1	4050	130	526 500	1 822 500	– 70.53	20 146 647.65
2	19 000	50	950 000	15 833 333.33	9.47	1 703 937.1
3	4500	33.33	149 985	2 500 000	26.14	3 074 848.2
4	– 1256.6	50	– 62 830	– 125 663.71	9.47	– 112 693.02
	$\Sigma A =$ 26 293.4		$\Sigma Ay =$ 1 563 655	$\Sigma I_{xx} =$ 20 030 169.62		$\Sigma A(\bar{y}-y)^2 =$ 24 812 739.93

For I_{yy}

Part No.	Area (mm^2)	X (mm)	$A \times X$ (mm^3)	I_{yy} (mm^4)	$(\bar{x}-x)$ (mm)	$A(\bar{x}-x)^2$ (mm^4)
1	4050	30	121 500	1 822 500	74.76	22 635 683.28
2	19 000	95	1 805 000	57 158 333.33	9.23	1 618 665.1
3	4500	220	990 000	2 025 000	– 115.77	60 312 118.05
4	– 1256.6	140	– 175 924	– 125 663.71	– 35.77	– 1 607 810.78
	$\Sigma A =$ 26 293.4		$\Sigma Ax =$ 2 740 576	$\Sigma I_{yy} =$ 60 880 169.62		$\Sigma A(\bar{x}-x)^2 =$ 82 958 655.65

$\bar{x}$ = 104.23 mm $\bar{y}$ = 59.47 mm I_{xx} = 44 842 909.55 mm^4 I_{yy} = 143 838 825.3 mm^4

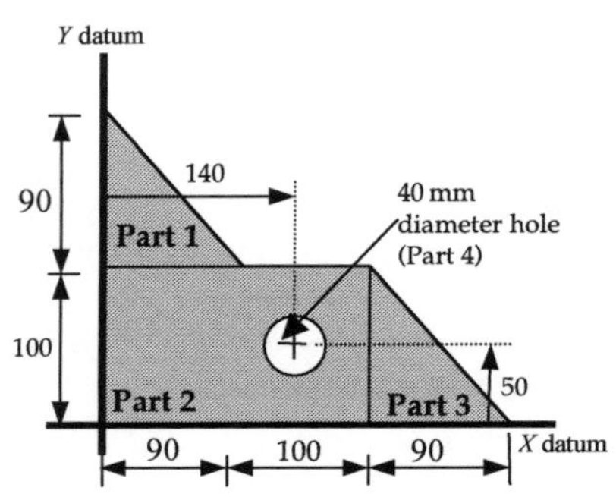

Figure Question 1

2. **For I_{xx}**

Part No.	Area (mm^2)	Y (mm)	$A \times Y$ (mm^3)	I_{xx} (mm^4)	$(\bar{y}-y)$ (mm)	$A(\bar{y}-y)^2$ (mm^4)
1	36 000	150	5 400 000	270 000 000	0	0
2	5400	260	1 404 000	4 320 000	–110	65 340 000
3	5400	40	216 000	4 320 000	110	65 340 000
	$\Sigma A =$ 46 800		$\Sigma Ay =$ 7 020 000	$\Sigma I_{xx} =$ 278 640 000		$\Sigma A(\bar{y}-y)^2 =$ 130 680 000

For I_{yy}

Part No.	Area (mm^2)	X (mm)	$A \times X$ (mm^3)	I_{yy} (mm^4)	$(\bar{x}-x)$ (mm)	$A(\bar{x}-x)^2$ (mm^4)
1	36 000	60	2 160 000	43 200 000	20.76	15 515 193.6
2	5400	150	810 000	2 430 000	–69.24	25 888 559.04
3	5400	150	810 000	2 430 000	–69.24	25 888 559.04
	$\Sigma A =$ 46 800		$\Sigma Ax =$ 3 780 000	$\Sigma I_{yy} =$ 48 060 000		$\Sigma A(\bar{x}-x)^2 =$ 67 292 311.68

$\bar{x}$ = 80.76 mm $\bar{y}$ = 150 mm I_{xx} = 409 320 000 mm^4 I_{yy} = 1 153 523 11.7 mm^4

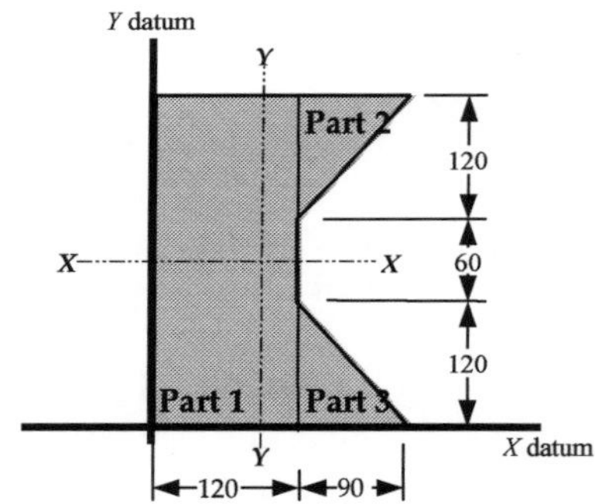

Figure Question 2

3. **For I_{yy}**

Part No.	Area (mm^2)	X (mm)	$A \times X$ (mm^3)	I_{yy} (mm^4)	$(\bar{x}-x)$ (mm)	$A(\bar{x}-x)^2$ (mm^4)
1	40 000	200	8 000 000	533 333 333.3	8.04	2 585 664
2	6000	66.67	400 020	3 333 333.33	141.37	119 912 851.4
3	6000	333.33	1 999 980	3 333 333.33	– 125.29	94 185 504.6
4	68 000	200	13 600 000	226 666 666.7	8.04	4 395 628.8
5	26 100	245	6 394 500	182 917 500	– 36.96	35 653 685.76
	$\Sigma A =$ 146 100		$\Sigma Ax =$ 30 394 500	$\Sigma I_{yy} =$ 949 584 166.7		$\Sigma A(\bar{x}-x)^2 =$ 256 733 344.6

$\bar{x}$ = 208.04 mm $\bar{y}$ = 292.502 mm I_{yy} = 1 206 317 511 mm^4

Moment of Resistance (extreme left) = $\sigma I/x$ = $(190 \times 1\,206\,317\,511)/(208.04 \times 10^6)$ = 1101.71 kN m

Moment of Resistance (extreme right) = $\sigma I/x$ = $(190 \times 1\,206\,317\,511)/(191.96 \times 10^6)$ = 1194 kN m

BEAM IS ADEQUATE since MoR (min.) = 1101.71 > 1100 kN m = M (max.)

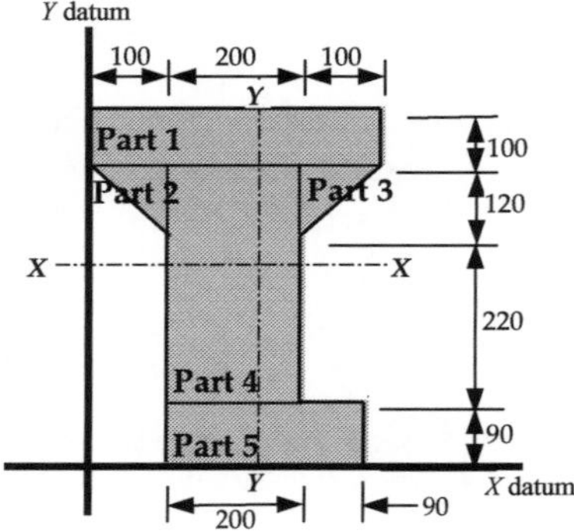

Figure Question 3

4. **For I_{xx}**

Part No.	Area (mm²)	Y (mm)	$A \times Y$ (mm³)	I_{xx} (mm⁴)	$(\bar{y}-y)$ (mm)	$A(\bar{y}-y)^2$ (mm⁴)
1	3600	157	565 200	33 700 000	−101.43	37 036 961.64
2	9000	15	135 000	675 000	40.57	14 813 324.1
	$\Sigma A =$ 12 600		$\Sigma Ay =$ 700 200	$\Sigma I_{xx} =$ 34 375 000		$\Sigma A(\bar{y}-y)^2 =$ 51 850 285.74

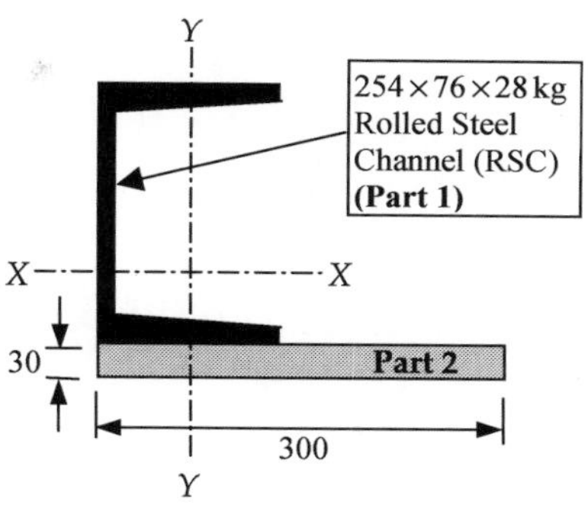

Figure Question 4

$\bar{y}$ = 55.57 mm **I_{xx} = 86 225 285.74 mm⁴**

Moment of Resistance (top) = $\sigma I/y_{top}$ = (165 × 86 225 285.74)/(228.43 × 10⁶)
= 62.28 kN m

Moment of Resistance (bottom) = $\sigma I/y_{bottom}$ = (165 × 86 225 285.74)/(55.57 × 10⁶)
= 256.02 kN m

Moment of Resistance (MoR) for compound section = 62.28 kN m

3. Analysis of Beams

> ***In this chapter we will learn about:***
> - ***Different types of load application***
> - ***Different support conditions***
> - ***Static indeterminacy***
> - ***Calculating beam reactions using the equations of statics***
> - ***Plotting Shear Force diagrams***
> - ***Plotting Bending Moment diagrams***

3.1 Introduction

We will now study the analysis of various beam systems. The beams considered here will be solved using the equations of statics alone, that is the equations given in Chapter 1, section 1.2.1 for the analysis of two-dimensional structures such that:

$$\Sigma F_x = 0,\ \ \Sigma F_y = 0,\ \ \Sigma M_z = 0 \qquad [1.5\,\text{a, b, c}]$$

Or, more familiarly the sum of the horizontal applied forces and horizontal reactions at the supports equals zero, the sum of the vertical applied forces and vertical reactions at the supports equals zero, the sum of all moments applied to the structure by the loads and the supports equals zero. This relationship can be extended to a three-dimensional structure as discussed in Chapter 1.

We will therefore only consider *statically determinate structures,* or structures that can be solved using the equations of statics alone (equations 1.5a, b, c). We will not, at this stage, extend our studies to include *statically indeterminate (or hyperstatic) structures.* These cannot be solved using the equations of statics alone and other equations are required which are formed by considering the overall response of the structure to the applied load and relating the movement of each member to all other members connected to it (Compatibility).

In order to ensure that the structure we are analysing is statically determinate, a simple test will be proposed.

First we will consider our simple beam system and the effects different types of applied loads and support have on our analysis. We will then look at a test for static determinacy. Having determined that our system can be solved using the equations of statics alone, we will then consider methods for solving these equations and ensuring equilibrium of the structure. Finally, we will plot simple graphs showing the distribution of both Shear Force and Bending Moment through our structural system. These graphs are known as Shear Force Diagrams and Bending Moment Diagrams and are used as a visual aid by designers in the analysis of various structural members. They are particularly useful in the design and detailing of concrete sections, where it is often necessary to vary the amount of reinforcement used in proportion to the shear and moment being resisted.

3.2 Types of Beams, Loads and Supports

Considered here are simple statically determinate beams and support systems. Beams are generally horizontal members. They can support applied vertical and horizontal loads along with applied moments; however we will only be analysing beams with a single span, that is either cantilevered beams (Figure 3.1a) or simply supported beams (Figure 3.1b). Beams can span over a number of supports, and these are commonly termed 'continuous beams', however we will not consider the analysis of this type of structure here. It is also important to note that the techniques developed can be applied to columns with similar support and load conditions.

Firstly we will consider the types of loads which can be applied to our structures.

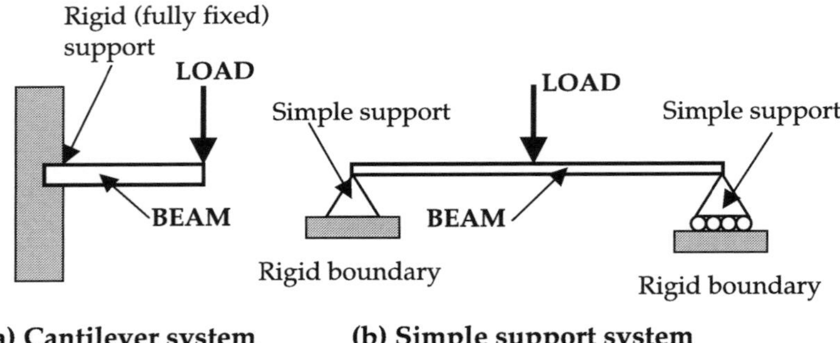

Figure 3.1: Example of simple beam, support and load systems

3.2.1 Types of Applied Loads

In Chapter 1, section 1.3 we looked at the determination of loads applied to a specific part of our structure. We looked at how we determine the amount of load applied and how this is distributed throughout a building. Consider now how these loads are analysed in a simple beam system.

The two types of applied loads we will consider are:

(1) Point loads
(2) Distributed loads.

Point loads

These are loads applied at a specific point on the structure. In terms of analysis this specific point can be identified discretely on our beam system. These loads are only applied at the point considered. Examples of this type of load are the connection of two beams (Figure 3.2a), the load transferred from a column to its supporting beam (Figure 3.2b), or the load applied by a person in the system of Figure 3.2c.

Obviously, all of the loading types do have some finite width, the beam connection for instance in Figure 3.2a may typically be 200 mm or more wide. However, with respect to the length of the beam or column, this width is small and we therefore consider the load to be applied at a specific point. Point loads are specified as the total load applied at the point under consideration. The units are therefore Newtons (N) or more typically kiloNewtons (kN).

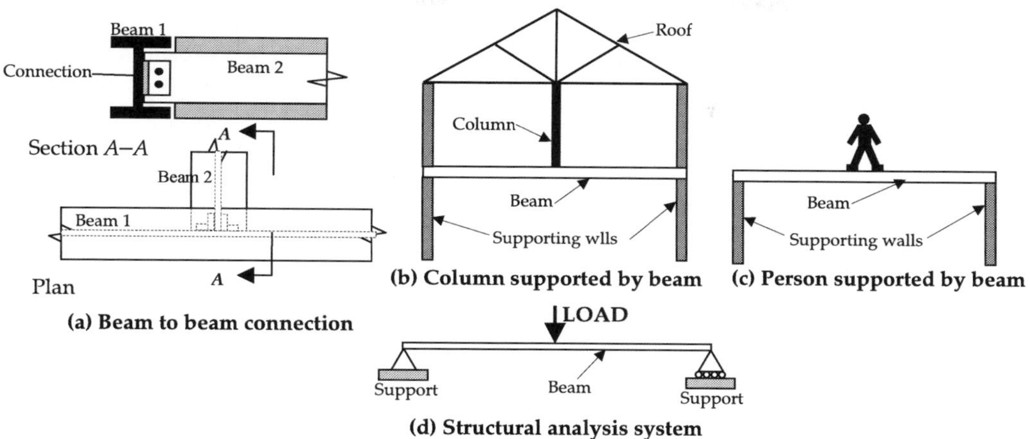

Figure 3.2: Examples of different types of point load

Distributed loads (DL)

These are loads applied along a specified length on the structure. In terms of our analysis we will need to know both the magnitude of the load and its length in order to calculate the total load applied. These loads may be uniform or vary in their distribution. Examples of this type of load are the load applied by a crowd on the beam of a grandstand at a football match (Figure 3.3a), the load from snow applied to a roof rafter (Figure 3.3b), or the load applied by the roof structure to the beam in the system of Figure 3.3c.

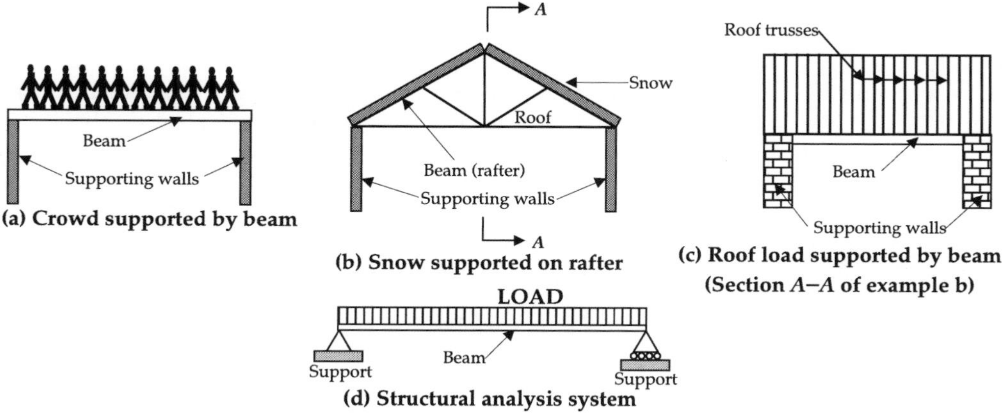

Figure 3.3: Examples of different types of uniformly distributed load (UDL)

Again, some of the loading types considered do have a finite width, the roof trusses supported by the beam in Figure 3.3c for example. However, in comparison to the length of the beam, the spacing between the trusses is small and we therefore consider the load to be evenly spread along the beam. The units for distributed loads are Newton per metre (N/m) or more typically kiloNewton per metre (kN/m). The total load may be calculated by multiplying the load per metre by the length of the load.

Distributed loads can also vary in magnitude along their length and these can be analysed using various techniques. For evenly distributed uniform loads we can calculate the total load by multiplying the load per metre by its length. The total load is considered to act at the load's centroid, which for a rectangle is located at half its length (see Figure 3.4a). If we have two rectangular loads these can be considered separately as shown in Figure 3.4b. For the triangular load of Figure 3.4c we must first calculate the total load. This will be found to be equivalent to the area of the 'load' triangle such that:

$$\text{Area of triangle} = 0.5 \times \text{base} \times \text{height} \qquad [3.1]$$

In this case the base is the length of the load and the height is the maximum intensity of the force per metre. This total load will again act at the centroid of the area which is located at two-thirds of the length of the base from the apex (see Chapter 2, Figure 2.8).

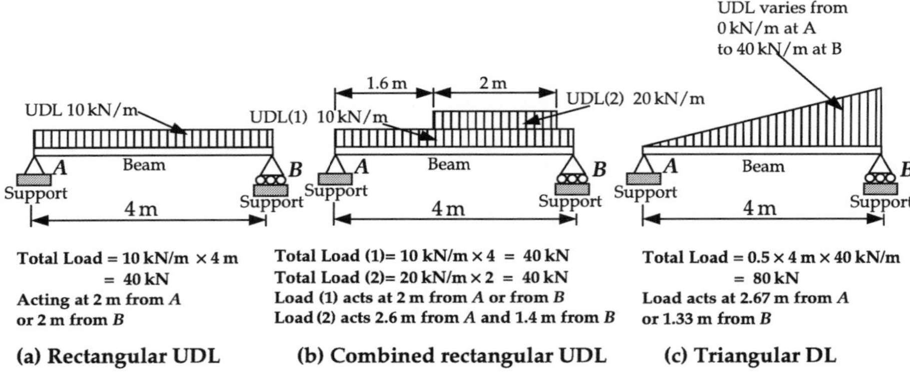

Figure 3.4: Calculation of total load and point of application for UDL

For Figure 3.4b we note that load (1) acts at half its length, 2 m from *A* or *B* similar to that in Figure 3.4a, however load (2) acts at 2.6 m from support *A* and 1.4 m from support *B*. This is easily determined since load (2) acts at half its length, as before, which is 2/2 m = 1 m. The distance from support *A* to the point of action is therefore 1.6 m + 2/2 m = 2.6 m. The distance to the point of action may be similarly measured from point *B* as 4 m – 2.6 m = 1.4 m or (4 m – 1.6 m – 2 m) + 2/2 m = 1.4 m. We must therefore treat each load separately in our analysis as we will discover later. The triangular load of Figure 3.4c acts at its centroid, which is located at two-thirds its base length from *A* (2.67 m) or one-third its length from *B* (1.33 m).

These are the most common types of Distributed Load (DL) that we will consider. The triangular distributed load is commonly found in the analysis of structures such as retaining walls and dams where the imposed load is due to soil or hydrostatic (water) pressure.

3.2.2 Types of Support

As stated in Chapter 1, section 1.3.1 loads can be classified as those applied to the system and those applied to the boundary of the system. All structures must be supported. The final support will be the ground on which the structure rests, but throughout the structural system each member must be supported by another in order that the load is finally transferred to the foundation.

We will now consider the three most common types of support:

(1) Fully fixed (or encastré) support
(2) Simple (or knife edge) support
(3) Roller support.

Other types of support do exist, such as spring support systems, however these form part of more advanced studies in structural mechanics.

Fully fixed support

With this type of support, illustrated in Figure 3.5c, the end of the beam is clamped by the member supporting it. As the boundary of the beam is considered to be unyielding, the beam is fully fixed at this position; it cannot move up or down (δy), left or right (δx) and, at its junction with the support, cannot rotate (θz). Typical examples of this type of joint would be a beam built into a wall, a column built into a foundation, or a suitably reinforced concrete beam-to-column joint (Figure 3.5a and 3.5b).

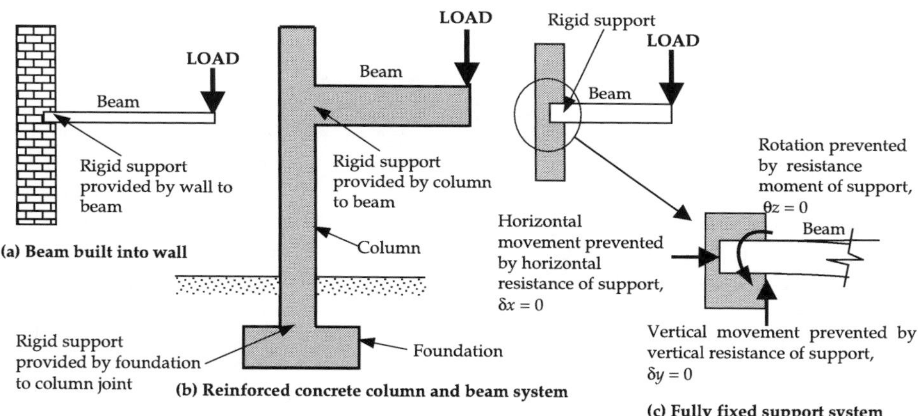

Figure 3.5: Examples of a fully fixed support

The fully fixed support can therefore provide resistance to vertical load, horizontal load and applied moment.

Simple support

This type of support, illustrated in Figure 3.6c. will not provide resistance to bending moments. The end of the beam generally rests on the supporting structure or is bolted with a single bolt. The boundary of the support is still considered to be unyielding. The beam is fixed in position and cannot move down (δy). Horizontal (δx) movement of the support is also resisted. Typical examples of this type of joint would be a beam resting on a wall, or connected by a simple fixing to a column (Figure 3.6a and b).

The simple support will provide resistance to vertical and horizontal load but has no resistance to an applied moment. The beam is therefore free to rotate at the support.

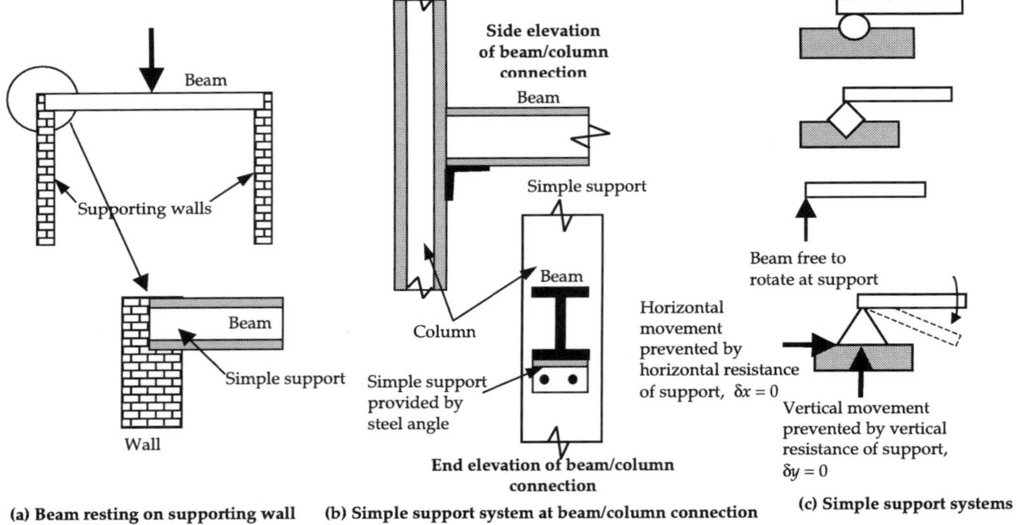

Figure 3.6: Examples of a simple support

Roller support

The roller support is a further modification to the simple support and is illustrated in Figure 3.7c. It will provide resistance to vertical movement only. It is unable to resist applied moments or horizontal force. The end of the beam generally rests on the supporting structure and may have a bearing plate inserted underneath to reduce resistance to movement in the horizontal direction. The boundary of the support is still considered to be unyielding and the beam is fixed in its vertical position (δy) and cannot move in a downward direction. Typical examples of this type of support would be the bearing of a bridge or the support of heating pipes, both of which require that some horizontal movement be allowed for thermal expansion and contraction (Figure 3.7a and b).

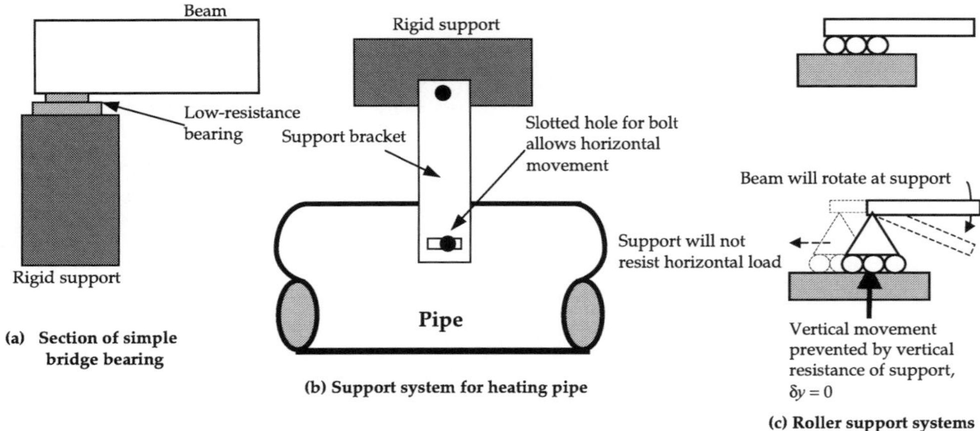

Figure 3.7: Examples of roller support systems

The roller support can only provide resistance to vertical load, it has no resistance to horizontal load or applied moment. The beam is therefore free to rotate at the support and the support will move under horizontal loading (much like a model car, which moves when pushed horizontally but can support vertical load if pressure is applied downwards!).

3.2.3 Types of Beam

In our analysis of structures we will be considering two different types of beams. These are:

(1) Simply supported beams
(2) Cantilevered beams.

Examples of these beams are shown in Figure 3.7a and b respectively.

Members may be orientated in either the horizontal (beam) or vertical (column) direction, however the analysis of each system is similar. Continuous beams will not be considered at this stage as generally they cannot be solved using the equations of statics alone.

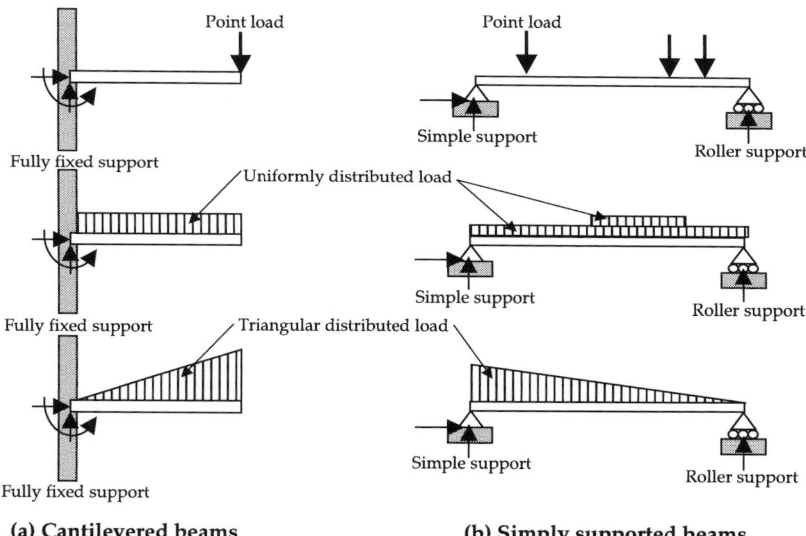

Figure 3.8: Examples of cantilever and simply supported beams

Each of the systems shown in Figure 3.8 is statically determinate. If, as is usual, the applied loads are known the only 'unknowns' will be the value of each of the reactive forces or 'reactants' at the supports; that is, the horizontal and vertical force and the resisting moment of the fully fixed support, the horizontal and vertical forces at the simple support, and the vertical force at the roller support. For equilibrium of a beam system we require that:

$\Sigma F_x = 0$ (summation of all horizontal forces must be zero)
$\Sigma F_y = 0$ (summation of all vertical forces must be zero)
$\Sigma M_z = 0$ (summation of all moments must be zero)

We can use this relationship to solve for the unknown values.

However, before we can commence the analysis of any structural system using the equations of statics alone, we must ensure that it is statically determinate. We will now consider a simple test to determine the static determinacy of various types of structure.

3.3 Determination of the Static Determinacy of Structures

A structure which can be solved using the equations of statics alone is classified as a *statically determinate structure*. If a structure cannot be solved using the equations of statics alone it is called a *statically indeterminate* or *hyperstatic structure*. A simple test exists in order to determine if a structure is statically determinate or not.

For plane frames, that is two-dimensional structures such as the portal frame shown in Figure 3.9 we know that for equilibrium ΣF_x, ΣF_y and ΣM_z must each equal zero. We also know that at joints B and C on the frame, any movement in the member AB must result in movement in the member BC if the connection at the joint is fixed. If the joint between the two members is considered to be rigid, then the angle between members AB and BC must remain constant (in this case at 90°). We can therefore relate all member movements at a joint and these must be 'compatible' with each other if the members are to remain connected. It is thus possible to relate both geometric properties and force equations in order to analyse a structure. Methods such as Moment Area, Stiffness and Flexibility have been developed to analyse structures that are indeterminate using the equations of statics and equations derived from geometric properties such as compatibility. However, at this stage we will only concern ourselves with structures that can be solved using the equations of statics alone and we will now consider a simple test that can be used to determine the static indeterminacy of various types of frame.

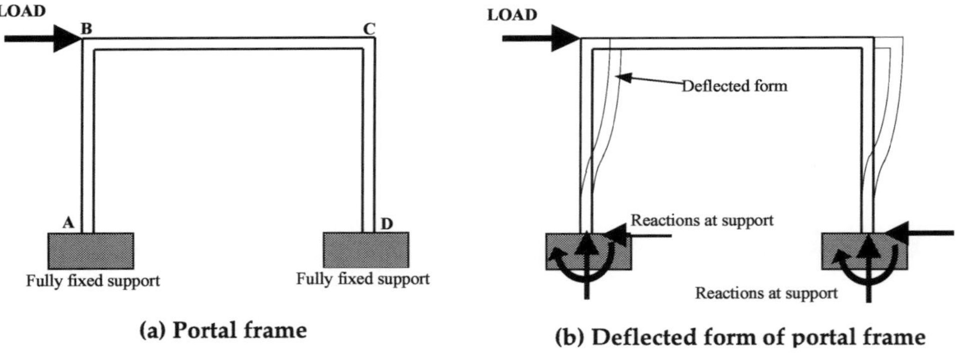

(a) Portal frame **(b) Deflected form of portal frame**

Figure 3.9: Simple portal frame structure

For any structure we will know the geometry and the applied load. Since the load is known we will be able to derive a set of three equations, based on the equations of statics, for each joint, to determine the vertical and horizontal forces along with the moments. However, these equations will contain references to the unknown forces in the members and at the reactions. Therefore, for each joint we will have three 'knowns' whilst for each member we will have three 'unknowns'. We will also have 'unknown' values for the forces and moments, or reactants, at the supports. This is shown in diagrammatic form for the portal frame in Figure 3.10.

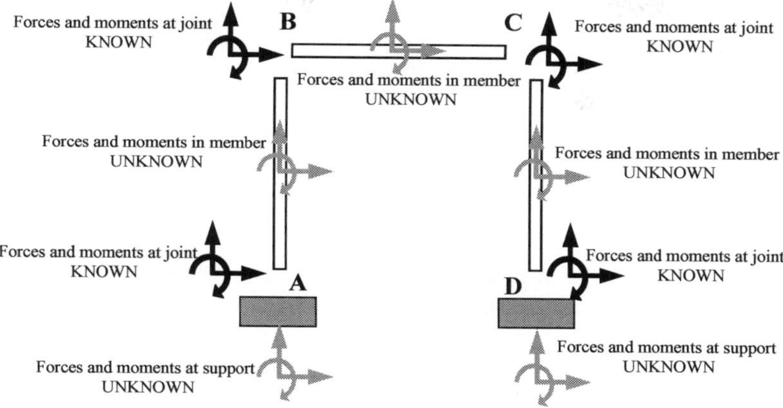

Figure 3.10: Forces and moments on portal frame

At this stage a simple test is presented which will enable us to determine whether a structure is statically determinate or not. If it is not statically determinate we will not be able to solve it using the methods discussed in this chapter and we will need to develop further, more advanced techniques.

We should also remember that the number of reactant forces will depend on the type of support used, as shown in Figure 3.11.

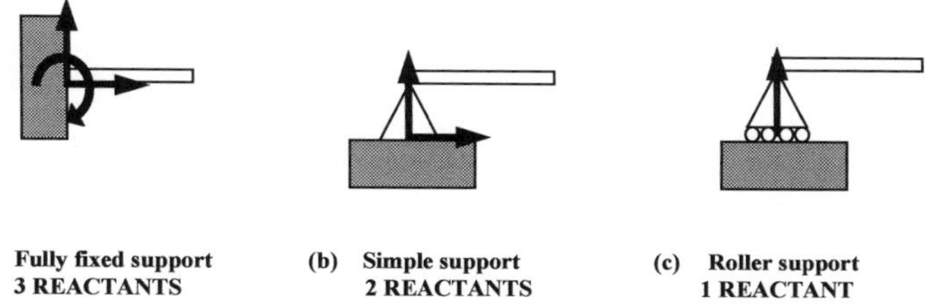

Figure 3.11: Reactant forces at different types of support

From the relationship of 'known' and 'unknown' forces and moments previously discussed we can derive an equation such that:

$$(m + r - f) - 2j = \text{degree of indeterminacy} \qquad [3.2]$$

where m = number of members
r = number of rigid joints (between members and supports)
f = roller releases
and j = number of joints (other than at supports).

Thus, applying this equation to the simple portal frame of Figure 3.9 we find that:

Number of members m = 3
Number of rigid joints r = 4 (including supports)
Number of roller releases f = 0
Number of joints j = 2 (not including supports)

Therefore, from equation [3.2]:

$$(m + r - f) - 2j = \text{degree of indeterminacy}$$
$$(3 + 4 - 0) - 2(2) = 7 - 4 = 3^\circ \text{ indeterminate}$$

Equation [3.2] can be applied to all two-dimensional structures and takes account of both the support conditions and also any internal releases contained within the structure. These internal releases, often referred to as pins, may be visualised as 'hinges' within the frame. These hinges provide a further equation for our solution since the sum of the moments at a hinge must equal zero. In order to visualise a pinned joint, consider the simple hinge joint of a door as shown in Figure 3.12, which has very similar structural properties. Also shown are typical pin joints that may be formed in a structure.

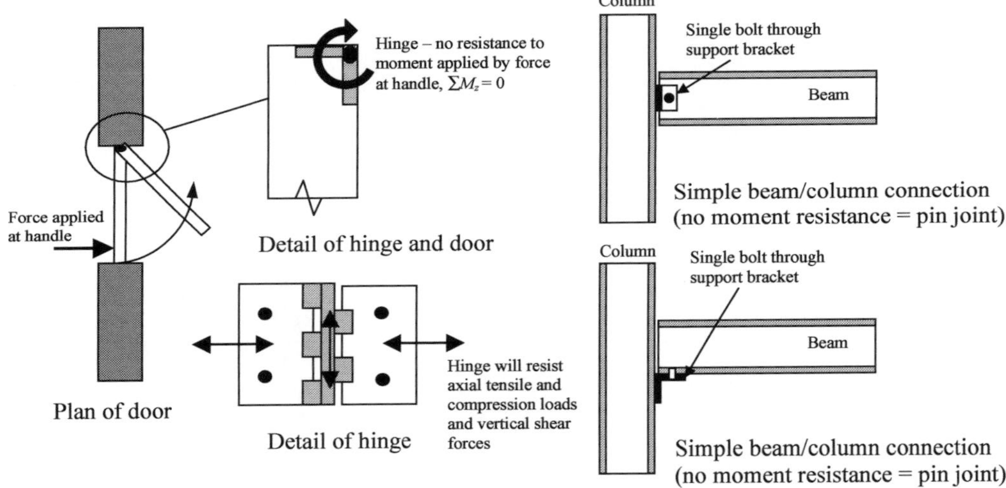

Figure 3.12: Detail of simple door hinge and internal pinned supports

Whilst some frictional resistance will be provided at the hinge or pin joint, this is small when compared to the applied forces and is normally ignored. These internal releases are accounted for in the determination of static determinacy as it is only the rigid joints that are included in the equation of [3.2].

◇

Example 3.1: Static determinacy of a 'one-pin' portal frame

Using equation [3.2], determine the static determinacy of the portal frame structure of Figure 3.13.

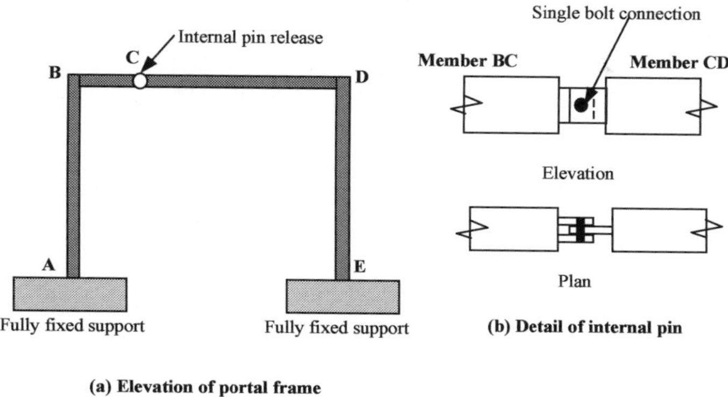

Figure 3.13: Details of portal frame – Example 3.1

First we number the joints and members as shown in Figure 3.14 noting that all the joints apart from that at C are rigid and fully fixed.

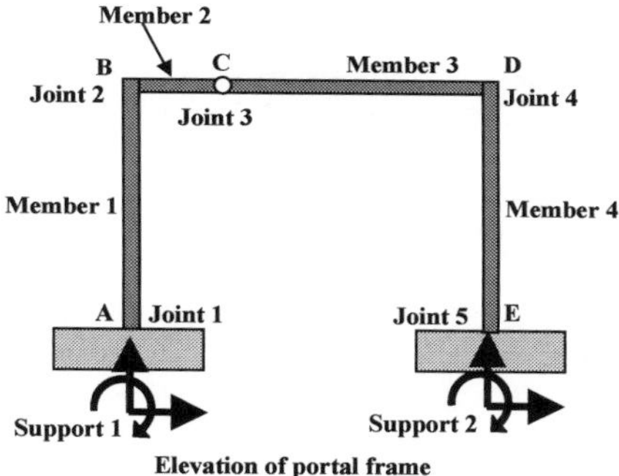

Figure 3.14: Analysis of portal frame – Example 3.1

From Figure 3.14 we find:

Number of members m	=	4
Number of rigid joints r	=	4 (including supports)
Number of roller releases f	=	0
Number of joints j	=	3 (not including supports)

Therefore, from equation [3.2]:

$$(m + r - f) - 2j = \text{degree of indeterminacy}$$
$$(4 + 4 - 0) - 2(3) = 8 - 6 = 2^\circ \text{ indeterminate}$$

The structure in Example 3.1 is therefore 2° indeterminate.

We will now consider some further examples. However it is important to note that if our final result is 0° indeterminate we have a statically determinate structure. If however the result is negative, the structure will be a mechanism. This type of structure is unstable and cannot resist applied forces. Such a structure is shown in Example 3.2.

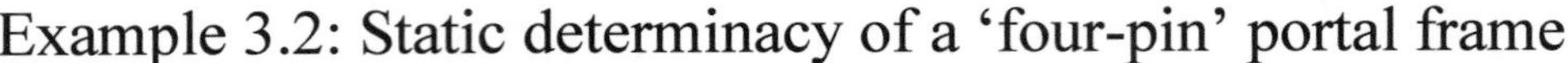

Example 3.2: Static determinacy of a 'four-pin' portal frame

Using equation [3.2], determine the static determinacy of the portal frame structure of Figure 3.15.

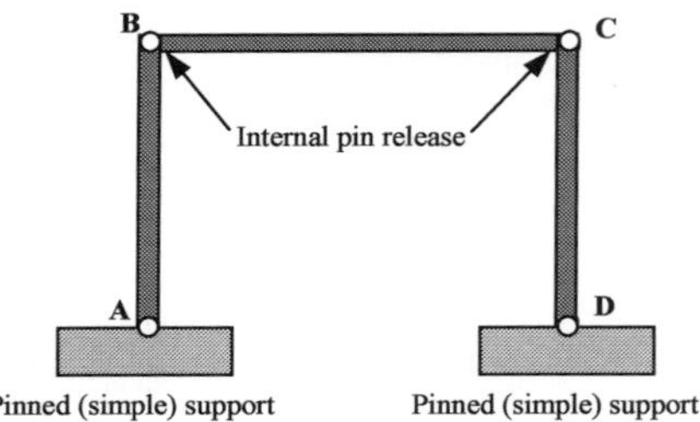

Figure 3.15: Portal frame – Example 3.2

Again we number all joints, members and reactions as if the structure were rigid with fully fixed supports.

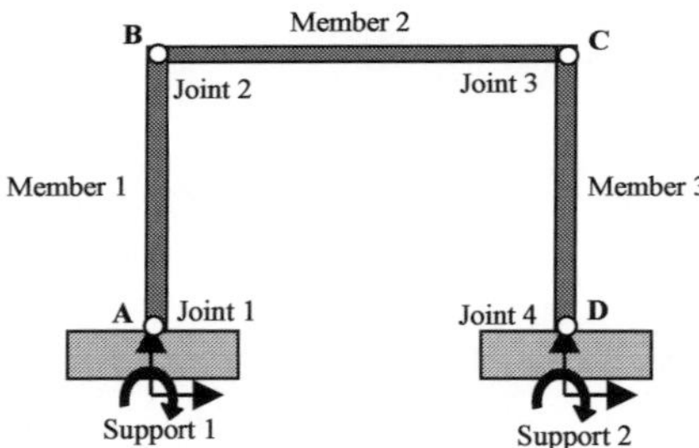

Figure 3.16: Analysis of portal frame – Example 3.2

From Figure 3.16 we find:

Number of members m	=	3	
Number of rigid joints r	=	0	(including supports)
Number of roller releases f	=	0	
Number of joints j	=	2	(not including supports)

Therefore, from equation [3.2]:

$$(m + r - f) - 2j = \text{degree of indeterminacy}$$
$$(3 + 0 - 0) - 2(2) = 3 - 4 = -1^\circ \text{ indeterminate}$$

From our analysis we find the structure in Example 3.2 is -1° indeterminate. It is therefore a mechanism and will collapse under applied load.

Example 3.3: Static determinacy of a rigid two-storey frame

Using equation [3.2], determine the static determinacy of the frame structure of Figure 3.17.

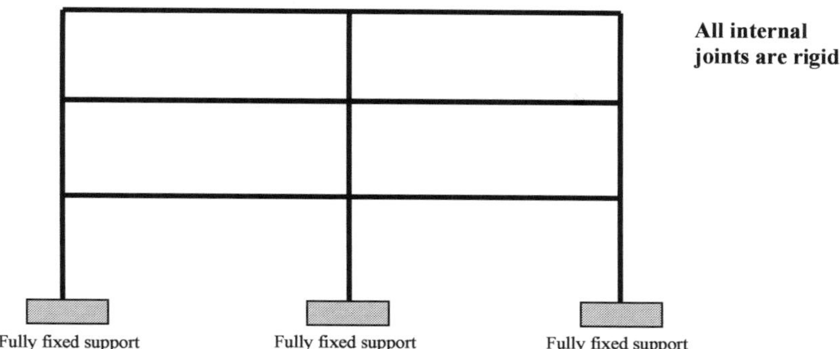

Figure 3.17: Elevation of frame structure – Example 3.3

First number joints, members and reactions as before (Figure 3.18).

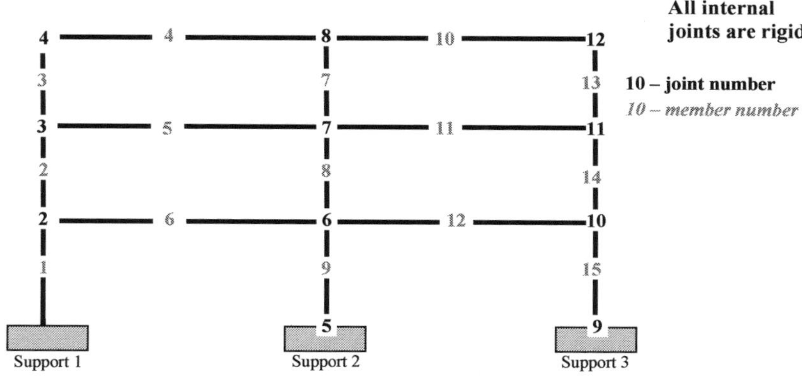

Figure 3.18: Analysis of frame structure – Example 3.3

From Figure 3.18 we find:

	Number of members m	=	15
	Number of rigid joints r	=	12 (including supports)
	Number of roller releases f	=	0
	Number of joints j	=	9 (not including supports)

Therefore, from equation [3.2]:

$$(m + r - f) - 2j = \text{degree of indeterminacy}$$
$$(15 + 12 - 0) - 2(9) = 27 - 18 = 9^\circ \text{ indeterminate}$$

The structure in Example 3.3 is therefore 9° indeterminate.

Example 3.4: Static determinacy of a non-symmetric frame

Using equation [3.2], determine the static determinacy of the frame structure of Figure 3.19, which has a fully fixed support at A and a pinned support at G. It also has an internal pin at joint C.

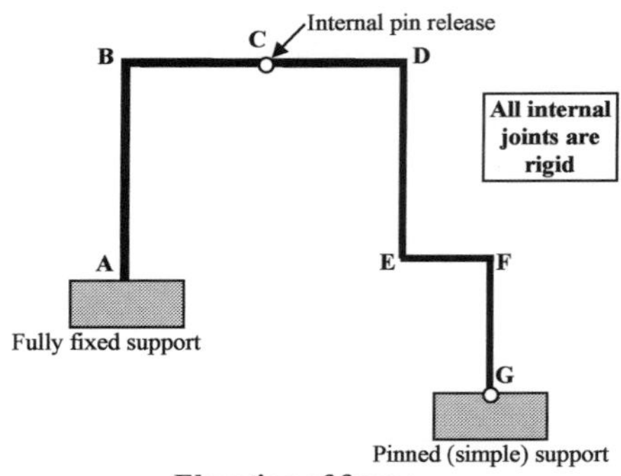

Figure 3.19: Details of frame – Example 3.4

We number the joints, members and reactions, as before.

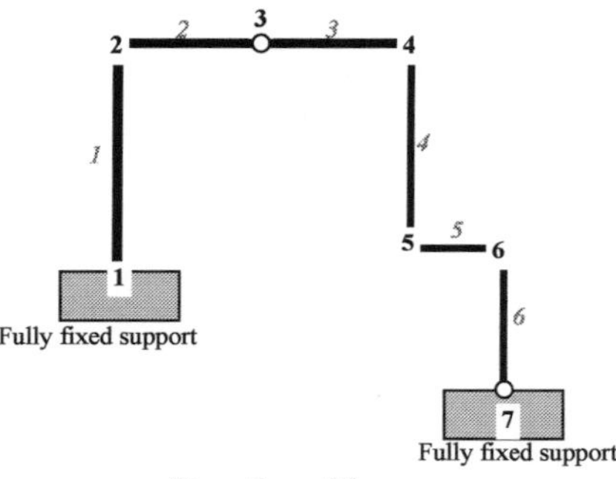

Figure 3.20: Analysis of frame structure – Example 3.4

From Figure 3.20 we find:			
	Number of members m	=	6
	Number of rigid joints r	=	5 (including supports)
	Number of roller releases f	=	0
	Number of joints j	=	5 (not including supports)

Therefore, from equation [3.2]:

$$(m + r - f) - 2j = \text{degree of indeterminacy}$$
$$(6 + 5 - 0) - 2(5) = 11 - 10 = 1^\circ \text{ indeterminate}$$

The structure in Example 3.4 is therefore 1° indeterminate.

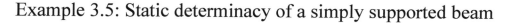

Example 3.5: Static determinacy of a simply supported beam

Using equation [3.2], calculate the static determinacy of the beam of Figure 3.21a, which is simply supported at A and has a roller support at B.

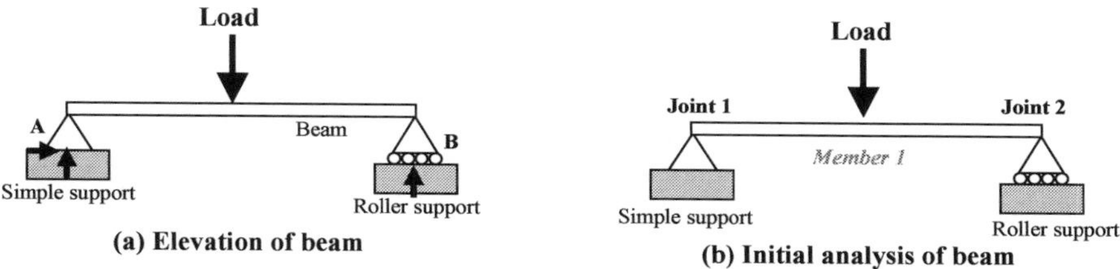

Figure 3.21: Details of beam – Example 3.5

From Figure 3.21b we find:

Number of members m	=	1	
Number of rigid joints r	=	0	(including supports)
Number of roller releases f	=	1	
Number of joints j	=	0	(not including supports)

Therefore, from equation [3.2]:

$$(m + r - f) - 2j = \text{degree of indeterminacy}$$
$$(1 + 0 - 1) - 2(0) = 1 - 1 = 0^{\circ} \text{ determinate}$$

The structure in Example 3.4 is therefore a determinate structure and can be solved using the equations of statics alone.

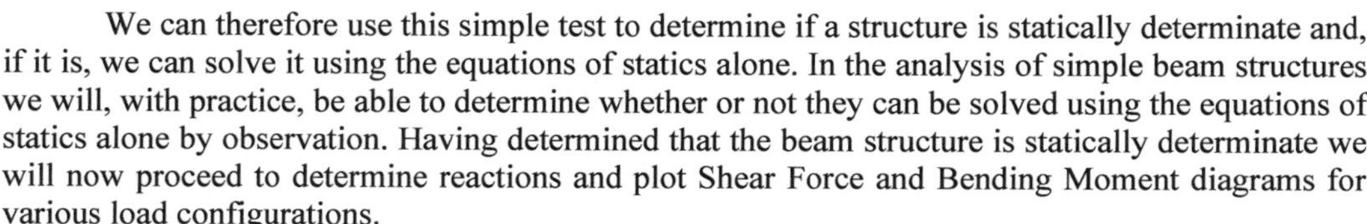

We can therefore use this simple test to determine if a structure is statically determinate and, if it is, we can solve it using the equations of statics alone. In the analysis of simple beam structures we will, with practice, be able to determine whether or not they can be solved using the equations of statics alone by observation. Having determined that the beam structure is statically determinate we will now proceed to determine reactions and plot Shear Force and Bending Moment diagrams for various load configurations.

3.4 Beam Reactions

In the examples that follow, the beams will be considered as weightless in order to simplify these early problems. We must however remember to include the self-weight of the beam in the final design calculations in order to complete an accurate analysis of the structure. This can be achieved by including a Distributed Load to represent the beam's own weight. It is also important to note that the beam is assumed to be a rigid structure. This again will simplify the analysis.

This simplification is permissible since, in comparison to the length (or span) of the beam, the deflection will be very small.

Assuming the beam to be rigid leads us to develop what is known as the **Rigid Stressed Body Paradox**. It is necessary, in the analysis techniques considered here, that the beam remains rigid (or deflections are small in comparison to length) and does not bend (deflect) under load. This is required because, if the beam were to bend excessively under load, secondary moments would be induced, as indicated in Figure 3.22b, which would further complicate the analysis.

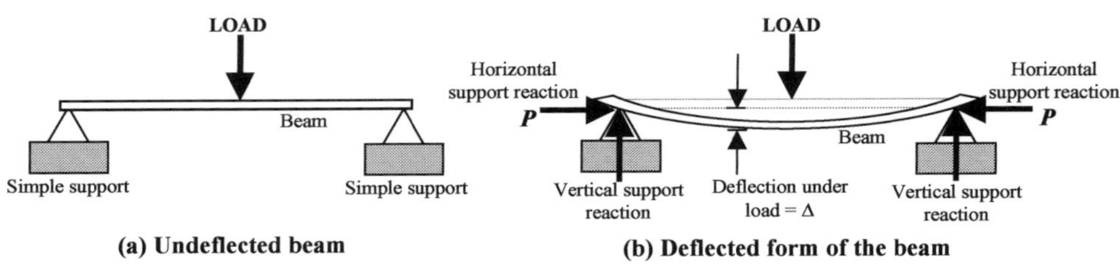

Figure 3.22: Deflection (bending) of beam under load

From Figure 3.22 we note that, as the beam is loaded it bends or deflects downward. The amount of deflection will be designated by Δ and the horizontal reactions induced designated by P. The maximum moment can be calculated at the mid-span of the beam using the rigid stressed body paradox, that is considering the beam does not deflect. However, in real structures, displacement does take place and a secondary moment is induced equivalent to the moment developed by the force P:

$$\text{Secondary moment} = P\Delta$$

In most cases this moment is small in comparison to the moment from the applied loads, since the deflection will be quite small when compared to the length of the beam, and is therefore considered to have a second-order (insignificant) effect. It may be important to consider this $P\Delta$ (P-Delta) effect in the design of some structures, particularly columns, and we should always be aware that it will have an influence, to a greater or lesser extent, on all structures. In the determination of the beam reactions, Shear Force and Bending Moment diagrams which follow, we will consider our beams to be both weightless and rigid.

In order to calculate beam reactions, we need to ensure that the three equations of statics are fully satisfied and that the summation of all the applied forces acting on the beam and all the external forces acting at the supports equate to zero. In order to determine this, we derive equations by summing all forces and moments on the structure such that:

$$\sum F_x = 0$$
$$\sum F_y = 0$$
$$\sum M_z = 0$$

If these equations can be satisfied, we can ensure equilibrium of our system. We will first consider the simplest system of a beam supporting a point load applied at mid-span as shown in Example 3.6.

Example 3.6: Beam reactions for a simply supported beam carrying a single point load at mid-span

For the beam shown in Figure 3.23, loaded at mid-span with a point load of 10 kN, determine the forces (reactions) applied at the supports A and B required to ensure equilibrium of the system.

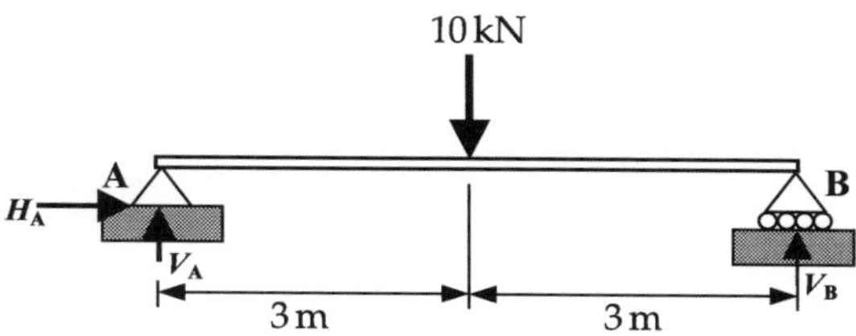

Figure 3.23: Beam system – Example 3.6

It should be possible, by observation, to conclude that the vertical forces (reactions) at each support will be equal to half the applied load – in this case 5 kN at each support, in an upwards direction. It should also be possible, by observation, to determine that the horizontal reaction H_A must be zero, as there is no horizontal applied load. However, we will proceed to determine the reactions using the three equations of statics. We do this by summing all the forces and moments in turn, external and internal, using normal Cartesian co-ordinates to determine the signs of forces and the clockwise positive convention for moments, as shown in Figure 3.24.

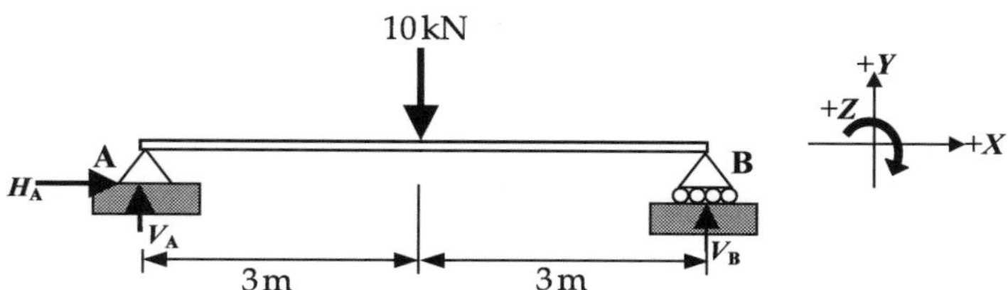

Figure 3.24: Co-ordinate system for beam analysis – Example 3.6

First we sum all horizontal forces, that is all horizontal loads and reactions (forces at supports), such that:

$$\Sigma F_x = H_A + 0 = 0 \qquad [1]$$

therefore, for equilibrium, from [1]:

$H_A = 0$ kN (note that H_B must be zero for a roller support)

We now consider all vertical forces and reactions:

$$\Sigma F_y = V_A + V_B - 10 \text{ kN} = 0$$

and we can simplify this equation to:

$$V_A + V_B = 10 \text{ kN} \qquad [2]$$

Note that we cannot solve equation [2] at this stage since no unique answer can be derived. Any solution, for instance 2 + 8 = 10 or even 3 + 7 = 10 will satisfy this single equation. To overcome this problem we require a further equation to solve simultaneously with equation [2]. This equation can be derived by taking moments along the beam.

In theory, we can take moments at any point along the beam; however we can simplify the problem by calculating moments using one of the reactions as our datum or origin, as shown in Figure 3.25 as we already know that the moments at this point must be zero (simple support). If we do this, we effectively reduce our equation to one 'unknown' since the summation of all moments of forces at the datum must be zero.

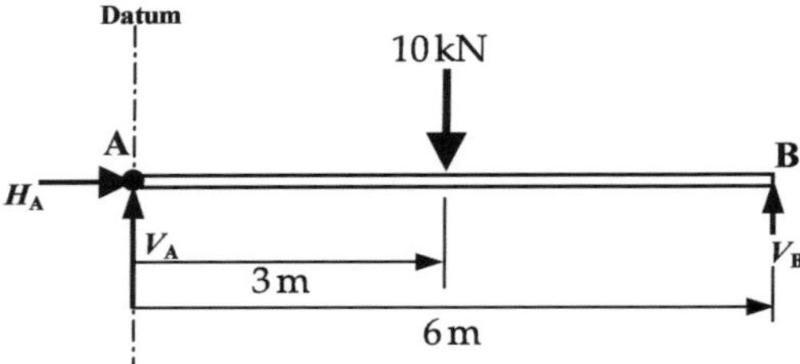

Figure 3.25: Position of datum for calculation of moments (support A) – Example 3.6

We move from our datum, along the beam, in this case from left to right, and calculate the moment for each force in turn. So taking moments about support A:

$$\Sigma M_z = (V_A \times 0) + (10\text{ kN} \times 3\text{m}) - (V_B \times 6\text{m}) = 0 \qquad [3]$$

Note that all distances are measured from the datum and that we follow the clockwise positive notation for our moments. Note also that the force H_A has a zero distance and therefore has no influence on these calculations. We can now simplify and solve equation [3] thus:

$$(30) - (6V_B) = 0$$
$$30 = 6V_B$$

which gives $$V_B = (30/6) = 5\text{ kN}$$

We can then back-substitute this answer into equation [2] to calculate V_A such that:

$$V_A + V_B = 10\text{ kN} \qquad [2]$$

But $V_B = 5$ kN hence: $$V_A + 5\text{ kN} = 10\text{ kN}$$
$$V_A = 10 - 5 = 5\text{ kN}$$

These simple steps may be followed in order to calculate the reactions for all types of loading on various statically determinate systems, as shown in the following examples.

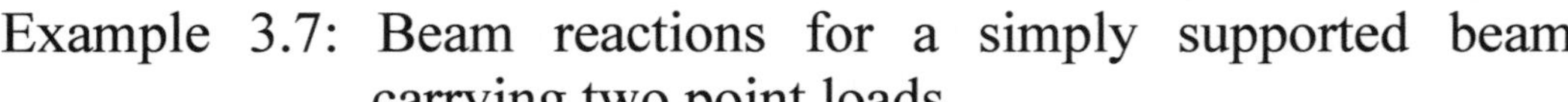

Example 3.7: Beam reactions for a simply supported beam carrying two point loads

For the beam shown in Figure 3.26, loaded with two point loads of 30 kN and 50 kN, determine the forces (reactions) at the supports A and B required to ensure equilibrium of the system.

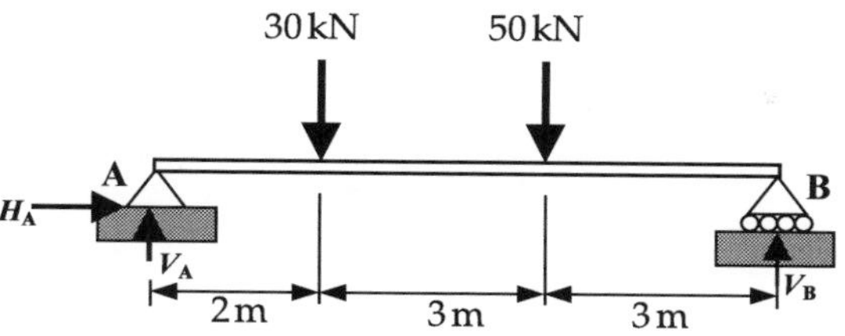

Figure 3.26: Beam system – Example 3.7

We will follow the same procedure as before:

Summation of all horizontal forces $= \Sigma F_x = H_A + 0 = 0$ [1]

$\therefore H_A = 0$ kN

Summation of all vertical forces $= \Sigma F_y = V_A + V_B - 30\text{ kN} - 50\text{ kN} = 0$

$V_A + V_B = 80$ kN [2]

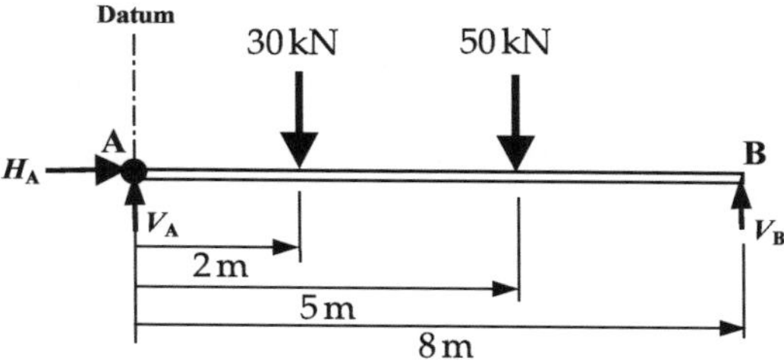

Figure 3.27: Position of datum for calculation of moments (support A) – Example 3.7

Taking moments about support A as shown in Figure 3.27:

Summation of all moments $= \Sigma M_z = (V_A \times 0) + (30\text{ kN} \times 2\text{ m}) + (50\text{ kN} \times 5\text{ m})$

$- (V_B \times 8\text{ m}) = 0$ [3]

$= (60) + (250) - 8V_B = 0$

$8V_B = 310$

$V_B = (310/8) = 38.75$ kN

Back-substituting the result for V_B into equation [2] we find:

$V_A + V_B = 80$ kN [2]

But $V_B = 38.75$ kN so: $V_A + 38.75\text{ kN} = 80$ kN

$V_A = 80 - 38.75 = 41.25$ kN

It is important to note that it does not matter which reaction we decide to take moments about. If we now take moments about support B as indicated in Figure 3.28 we will find:

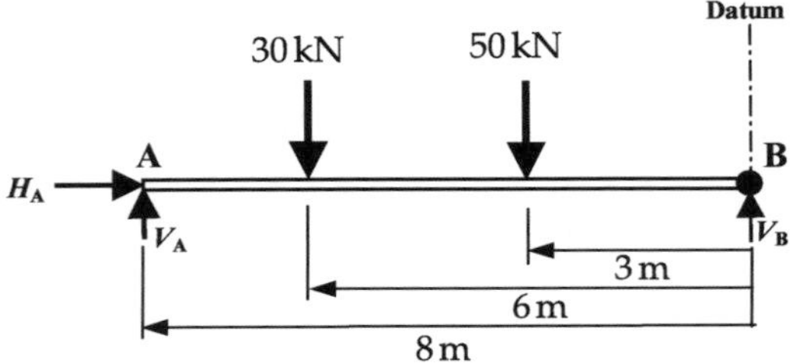

Figure 3.28: Position of datum for calculation of moments (support B) – Example 3.7

Taking moments about support B:

Summation of all moments $= \sum M_z = (V_B \times 0) - (50 \text{ kN} \times 3 \text{ m}) - (30 \text{ kN} \times 6 \text{ m}) + (V_A \times 8 \text{ m}) = 0$ [3]

$= -(150) - (180) + 8V_A = 0$

$8V_A = 330$

$V_A = (330/8) = 41.25$ kN (as before).

The final beam forces and reactions are shown in Figure 3.29.

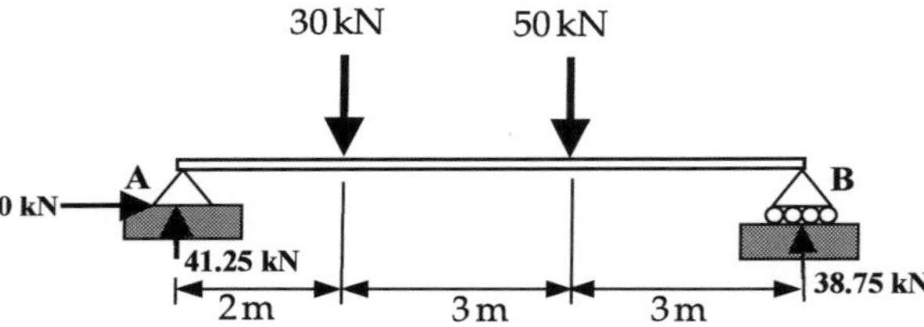

Figure 3.29: Final forces applied to beam – Example 3.7

◇

In the previous examples, the supports have been located at the ends of the beam. We will now consider the situation where one of the supports is not located at the end.

◇

Example 3.8: Beam reactions for a simply supported beam with a cantilevered end carrying two point loads

For the beam shown in Figure 3.30, loaded with two point loads of 60 kN and 8 kN, determine the forces (reactions) applied at the supports A and B required to ensure equilibrium of the system.

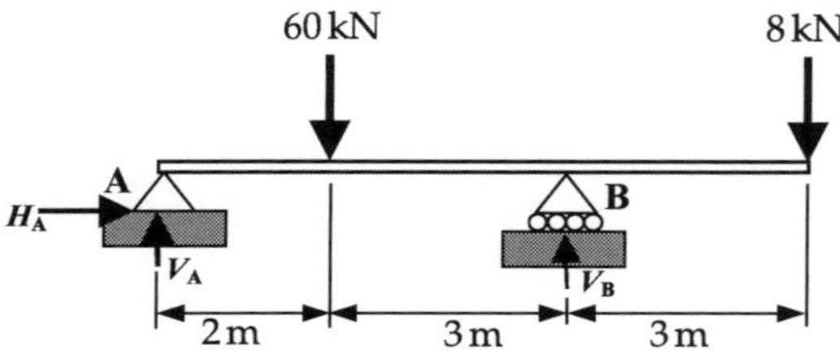

Figure 3.30: Beam system – Example 3.8

Summation of all horizontal forces $= \Sigma F_x = H_A + 0 = 0$ [1]

$\therefore H_A = 0$ kN

Summation of all vertical forces $= \Sigma F_y = V_A + V_B - 60 \text{ kN} - 8 \text{ kN} = 0$

$V_A + V_B = 68$ kN [2]

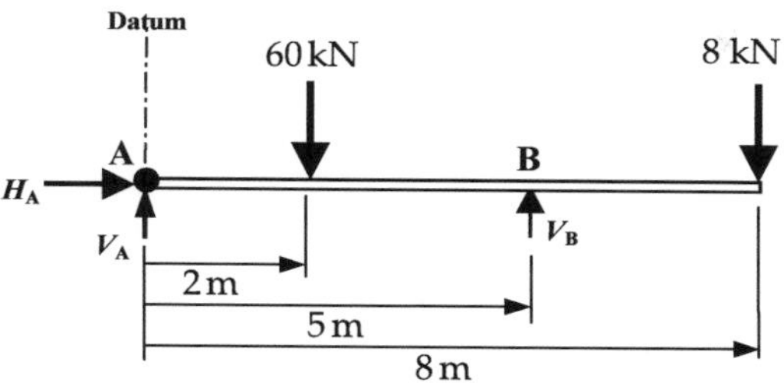

Figure 3.31: Position of datum for calculation of moments (support A) – Example 3.8

Taking moments about support A as shown in Figure 3.31:

Summation of all moments $= \Sigma M_z = (V_A \times 0) + (60 \text{ kN} \times 2 \text{ m}) - (V_B \times 5 \text{ m}) + (8 \text{ kN} \times 8 \text{ m}) = 0$ [3]

$= (120) - 5V_B + (64) = 0$

$5V_B = 184$

$V_B = (184/5) = 36.8$ kN

Back-substituting the result for V_B into equation [2] we find:

$V_A + V_B = 68$ kN [2]

But since $V_B = 36.8$ kN then: $V_A + 36.8 \text{ kN} = 68$ kN

$V_A = 68 - 36.8 = 31.2$ kN

Whilst calculating the moments about support B might be considered slightly more difficult, the same rules apply equally well. The only problem stems from having to determine whether the moments are clockwise or anticlockwise about the datum. The conventions to be adopted in the analysis of the beam, and the calculation of moments about support B, are shown in Figure 3.32.

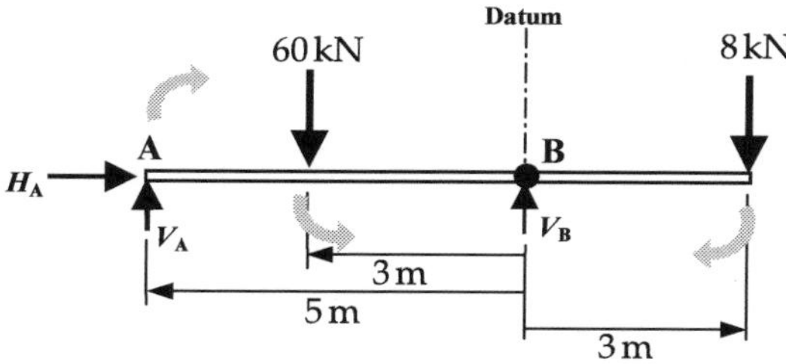

Figure 3.32: Position of datum for calculation of moments (support B) – Example 3.8

It can be seen from Figure 3.32 that the moments from the 8 kN load and the support V_A will induce a clockwise rotation about support B, whilst the moment from the 60 kN load will be anticlockwise. This is easily introduced into our calculations such that, taking moments about support B we find:

Taking moments about support B:

Summation of all moments $= \Sigma M_z = (V_B \times 0) - (60\text{ kN} \times 3\text{ m}) + (V_A \times 5\text{ m}) + (8 \times 3) = 0$ [3]

$= -(180) + 5V_A + (24) = 0$

$= 5V_A - 156$

$5V_A = 156$

$V_A = (156/5) = 31.2$ kN (as before)

It is therefore important to consider the direction of the moments about the point at which the moments are calculated and to ensure that the sign convention of clockwise positive, anticlockwise negative is adhered to.

Having considered some elementary beams with applied point loads, we will now extend our studies to beams supporting Distributed Loads (DL).

Example 3.9: Beam reactions for a simply supported beam carrying a uniformly distributed load (UDL)

For the beam shown in Figure 3.33, loaded along its entire length with a UDL of 5 kN/m, determine the forces (reactions) applied at the supports A and B required to ensure equilibrium of the system.

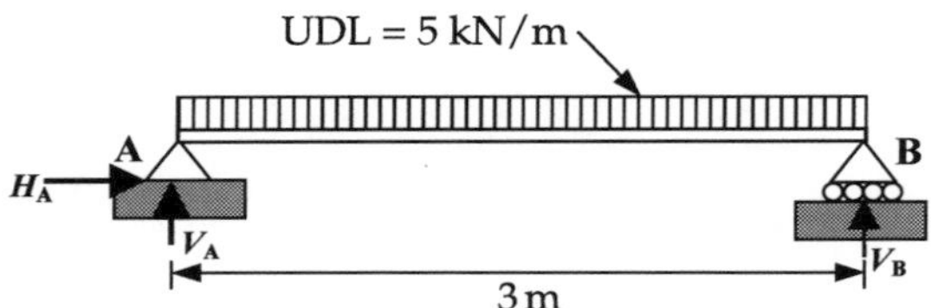

Figure 3.33: Beam system – Example 3.9

Here we note that the load is 'spread' over the entire length of the beam. The UDL is specified as 5 kN/m and it is 3 m long, so therefore the total load is:

$$\text{Total load} = 5\text{ kN/m} \times 3\text{ m} = 15\text{ kN}$$

It should again be possible, by observation, to conclude that the vertical forces (reactions) at the supports will be equal and opposite to half the total applied load, 7.5 kN each upwards. It should also be possible, by observation, to determine that the horizontal reaction H_A is zero since there is no horizontal applied load. We will now determine the reactions using the three equations of statics which we do by summing all the forces and moments, in turn, external and internal using normal Cartesian co-ordinates to determine the signs of forces and the clockwise positive convention for moments, as shown previously in Figure 3.24. However, in this case, we need to calculate the position of the centroid of the load in order to calculate its moment. For a rectangular load, such as that in Figure 3.33, the centroid is located at half its length, that is, 1.5 m from support A or B. The equations for horizontal and vertical equilibrium may be determined in a similar manner to that used for a point load. Hence:

Summation of all horizontal forces $= \Sigma F_x = H_A + 0 = 0$ [1]

$\therefore H_A = 0$ kN

Summation of all vertical forces $= \Sigma F_y = V_A + V_B - (5 \text{ kN/m} \times 3 \text{ m}) = 0$

$$V_A + V_B = 15 \text{ kN} \qquad [2]$$

We can now determine the moments about support A.

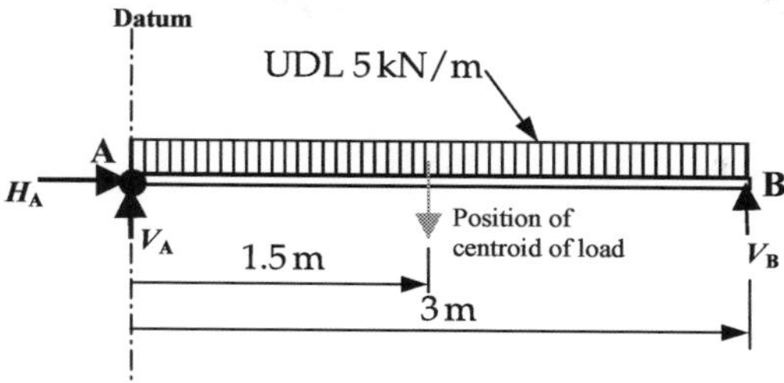

Figure 3.34: Position of datum for calculation of moments (support A) – Example 3.9

We move from our datum, along the beam, in this case from left to right, and calculate the moment for the UDL by calculating the total load and the distance of the centroid of the load from the datum (known as the 'lever arm').

Taking moments about support A we find:

$$\Sigma M_z = (V_A \times 0) + (\underbrace{(5 \text{ kN/m} \times 3\text{ m})}_{\text{Total load}} \times \underbrace{3\text{ m}/2}_{\text{lever arm}}) - (V_B \times 3\text{ m}) = 0 \qquad [3]$$

Note that all distances are measured from the datum and that we follow the clockwise positive convention for moments. We can simplify and solve equation [3] thus:

$$(22.5) - (3V_B) = 0$$
$$22.5 = 3V_B$$

which gives $V_B = (22.5/3) = 7.5 \text{ kN}$

We can now back-substitute this answer into equation [2] to calculate V_A such that:

$$V_A + V_B = 15 \text{ kN} \qquad [2]$$

But since $V_B = 7.5$ kN then: $V_A + 7.5 \text{ kN} = 15 \text{ kN}$

$$V_A = 15 - 7.5 = 7.5 \text{ kN}$$

◇

Therefore, calculation of reactions for beams supporting DLs varies little from that for point loads, except we must ensure that we calculate the total load which acts at its centroid position. These simple steps may be followed in order to calculate the reactions for all types of DL on various statically determinate structures, as shown in the following examples.

◇

Example 3.10: Beam reactions for a simply supported beam supporting two uniformly distributed loads

For the beam shown in Figure 3.35, loaded with two UDLs, one of 10 kN/m applied along the entire length of the beam and the other 30 kN/m applied over a length of 3 m, determine the forces (reactions) applied at the supports A and B required to ensure equilibrium of the system.

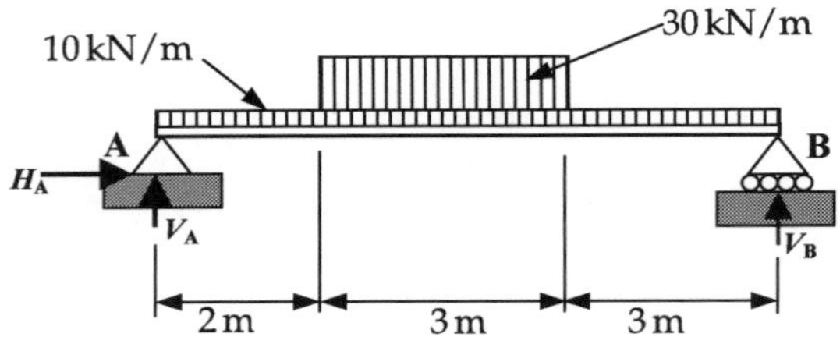

Figure 3.35: Beam system – Example 3.10

We will follow the same procedure as before:

Summation of all horizontal forces $= \sum F_x = H_A + 0 = 0$ [1]

$\therefore H_A = 0$ kN

Summation of all vertical forces $= \sum F_y = V_A + V_B - (10 \text{ kN/m} \times (2\,\text{m} + 3\,\text{m} + 3\,\text{m})) - (30 \text{ kN/m} \times 3\,\text{m}) = 0$

$V_A + V_B - 80 \text{ kN} - 90 \text{ kN} = 0$

$V_A + V_B = 170 \text{ kN}$ [2]

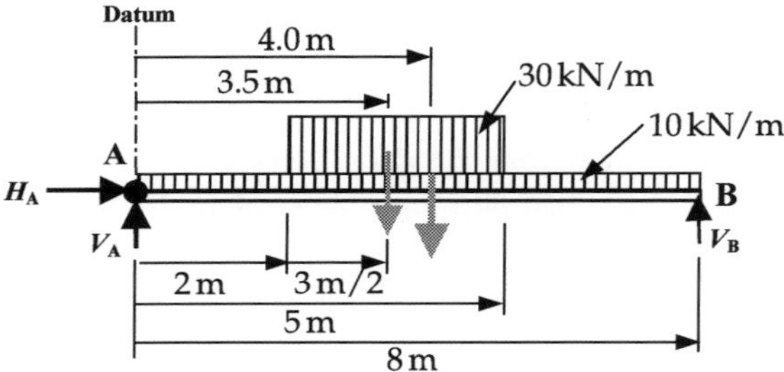

Figure 3.36: Position of datum for calculation of moments (support A) – Example 3.10

Taking moments about support A as shown in Figure 3.36:

Summation of all moments $= \sum M_z = (V_A \times 0) + ((10 \text{ kN/m} \times 8\,\text{m}) \times 8\,\text{m}/2) + ((30 \text{ kN} \times 3\,\text{m}) \times (2\,\text{m} + 3\,\text{m}/2)) - (V_B \times 8\,\text{m}) = 0$ [3]

$= (320) + (315) - 8V_B = 0$

$8V_B = 635$

$V_B = (635/8) = 79.375 \text{ kN}$

Back-substituting the result for V_B into equation [2] we find:

$V_A + V_B = 170 \text{ kN}$ [2]

But as $V_B = 79.375$ kN then: $V_A + 79.375 \text{ kN} = 170 \text{ kN}$

$V_A = 170 - 79.375 = 90.625 \text{ kN}$

Again it does not matter which reaction we decide to take moments about. If we now take moments about support B as indicated in Figure 3.37 we will find:

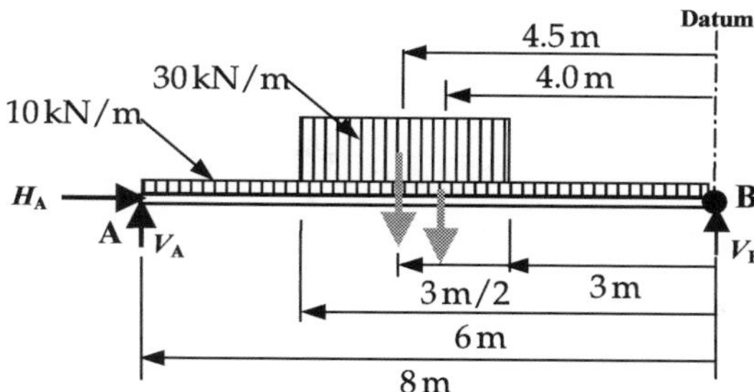

Figure 3.37: Position of datum for calculation of moments (support B) – Example 3.10

Taking moments about support B:

Summation of all moments $= \Sigma M_z = (V_B \times 0) - ((10\text{ kN/m} \times 8\text{ m}) \times 8\text{ m}/2)$
$- ((30\text{ kN} \times 3\text{ m}) \times (3\text{ m} + 3\text{ m}/2))$
$+ (V_A \times 8\text{ m}) = 0$ [3]

$= -(320) - (405) + 8V_A = 0$

$8V_A = 725$

$V_A = (725/8) = 90.625\text{ kN}$ (as before)

The final beam forces and reactions are shown in Figure 3.38.

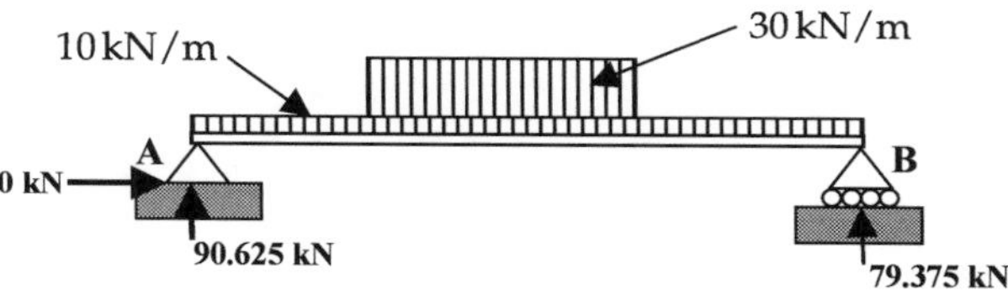

Figure 3.38: Final forces applied to beam – Example 3.10

◇

We will now consider the analysis of a beam where only one support is at the end.

◇

Example 3.11: Beam reactions for a simply supported beam with a cantilevered end carrying two distributed loads

For the beam shown in Figure 3.39, loaded with two UDLs, one of 25 kN/m applied along the entire length of the beam and the other of 12 kN/m applied over a length of 6 m, determine the forces (reactions) applied at the supports A and B required to ensure equilibrium of the system.

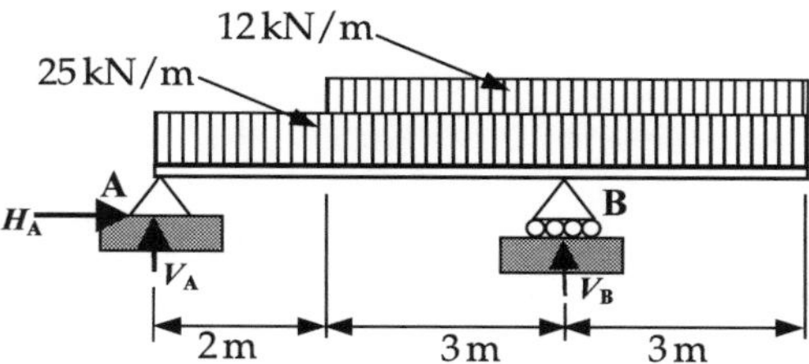

Figure 3.39: Beam system – Example 3.11

Summation of all horizontal forces $= \sum F_x = H_A + 0 = 0$ [1]

$\therefore H_A = 0$ kN

Summation of all vertical forces $= \sum F_y = V_A + V_B - (25 \text{ kN/m} \times (2\text{ m} + 3\text{ m} + 3\text{ m})) - (12 \text{ kN/m} \times (3\text{ m} + 3\text{ m})) = 0$

$V_A + V_B - (200) - (72) = 0$

$V_A + V_B = 272$ kN [2]

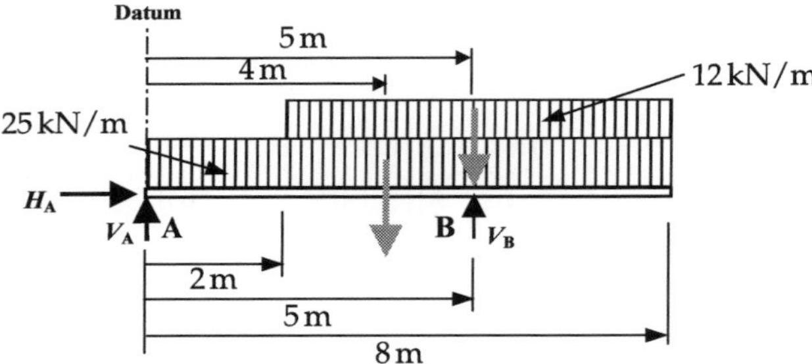

Figure 3.40: Position of datum for calculation of moments (support A) – Example 3.11

Taking moments about support A as shown in Figure 3.40:

Summation of all moments $= \sum M_z = (V_A \times 0) + ((25 \text{ kN/m} \times 8\text{ m}) \times 8/2) - (V_B \times 5\text{ m}) + ((12 \text{ kN/m} \times 6\text{ m}) \times (2 + 6/2)) = 0$ [3]

$= (800) - 5V_B + (360) = 0$

$5V_B = 1160$

$V_B = (1160/5) = 232$ kN

Back-substituting the result for V_B into equation [2] we find:

$V_A + V_B = 272$ kN [2]

But since $V_B = 232$ kN then: $V_A + 232 \text{ kN} = 272$ kN

$V_A = 272 - 232 = 40$ kN

If we wished to calculate moments about support B rather than support A, we could split the UDL into that acting to the right and that acting to the left of the support; however there is no real need to complicate the problem since the calculations for the lever arm can be completed as before.

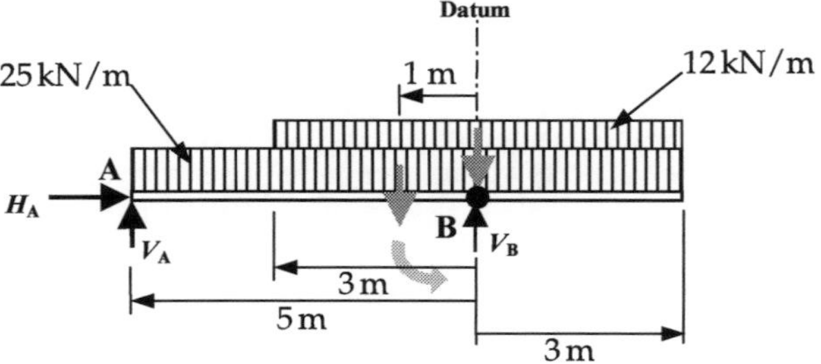

Figure 3.41: Position of datum for calculation of moments (support B) – Example 3.11

We note from Figure 3.41 that the centroid of the 12 kN/m load is directly over the support B and along the line of the datum. The distance to the centroid is therefore 0 m and we could omit the load from the calculation, but for the sake of completeness it has been included in equation [3].

Taking moments about support B:
Summation of all moments $= \Sigma M_z = (V_B \times 0) - ((25\,\text{kN/m} \times 8\,\text{m}) \times (5 - 8/2))$
$+ (V_A \times 5\,\text{m}) - ((12\,\text{kN/m} \times 6) \times 0) = 0$ [3]
$= -(200) + 5V_A + (0) = 0$
$= 5V_A - 200 = 0$
$5V_A = 200$
$V_A = (200/5) = 40\,\text{kN}$ (as before).

◇

It is therefore important to pick carefully which support we take moments about in order to make the problem as easy as possible. However, if care is taken in determining the direction of the moment, the results will be the same whichever support is used.

We have, so far, only considered beams loaded with either point loads or uniformly distributed loads. In real structures we may have a combination of both. Before we commence the analysis of more complex systems we will consider a method for simplifying the problem by applying the **Law of Linear Superposition**.

3.4.1 Law of Linear Superposition

Before we consider structures of increased complexity it is worth noting that, so long as the $P\Delta$ (P-Delta) effect discussed earlier does not exist, we can separate a difficult problem into parts, summing the results for each part to find the solution to the original system.

Consider the beam shown in Figure 3.42. It is supported at points A and B and carries both a UDL and point loads. Calculate the forces at the supports.

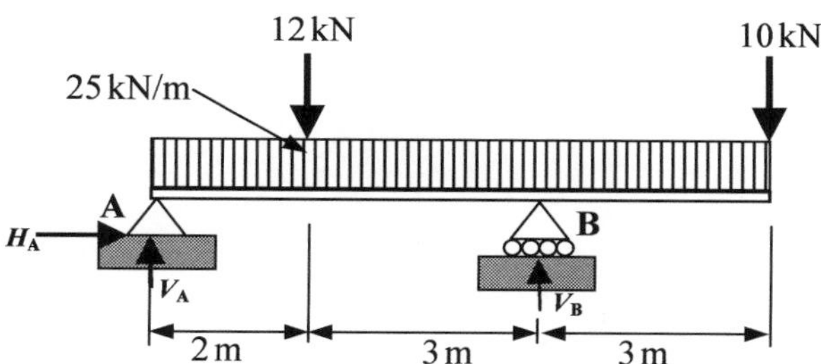

Figure 3.42: Beam carrying both uniformly distributed and point loads

We can solve this problem using the equations of statics as before. However, rather than solving the problem as a whole we can split it into separate parts. We can then sum the reaction forces determined from each part in order to derive the solution to the entire problem. The separate parts of the solution to the beam of Figure 3.42 are shown in Figure 3.43.

We will solve for each separate part, noting that no horizontal load is applied to either the original beam, or therefore to any of its parts. Hence, ΣF_x will be zero in each case and it will not be necessary to include this equation in the calculations.

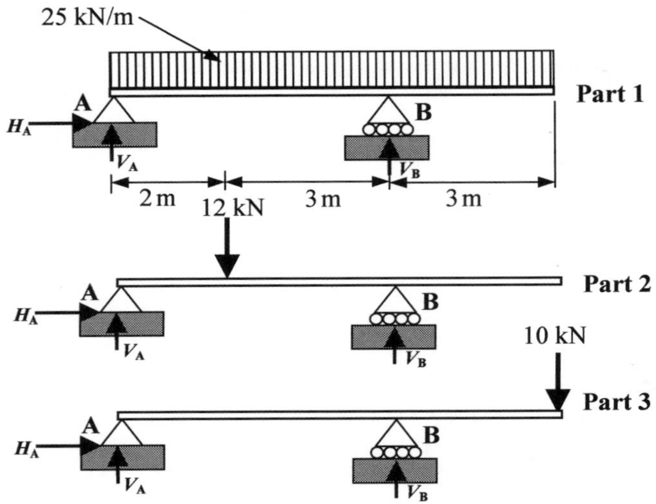

Figure 3.43: Separated parts of beam to be analysed

For Part 1:

$\sum F_y = V_A + V_B - (25\,\text{kN/m} \times (2\,\text{m} + 3\,\text{m} + 3\,\text{m})) = 0$

$= V_A + V_B = 200\,\text{kN}$ [1]

Moments taken about support A.

$\sum M_z = ((25\,\text{kN/m} \times 8\,\text{m} \times (8\,\text{m}/2)) - V_B \times 5\,\text{m} = 0$

$= 800 - 5V_B = 0$

$\therefore\ 800 = 5V_B$

$V_B = (800/5) = 160\,\text{kN}$ [2]

Back-substitute result from [2] above into [1] above:

$V_A + V_B = 200\,\text{kN}$ [1] but, $V_B = 160\,\text{kN}$

$V_A + 160\,\text{kN} = 200\,\text{kN}$

$V_A = 200 - 160 = 40\,\text{kN}$

For Part 2:

$\sum F_y = V_A + V_B - 12\,\text{kN} = 0$

$= V_A + V_B = 12\,\text{kN}$ [1]

Moments taken about support A.

$\sum M_z = (12\,\text{kN} \times 2\,\text{m}) - V_B \times 5\,\text{m} = 0$

$= 24 - 5V_B = 0$

$\therefore\ 24 = 5V_B$

$V_B = (24/5) = 4.8\,\text{kN}$ [2]

Back-substitute result from [2] above into [1] above:

$V_A + V_B = 12\,\text{kN}$ [1] but, $V_B = 4.8\,\text{kN}$

$V_A + 4.8\,\text{kN} = 12\,\text{kN}$

$V_A = 12 - 4.8 = 7.2\,\text{kN}$

For Part 3:

$\sum F_y = V_A + V_B - 10\text{kN} = 0$

$= V_A + V_B = 10\text{kN}$ [1]

Moments taken about support A.

$\sum M_z = (10\,\text{kN} \times 8\,\text{m}) - V_B \times 5\,\text{m} = 0$

$= 80 - 5V_B = 0$

$\therefore\ 80 = 5V_B$

$V_B = (80/5) = 16\,\text{kN}$ [2]

Back-substitute result from [2] above into [1] above:

$V_A + V_B = 10\,\text{kN}$ [1] but, $V_B = 16\,\text{kN}$

$V_A + 16\,\text{kN} = 10\,\text{kN}$

$V_A = 10 - 16 = -6\,\text{kN}$

We first note that the solution to Part 3 yields a negative result for the reaction at support A. This negative sign indicates that the vertical reaction at support A is in a downward direction, effectively holding the beam down as it attempts to rotate about support B. The type of support indicated at A would not normally be able to exert this downward force, however when we derive the solution for the entire system we will find that the total resultant force is upward such that the forces at the supports for the original system will be:

Vertical force at support A for beam of Figure 3.42 $= V_A(\text{Part 1}) + V_A(\text{Part 2}) + V_A(\text{Part 3})$
$= 40\text{ kN} + 7.2\text{ kN} - 6\text{ kN}$
$V_A = 41.2\text{ kN}$

Vertical force at support B for beam of Figure 3.42 $= V_B(\text{Part 1}) + V_B(\text{Part 2}) + V_B(\text{Part 3})$
$= 160\text{ kN} + 4.8\text{ kN} + 16\text{ kN}$
$V_B = 180.8\text{ kN}$

The final forces at the reactions are shown in Figure 3.44.

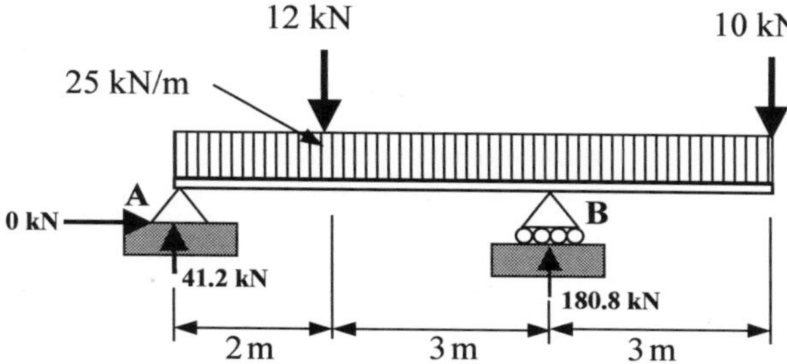

Figure 3.44: Summary of forces on beam

Whilst the problem considered here could have been solved by simply forming each of the equations of statics in turn, considering all applied loads at the same time, as structures and loading become more complex it will sometimes be easier to separate the system into more manageable parts and to sum the results to find the solution for the entire problem. This method of analysis can be adopted for any beam so long as the $P\Delta$ effect does not predominate.

We will now apply the equations of statics to some further problems.

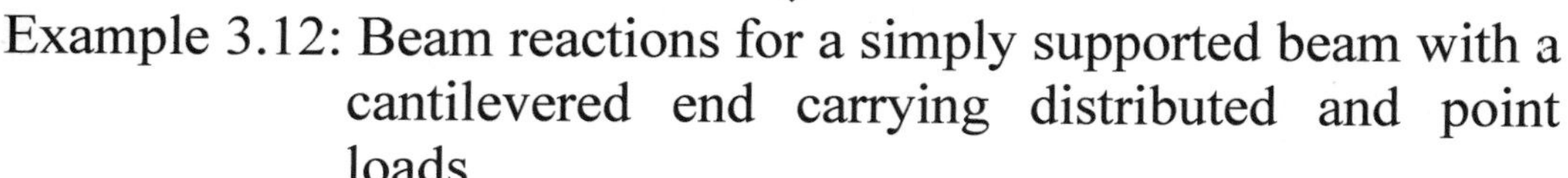

Example 3.12: Beam reactions for a simply supported beam with a cantilevered end carrying distributed and point loads

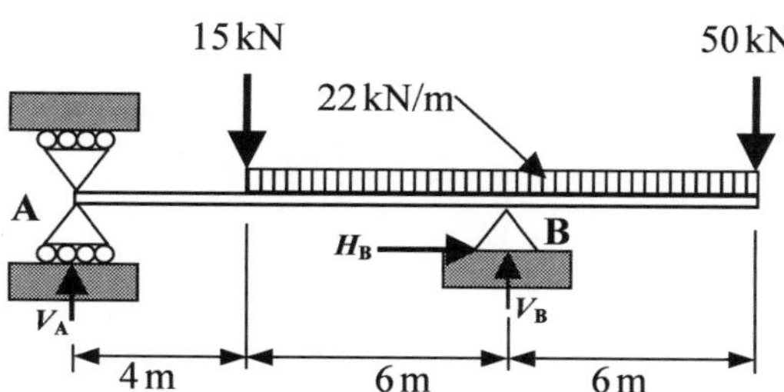

Figure 3.45: Beam system – Example 3.12

For the beam of Figure 3.46, supported at points A and B, determine the reactions at the supports V_A, V_B and H_B as shown.

We first note that we appear to have three supports, not two! This is not in fact the case. The support shown at A is a simple support which will support either an upward or a downward reaction depending on the results of our analysis. For the purpose of forming our equations we will still consider V_A to be upwards. If this is the case, the support at the bottom of the beam will carry the load. If, however, the analysis reveals a negative result for the reaction V_A then it will be the top reaction that supports the load. Our initial analysis will therefore consider the system of Figure 3.46, ignoring the top support at A.

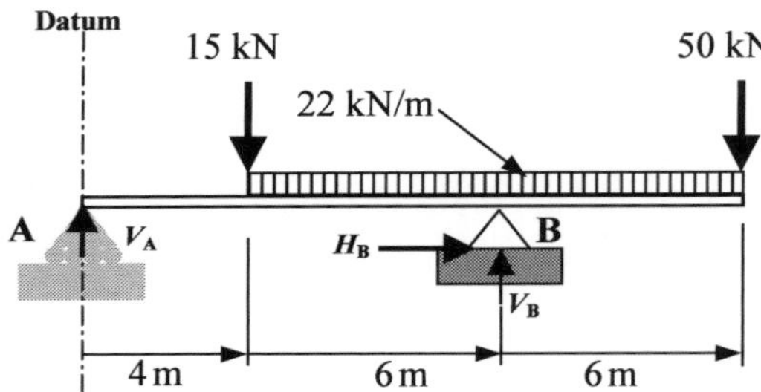

Figure 3.46: Initial analysis of beam system – Example 3.12

Resolving horizontally: $\Sigma F_x = H_B - 0 = 0 \quad \therefore \quad H_B = 0$

Resolving vertically: $\Sigma F_y = V_A + V_B - 15\,\text{kN} - (22\,\text{kN/m} \times (6\,\text{m}+6\,\text{m})) - 50\,\text{kN} = 0$

$= V_A + V_B - 15\,\text{kN} - 264\,\text{kN} - 50\,\text{kN} = 0$

$= V_A + V_B = 329\,\text{kN}$ [1]

Taking moments about A: $\Sigma M_z = (15\,\text{kN} \times 4\,\text{m}) + (22\,\text{kN/m} \times 12\,\text{m} \times (4\,\text{m} + 12\,\text{m}/2))$

$+ (50\,\text{kN} \times 16\,\text{m}) - (V_B \times (4\,\text{m}+6\,\text{m})) = 0$

$= 60 + 2640 + 800 - 10V_B = 0$

$V_B = (3500/10) = 350\,\text{kN}$

Back-substituting V_B into [1] gives:

$V_A + V_B = 329\,\text{kN}$

But $V_B = 350\,\text{kN}$ hence: $V_A + 350 = 329$

And so: $V_A = 329 - 350 = -21\,\text{kN}$ (negative result therefore downwards)

Our final load/reaction system is indicated in Figure 3.47.

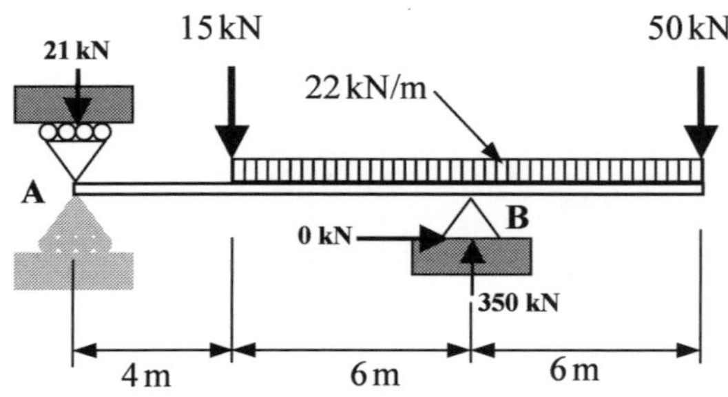

Figure 3.47: Final forces and reactions for beam system – Example 3.12

Example 3.13: Beam reactions for a simply supported beam with both ends cantilevered carrying distributed and point loads

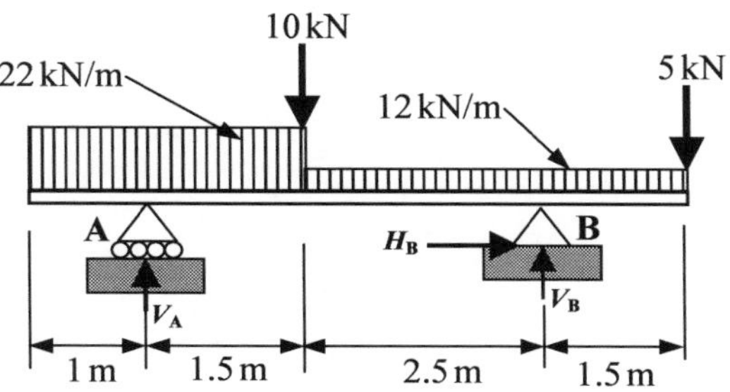

Figure 3.48: Beam system – Example 3.13

For the beam system of Figure 3.48 calculate the reactions V_A, V_B and H_B at the supports A and B.

Resolving horizontally: $\Sigma F_x = H_B - 0 = 0$ $\therefore H_B = 0$

Resolving vertically:

$$\Sigma F_y = V_A + V_B - (22\,\text{kN/m} \times (1\,\text{m}+1.5\,\text{m})) - 10\,\text{kN} - (12\,\text{kN/m} \times (2.5\,\text{m} + 1.5\,\text{m})) - 5\,\text{kN} = 0$$

$$= V_A + V_B - 55\,\text{kN} - 10\,\text{kN} - 48\,\text{kN} - 5\,\text{kN} = 0$$

$$= V_A + V_B = 118\,\text{kN} \qquad [1]$$

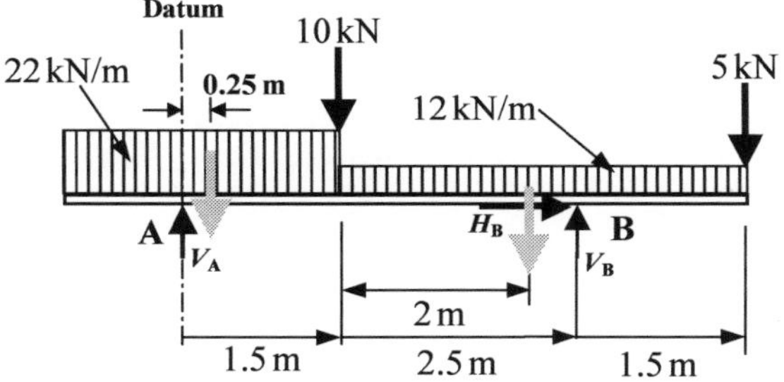

Figure 3.49: Moments about support A for beam system – Example 3.13

Taking moments about A as shown in Figure 3.49.

Taking moments about A:

$$\Sigma M_z = (22\,\text{kN/m} \times 2.5\,\text{m} \times (2.5\,\text{m}/2 - 1\,\text{m})) + (10\,\text{kN} \times 1.5\,\text{m}) + (12\,\text{kN/m} \times 4\,\text{m} \times (1.5\,\text{m} + 4\,\text{m}/2)) + (5\,\text{kN} \times 5.5\,\text{m}) - (V_B \times (1.5\,\text{m} + 2.5\,\text{m})) = 0$$

$$= 13.75 + 15 + 168 + 27.5 - 4V_B = 0$$

$$V_B = (224.25/4) = 56.0625\,\text{kN}$$

Back-substituting V_B into [1] gives:

$V_A + V_B = 118\,\text{kN}$

But $V_B = 56.0625\,\text{kN}$ hence: $V_A + 56.0625 = 118$

And so: $V_A = 118 - 56.0625 = 61.9375\,\text{kN}$

◇

Example 3.14: Beam reactions for a simply supported beam with a cantilevered end supporting two inclined point loads

In the beam system of Figure 3.50 we have a structure which supports two inclined point loads. In order to apply our method of analysis to this system and determine the reactions at the supports we must first resolve both forces to find their horizontal and vertical components.

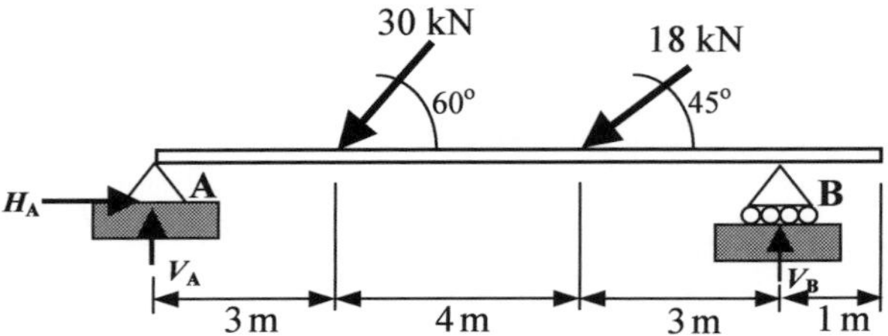

Figure 3.50: Beam system – Example 3.14

First we resolve each of the inclined forces as shown in Figure 3.51.

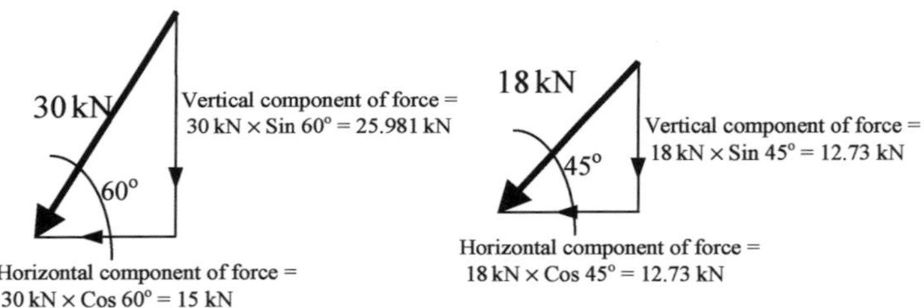

Figure 3.51: Resolution of forces – Example 3.14

Having resolved the forces into their horizontal and vertical components we can replace the inclined loads with their components and analyse the system shown in Figure 3.52.

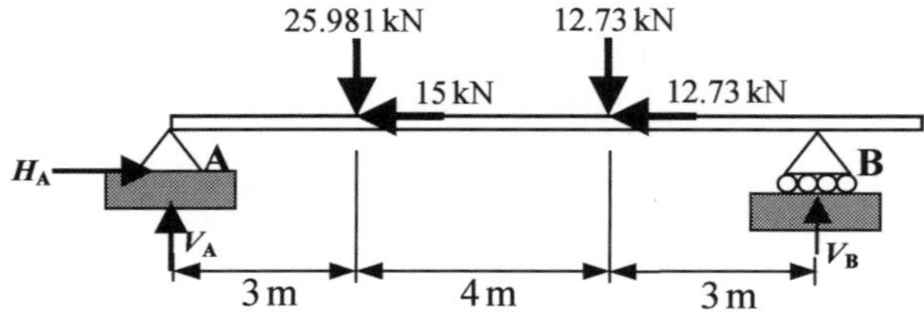

Figure 3.52: Modified beam system – Example 3.14

Resolving horizontally: $\Sigma F_x = H_A - 15\,\text{kN} - 12.7\,\text{kN} = 0 \qquad \therefore H_A = 27.73\,\text{kN}$

Resolving vertically: $\Sigma F_y = V_A + V_B - 25.981\,\text{kN} - 12.73\,\text{kN} = 0$
$= V_A + V_B - 38.711\,\text{kN} = 0$
$= V_A + V_B = 38.711\,\text{kN}$ [1]

Taking moments about A: $\Sigma M_z = (25.981\,\text{kN} \times 3\,\text{m}) + (12.73\,\text{kN} \times (3\,\text{m} + 4\,\text{m})) - V_B \times 10\,\text{m} = 0$
$= 77.943 + 89.11 - 10V_B = 0$
$V_B = (167.053/10) = 16.7053\,\text{kN}$

Back-substituting V_B into [1] gives:
$V_A + V_B = 38.711\,\text{kN}$

But $V_B = 16.7053\,\text{kN}$ hence: $V_A + 16.7053 = 38.711$

And so: $V_A = 38.711 - 16.7053 = 22.658\,\text{kN}$

The final beam forces are shown in Figure 3.53.

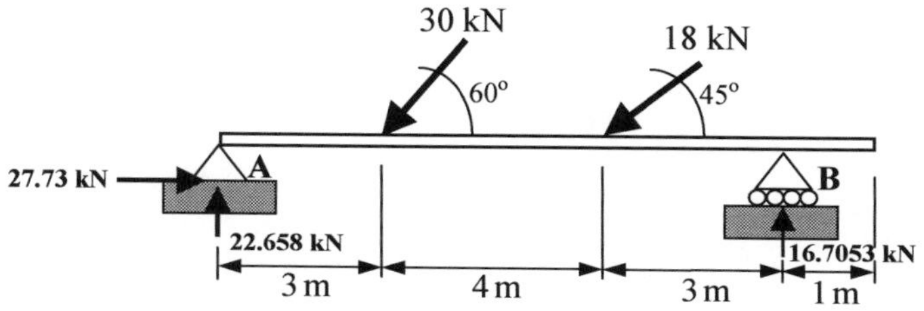

Figure 3.53: Final beam forces and reactions – Example 3.14

Note that the overhanging part of the beam to the right of support B has no effect on our calculations as there is no applied load or reaction on it.

◇

Example 3.15: Beam reactions for a simply supported beam with a cantilevered end carrying a trapezoidal distributed load

In the beam system of Figure 3.54 we have a structure which supports a trapezoidal load. This load can be separated into both a uniformly and a triangular distributed load as shown. We note that the triangular load varies in intensity from 0 kN/m at 2 m to the left of support A to a maximum of 22 kN/m at support B. Calculate the forces at the supports V_A, V_B and H_B.

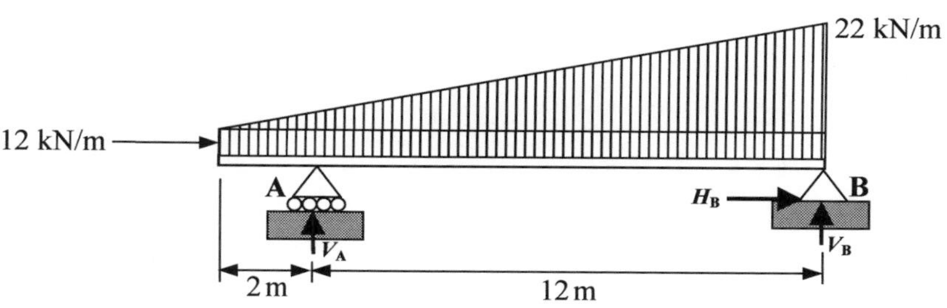

Figure 3.54: Beam system – Example 3.15

We can analyse this system as before such that:

Resolving horizontally: $\Sigma F_x = H_B - 0 = 0 \quad \therefore H_B = 0\,\text{kN}$

To resolve vertically we need to calculate the total load, noting that the area of a triangle is given by:

Area ◺ $= 0.5 \times \text{base} \times \text{height}$

Resolving vertically:

$$\begin{aligned}\Sigma F_y &= V_A + V_B - (12\,\text{kN/m} \times (2\,\text{m} + 12\,\text{m})) \\ &\quad - (0.5 \times (2 + 12\,\text{m}) \times 22\,\text{kN/m}) = 0 \\ &= V_A + V_B - 168\,\text{kN} - 154\,\text{kN} = 0 \\ &= V_A + V_B = 322\,\text{kN} \qquad [1]\end{aligned}$$

In order to calculate the moments we first note that the centroid of the triangle is located at two-thirds the length of the base from the apex, or one-third the length of the base from support B as shown in Figure 3.55. We also note that, in this case, it is easier to calculate the moments about support B as this is located at the end of the beam.

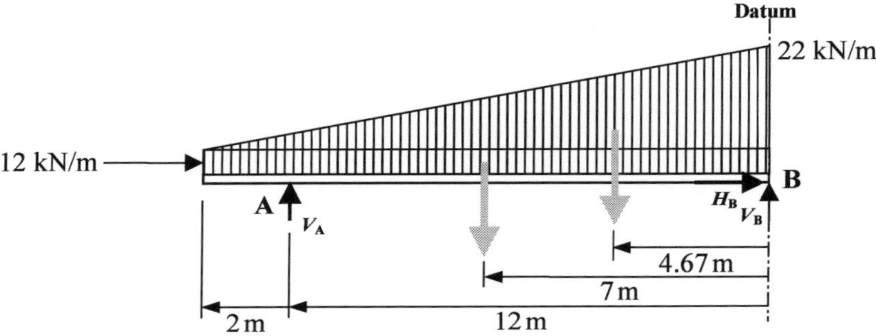

Figure 3.55: Moments about support B for beam system – Example 3.15

Taking moments about B:

$$\begin{aligned}\Sigma M_z &= (12\,\text{kN} \times 14\,\text{m} \times (14\,\text{m}/2)) + ((0.5 \times 14\,\text{m} \times 22\,\text{kN/m}\,) \times 14/3)) \\ &\quad - V_A \times 12\,\text{m} = 0 \\ &= 1176 + 719.18 - 12V_A = 0 \\ V_A &= (1895.18/12) = 153.932\,\text{kN}\end{aligned}$$

Back-substituting V_A into [1] gives:

$$V_A + V_B = 322\,\text{kN}$$

But $V_A = 153.932\,\text{kN}$ hence: $153.932 + V_B = 322$

And so: $V_B = 322 - 153.932 = 168.068\,\text{kN}$

Example 3.16: Beam reactions for a simply supported beam with both ends cantilevered carrying distributed and point loads

In the beam system of Figure 3.56 we have a structure that supports both uniformly distributed and point loads. Calculate the horizontal and vertical forces at the reactions as indicated.

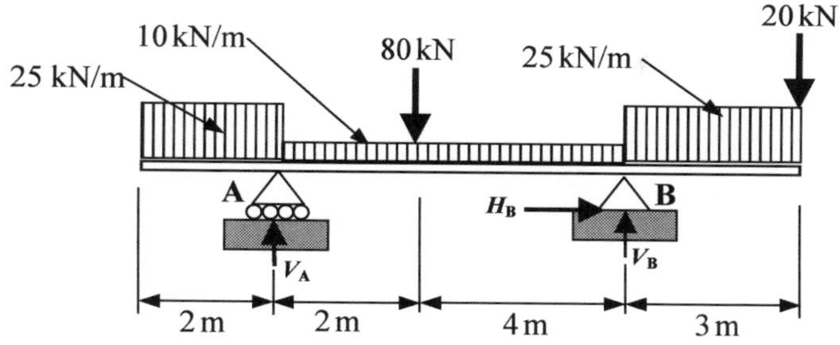

Figure 3.56: Beam system – Example 3.16

Resolving horizontally: $\Sigma F_x = H_B - 0 = 0 \quad \therefore H_B = 0\,\text{kN}$

Resolving vertically: $\Sigma F_y = V_A + V_B - (25\,\text{kN/m} \times 2\,\text{m}) - (10\,\text{kN/m} \times (2\,\text{m} + 4\,\text{m}))$

$$- 80\,\text{kN} - (25\,\text{kN/m} \times 3\,\text{m}) - 20\,\text{kN} = 0$$
$$= V_A + V_B - 50\,\text{kN} - 60\,\text{kN} - 80\,\text{kN} - 75\,\text{kN} - 20\,\text{kN} = 0$$
$$= V_A + V_B = 285\,\text{kN} \qquad [1]$$

In this system, careful consideration must be given to the calculation of moments. We note that neither support is located at the end position of the beam and that there is no obvious advantage to be gained from taking moments about support A rather than support B. As the selection of the origin (datum) is arbitrary we will, in this case, take moments about support A as shown in Figure 3.57.

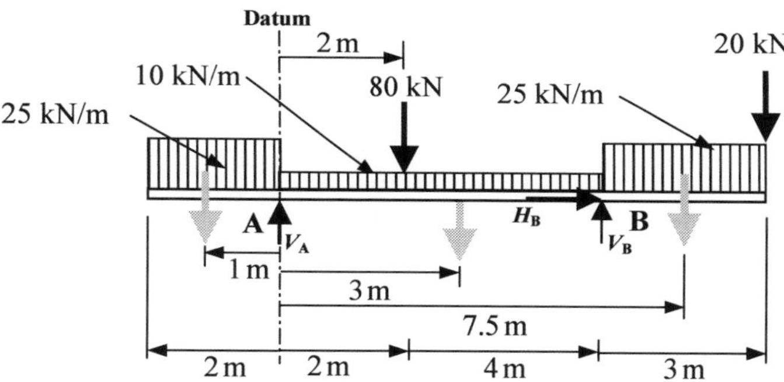

Figure 3.57: Moments about support A for beam system – Example 3.16

We note that the moment of the load to the left of support A is anticlockwise and therefore negative.

Taking moments about A: $\Sigma M_z = -(25\,\text{kN/m} \times 2\,\text{m} \times (2\,\text{m}/2)) + (10\,\text{kN/m} \times 6\,\text{m} \times (6\,\text{m}/2))$

$$+ (80\,\text{kN} \times 2\,\text{m}) + (25\,\text{kN/m} \times 3\,\text{m} \times (6\,\text{m} + 3\,\text{m}/2))$$
$$+ (20\,\text{kN} \times 9\,\text{m}) - V_B \times 6\,\text{m} = 0$$
$$= -50 + 180 + 160 + 562.5 + 180 - 6V_B = 0$$
$$V_B = (1032.5/6) = 172.083\,\text{kN}$$

Back-substituting V_B into [1] gives:

$$V_A + V_B = 285\,\text{kN}$$

But $V_B = 172.083\,\text{kN}$ hence: $V_A + 172.083 = 285$

And so: $V_A = 285 - 185.417 = 112.917\,\text{kN}$

We have now analysed various simply supported beam systems. Before we move on to plotting Shear Force and Bending Moment diagrams, we will consider two cantilevered beam problems.

Example 3.17: Beam reactions for a cantilever beam carrying uniformly distributed and point loads

In the beam system of Figure 3.58 we have a cantilevered beam supporting a vertical and horizontal point load and a UDL. The beam is fully fixed at the support A.

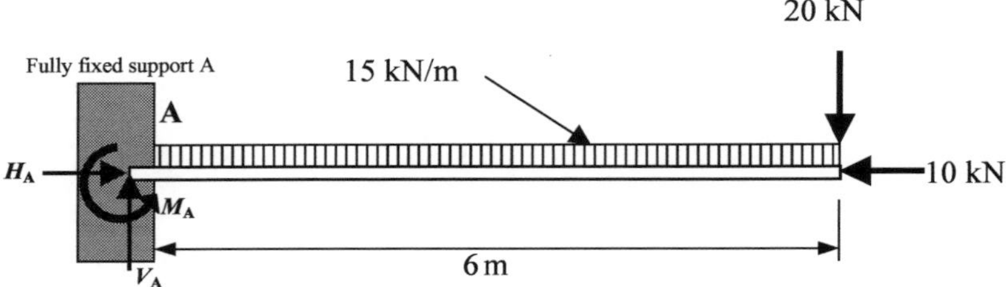

Figure 3.58: Beam system – Example 3.17

We first note that, with a cantilevered beam system, the applied moments due to the applied loads must be resisted by a resisting moment at the support M_A. We can apply the equations of statics as before.

Resolving horizontally: $\Sigma F_x = H_A - 10\,\text{kN} = 0 \qquad \therefore H_A = 10\,\text{kN}$

Resolving vertically: $\Sigma F_y = V_A - (15\,\text{kN/m} \times 6\,\text{m}) - 20\,\text{kN}$

$= V_A - 90\,\text{kN} - 20\,\text{kN} = 0$

$= V_A = 110\,\text{kN}$ [1]

Taking moments about A: $\Sigma M_z = (15\,\text{kN/m} \times 6\,\text{m} \times 6\,\text{m}/2) + (20 \times 6\,\text{m}) - M_A = 0$

$= 270 + 180 - M_A = 0$

$M_A = 450\,\text{kN m}.$

Example 3.18: Beam reactions for a cantilever beam carrying triangular distributed and point loads

In the beam system of Figure 3.59 we have a cantilevered beam supporting a vertical point load and a triangular distributed load varying from 0 kN/m at the support to 20 kN/m at 8 m from the support. The beam is fully fixed at support A. Calculate the reactions H_A, V_A and M_A.

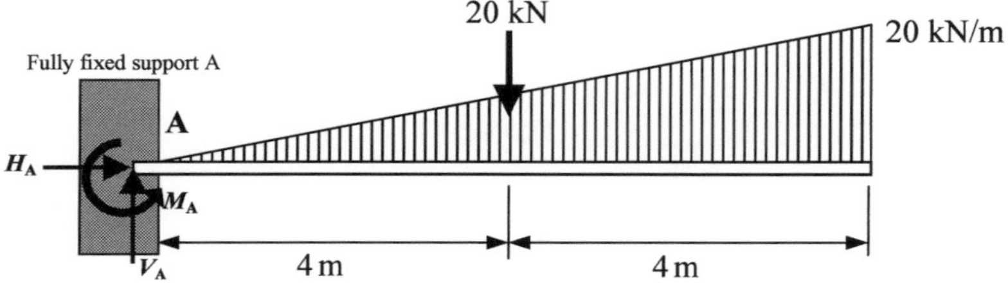

Figure 3.59: Beam system – Example 3.18

Resolving horizontally: $\Sigma F_x = H_A - 0\,\text{kN} = 0 \qquad \therefore H_A = 0\,\text{kN}$

Resolving vertically: $\Sigma F_y = V_A - (0.5 \times 20\,\text{kN/m} \times 8\,\text{m}) - 20\,\text{kN}$

$= V_A - 80\,\text{kN} - 20\,\text{kN} = 0$

$= V_A = 100\,\text{kN}$

Noting that the centroid for the triangular load is located at two-thirds the length of the base from support A, we have:

Taking moments about A: $\Sigma M_z = ((0.5 \times 20\,\text{kN/m} \times 8\,\text{m}) \times (2/3 \times 8\,\text{m})) + (20 \times 4\,\text{m}) - M_A = 0$

$= 426.667 + 80 - M_A = 0$

$M_A = 506.667\,\text{kN\,m}$

We can now calculate the reactions for various statically determinate beam systems, ensuring equilibrium using the equations of statics. We will now proceed to calculate the distribution of forces and moments along various beam systems and plot the calculated values in the form of graphs. These graphs are more familiarly known to engineers as Shear Force and Bending Moment diagrams.

3.5 Shear Force and Bending Moment Diagrams

Shear Force and Bending Moment diagrams provide useful graphs which plot the forces and moments on a structure. They enable the designer to easily visualise the distribution of forces and moments and to determine which parts of a member require detailed consideration during design. These diagrams are very useful for the final design of structures, particularly those constructed in reinforced concrete where the layout for the steel bars is critical. In such structures the reinforcement is designed, in conjunction with the concrete, to resist the forces and moments applied to the structure. Figure 3.60 shows a typical section through a concrete simply supported beam member supporting a vertical downward UDL on the top face.

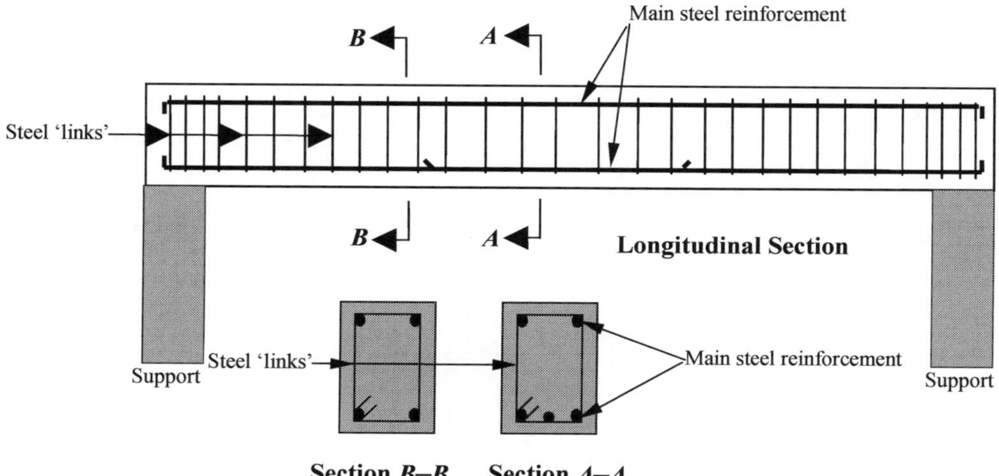

Figure 3.60: Section through a typical reinforced concrete beam

The beam reinforcement is determined by the amount of force or moment it must resist. We note that the main reinforcement consists of three bars in the bottom of section *A–A*. This may be considered to be at the point at which the maximum bending moment will occur and therefore the maximum tension force in the bottom of the beam must be resisted. As the moment decreases from its maximum value, so the amount of reinforcement in the tension zone at the bottom of the beam can be reduced as shown in section *B–B*. The steel 'links' of the beam are designed to improve the beam's resistance to shear force. Hence they are placed close together where the shear is high, in this case at the supports, and spread further apart as shear force reduces. It should be noted that, in this example, reinforcement is shown in the compression zone, that is at the top of the beam. The steel will normally be placed to resist tensile forces but it is sometimes necessary to provide nominal steel reinforcement in the compression zone in order to form the cage that provides support to the steel links.

Shear force and bending moment diagrams are of great value in designing such members. From the graphs the designer can determine the spacing for shear reinforcement links and decide where to terminate (or 'curtail') the main bending reinforcement. It is also possible to determine where the tension zones are located in the structure. This will aid in designing the most efficient and cost-effective structure possible.

It will also be necessary to be able to plot shear force and bending moment diagrams in order to take advantage of more advanced analysis techniques. Commonly, methods such as flexibility and virtual work are used to analyse statically indeterminate structures, and a good understanding of how to draw these simple diagrams will greatly enhance your ability to master these techniques.

We will initially consider the construction of shear force diagrams separately from bending moment diagrams, however it should be noted that both are normally required to complete the design of a structural member.

3.5.1 Shear Force Diagrams

We have already defined shear force in Chapter 1, and we will now consider how this force is distributed along a simple section. The applied forces and reactions of the system as a whole are balanced using the equations of statics to ensure the structure is in equilibrium. The forces themselves must be resisted by the structural member, and the amount of resistance is governed by its material properties and service requirements. We already know that a material will have a limiting stress above which the material will fail or become unserviceable. The member must therefore be designed to provide the required resistance to forces and moments and, in order to do this, we need to determine how these are distributed along its length.

The method used to determine the amount of shear force a beam must resist at a specific point is to consider a section of the member and then to sum all vertical forces and reactions applied to that section. The residual amount will be the vertical force to be resisted by the beam. Put simply, we will divide the member into discrete sections and will then sum all vertical forces to one side of a datum located at the division. The resultant will be the shear force resisted by the member and can be plotted in the form of a graph in order to show its distribution along the member.

We will now determine the shear force distribution in the beam of Example 3.19.

Example 3.19: Shear force diagram for a simply supported beam carrying a point load

For the beam shown in Figure 3.61, which is simply supported at points A and B, determine:

(a) the reactions at the supports V_A, H_A and V_B;

and (b) the shear force at 1 m intervals along the beam.

Plot the results of (b) in the form of a shear force diagram.

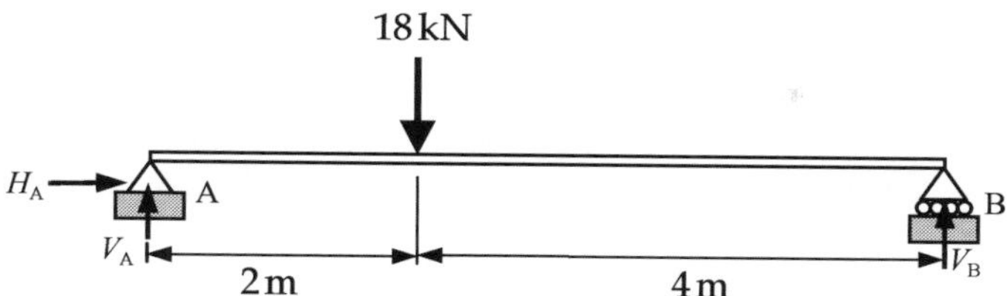

Figure 3.61: Beam system – Example 3.19

First, we must calculate the reactions using the procedure detailed earlier:

Resolving horizontally: $\Sigma F_x = H_A - 0 = 0 \qquad \therefore H_A = 0$

Resolving vertically: $\Sigma F_y = V_A + V_B - 18\,\text{kN} = 0$

$= V_A + V_B = 18\,\text{kN}$ [1]

Taking moments about A: $\Sigma M_z = (18\,\text{kN} \times 2\,\text{m}) - (V_B \times (2\,\text{m}+4\,\text{m})) = 0$

$= 36 - 6V_B = 0$

$V_B = (36/6) = 6\,\text{kN}$

Back-substituting V_B into [1] gives:

$V_A + V_B = 18\,\text{kN}$

But $V_B = 6\,\text{kN}$ hence: $V_A + 6\,\text{kN} = 18\,\text{kN}$

And so: $V_A = 18 - 6 = 12\,\text{kN}$

Having calculated the reactions, we now proceed to calculate the shear force along the beam. The shear forces will be calculated at 1-m intervals commencing at support A and proceeding to support B. In order to clearly identify the forces being considered, we will produce a drawing showing all forces to the left of the point under consideration. In the early stages of your studies, it is always advisable to draw such a sketch to show all forces to be considered, or to simply cover with your hand that portion of the beam not being included!

We commence our calculations at support A:

$H_A = 0\,\text{kN}$

Shear force = 12 kN

$V_A = 12\,\text{kN}$

We note that, for the entire system, the vertical forces are in equilibrium and therefore summate to zero. Then, using zero as our starting point, the vertical force at the support is 12 kN upwards and therefore positive. This will induce a shear force in the beam at support A equivalent to the force at the reaction. We also note that the horizontal force does not cause shear in the beam member. It can therefore be omitted from our present analysis. We will however consider this type of 'axial' force later in Chapter 4.

Shear force at 1 m from support A:

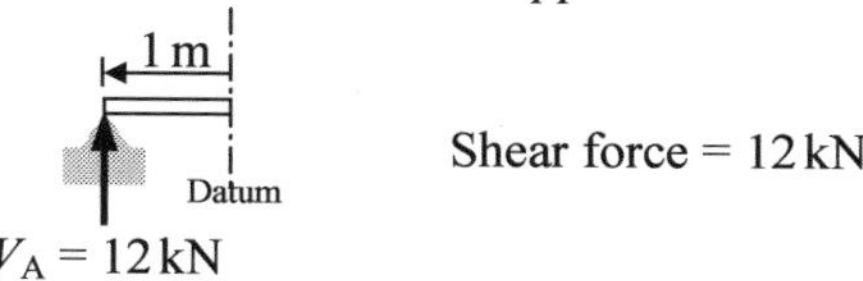

Shear force = 12 kN

Noting that no further vertical forces are applied to the beam between the datum at 1 m and support A, the shear force remains the same. This will always be the case since if no new force is applied to the member, no change in shear force can take place!

Shear force at 2 m from support A:

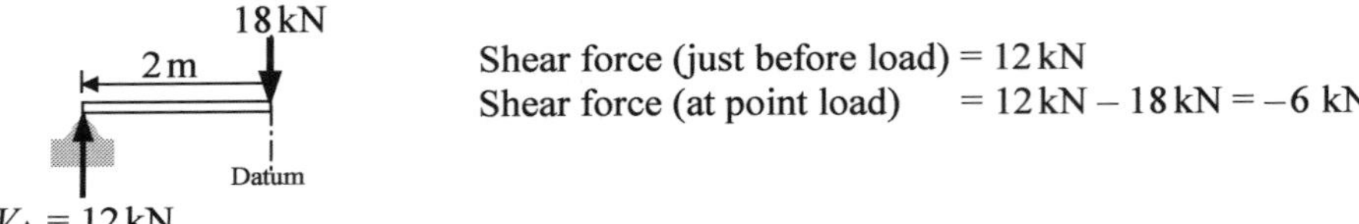

Shear force (just before load) = 12 kN
Shear force (at point load) = 12 kN – 18 kN = –6 kN

Here we first calculate the shear just before the applied load and note that no change has occurred. At the point at which the point load of 18 kN is applied, the effect takes place and the shear force is reduced to a value of –6 kN. The negative sign indicates that the residual force is in a downward direction. It will be necessary in all calculations for shear force to remember that the effect of a point load takes place at its point of application. The shear force just prior to this point is not affected by the load.

Shear force at 3 m from support A:

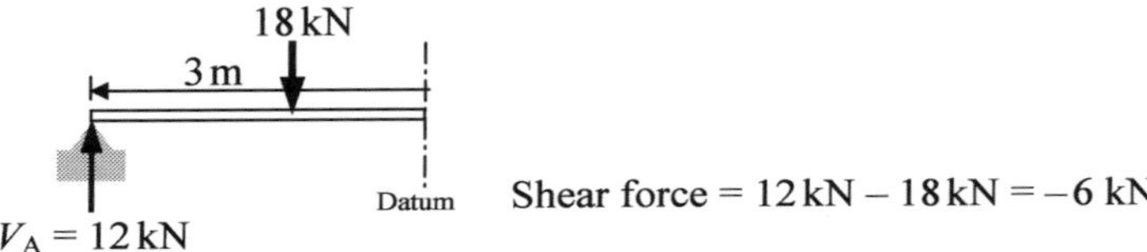

Shear force = 12 kN – 18 kN = –6 kN

Again no vertical forces are applied between the point load of 18 kN and the datum and so the shear force remains the same.

Shear force at 4 m from support A:

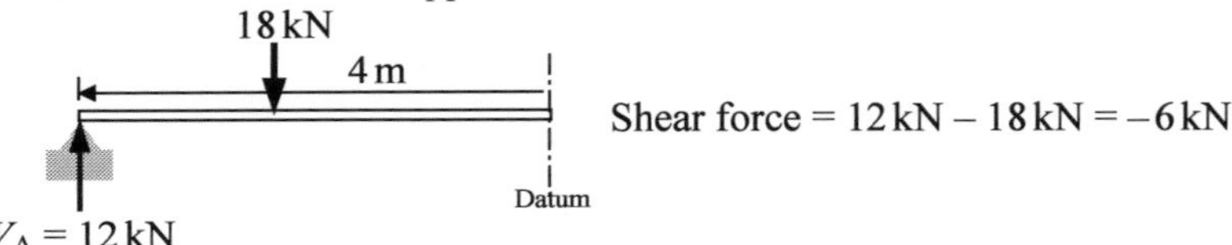

Shear force = 12 kN – 18 kN = –6 kN

Shear force at 5 m from support A:

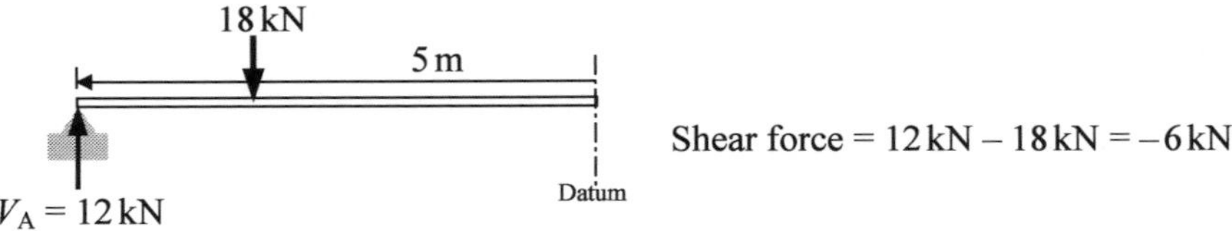

Shear force = 12 kN – 18 kN = –6 kN

Shear force at 6 m from support A, i.e. at support B:

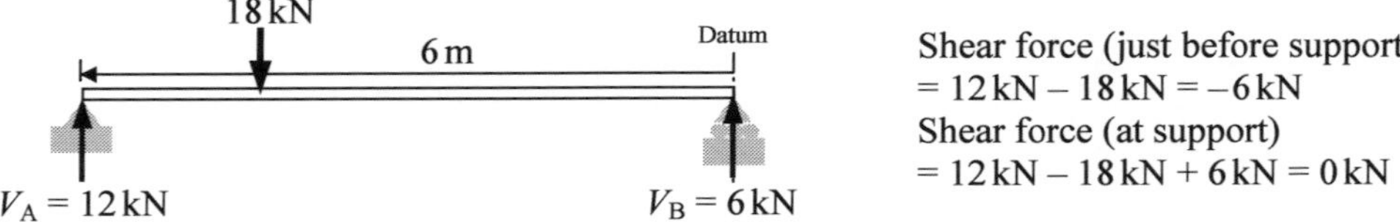

Shear force (just before support)
= 12 kN – 18 kN = –6 kN
Shear force (at support)
= 12 kN – 18 kN + 6 kN = 0 kN

Here again, just before support B there is no change in shear force. At the point of application, the vertical force at the support reduces the resultant shear force to zero.

Having calculated the shear force at 1-m intervals along the beam, we can record the results in a table as shown in Figure 3.62.

Distance along beam from support A	Shear force
0 m (at support A)	0 kN (start point) 12 kN
1 m	12 kN
2 m	12 kN (just before point load) −6 kN (at point load)
3 m	−6 kN
4 m	−6 kN
5 m	−6 kN
6 m (at support B)	−6 kN (just before support) 0 kN (end point)

Figure 3.62: Shear force along beam – Example 3.19

We can also use these results to plot a graph for the shear force as shown in Figure 3.63.

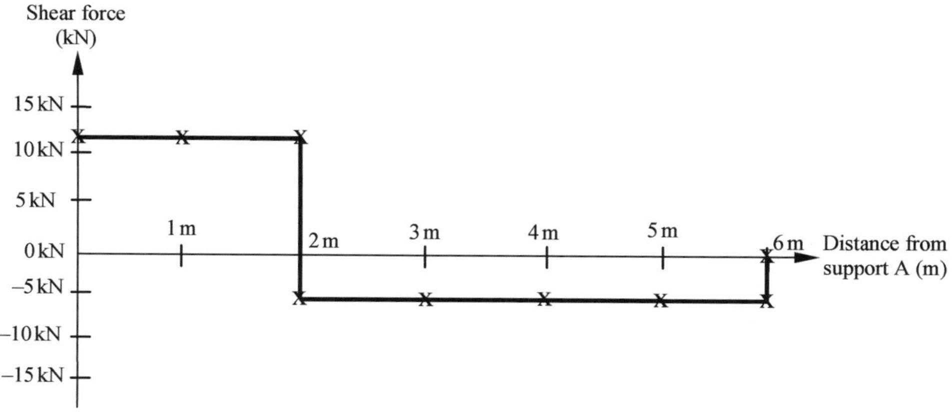

Figure 3.63: Plot of distribution of shear forces along beam – Example 3.19

The resultant graph shows the distribution of shear forces along the beam of Example 3.19. It is plotted in normal Cartesian co-ordinates with shear force values plotted along the y axis and the distance along the beam along the x axis.

Engineers tend to simplify these graphs by marking the principal values on a shear force diagram thus omitting the need for the y axis and in this case by annotating the diagram to show the position of the supports. The diagrams are often plotted 'proportionally' or using a suitable scale. The shear force diagram for the beam of Example 3.19 may therefore be drawn as shown in Figure 3.64.

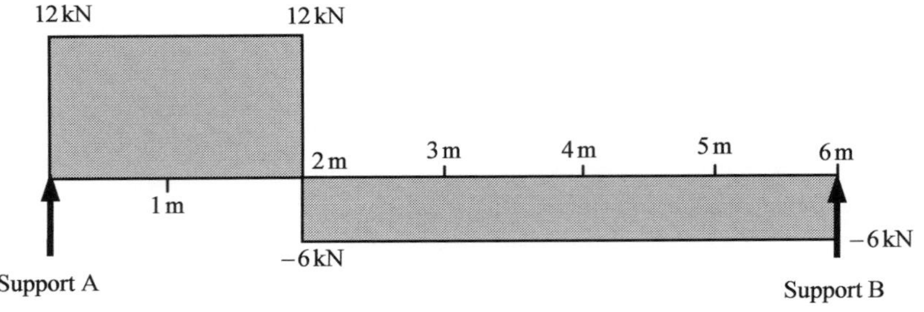

Figure 3.64: Shear force diagram for beam – Example 3.19

We can follow this procedure to construct shear force diagrams for various beam configurations and loading.

Example 3.20: Shear force diagram for a simply supported beam carrying two point loads

For the beam shown in Figure 3.65, which is simply supported at points A and B, determine:
(a) the reactions at the supports V_A, H_A and V_B;
and (b) the shear force at 1-m intervals along the beam.

Plot the results of (b) in the form of a shear force diagram.

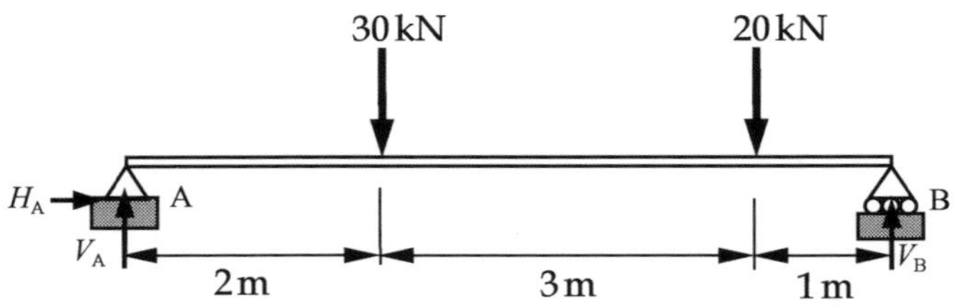

Figure 3.65: Beam system – Example 3.20

First, we must calculate the reactions:

Resolving horizontally: $\Sigma F_x = H_A - 0 = 0 \qquad \therefore H_A = 0$

Resolving vertically: $\Sigma F_y = V_A + V_B - 30\,\text{kN} - 20\,\text{kN} = 0$

$= V_A + V_B = 50\,\text{kN}$ [1]

Taking moments about A: $\Sigma M_z = (30\,\text{kN}\times 2\,\text{m}) + (20\,\text{kN}\times(2\,\text{m}+3\,\text{m})) - (V_B\times(2\,\text{m}+3\,\text{m}+1\,\text{m})) = 0$

$= 60 + 100 - 6V_B = 0$

$V_B = (160/6) = 26.67\,\text{kN}$

Back-substituting V_B into [1] gives:

$V_A + V_B = 50\,\text{kN}$

But $V_B = 26.67\text{kN}$ hence: $V_A + 26.67\,\text{kN} = 50\,\text{kN}$

And so: $V_A = 50 - 26.67 = 23.33\,\text{kN}$

We can now calculate the shear force at 1-m intervals as before.

We commence our calculations at the support A:

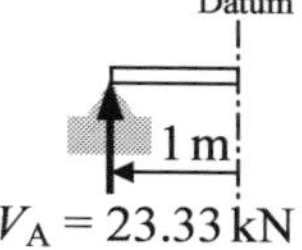

Shear force = 23.33 kN

Shear force at 1 m from support A:

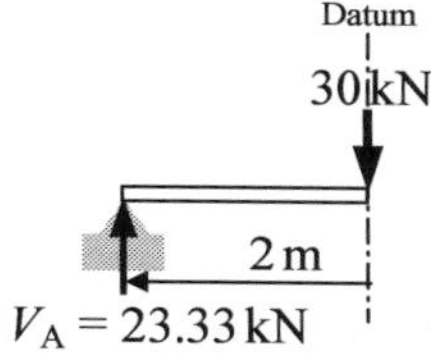

Shear force = 23.33 kN

Shear force at 2 m from support A:

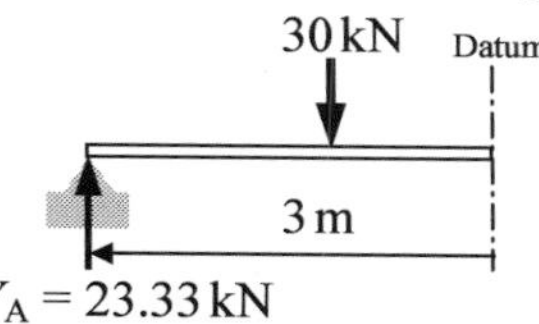

Shear force (just before load) = 23.33 kN
Shear force (at point load) = 23.33 kN – 30 kN = –6.67 kN

Shear force at 3 m from support A:

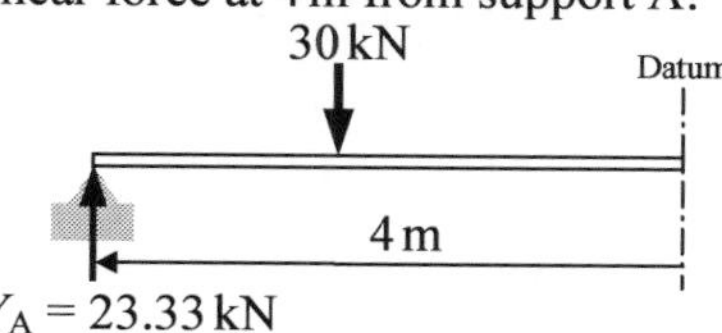

Shear force = 23.33 kN – 30 kN = –6.67 kN

Shear force at 4 m from support A:

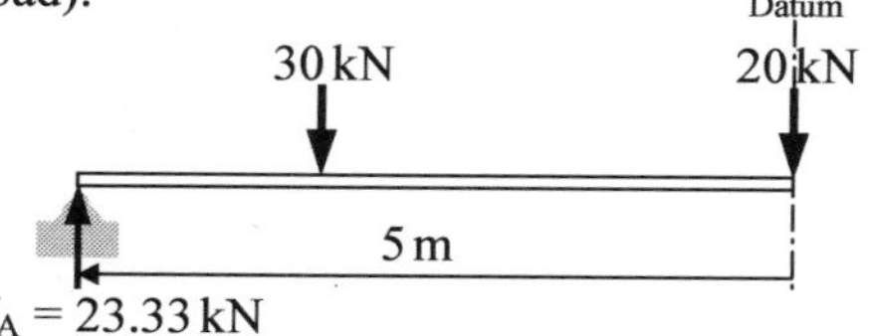

Shear force = 23.33 kN – 30 kN = –6.67 kN

Shear force at 5 m from support A (here we note that we must take into account our second point load):

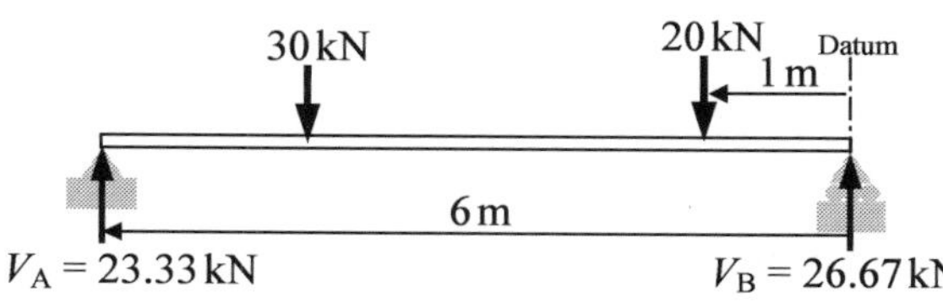

Shear force (just before load) = 23.33 kN – 30 kN
= –6.67 kN
Shear force (at point load) = 23.33 kN – 30 kN – 20 kN
= –26.67 kN

Shear force at 6 m from support A, i.e. at support B:

Shear force (just before support)
= 23.33 kN – 30 kN – 20 kN = –26.67 kN
Shear force (at support)
= 23.33 kN – 30 kN – 20 kN + 26.67 kN
= 0 kN

Having calculated the shear force at 1-m intervals we can record the results in a table as shown in Figure 3.66.

Distance along beam from support A	Shear force
0 m (at support A)	0 kN (start point) 23.33 kN
1 m	23.33kN
2 m	23.33 kN (just before point load) −6.67 kN (at point load)
3 m	−6.67 kN
4 m	−6.67 kN
5 m	−6.67 kN (just before point load) −26.67 kN (at point load)
6 m (at support B)	−26.67 kN (just before support) 0 kN (end point)

Figure 3.66: Shear force along beam – Example 3.20

The Shear force diagram will be that shown in Figure 3.67.

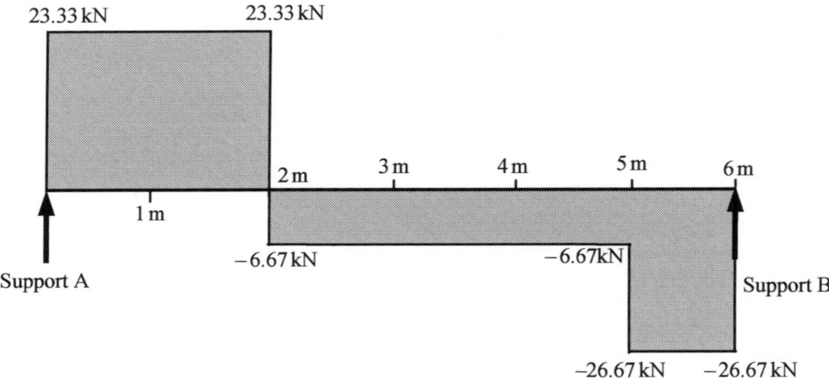

Figure 3.67: Shear force diagram for beam – Example 3.20

We will now consider a further example of a simply supported beam with a cantilevered end.

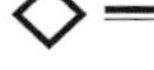

Example 3.21: Shear force diagram for a simply supported beam with a cantilevered end carrying two point loads

For the beam shown in Figure 3.68, which is simply supported at points A and B, determine:
(a) the reactions at the supports V_A, V_B and H_B;
and (b) the shear force at 1-m intervals along the beam.

Plot the results of (b) in the form of a shear force diagram.

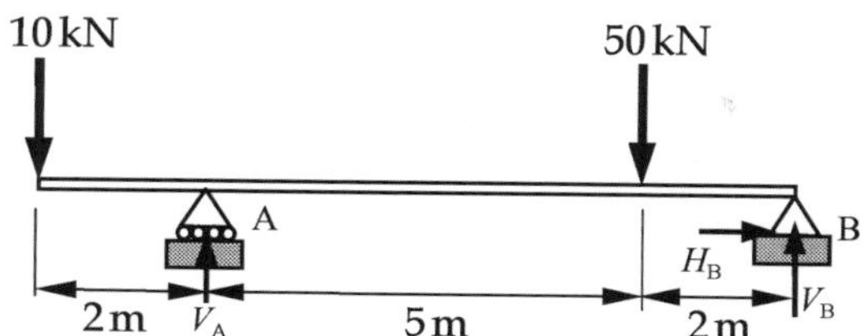

Figure 3.68: Beam system – Example 3.21

We must first calculate the reactions at the supports:

Resolving horizontally: $\Sigma F_x = H_B - 0 = 0 \qquad \therefore H_B = 0$

Resolving vertically: $\Sigma F_y = V_A + V_B - 10\,\text{kN} - 50\,\text{kN} = 0$

$= V_A + V_B = 60\,\text{kN}$ [1]

Taking moments about B: $\Sigma M_z = -(50\,\text{kN} \times 2\,\text{m}) + (V_A \times (2\,\text{m} + 5\,\text{m})) - (10\,\text{kN} \times (2\,\text{m} + 5\,\text{m} + 2\,\text{m})) = 0$

$= -100 + 7V_A - 90 = 0$

$V_A = (190/7) = 27.143\,\text{kN}$

Back-substituting V_B into [1] gives:

$V_A + V_B = 60\,\text{kN}$

But $V_B = 27.143$ kN hence: $27.143\,\text{kN} + V_B = 60\,\text{kN}$

And so: $V_B = 60 - 27.143 = 32.857\,\text{kN}$

We can now calculate the shear forces at 1-m intervals.

We commence our calculations at the left end of the beam, 2 m to the left of support A:

10 kN

Shear force (start point) = 0 kN
Shear force = – 10 kN (note negative sign, to indicate downward force)

Shear force at 1 m from left end of beam:

Shear force = – 10 kN

Shear force at 2 m from left end of beam:

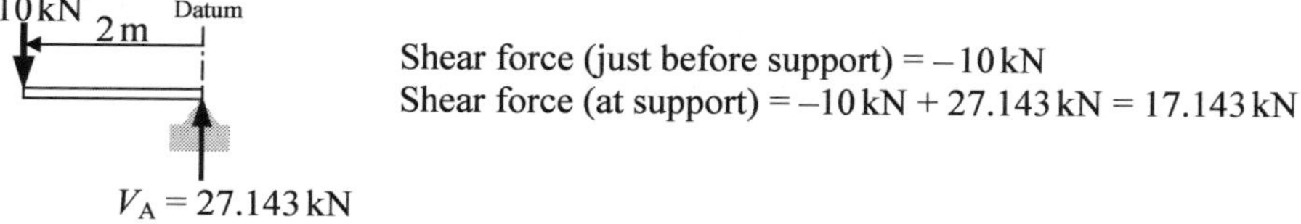

Shear force (just before support) = – 10 kN
Shear force (at support) = –10 kN + 27.143 kN = 17.143 kN

Shear force at 3 m from left end of beam:

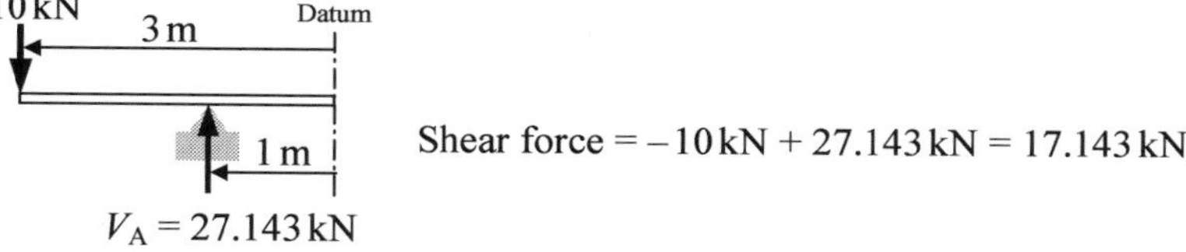

Shear force = – 10 kN + 27.143 kN = 17.143 kN

We note again that there is no change in applied loads or support reactions until the 50 kN point load located 7 m from the left end of the beam. The shear force will therefore remain constant to 7 m and we will simplify our calculations by omitting the intermediate points.

Shear force at 7 m from the left end of the beam:

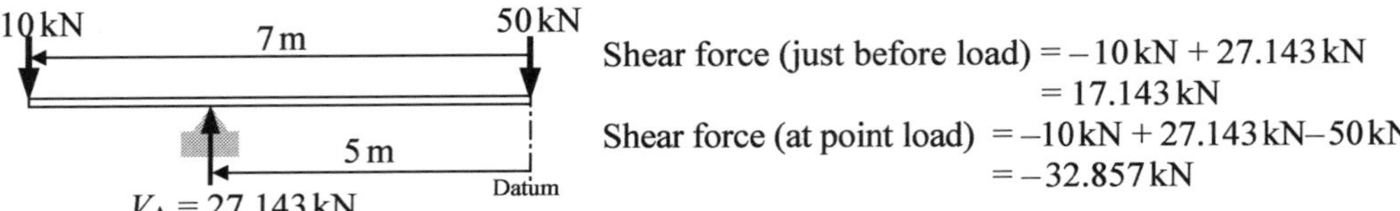

As there is no change in load between the application of the 50 kN point load and the reaction at support B, the shear force will remain constant. We will therefore move directly to the end of the beam at support B.

Shear force at 9 m from support A, i.e. at support B:

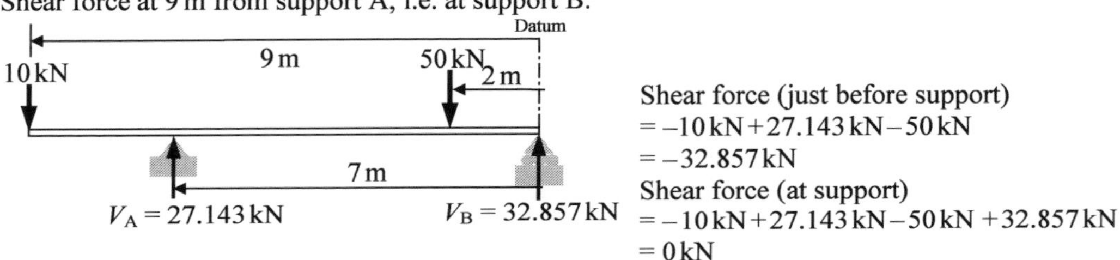

Having calculated the shear force we can record the results, at 1-m intervals, in a table as shown in Figure 3.69.

Distance along beam from left end	Shear force
0 m	0 kN (start point) – 10 kN
1 m	– 10 kN
2 m (at support A)	– 10 kN (just before support) 17.143 kN (at support)
3 m	17.143 kN
4 m	17.143 kN
5 m	17.143 kN
6 m	17.143 kN
7 m	17.143 kN (just before point load) –32.857 kN (at point load)
8 m	–32.857 kN
9 m (at support B)	–32.857 kN (just before support) 0 kN (end point)

Figure 3.69: Shear force along beam – Example 3.21

The Shear force diagram is shown in Figure 3.70.

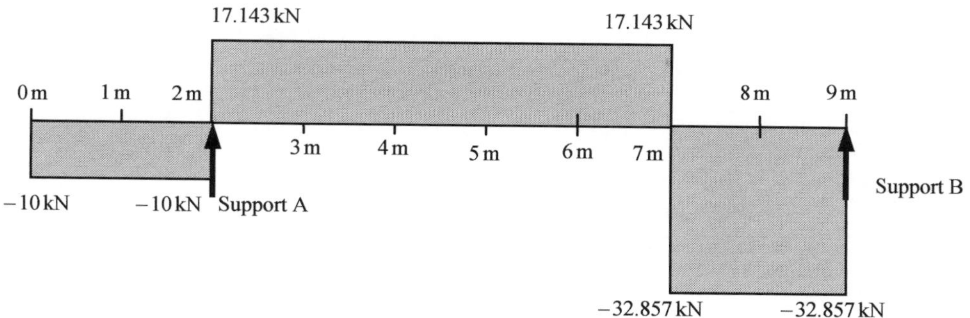

Figure 3.70: Shear force diagram for beam – Example 3.21

Having constructed the shear force diagrams for simply supported beams carrying point loads, we will now extend our studies to those beams supporting distributed loads (DLs)

Example 3.22: Shear force diagram for a simply supported beam carrying a uniformly distributed load (UDL)

For the beam shown in Figure 3.71, which is simply supported at points A and B, determine:
(a) the reactions at the supports V_A, H_A and V_B;
and (b) the shear force at 1-m intervals along the beam.

Plot the results of (b) in the form of a shear force diagram.

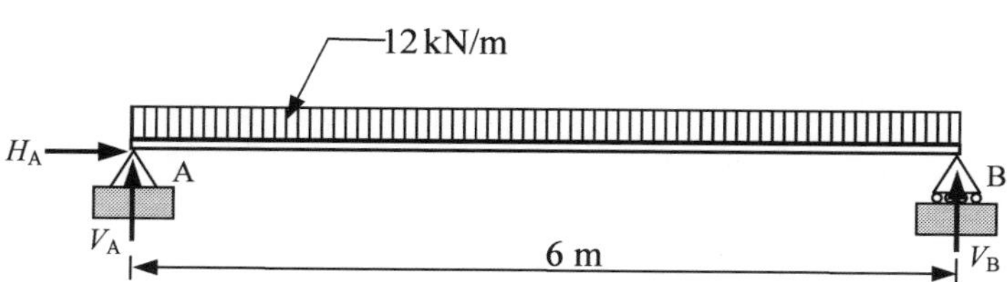

Figure 3.71: Beam supporting a uniformly distributed load – Example 3.22

Resolving horizontally: $\sum F_x = H_A - 0 = 0 \qquad \therefore H_A = 0$

Resolving vertically: $\sum F_y = V_A + V_B - (12\,\text{kN/m} \times 6\,\text{m}) = 0$

$= V_A + V_B = 72\,\text{kN}$ [1]

Taking moments about A: $\sum M_z = -(12\,\text{kN/m} \times 6\,\text{m} \times (6\,\text{m}/2)) + (V_B \times 6\,\text{m}) = 0$

$= -216 + 6V_B = 0$

$V_B = (216/6) = 36\,\text{kN}$

Back-substituting V_B into [1] gives:

$V_A + V_B = 72\,\text{kN}$

But $V_B = 36\text{kN}$ hence: $V_A + 36 = 72\,\text{kN}$

And so: $V_A = 72 - 36 = 36\,\text{kN}$

We note that the reactions of the beam could well have been predicted by observation since it is symmetrically loaded.

We can now calculate the shear forces at 1-m intervals as before by summing the total load applied to the left of our datum.

We commence our calculations at support A:

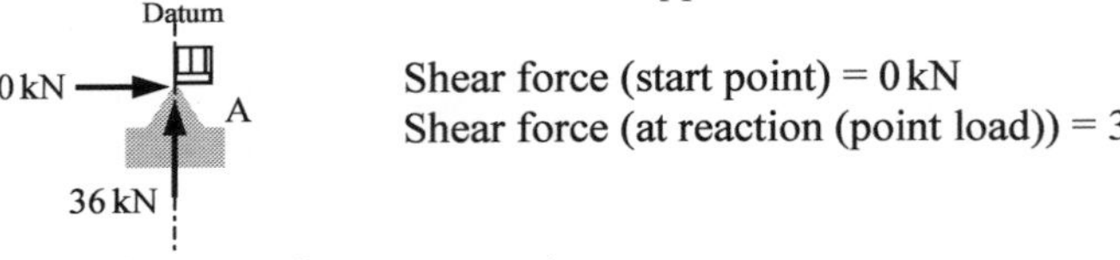

Shear force (start point) = 0 kN
Shear force (at reaction (point load)) = 36 kN

Shear force 1 m from support A:

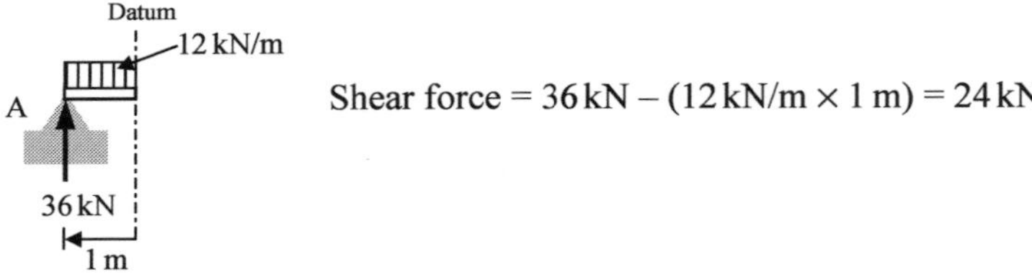

Shear force = 36 kN – (12 kN/m × 1 m) = 24 kN

Shear force 2 m from support A:

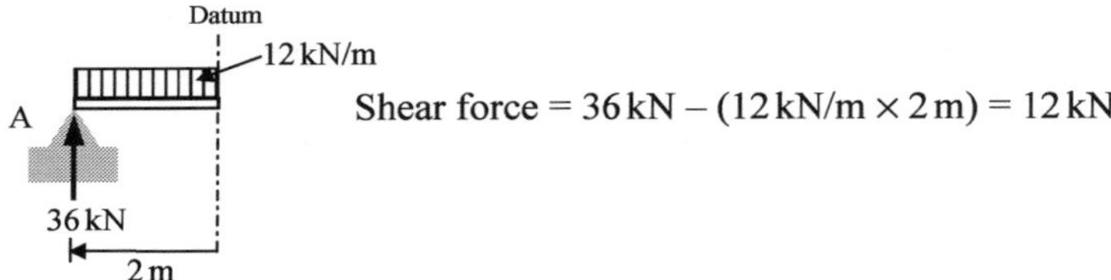

Shear force = 36 kN – (12 kN/m × 2 m) = 12 kN

Shear force 3 m from support A:

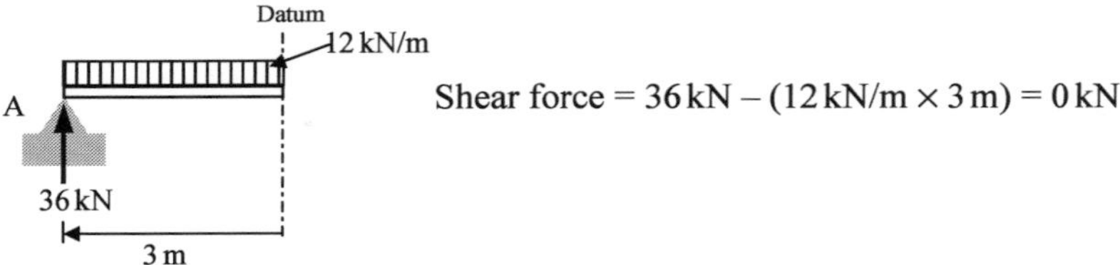

Shear force = 36 kN – (12 kN/m × 3 m) = 0 kN

Shear force 4 m from support A:

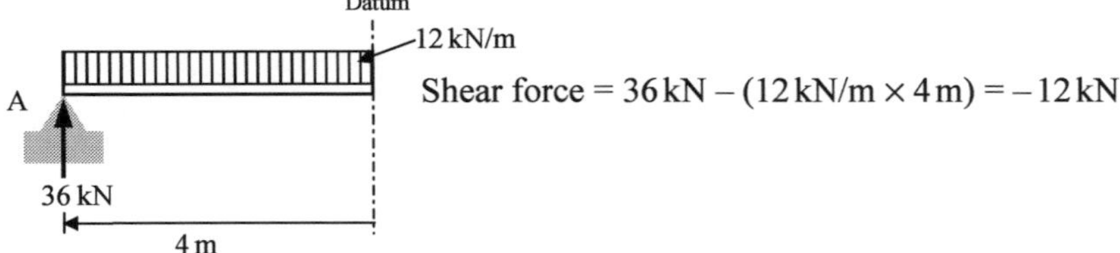

Shear force = 36 kN – (12 kN/m × 4 m) = – 12 kN

Shear force 5 m from support A:

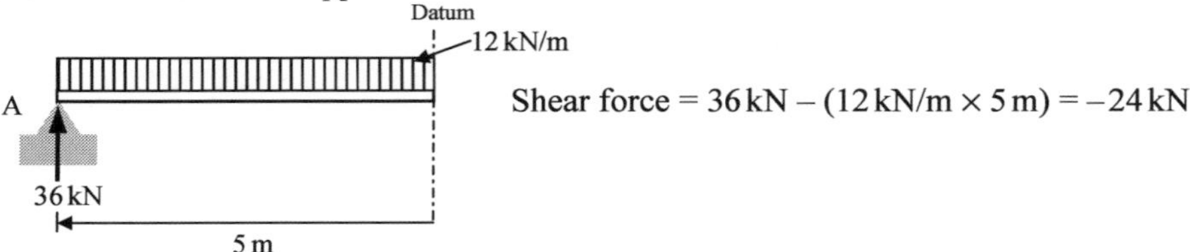

Shear force = 36 kN – (12 kN/m × 5 m) = –24 kN

Shear force 6 m from support A (at support B):

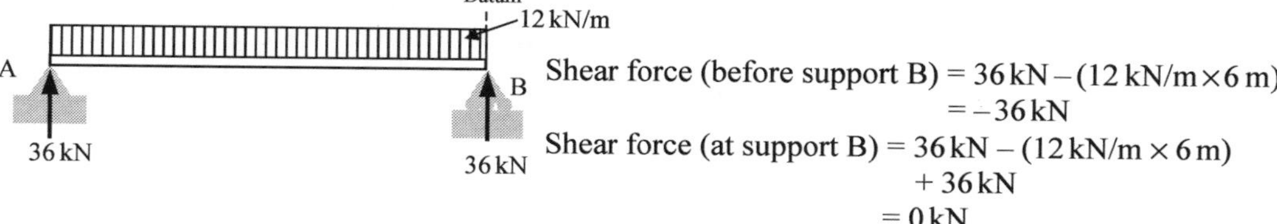

$$\text{Shear force (before support B)} = 36\,\text{kN} - (12\,\text{kN/m} \times 6\,\text{m}) = -36\,\text{kN}$$

$$\text{Shear force (at support B)} = 36\,\text{kN} - (12\,\text{kN/m} \times 6\,\text{m}) + 36\,\text{kN} = 0\,\text{kN}$$

Having calculated the shear force we can record the results, at 1-m intervals, in a table as shown in Figure 3.72.

Distance along beam from support A	Shear force
0 m (at support A)	0 kN (start point) 36 kN (at support)
1 m	24 kN
2 m	12 kN
3 m	0 kN
4 m	−12 kN
5 m	−24 kN
6 m (at support B)	−36 kN (just before support) 0 kN (end point)

Figure 3.72: Shear force along beam – Example 3.22

The Shear force diagram will be that shown in Figure 3.73.

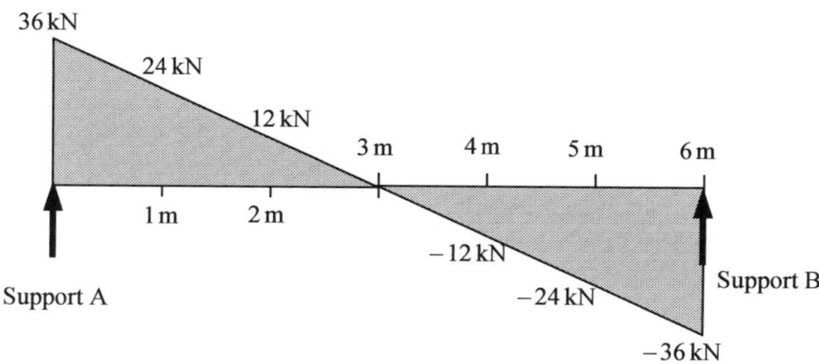

Figure 3.73: Shear force diagram for beam – Example 3.22

For the UDL of Example 3.22, the shear force will vary linearly along the beam causing a distribution in the triangular form indicated in Figure 3.73. The knowledge of this fact will aid us in our analysis, as it will enable us to be able to interpolate values at intermediate points along the beam. It is important to note the difference between this shear force diagram and that for a point load. For the point load, the shear force will only vary at the point at which the load is applied. For a uniformly distributed load, the shear force will vary along its length.

Consider Example 3.23 with two different DLs applied to the beam.

Example 3.23: Shear force diagram for a simply supported beam carrying two uniformly distributed loads

For the beam shown in Figure 3.74, which is simply supported at A and B, determine:
(a) the reactions at the supports V_A, H_A and V_B;
and (b) the shear force at 1-m intervals along the beam.

Plot the results of (b) in the form of a shear force diagram.

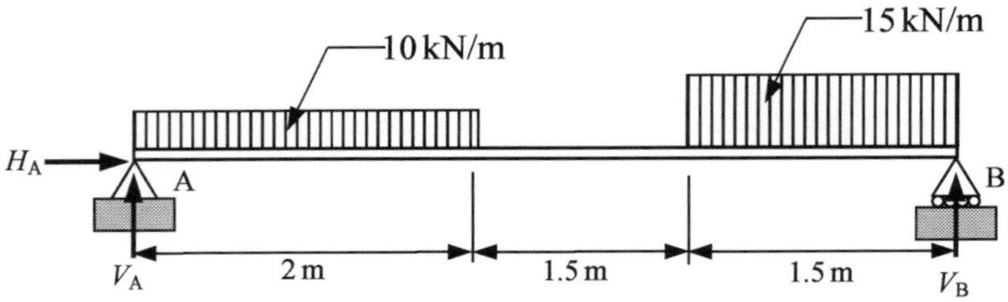

Figure 3.74: Beam supporting uniformly distributed load – Example 3.23

Resolving horizontally: $\Sigma F_x = H_A - 0 = 0 \quad \therefore H_A = 0$

Resolving vertically: $\Sigma F_y = V_A + V_B - (10\,\text{kN/m} \times 2\,\text{m}) - (15\,\text{kN/m} \times 1.5\,\text{m}) = 0$

$= V_A + V_B = 42.5\,\text{kN}$ [1]

Taking moments about A: $\Sigma M_z = (10\,\text{kN/m} \times 2\,\text{m} \times (2\,\text{m}/2))$

$+ (15\,\text{kN/m} \times 1.5\,\text{m} \times (2\,\text{m} + 1.5\,\text{m} + (1.5\,\text{m}/2)))$

$- (V_B \times (2\,\text{m} + 1.5\,\text{m} + 1.5\,\text{m})) = 0$

$= 20 + 95.625 - 5V_B = 0$

$V_B = (115.625/5) = 23.125\,\text{kN}$

Back-substituting V_B into [1] gives:

$V_A + V_B = 42.5\,\text{kN}$

But $V_B = 23.125\,\text{kN}$ hence $V_A + 23.125 = 42.5\,\text{kN}$

And so: $V_A = 42.5 - 23.125 = 19.375\,\text{kN}$

We can now calculate the shear forces at 1-m intervals as before by summing the total load applied to the left of our datum, however we note that the loads are not spaced at 'whole' metre intervals along the beam. It will therefore be necessary to calculate an extra point at 3.5 m from support A, in order to accurately plot the shear force diagram.

We commence our calculations at support A:

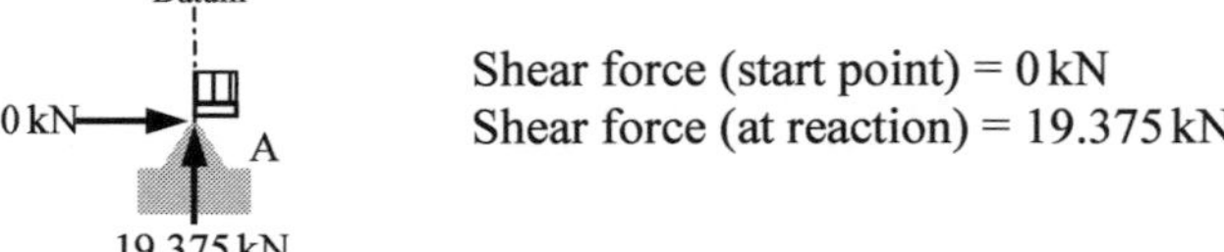

Shear force (start point) = 0 kN
Shear force (at reaction) = 19.375 kN

Shear force 1 m from support A:

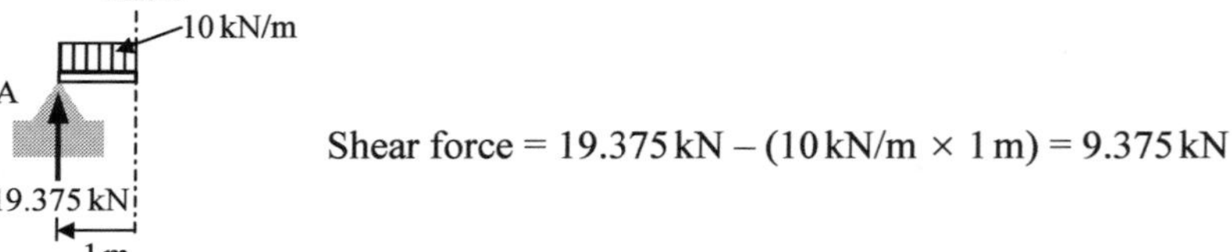

Shear force = 19.375 kN – (10 kN/m × 1 m) = 9.375 kN

Shear force 2 m from support A:

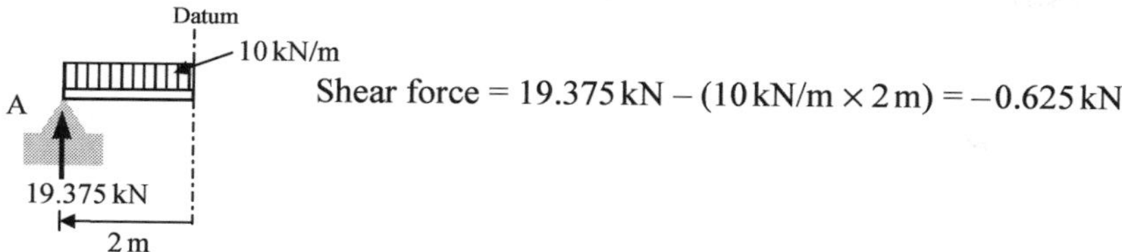

Shear force = 19.375 kN – (10 kN/m × 2 m) = –0.625 kN

Shear force 3 m from support A:

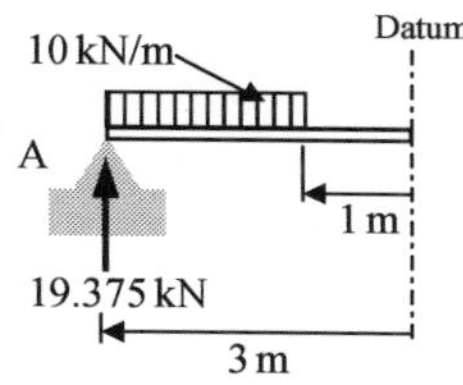

Shear force = 19.375 kN – (10 kN/m × 2 m) = –0.625 kN

Shear force 3.5 m from support A:

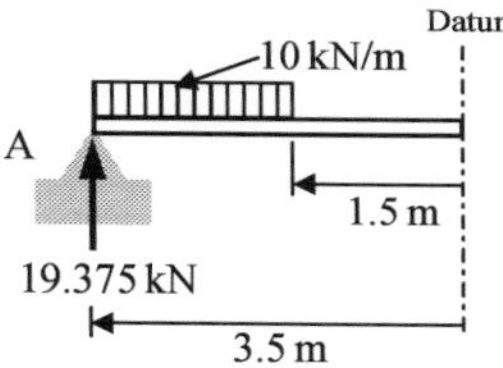

Shear force = 19.375 kN – (10 kN/m × 2 m) = –0.625 kN

Shear force 4 m from support A:

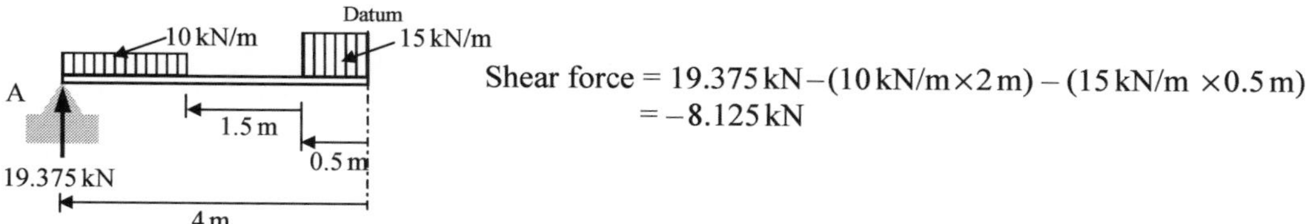

Shear force = 19.375 kN – (10 kN/m × 2 m) – (15 kN/m × 0.5 m)
= –8.125 kN

Shear force 5 m from support A (at support B):

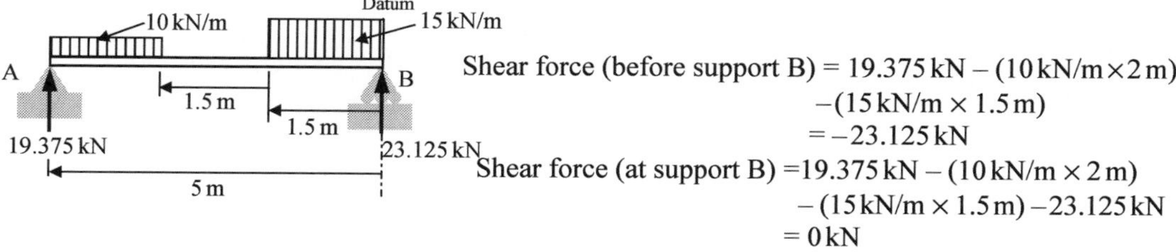

Shear force (before support B) = 19.375 kN – (10 kN/m × 2 m)
–(15 kN/m × 1.5 m)
= –23.125 kN

Shear force (at support B) = 19.375 kN – (10 kN/m × 2 m)
– (15 kN/m × 1.5 m) – 23.125 kN
= 0 kN

The calculated shear forces are tabulated in the table of Figure 3.75 for the beam of Example 3.23.

Distance along beam from support A	Shear force
0 m (at support A)	0 kN (start point) 19.375 kN (at support)
1 m	9.375 kN
2 m	−0.625 kN
3 m	−0.625 kN
3.5 m	−0.625 kN
4 m	−8.125 kN
5 m (at support B)	−23.125 kN (just before support) 0 kN (end point)

Figure 3.75: Shear force along beam – Example 3.23

The Shear force diagram will be that shown in Figure 3.76.

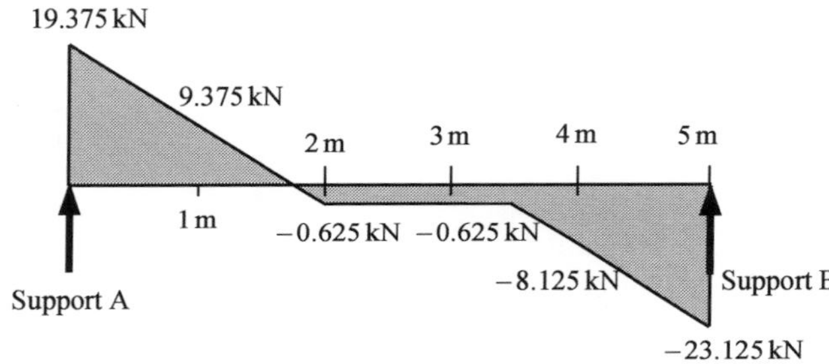

Figure 3.76: Shear force diagram for beam – Example 3.23

We will now consider beams with a combination of point and uniformly distributed loads

---

Example 3.24: Shear force diagram for a simply supported beam carrying uniformly distributed and point loads

For the beam shown in Figure 3.77, which is simply supported at points A and B:
(a) determine the reactions at the supports V_A, H_A and V_B;
and (b) plot the shear force diagram for the beam.

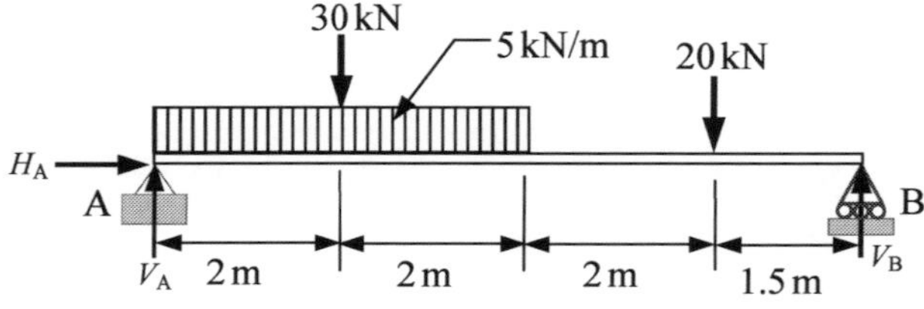

Figure 3.77: Beam system – Example 3.24

Resolving horizontally: $\Sigma F_x = H_A - 0 = 0 \qquad \therefore H_A = 0$

Resolving vertically: $\Sigma F_y = V_A + V_B - 30\,\text{kN} - (5\,\text{kN/m} \times (2\,\text{m} + 2\,\text{m})) - 20\,\text{kN} = 0$

$= V_A + V_B - 30\,\text{kN} - 20\,\text{kN} - 20\,\text{kN} = 0$

$V_A + V_B = 70\,\text{kN}$ [1]

Taking moments about A: $\Sigma M_z = (30\,\text{kN} \times 2\,\text{m}) + (5\,\text{kN/m} \times 4\,\text{m} \times (4\,\text{m}/2))$

$+ (20\,\text{kN} \times (2\,\text{m} + 2\,\text{m} + 2\,\text{m}) - (V_B \times (2\,\text{m} + 2\,\text{m} + 2\,\text{m} + 1.5\,\text{m})) = 0$

$= 60 + 40 + 120 - 7.5 V_B = 0$

$V_B = (220/7.5) = 29.33\,\text{kN}$

Back-substituting V_B into [1] gives:

$V_A + V_B = 70\,\text{kN}$

But $V_B = 29.33$ kN hence: $V_A + 29.33\,\text{kN} = 70\,\text{kN}$

And so: $V_A = 70 - 29.33 = 40.67\,\text{kN}$

We can now calculate the shear forces along the beam, ensuring that we select enough points to produce an accurate plot.

We commence our calculations at the support A:

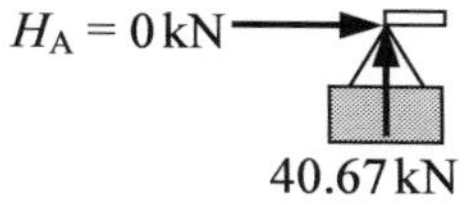

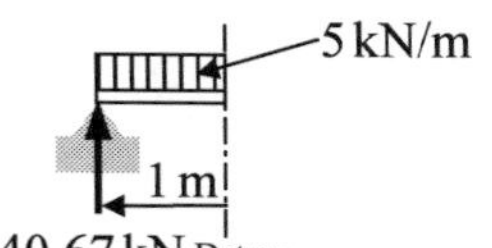

Shear force (start point) = 0 kN
Shear force (at reaction) = 40.67 kN

Shear force at 1 m from support A:

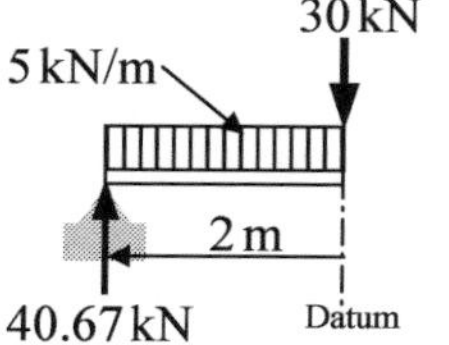

Shear force = 40.67 kN – (5 kN/m × 1 m) = 35.67 kN

Shear force at 2 m from support A:

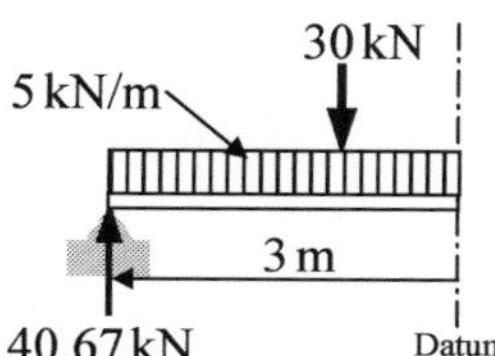

Shear force (just before point load) = 40.67 kN – (5 kN/m × 2 m)
= 30.67 kN

Shear force (at point load) = 40.67 kN – (5 kN/m × 2 m) – 30 kN
= 0.67 kN

Shear force at 3 m from support A:

30 kN
5 kN/m
3 m
40.67 kN
Datum

Shear force = 40.67 kN – (5 kN/m × 3 m) – 30 kN = –4.33 kN

Shear force at 4 m from support A:

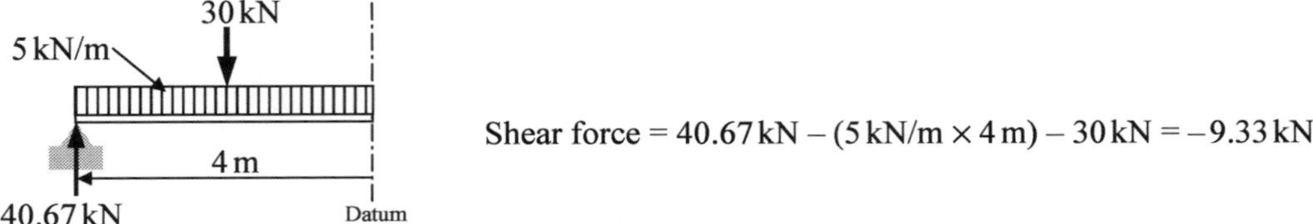

Shear force = 40.67 kN – (5 kN/m × 4 m) – 30 kN = –9.33 kN

We now note that no further loads are applied to the beam until the point load of 20 kN at 6 m from support A (therefore, there will be no change in shear force between 4 and 6 m).

Shear force at 6 m from support A:

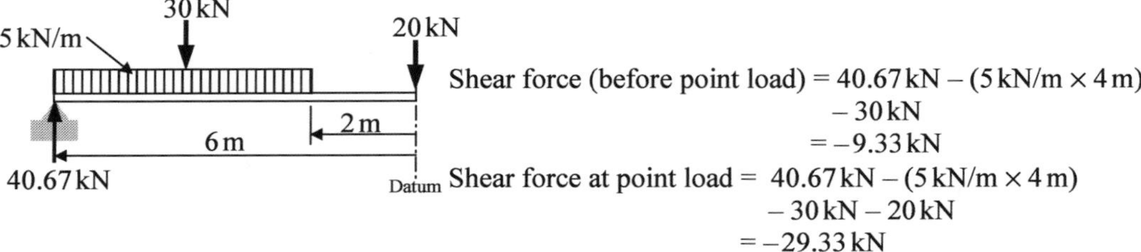

Shear force (before point load) = 40.67 kN – (5 kN/m × 4 m) – 30 kN = –9.33 kN

Shear force at point load = 40.67 kN – (5 kN/m × 4 m) – 30 kN – 20 kN = –29.33 kN

Again we have no loads applied until support B (and again, no change in shear force).

Shear force at 7.5 m from support A, i.e. at support B:

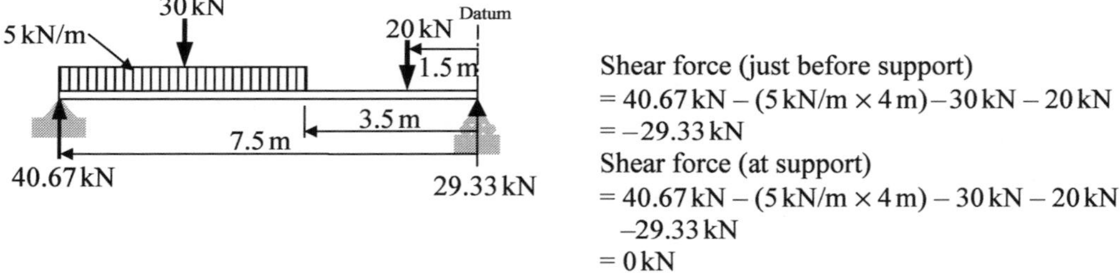

Shear force (just before support)
= 40.67 kN – (5 kN/m × 4 m) – 30 kN – 20 kN
= –29.33 kN

Shear force (at support)
= 40.67 kN – (5 kN/m × 4 m) – 30 kN – 20 kN –29.33 kN
= 0 kN

The final results are shown in a tabular form in Figure 3.78.

Distance along beam from support A	**Shear force**
0 m (at support A)	0 kN (start point) 40.67 kN
1 m	35.67 kN
2 m	30.67 kN (just before point load) 0.67 kN (at point load
3 m	–4.33 kN
4 m	–9.33 kN
6 m	–9.33 kN (just before point load) –29.33 kN (at point load)
7.5 m (at support B)	–29.33 kN (just before support) 0 kN (end point)

Figure 3.78: Shear force along beam – Example 3.24

The Shear force diagram will be that shown in Figure 3.79.

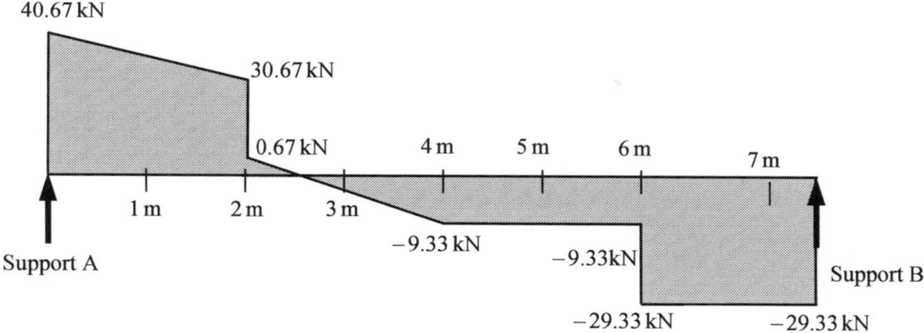

Figure 3.79: Shear force diagram for beam – Example 3.24

Example 3.25: Shear force diagram for a simply supported beam with a cantilevered end carrying distributed and point loads

For the beam shown in Figure 3.80, which is simply supported at A and B:
(a) determine the reactions at the supports V_A, V_B and H_B;
and (b) plot the shear force diagram for the beam.

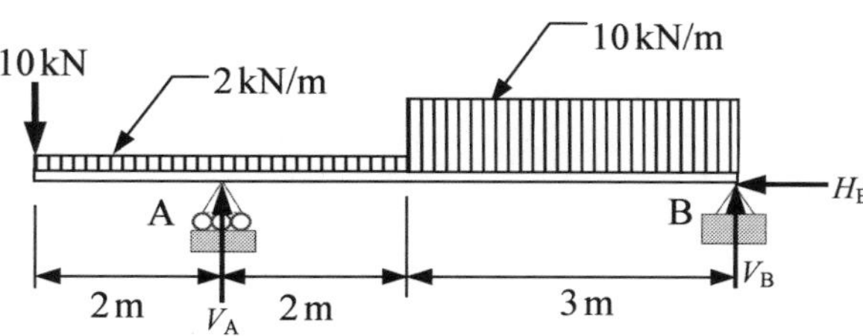

Figure 3.80: Beam system – Example 3.25

We must first calculate the reactions.

Resolving horizontally: $\Sigma F_x = H_B - 0 = 0 \qquad \therefore H_B = 0$

Resolving vertically: $\Sigma F_y = V_A + V_B - 10\,\text{kN} - (2\,\text{kN/m} \times (2\,\text{m} + 2\,\text{m})) - (10\,\text{kN/m} \times 3\,\text{m}) = 0$

$$= V_A + V_B - 10\,\text{kN} - 8\,\text{kN} - 30\,\text{kN}$$

$$V_A + V_B = 48\,\text{kN} \qquad [1]$$

Taking moments about B: $\Sigma M_z = -(10\,\text{kN/m} \times 3\,\text{m} \times (3\,\text{m}/2)) - (2\,\text{kN/m} \times 4\,\text{m} * (3\,\text{m} + 4\,\text{m}/2))$

$$+ (V_A \times (2\,\text{m} + 3\,\text{m})) - (10\,\text{kN} \times (3\,\text{m} + 2\,\text{m} + 2\,\text{m})) = 0$$

$$= -45 - 40 + 5V_A - 70 = 0$$

$$V_A = (155/5) = 31\,\text{kN}$$

Back-substituting V_B into [1] gives:

$$V_A + V_B = 48\,\text{kN}$$

But $V_B = 31\,\text{kN}$ hence: $31\,\text{kN} + V_B = 48\,\text{kN}$

And so: $V_B = 48 - 31 = 17\,\text{kN}$

We can now calculate the shear force, in this case at 1-m intervals, as before.

We commence our calculations at the left end of the beam, 2 m to the left of support A:

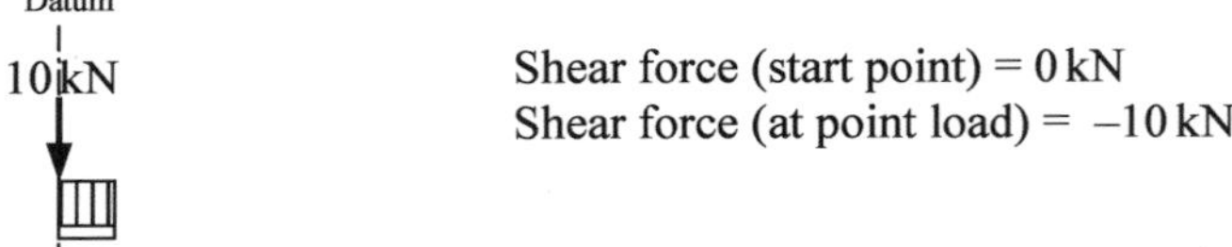

Shear force (start point) = 0 kN
Shear force (at point load) = –10 kN

Shear force at 1 m from left end of beam:

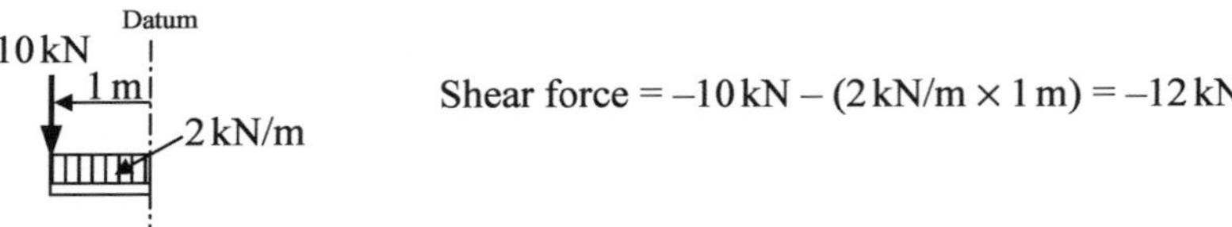

Shear force = –10 kN – (2 kN/m × 1 m) = –12 kN

Shear force at 2 m from left end of beam:

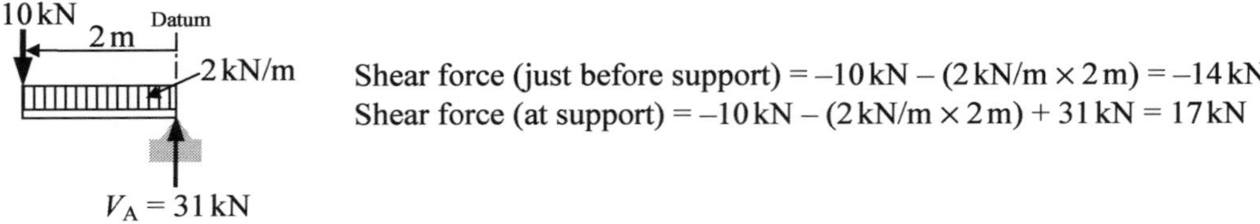

Shear force (just before support) = –10 kN – (2 kN/m × 2 m) = –14 kN
Shear force (at support) = –10 kN – (2 kN/m × 2 m) + 31 kN = 17 kN

We again note that the shear force will vary linearly from support A to the junction between the 2 kN/m UDL and the 10 kN/m UDL. We can therefore take our next section at 4 m from the left of the beam. However, in order to provide a complete solution to the problem the shear force at 3 m is calculated below.

Shear force at 3 m from left end of beam = –10 kN – (2 kN/m × 3 m) + 31 kN = 15 kN

Shear force at 4 m from left end of beam:

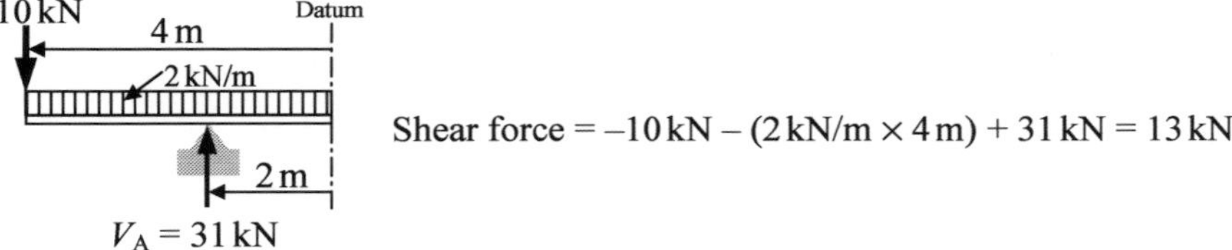

Shear force = –10 kN – (2 kN/m × 4 m) + 31 kN = 13 kN

Shear force at 5 m from left end of beam:

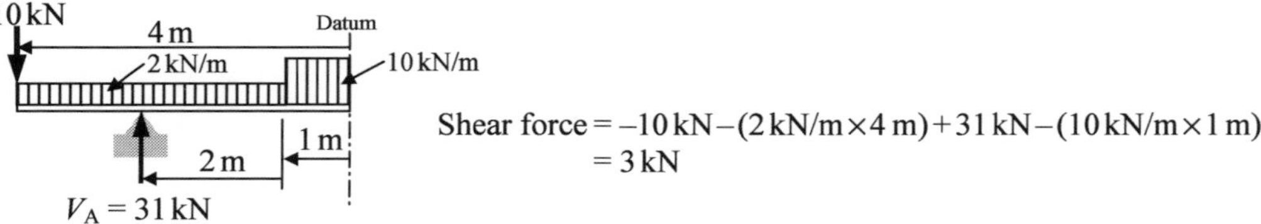

Shear force = –10 kN – (2 kN/m × 4 m) + 31 kN – (10 kN/m × 1 m)
= 3 kN

The shear force at 6 m can again be determined without a calculation since the increase in shear force is linear such that:

Shear force at 6 m from left end of beam = –10 kN – (2 kN/m × 4 m) + 31 kN – (10 kN/m × 2 m)
= –7 kN

Shear force at 7 m from the left end of the beam (at support B):

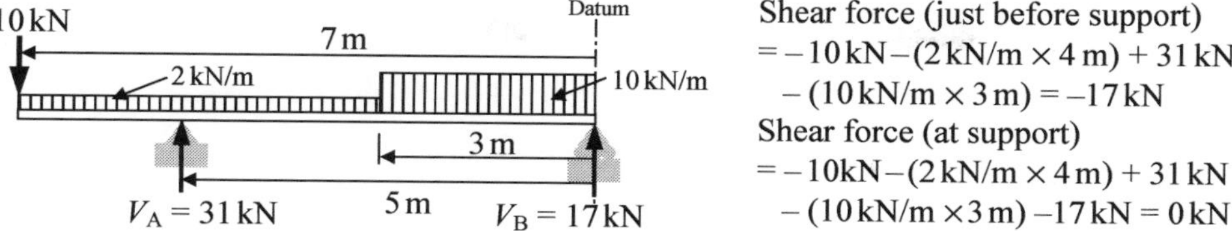

Shear force (just before support)
$= -10\,\text{kN} - (2\,\text{kN/m} \times 4\,\text{m}) + 31\,\text{kN} - (10\,\text{kN/m} \times 3\,\text{m}) = -17\,\text{kN}$

Shear force (at support)
$= -10\,\text{kN} - (2\,\text{kN/m} \times 4\,\text{m}) + 31\,\text{kN} - (10\,\text{kN/m} \times 3\,\text{m}) - 17\,\text{kN} = 0\,\text{kN}$

The shear forces for the beam of Example 3.25 are tabulated in Figure 3.81.

Distance along beam from support A	Shear force
0 m	0 kN (start point) −10 kN
1 m	−12 kN
2 m (at support A)	−14 kN (just before support) 17 kN (at support)
3 m	15 kN
4 m	13 kN
5 m	3 kN
6 m	−7 kN
7 m (at support B)	−17 kN (just before support) 0 kN (end point)

Figure 3.81: Shear force along beam – Example 3.25

The Shear force diagram is shown in Figure 3.82.

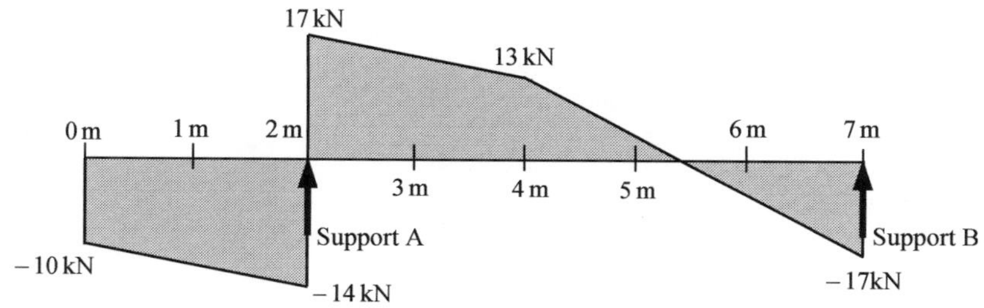

Figure 3.82: Shear force diagram for beam – Example 3.25

◇

Having considered, in detail, the construction of shear force diagrams, we will now discuss the plotting of bending moment diagrams using the beam examples from the previous section.

3.5.2 Bending Moment Diagrams

The method of determining the amount of bending moment that the member must resist at a specific point is to consider a section of the member and then to sum all bending moments, clockwise and anticlockwise, applied to the left (or right) of an imaginary cut. The residual amount will be the moment to be resisted by the structure. Put simply, we will divide the member into discrete sections, just as we did when calculating the shear force and we will then sum all moments to one side of a datum located at the division. The resultant will be the moment resisted by the member and can be plotted on a graph in order to show its distribution through the structure.

We will now consider the beam of Example 3.19 here shown as Example 3.26.

Example 3.26: Bending moment diagram for a simply supported beam carrying a point load

For the beam shown in Figure 3.83, which is simply supported at points A and B, determine:
(a) the reactions at the supports V_A, H_A and V_B;
and (b) the bending moments at 1-m intervals along the beam.
Plot the results of (b) in the form of a bending moment diagram.

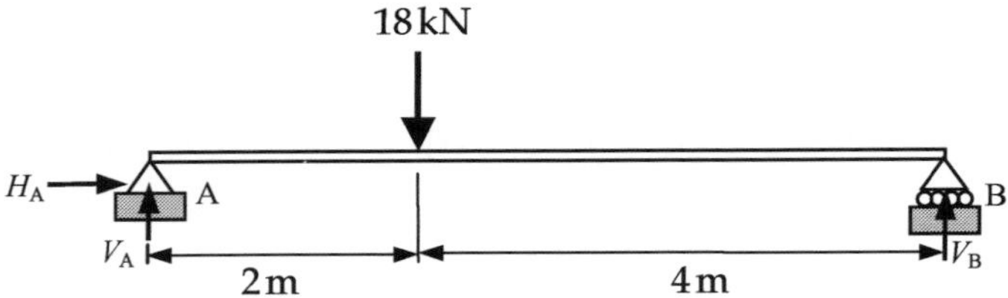

Figure 3.83: Beam system – Example 3.26

We have already calculated the reactions for this beam as:

$H_A = 0$, $V_A = 12\,\text{kN}$ and $V_B = 6\,\text{kN}$

Having calculated the reactions, we now proceed to calculate the moments along the beam at 1-m intervals commencing at support A and proceeding to support B. We will again use the sketches previously produced to assist in calculating the shear force as these will enable us to clearly visualise forces and distances to be considered. It is important to emphasise here that all moments are calculate about the datum therefore **ALL DISTANCES ARE MEASURED FROM THE DATUM TO THE POINT OF APPLICATION!**

We commence calculations at support A.

$H_A = 0\,\text{kN}$

$V_A = 12\,\text{kN}$

Bending moment $= 12\,\text{kN} \times 0\,\text{m} = 0\,\text{kNm}$

Again, the horizontal force can be omitted from these calculations as its distance, in the horizontal direction, from the datum will always be zero. We also note that clockwise moments will be taken to be positive and anticlockwise moments will be negative.

Bending moment at 1 m from support A:

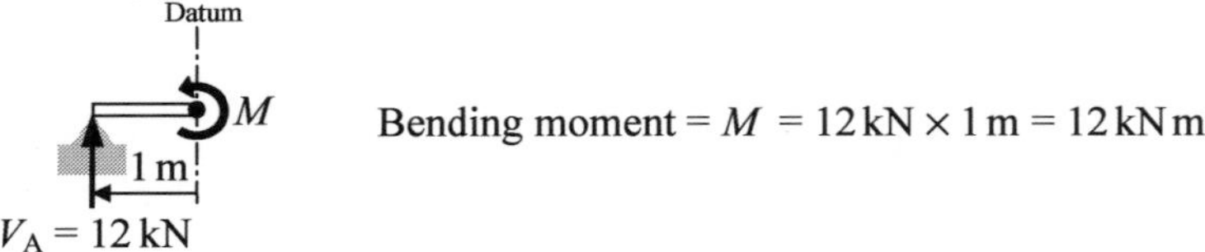

Bending moment $= M = 12\,\text{kN} \times 1\,\text{m} = 12\,\text{kNm}$

Here we note that as the distance to the datum increases, so does the moment. It will therefore be necessary, at this stage, to calculate moments along the entire length of the beam.

Bending moment at 2 m from support A:

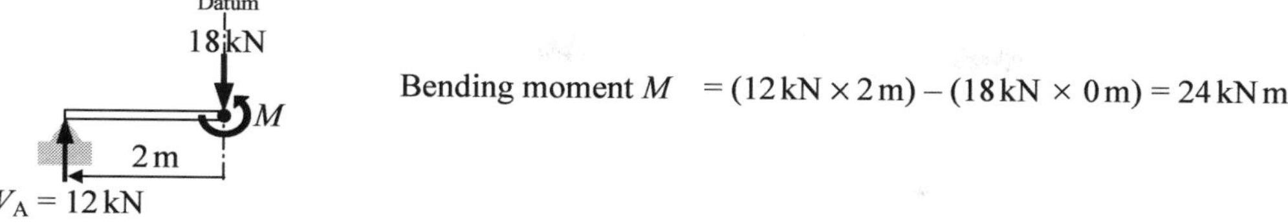

Here we include the point load of 18 kN; however as it is located at the datum it has zero distance and therefore zero moment.

Bending moment at 3 m from support A:

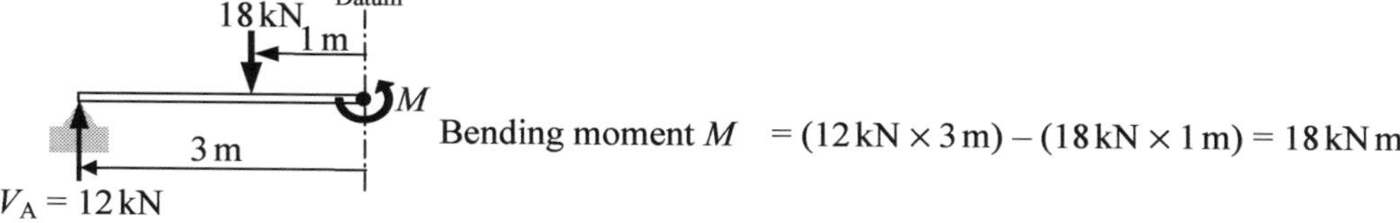

Bending moment at 4 m from support A:

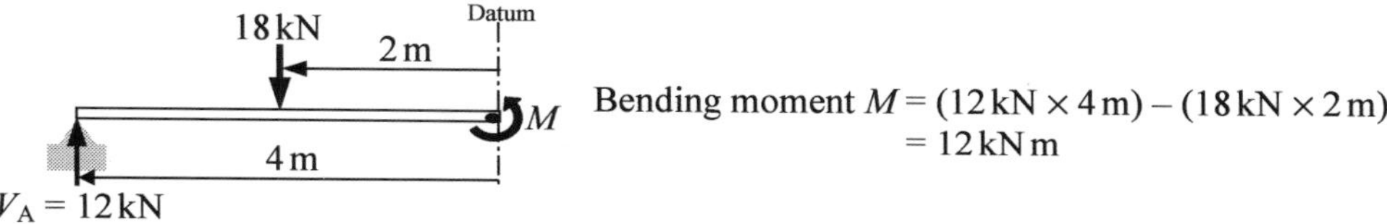

Bending moment at 5 m from support A:

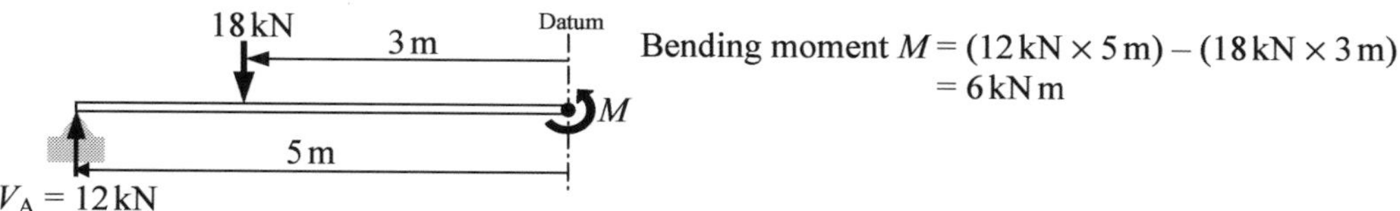

Bending moment at 6 m from support A, i.e. at support B:

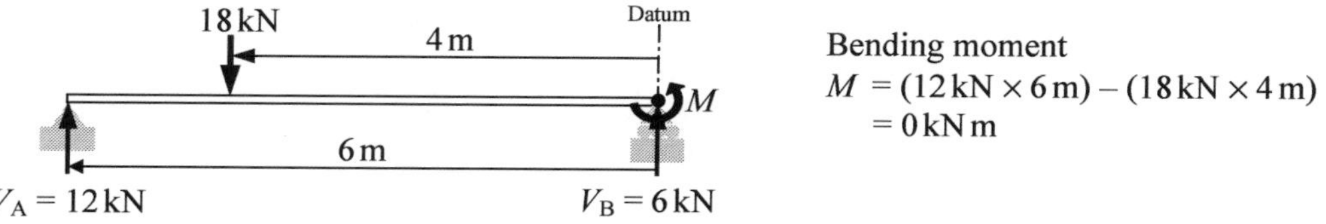

The bending moment at either end of the beam MUST be zero since there is no resistance to moment at this point for a simply supported beam and the section has zero length. This is an important check when we calculate bending moments on simply supported beams. We will consider beams with fixed supports later in this chapter.

Having calculated the bending moments at 1-m intervals, we can record the results in a table as shown in Figure 3.84.

Distance along beam from support A	Bending moment
0 m (at support A)	0 kN m
1 m	12 kN m
2 m	24 kN m
3 m	18 kN m
4 m	12 kN m
5 m	6 kN m
6 m (at support B)	0 kN m

Figure 3.84: Bending moments along beam – Example 3.26

We can also use these results to plot a graph of the bending moments as shown in Figure 3.85.

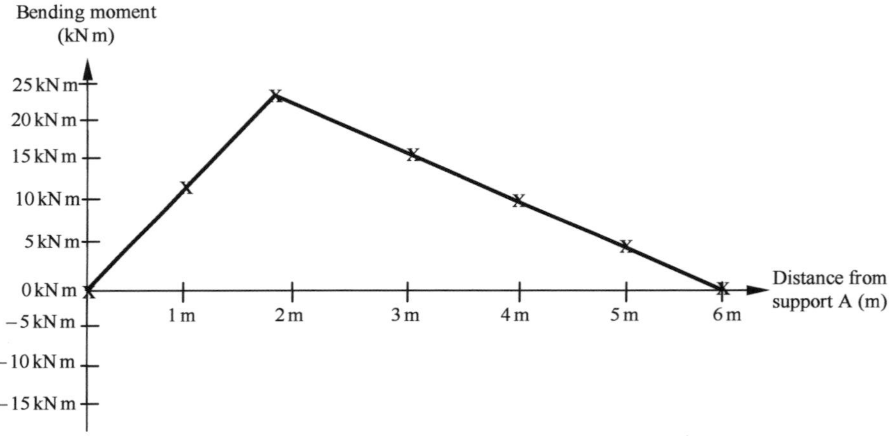

Figure 3.85: Graph of distribution of bending moments along beam – Example 3.26

The resultant graph shows the distribution of bending moments along the beam of Example 3.26. It is plotted in normal Cartesian co-ordinates with bending moment values plotted along the *y* axis and the distance along the beam along the *x* axis. As the values in these calculations yielded clockwise, positive moments, we plot the graph as shown. However, had we calculated bending moments commencing at support B, the results would have been negative. For example, consider the calculation of bending moments at 2 m from support B:

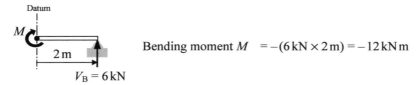

Datum

M

2 m

$V_B = 6\,kN$

Bending moment $M = -(6\,kN \times 2\,m) = -12\,kN\,m$

The results from calculating the moments commencing at support B would therefore have the same value as those previously calculated, but would have the opposite sign as shown in the table of Figure 3.86.

Distance along beam from support B	Bending moment
0 m (at support B)	0 kN m
1 m	−6 kN m
2 m	−12 kN m
3 m	−18 kN m
4 m	−24 kN m
5 m	−12 kN m
6 m (at support A)	0 kN m

Figure 3.86: Bending moments along beam – Example 3.26

And the graph for the bending moments would take the form of Figure 3.87.

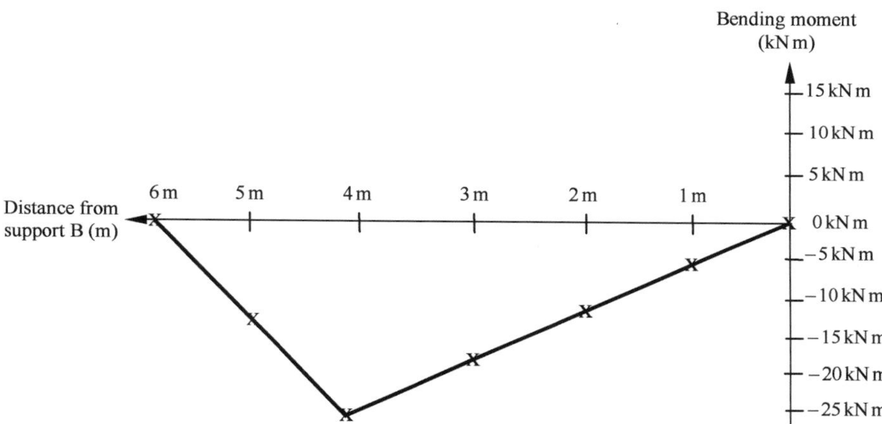

Figure 3.87: Plot of distribution of bending moments along beam – Example 3.26

It can be seen that both sets of calculations yield the same shaped graphs with the same principal values and, in fact, both solutions are correct. However, in order to take advantage of more advanced analysis techniques and in order that a standard convention is adopted, we will always plot our bending moment diagrams on the **tension face** of the structure under analysis. This requires us to determine which side of the beam will be in tension. Consider again the bending moment at 2 m from support A.

Bending moment at 2 m from support A:

2 m Datum

Tension M

$V_A = 12\,\text{kN}$

Bending moment $M \;= (12\,\text{kN} \times 2\,\text{m}) - (18\,\text{kN} \times 0\,\text{m}) = 24\,\text{kN m}$

At the datum, the beam is effectively 'clamped' or fixed in position. The effect of the load at the support is to push the end of the beam upwards and to stretch the bottom face, thus creating tension. We will therefore plot the bending moments on the underside of our x axis. If you are in doubt as to which side is the tension face, a simple test can be undertaken using a flexible ruler. Clamp the ruler with one hand at some distance from the end and then, with the finger of the other hand, apply a force in the direction of the load.

Engineers again tend to simplify these graphs by marking the principal values on the diagram, thus omitting the need for the y axis and, in this case, by annotating the diagram to show the position of the supports. The diagrams are often plotted 'proportionally' or using a suitable scale. The bending moment diagram for the beam of Example 3.26 is shown in Figure 3.88.

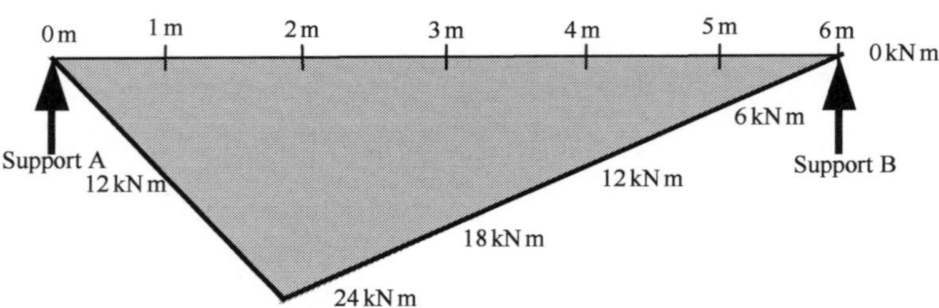

Figure 3.88: Bending moment diagram for beam – Example 3.26

Note that no signs are attributed to the values shown on the diagram.

We can now construct the bending moment diagrams for the remaining beam examples.

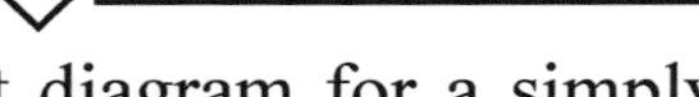

Example 3.27: Bending moment diagram for a simply supported beam carrying two point loads

For the beam shown in Figure 3.89, which is simply supported at A and B, determine:

(a) the reactions at the supports V_A, H_A and V_B;

and (b) the bending moments at 1-m intervals along the beam.

Plot the results of (b) in the form of a bending moment diagram.

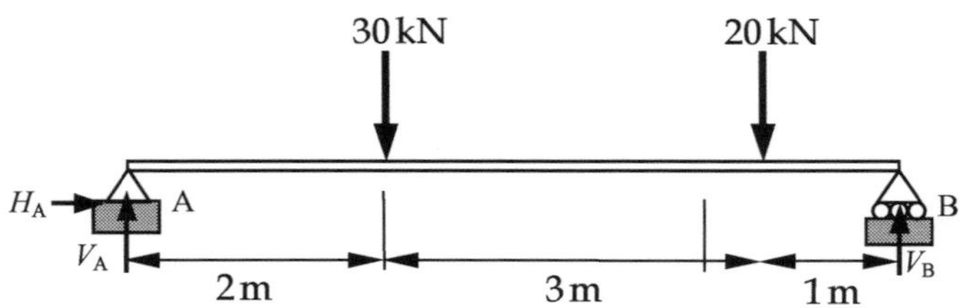

Figure 3.89: Beam system – Example 3.27

We have already calculated the reactions as:

$$H_A = 0, \; V_A = 23.33\,\text{kN} \text{ and } V_B = 26.67\,\text{kN}$$

We can now calculate the bending moments at 1-m intervals as before.

We commence our calculations at the support A:

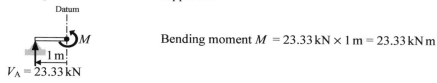

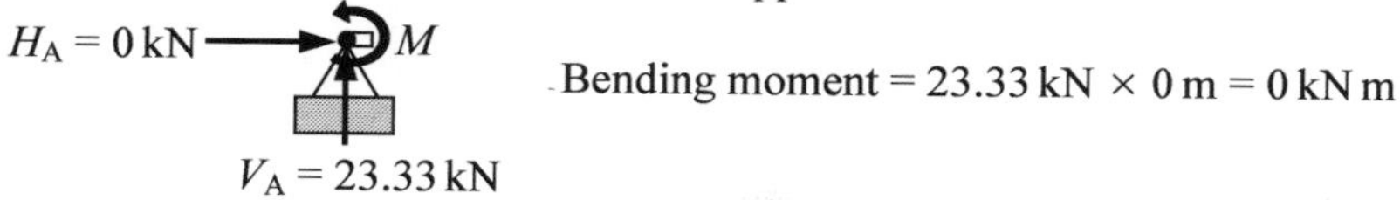

Bending moment = 23.33 kN × 0 m = 0 kN m

Bending moment at 1 m from support A:

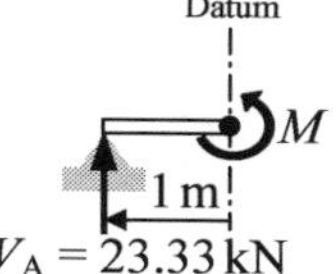

Bending moment $M = 23.33\,\text{kN} \times 1\,\text{m} = 23.33\,\text{kN m}$

Bending moment at 2 m from support A:

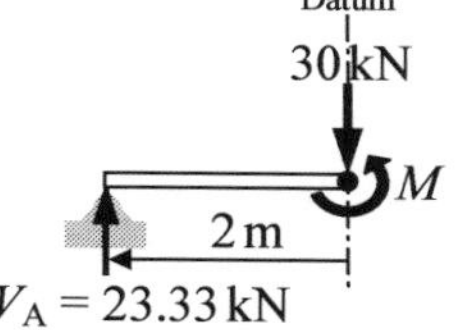

Bending moment $M = 23.33\,\text{kN} \times 2\,\text{m} - 30\,\text{kN} \times 0\,\text{m} = 46.66\,\text{kN m}$

Bending moment at 3 m from support A:

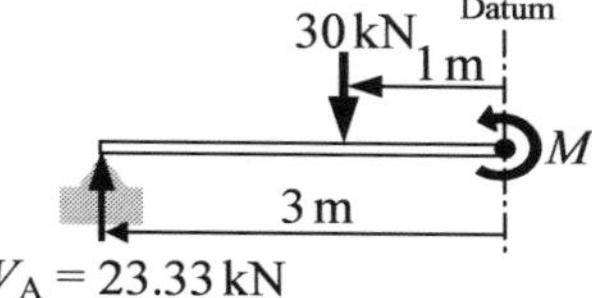

Bending moment $M = 23.33\,\text{kN} \times 3\,\text{m} - 30\,\text{kN} \times 1\,\text{m} = 39.99\,\text{kN m}$

Bending moment at 4 m from support A:

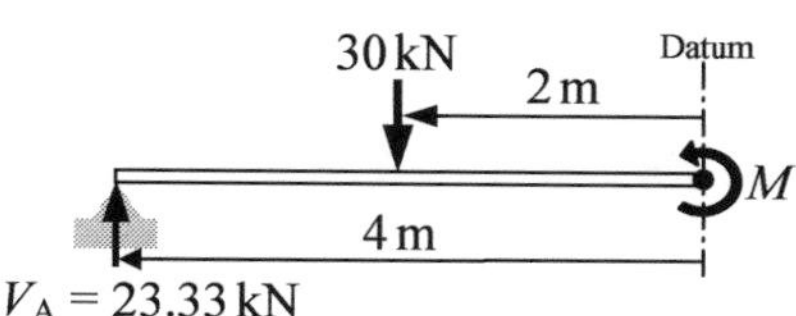

Bending moment $M = 23.33\,\text{kN} \times 4\,\text{m} - 30\,\text{kN} \times 2\,\text{m}$
$= 33.33\,\text{kN m}$

Bending moment at 5 m from support A:

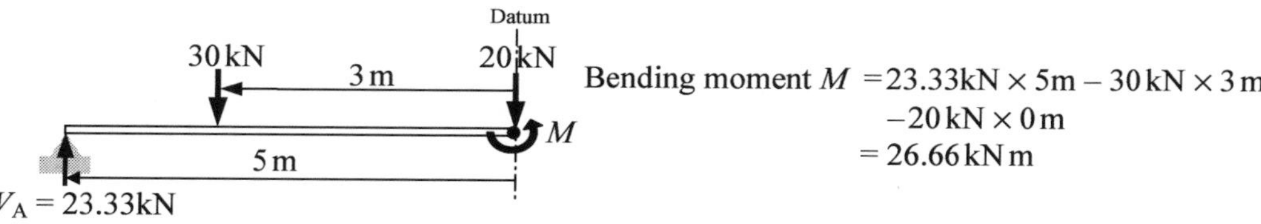

Bending moment $M = 23.33\text{kN} \times 5\text{m} - 30\,\text{kN} \times 3\,\text{m}$
$-20\,\text{kN} \times 0\,\text{m}$
$= 26.66\,\text{kN m}$

Bending moment at 6 m from support A, i.e. at support B:

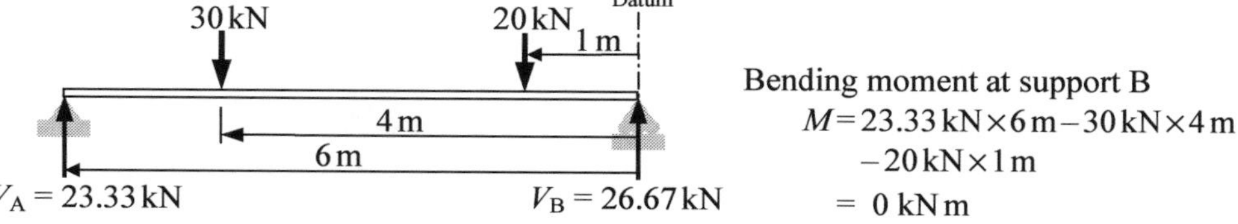

Bending moment at support B
$M = 23.33\,\text{kN} \times 6\,\text{m} - 30\,\text{kN} \times 4\,\text{m}$
$-20\,\text{kN} \times 1\,\text{m}$
$= 0\,\text{kN m}$

Having calculated the bending moments at 1-m intervals we can tabulate the values as shown in Figure 3.90.

Distance along beam from support A	Bending moment
0 m (at support A)	0 kN m
1 m	23.33 kN m
2 m	46.66 kN m
3 m	39.99 kN m
4 m	33.33 kN m
5 m	26.66 kN m
6 m (at support B)	0 kN m

Figure 3.90: Bending moments along beam – Example 3.27

Noting that the tension face is at the bottom of the beam, the bending moment diagram will therefore be that shown in Figure 3.91.

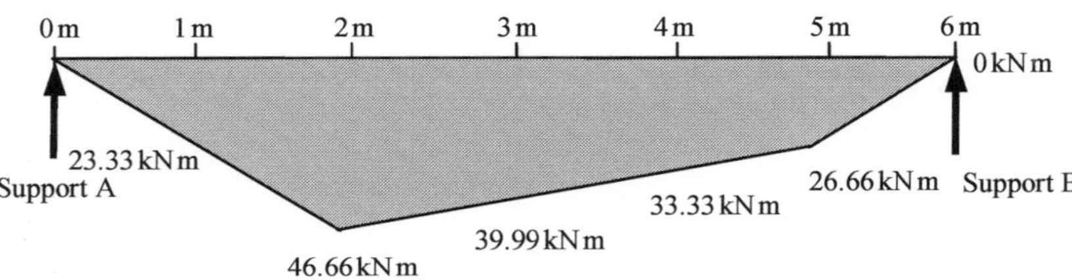

Figure 3.91: Bending moment diagram for beam – Example 3.27

◇

We will now consider an example of a simply supported beam with a cantilevered end.

◇

Example 3.28: Bending moment diagram for a simply supported beam with a cantilevered end carrying two point loads

For the beam shown in Figure 3.92, which is simply supported at A and B, determine:

(a) the reactions at the supports V_A, V_B and H_B;

and (b) the bending moments at 1-m intervals along the beam.

Plot the results of (b) in the form of a bending moment diagram.

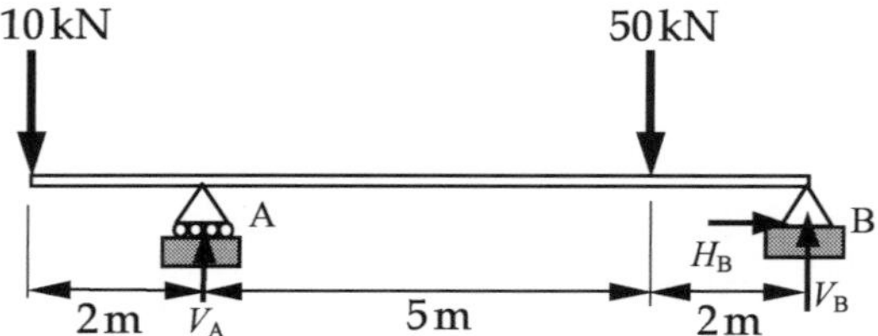

Figure 3.92: Beam system – Example 3.28

The reactions as previously calculated are: $H_A = 0$, $V_A = 27.143\,\text{kN}$ and $V_B = 32.857\,\text{kN}$.

We can now calculate the bending moments at 1-m intervals.

We commence our calculations at the left end of the beam, 2 m to the left of support A:

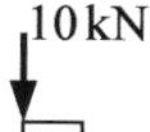

Bending moment $M = -10\,\text{kN} \times 0\,\text{m} = 0\,\text{kN m}$

Bending moment at 1 m from left end of beam:

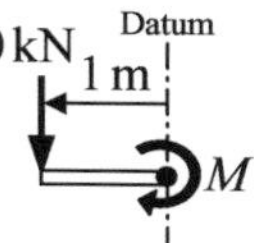

Bending moment $M = -10\,\text{kN} \times 1\,\text{m} = -10\,\text{kN m}$ (anticlockwise)

Bending moment at 2 m from left end of beam:

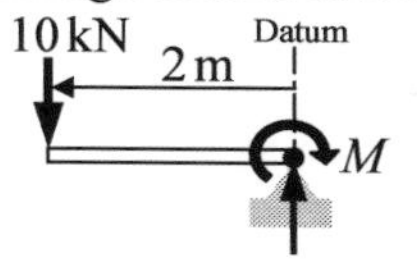

Bending moment $M = -10\,\text{kN} \times 2\,\text{m} = -20\,\text{kN m}$

Bending moment at 3 m from left end of beam:

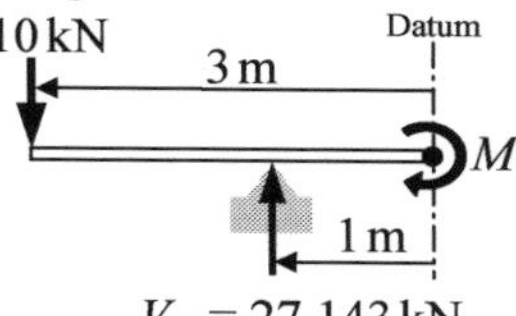

Bending moment $M = -10\,\text{kN} \times 3\,\text{m} + 27.143\,\text{kN} \times 1\,\text{m} = -2.857\,\text{kN m}$

Bending moment at 4 m from left end of beam:

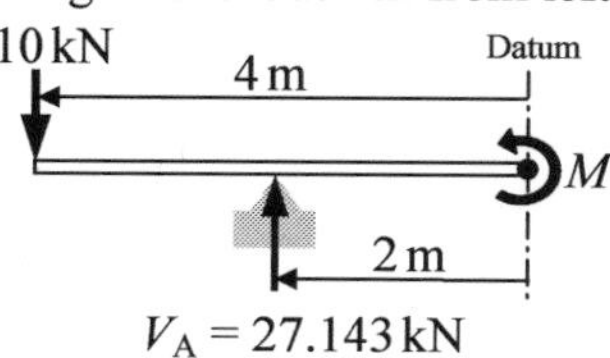

Bending moment

$$M = -10\,\text{kN} \times 4\,\text{m} + 27.143\,\text{kN} \times 2\,\text{m} = 16.143\,\text{kN m}$$

Bending moment at 5 m from left end of beam:

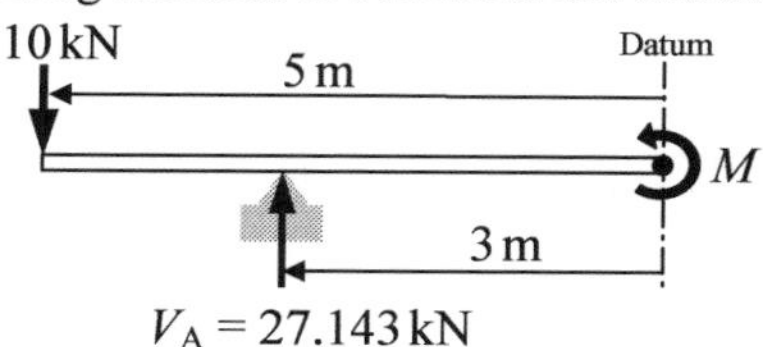

Bending moment

$$M = -10\,\text{kN} \times 5\,\text{m} + 27.143\,\text{kN} \times 3\,\text{m} = 31.429\,\text{kN m}$$

Bending moment at 6 m from left end of beam:

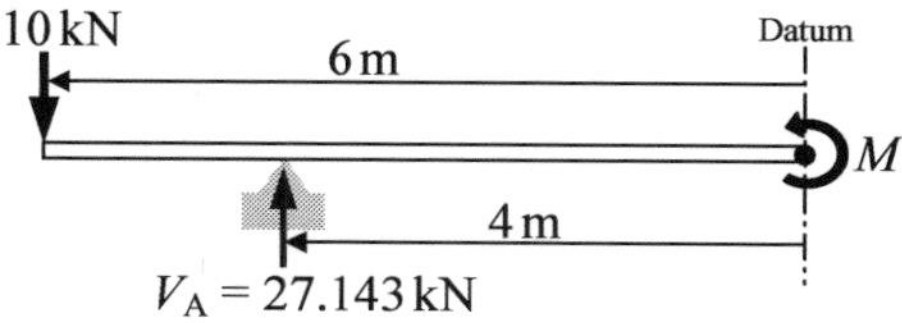

Bending moment

$$M = -10\,\text{kN} \times 6\,\text{m} + 27.143\,\text{kN} \times 4\,\text{m} = 48.572\,\text{kN m}$$

Bending moment at 7 m from the left end of the beam:

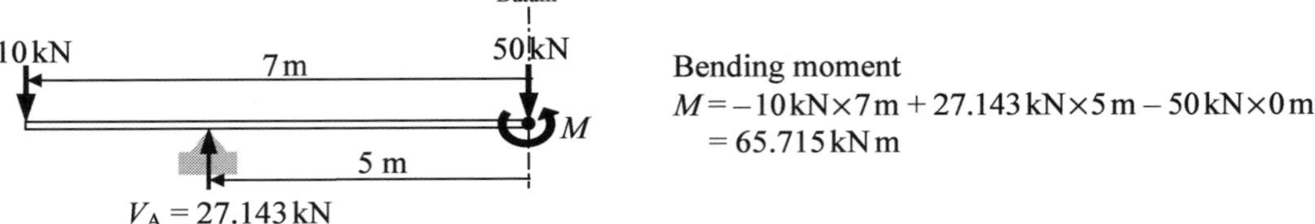

Bending moment at 8 m from the left end of the beam:

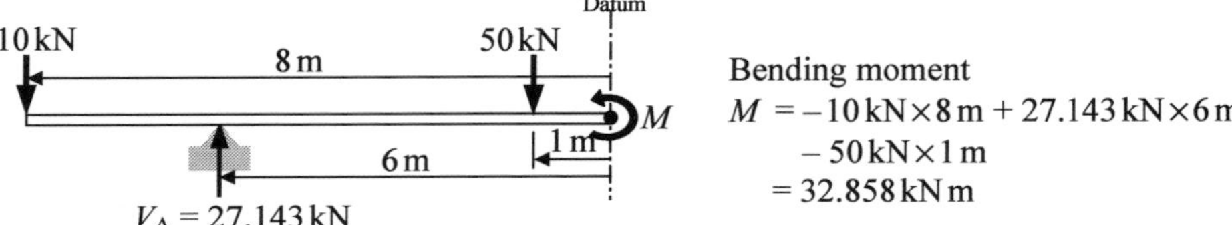

Bending moment at 9 m from support A, i.e. at support B:

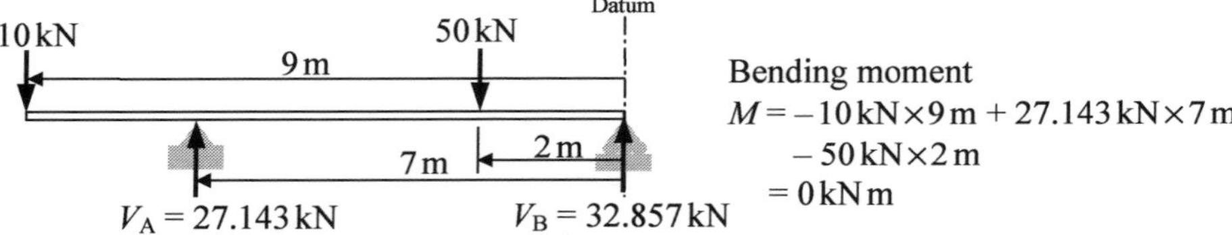

The bending moments for the beam of Example 3.28 are shown in Figure 3.93.

Distance along beam from left end	**Bending moment**
0 m	0 kN m
1 m	−10 kN m
2 m (at support A)	−20 kN m
3 m	−2.857 kN m
4 m	16.143 kN m
5 m	31.429 kN m
6 m	48.572 kN m
7 m	61.715 kN m
8 m	32.858 kN m
9 m (at support B)	0 kN m

Figure 3.93: Bending moment along beam – Example 3.28

Noting that, for the cantilever, the tension face is at the top of the beam. The bending moment diagram is shown in Figure 3.94. We also note that the maximum bending moment on the beam is 61.715 kN m and that it is located 7 m from the left end. With reference to Figure 3.70, we see that this corresponds to the point where the shear force passes through zero. It is useful to remember that maximum bending moment will always occur at the point where shear force is zero.

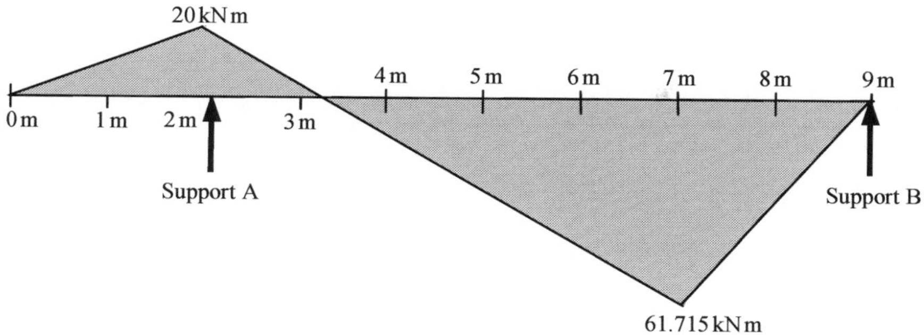

Figure 3.94: Bending moment diagram for beam – Example 3.28

Having constructed the bending moment diagrams for various simply supported beams carrying point loads we will now extend our studies to those beams supporting distributed loads (DLs).

Example 3.29: Bending moment diagram for a simply supported beam carrying a uniformly distributed load

For the beam shown in Figure 3.95, which is simply supported at A and B, determine:

(a) the reactions at the supports V_A, H_A and V_B;

and (b) the bending moments at 1-m intervals along the beam.

Plot the results of (b) in the form of a bending moment diagram.

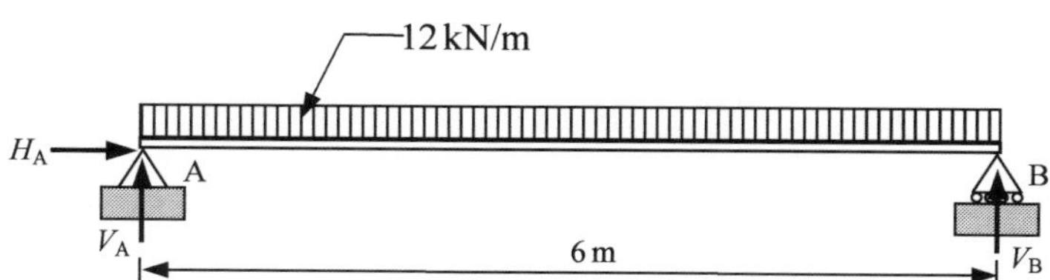

Figure 3.95: Beam supporting uniformly distributed load – Example 3.29

The reactions previously calculated are: $H_A = 0$, $V_A = 36\,\text{kN}$ and $V_B = 36\,\text{kN}$.

We commence our calculations at support A:

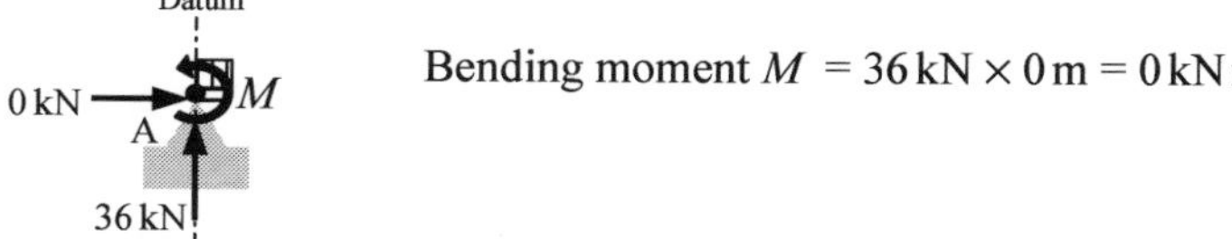

Bending moment $M = 36\,\text{kN} \times 0\,\text{m} = 0\,\text{kN m}$

Bending moment 1 m from support A:

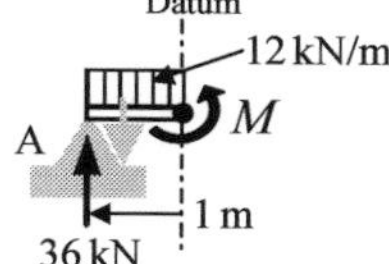

Bending moment $M = 36\,\text{kN} \times 1\,\text{m} - (12\,\text{kN/m} \times 1\,\text{m} \times 0.5\,\text{m}) = 30\,\text{kN m}$

(Note that the centroid of the UDL is located at 0.5 m from the datum.)

Bending moment 2 m from support A:

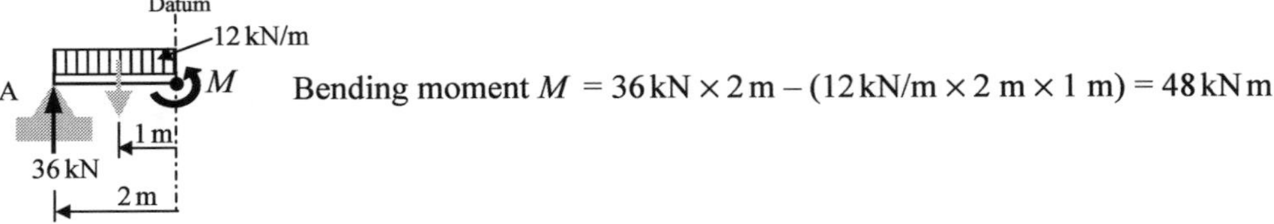

Bending moment $M = 36\,\text{kN} \times 2\,\text{m} - (12\,\text{kN/m} \times 2\,\text{m} \times 1\,\text{m}) = 48\,\text{kN m}$

Bending moment 3 m from support A:

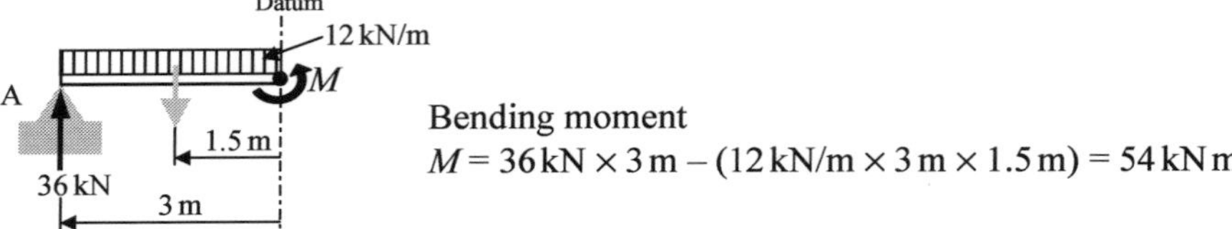

Bending moment
$M = 36\,\text{kN} \times 3\,\text{m} - (12\,\text{kN/m} \times 3\,\text{m} \times 1.5\,\text{m}) = 54\,\text{kN m}$

Bending moment 4 m from support A:

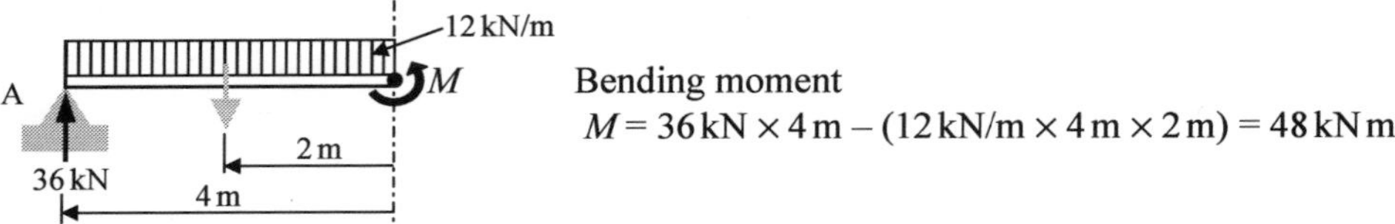

Bending moment
$M = 36\,\text{kN} \times 4\,\text{m} - (12\,\text{kN/m} \times 4\,\text{m} \times 2\,\text{m}) = 48\,\text{kN m}$

Bending moment 5 m from support A:

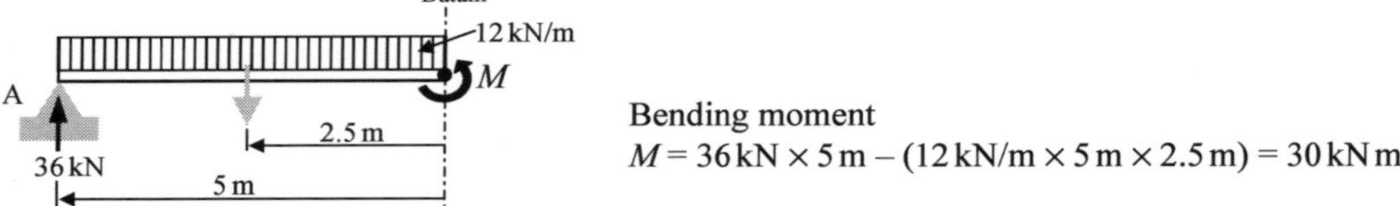

Bending moment
$M = 36\,\text{kN} \times 5\,\text{m} - (12\,\text{kN/m} \times 5\,\text{m} \times 2.5\,\text{m}) = 30\,\text{kN m}$

Bending moment 6 m from support A (at support B):

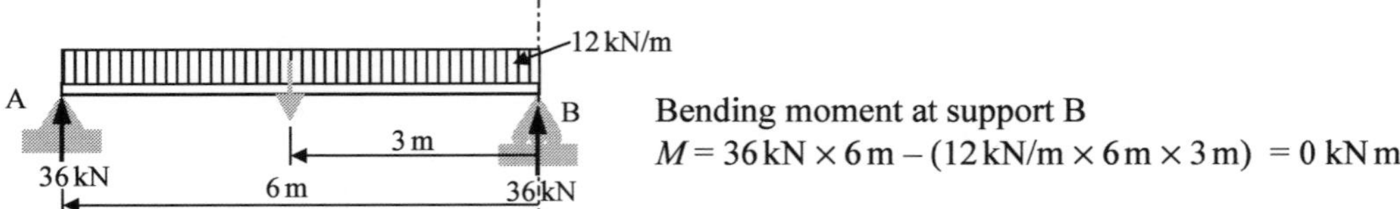

Bending moment at support B
$M = 36\,\text{kN} \times 6\,\text{m} - (12\,\text{kN/m} \times 6\,\text{m} \times 3\,\text{m}) = 0\,\text{kN m}$

Having calculated the bending moments we can record the results, at 1-m intervals, in a table as shown in Figure 3.96.

Distance along beam from support A	**Bending moment**
0 m (at support A)	0 kN m
1 m	30 kN m
2 m	48 kN m
3 m	54 kN m
4 m	48 kN m
5 m	30 kN m
6 m (at support B)	0 kN m

Figure 3.96: Bending moment along beam – Example 3.29

The bending moment diagram is shown in Figure 3.97.

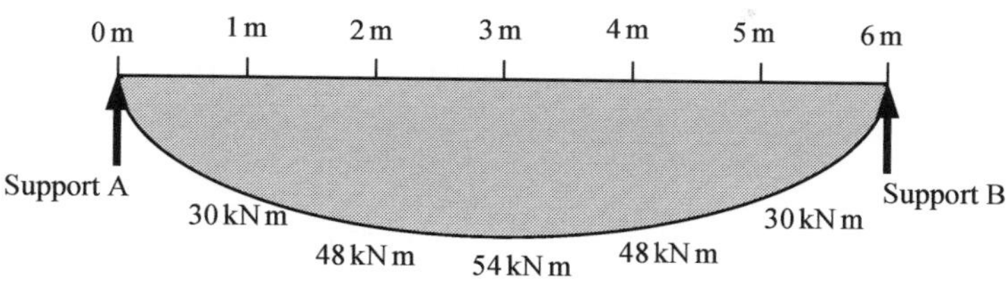

Figure 3.97: Bending moment diagram for beam – Example 3.29

For the UDL of Example 3.29, the bending moment will vary linearly along the beam resulting in a distribution in the curved (parabolic) form indicated in Figure 3.97. It will again not be necessary to calculate all the bending moments at all points along the beam as we will be able to interpolate the intermediate values. It is important to note the difference between this bending moment diagram and that for a point load. For the point load, the bending moment varies linearly and produces a straight line graph. For a uniformly distributed load, the bending moment will produce a parabolic plot.

Example 3.30 has two separate UDLs applied to the beam.

Example 3.30: Bending moment diagram for a simply supported beam carrying two distributed loads

For the beam shown in Figure 3.98, which is simply supported at A and B, determine:
(a) the reactions at the supports V_A, H_A and V_B;
and (b) the bending moments at 1-m intervals along the beam.
Plot the results of (b) in the form of a bending moment diagram.

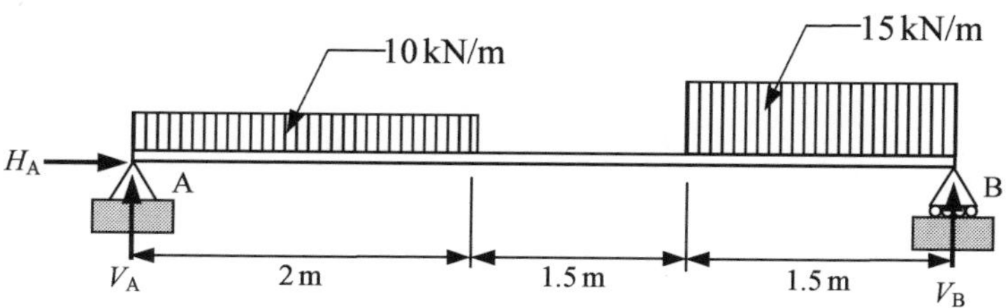

Figure 3.98: Beam supporting uniformly distributed load – Example 3.30

The reactions previously calculated are: $H_A = 0$, $V_A = 19.375$ kN and $V_B = 23.125$ kN.

We will calculate the bending moments at 1-m intervals as before; however we note that the loads are not spaced at increments of 1-m along the beam. It will therefore be necessary to calculate an extra point at 3.5 m from support A, in order to accurately plot the bending moment diagram.

We commence our calculations at support A:

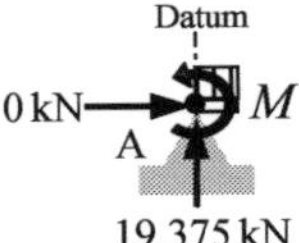

Bending moment $M = 19.375\,\text{kN} \times 0\,\text{m} = 0\,\text{kN m}$

Bending moment 1 m from support A:

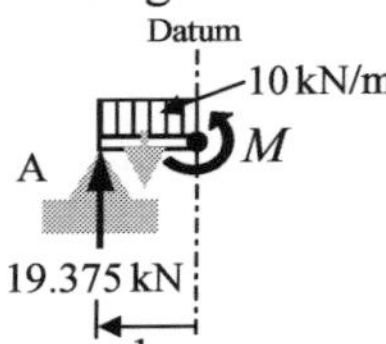

Bending moment $M = 19.375\,\text{kN} - (10\,\text{kN/m} \times 1\,\text{m} \times 0.5\,\text{m}) = 14.375\,\text{kN m}$

Bending moment 2 m from support A:

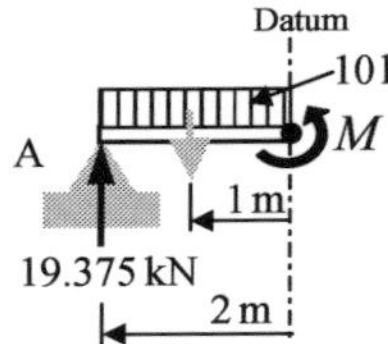

Bending moment $M = 19.375\,\text{kN} \times 2\,\text{m} - (10\,\text{kN/m} \times 2\,\text{m} \times 1\,\text{m}) = 28.75\,\text{kN m}$

Bending moment 3 m from support A:

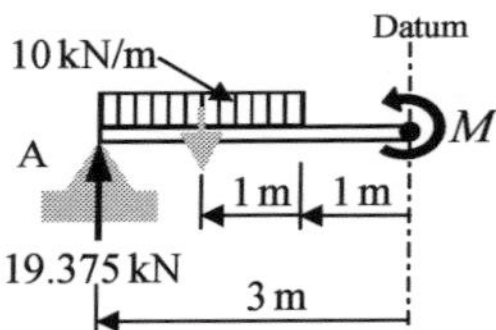

Bending moment
$M = 19.375\,\text{kN} \times 3\,\text{m} - (10\,\text{kN/m} \times 2\,\text{m} \times (1\,\text{m} + 1\,\text{m})) = 18.125\,\text{kN m}$

Bending moment 3.5 m from support A:

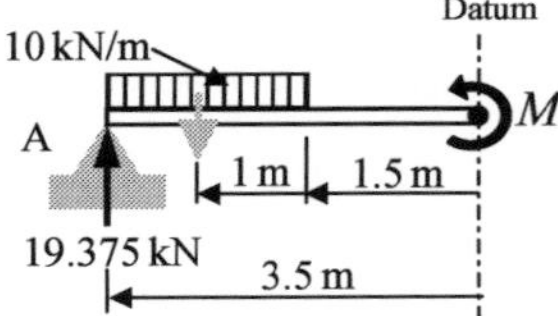

Bending moment
$M = 19.375\,\text{kN} \times 3.5\,\text{m} - (10\,\text{kN/m} \times 2\,\text{m} \times (1\,\text{m} + 1.5\,\text{m}))$
$= 17.8125\,\text{kN m}$

Bending moment 4 m from support A:

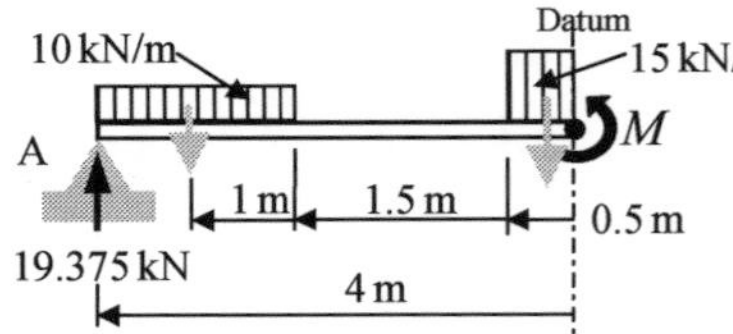

Bending moment
$M = 19.375\,\text{kN} \times 4\,\text{m}$
$- (10\,\text{kN/m} \times 2\,\text{m} \times (1\,\text{m} + 1.5\,\text{m} + 0.5\,\text{m}))$
$- (15\,\text{kN/m} \times 0.5\,\text{m} \times 0.25\,\text{m}) = 15.625\,\text{kN m}$

Bending moment 5 m from support A (at support B):

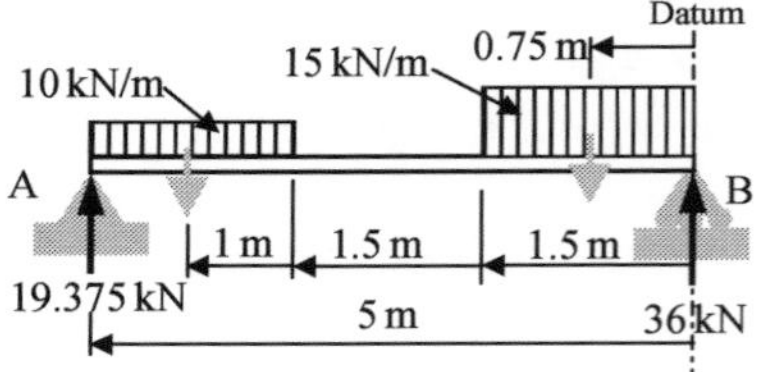

Bending moment (at support B)
$M = 19.375\,\text{kN} \times 5\,\text{m}$
$- (10\,\text{kN/m} \times 2\,\text{m} \times (1\,\text{m} + 1.5\,\text{m} + 1.5\,\text{m}))$
$- (15\,\text{kN/m} \times 1.5\,\text{m} \times 0.75\,\text{m}) = 0\,\text{kN m}$

The calculated bending moments are tabulated in the table of Figure 3.99.

Distance along beam from support A	**Bending moment**
0 m (at support A)	0 kN m
1 m	14.375 kN m
2 m	28.75 kN m
3 m	18.125 kN m
3.5 m	17.8125 kN m
4 m	15.625 kN m
5 m (at support B)	0 kN m

Figure 3.99: Bending moments along beam – Example 3.30

The bending moment diagram will be that shown in Figure 3.100.

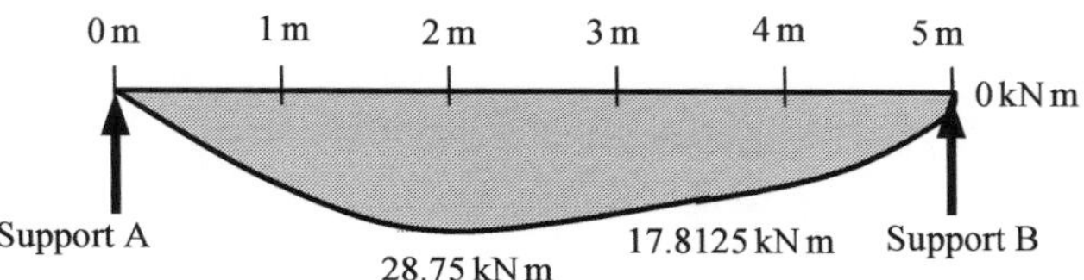

Figure 3.100: Bending moment diagram for beam – Example 3.30

We will now consider beams with a combination of point and uniformly distributed loads.

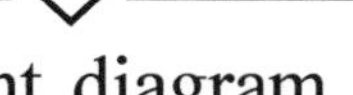

Example 3.31: Bending moment diagram for a simply supported beam carrying uniformly distributed and point loads

For the beam shown in Figure 3.101, which is simply supported at A and B:
(a) determine the reactions at the supports V_A, H_A and V_B;
and (b) plot the bending moment diagram for the beam.

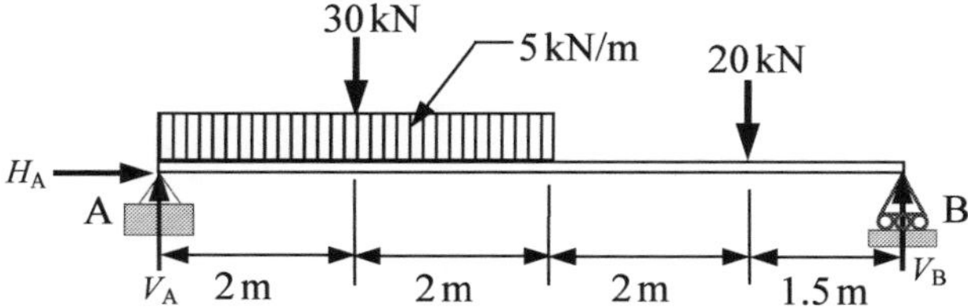

Figure 3.101: Beam system – Example 3.31

The reactions are as previously calculated: $H_A = 0$, $V_A = 40.67$ kN and $V_B = 29.33$ kN.

We shall now calculate the bending moments along the beam, ensuring that we select enough points to produce an accurate plot.

We commence our calculations at the support A:

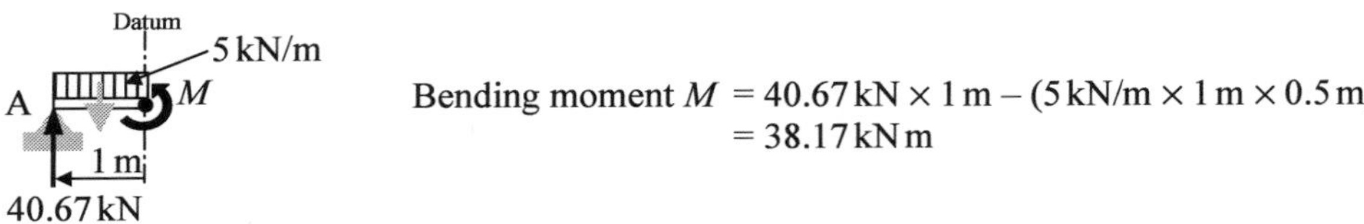

Bending moment $M = 40.67\,\text{kN} \times 0\,\text{m} = 0\,\text{kN m}$

Bending moment at 1 m from support A:

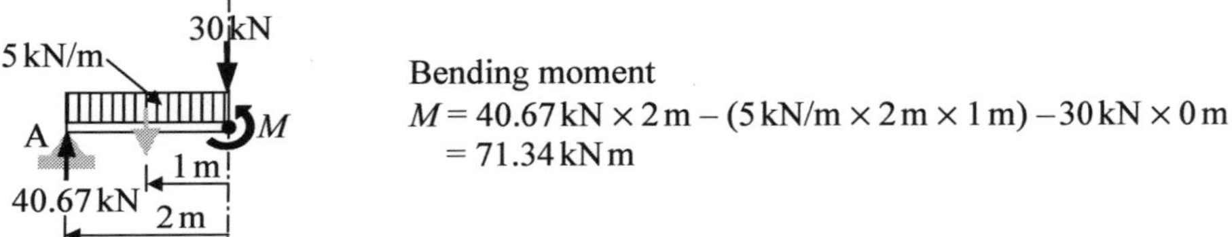

Bending moment $M = 40.67\,\text{kN} \times 1\,\text{m} - (5\,\text{kN/m} \times 1\,\text{m} \times 0.5\,\text{m})$
$= 38.17\,\text{kN m}$

Bending moment at 2 m from support A:

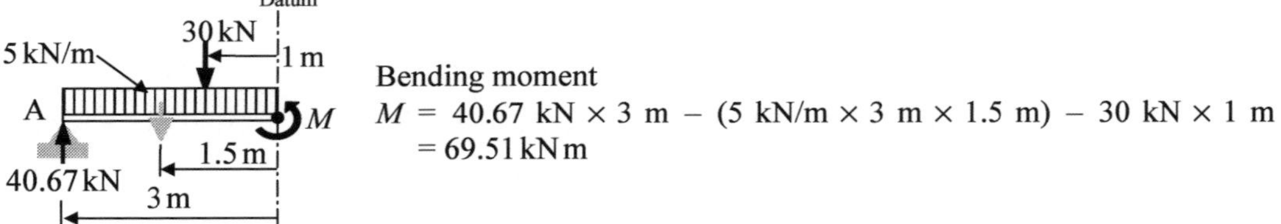

Bending moment
$M = 40.67\,\text{kN} \times 2\,\text{m} - (5\,\text{kN/m} \times 2\,\text{m} \times 1\,\text{m}) - 30\,\text{kN} \times 0\,\text{m}$
$= 71.34\,\text{kN m}$

Bending moment at 3 m from support A:

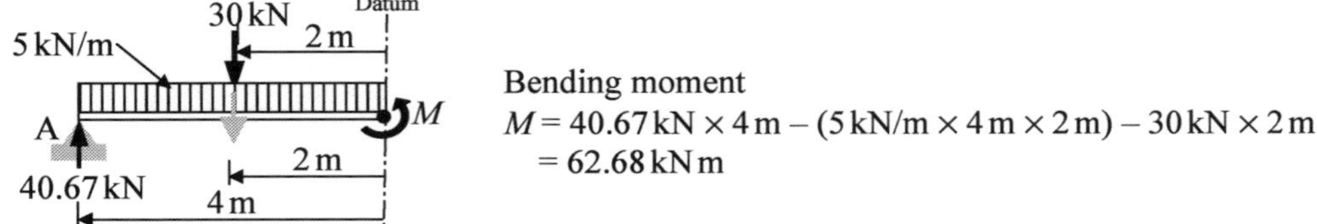

Bending moment
$M = 40.67\,\text{kN} \times 3\,\text{m} - (5\,\text{kN/m} \times 3\,\text{m} \times 1.5\,\text{m}) - 30\,\text{kN} \times 1\,\text{m}$
$= 69.51\,\text{kN m}$

Bending moment at 4 m from support A:

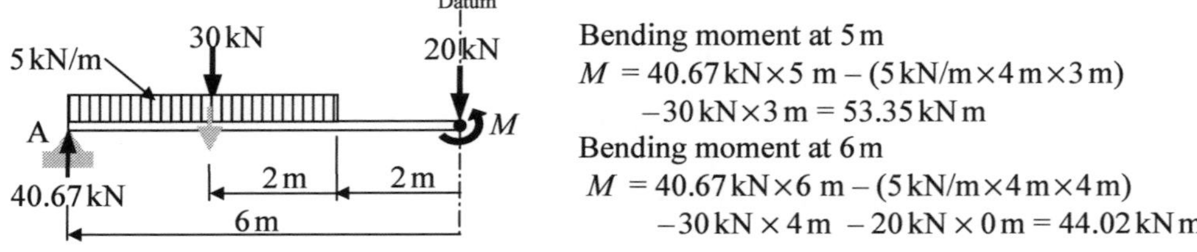

Bending moment
$M = 40.67\,\text{kN} \times 4\,\text{m} - (5\,\text{kN/m} \times 4\,\text{m} \times 2\,\text{m}) - 30\,\text{kN} \times 2\,\text{m}$
$= 62.68\,\text{kN m}$

Bending moment at 5 m and at 6 m from support A:

Datum

30 kN 20 kN

5 kN/m

A M

40.67 kN 2 m 2 m

6 m

Bending moment at 5 m
$M = 40.67\,\text{kN} \times 5\,\text{m} - (5\,\text{kN/m} \times 4\,\text{m} \times 3\,\text{m})$
$-30\,\text{kN} \times 3\,\text{m} = 53.35\,\text{kN m}$

Bending moment at 6 m
$M = 40.67\,\text{kN} \times 6\,\text{m} - (5\,\text{kN/m} \times 4\,\text{m} \times 4\,\text{m})$
$-30\,\text{kN} \times 4\,\text{m} - 20\,\text{kN} \times 0\,\text{m} = 44.02\,\text{kN m}$

Bending moment at 7.5 m from support A, i.e. at support B:

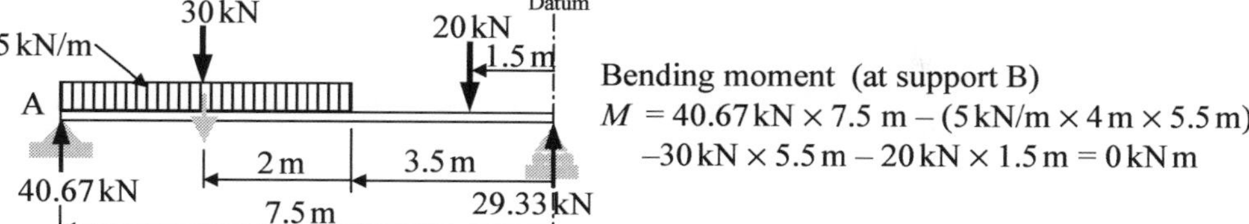

Bending moment (at support B)

$$M = 40.67\,\text{kN} \times 7.5\text{ m} - (5\,\text{kN/m} \times 4\,\text{m} \times 5.5\,\text{m})$$
$$-30\,\text{kN} \times 5.5\,\text{m} - 20\,\text{kN} \times 1.5\,\text{m} = 0\,\text{kN m}$$

The final results are shown in Figure 3.102.

Distance along beam from support A	**Bending moment**
0 m (at support A)	0 kN m
1 m	38.17 kN m
2 m	71.34 kN m
3 m	69.51 kN m
4 m	62.68 kN m
5 m	53.35 kN m
6 m	44.02 kN m
7.5 m (at support B)	0 kN m

Figure 3.102: Bending moment along beam – Example 3.31

The bending moment diagram will be that shown in Figure 3.103.

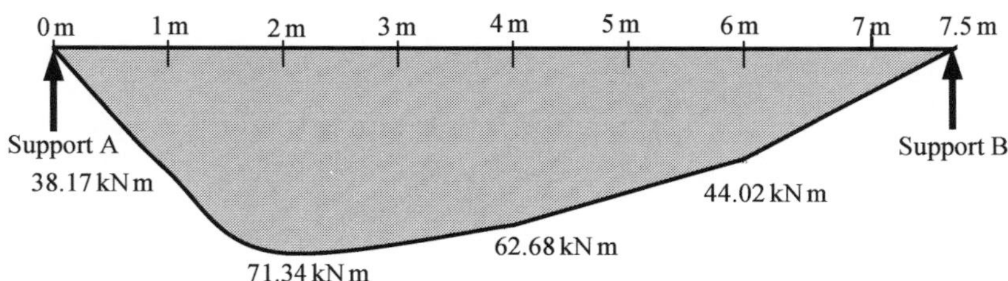

Figure 3.103: Bending moment diagram for beam – Example 3.31

◇

Example 3.32: Bending moment diagram for a simply supported beam carrying distributed and point loads

For the beam shown in Figure 3.104, which is simply supported at A and B:

(a) determine the reactions at the supports V_A, V_B and H_B;

and (b) plot the bending moment diagram for the beam.

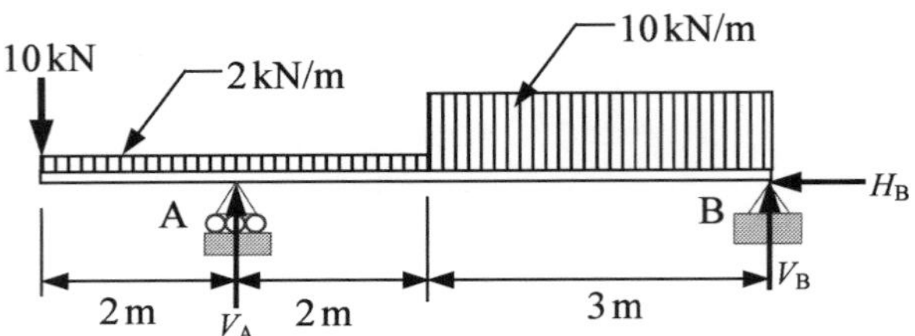

Figure 3.104: Beam system – Example 3.32

The reactions for this system are: $H_A = 0$, $V_A = 31\,kN$ and $V_B = 17\,kN$.

We can now calculate the bending moment, in this case at 1-m intervals, as before.

We commence calculations at the left end of the beam, 2 m to the left of support A:

10 kN

M

Bending moment $= -10\,kN \times 0\,m = 0\,kN\,m$

Bending moment at 1 m from left end of beam:

Datum

10 kN 1 m 2 kN/m

M

Bending moment
$M = -10\,kN \times 1\,m - (2\,kN/m \times 1\,m \times 0.5\,m) = -11\,kN\,m$

Bending moment at 2 m from left end of beam:

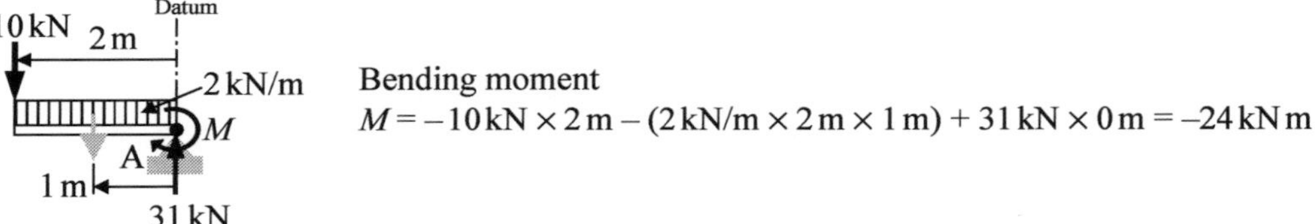

Bending moment
$M = -10\,kN \times 2\,m - (2\,kN/m \times 2\,m \times 1\,m) + 31\,kN \times 0\,m = -24\,kN\,m$

Bending moment at 3 m from left end of beam:

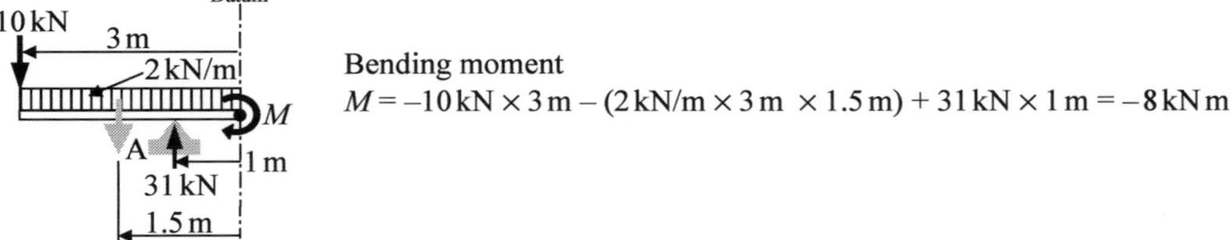

Bending moment
$M = -10\,kN \times 3\,m - (2\,kN/m \times 3\,m \times 1.5\,m) + 31\,kN \times 1\,m = -8\,kN\,m$

Bending moment at 4 m from left end of beam:

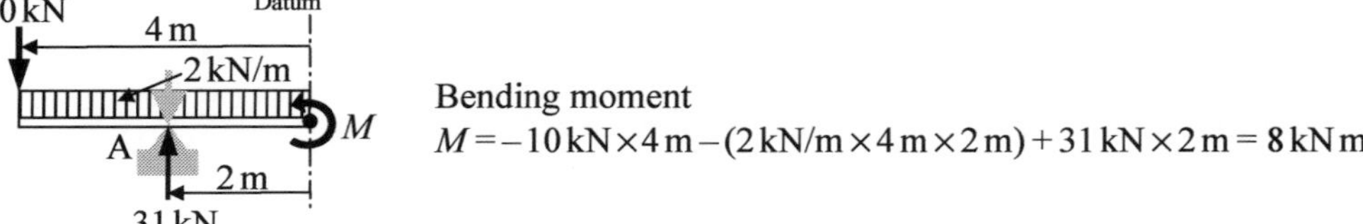

Bending moment
$M = -10\,kN \times 4\,m - (2\,kN/m \times 4\,m \times 2\,m) + 31\,kN \times 2\,m = 8\,kN\,m$

Bending moment at 5 m from left end of beam:

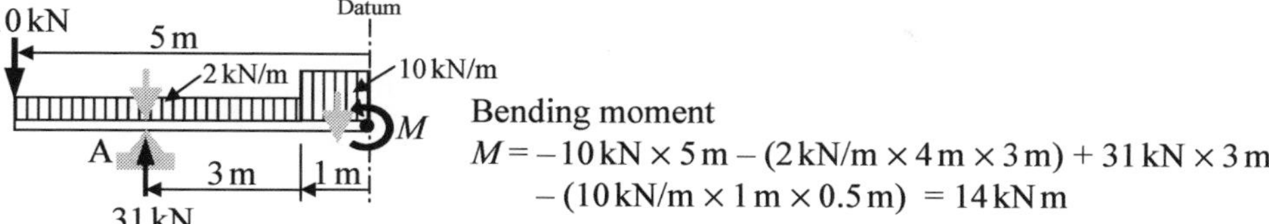

Bending moment

$M = -10\,\text{kN} \times 5\,\text{m} - (2\,\text{kN/m} \times 4\,\text{m} \times 3\,\text{m}) + 31\,\text{kN} \times 3\,\text{m}$
$\quad - (10\,\text{kN/m} \times 1\,\text{m} \times 0.5\,\text{m}) = 14\,\text{kN m}$

Bending moment at 6 m from left end of beam:

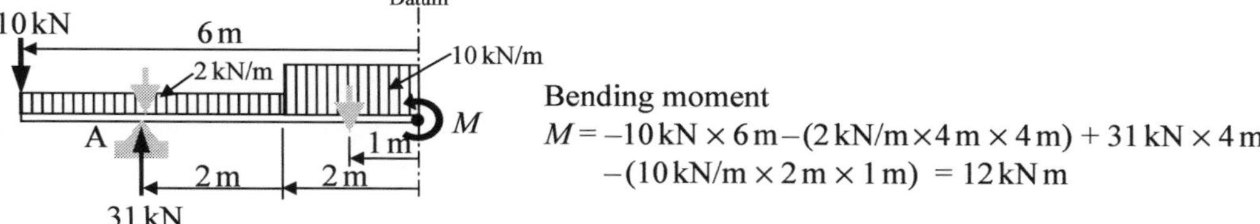

Bending moment

$M = -10\,\text{kN} \times 6\,\text{m} - (2\,\text{kN/m} \times 4\,\text{m} \times 4\,\text{m}) + 31\,\text{kN} \times 4\,\text{m}$
$\quad - (10\,\text{kN/m} \times 2\,\text{m} \times 1\,\text{m}) = 12\,\text{kN m}$

Bending moment at 7 m from the left end of the beam (at support B):

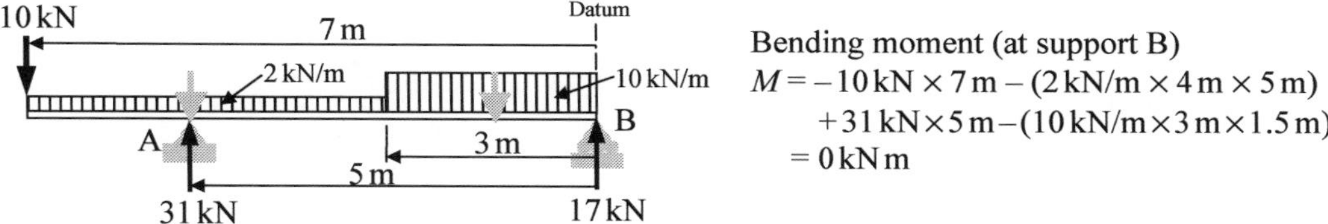

Bending moment (at support B)

$M = -10\,\text{kN} \times 7\,\text{m} - (2\,\text{kN/m} \times 4\,\text{m} \times 5\,\text{m})$
$\quad + 31\,\text{kN} \times 5\,\text{m} - (10\,\text{kN/m} \times 3\,\text{m} \times 1.5\,\text{m})$
$\quad = 0\,\text{kN m}$

The bending moments for the beam of Example 3.32 are tabulated in Figure 3.105.

Distance along beam from support A	**Bending moment**
0 m	0 kN m
1 m	−11 kN m
2 m (at support A)	−24 kN m
3 m	−8 kN m
4 m	8 kN m
5 m	14 kN m
6 m	12 kN m
7 m (at support B)	0 kN m

Figure 3.105: Bending moment along beam – Example 3.32

The bending moment diagram is shown in Figure 3.106.

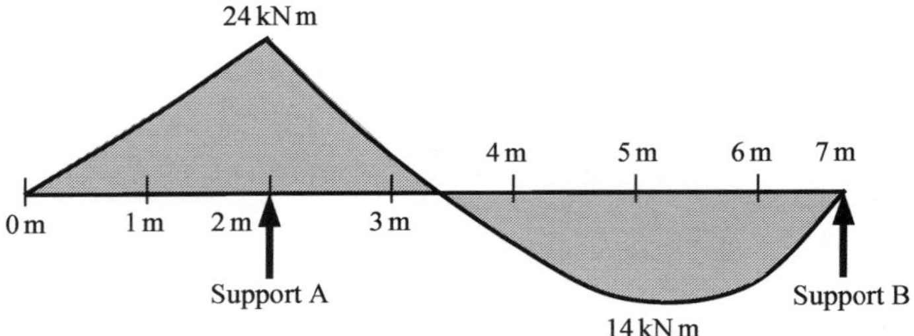

Figure 3.106: Bending moment diagram for beam – Example 3.32

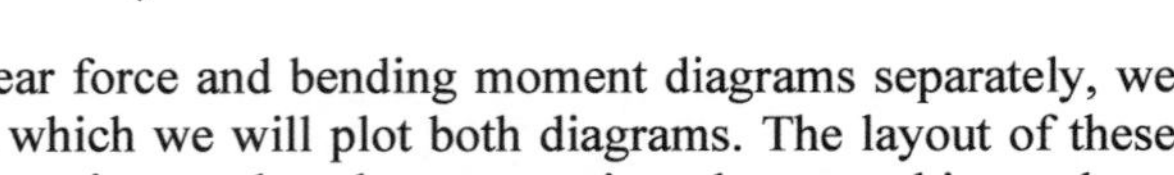

Having discussed the plotting of shear force and bending moment diagrams separately, we will now look at some further examples in which we will plot both diagrams. The layout of these problems is important especially as the information needs to be communicated as unambiguously as possible. The layout of the following examples is a standard format which it is suggested should be adopted for all such problems.

Example 3.33: Simply supported cantilevered beam

For the beam of Example 3.33 shown below:

(a) calculate the reactions at the supports V_A, V_B and H_B;

(b) plot the shear force and bending moment diagrams;

(c) determine the position and magnitude of the maximum bending moment.

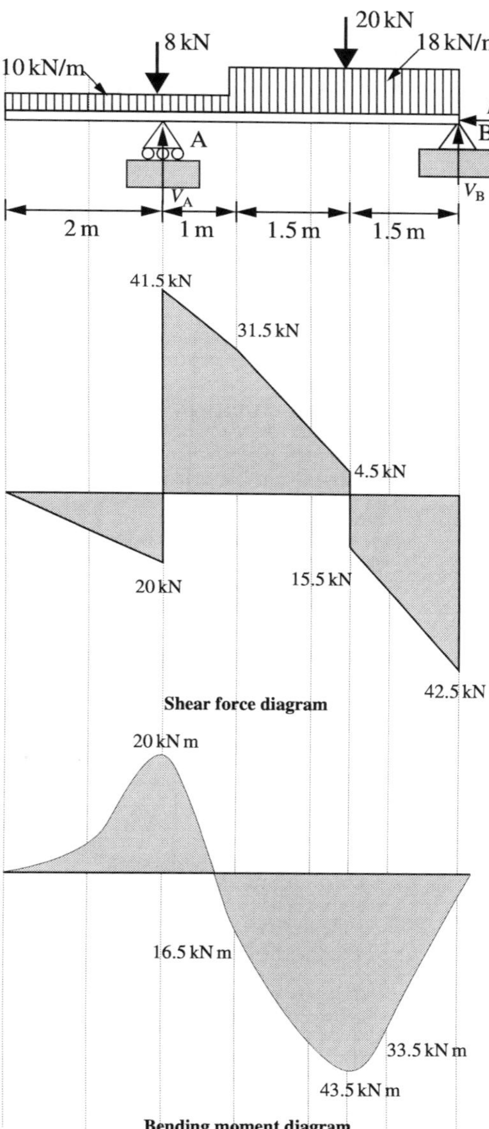

Calculate reactions:

$\Sigma F_x \quad H_B = 0\,\text{kN}$

$$\Sigma F_y \quad V_A + V_B - (10\,\text{kN/m} \times 3\,\text{m}) - 8\,\text{kN} - (18\,\text{kN/m} \times 3\,\text{m}) - 20\,\text{kN} = 0\,\text{kN}$$

$$V_A + V_B - 30 - 8 - 54 - 20 = 0\,\text{kN}$$

$$V_A + V_B = 112\,\text{kN} \qquad [1]$$

Take moments about support B:

$$\Sigma M_z = -(18\,\text{kN/m} \times 3\,\text{m} \times 3\,\text{m}/2) - (20\,\text{kN} \times 1.5\,\text{m}) - (10\,\text{kN/m} \times 3\,\text{m} \times 4.5\,\text{m}) - (8\,\text{kN} \times 4\,\text{m}) + V_A \times 4\,\text{m} = 0\,\text{kN m}$$

$$4\,V_A = 278 \quad \therefore \quad V_A = 69.5\,\text{kN}$$

Back-substituting into [1] gives $V_B = 42.5\,\text{kN}$

Calculate shear force along beam:

Distance from left end		Shear force (kN)
0 m	0	0
1 m	–10 kN/m × 1 m	–10
2 m	–10 kN/m × 2 m	–20
	–10 kN/m × 2 m – 8 kN + 69.5 kN	41.5
3 m	–10 kN/m × 3 m – 8 kN + 69.5 kN	31.5
4 m	–10 kN/m × 3 m – 8 kN + 69.5 kN –18 kN/m × 1 m	13.5
4.5 m	–10 kN/m × 3 m – 8 kN + 69.5 kN –18 kN/m × 1.5 m	4.5
	–10 kN/m × 3 m – 8 kN + 69.5 kN –18 kN/m × 1.5 m – 20 kN	15.5
5 m	–10 kN/m × 3 m – 8 kN + 69.5 kN –18 kN/m × 2 m – 20 kN	24.5
6 m	–10 kN/m × 3 m – 8 kN + 69.5 kN –18 kN/m × 3 m – 20 kN	42.5
	–10 kN/m × 3 m – 8 kN + 69.5 kN –18 kN/m × 3 m – 20 kN + 42.5 kN	0

Calculate bending moments along beam:

Distance from left end		Bending moment (kN m)
0 m	0	0
1 m	–10 kN/m × 1 m × 0.5 m	–5
2 m	–10 kN/m × 2 m ×1 m	–20
3 m	–10 kN/m × 3 m × 1.5 m – 8 kN × 1 m +69.5 kN/m × 1 m	16.5
4 m	–10 kN/m × 3 m × 2.5 m – 8 kN × 2 m +69.5 kN/m × 2 m – 18 kN/m × 1 m × 0.5 m	39
4.5 m	–10 kN/m × 3 m × 3 m – 8 kN × 2.5 m +69.5 kN/m × 2.5 m – 18 kN/m × 1.5 m × 0.75 m	43.5
5 m	–10 kN/m × 3 m × 3.5 m – 8 kN × 3 m +69.5 kN/m × 3 m – 18 kN/m × 2 m × 1 m –20 kN × 0.5 m	33.5
6 m	–10 kN/m × 3 m × 4.5 m – 8 kN × 4 m +69.5 kN/m × 4 m – 18 kN/m × 3 m × 1.5 m –20 kN × 1.5 m	0

Maximum bending moment will occur 4.5 m from left end of beam = 43.5 kN m

Example 3.36: Cantilevered beam

For the beam of Example 3.36 shown below:

(a) calculate the reactions at the supports; V_A, H_A and M_A;

(b) plot the shear force and bending moment diagrams;

(c) determine the position and magnitude of the maximum bending moment.

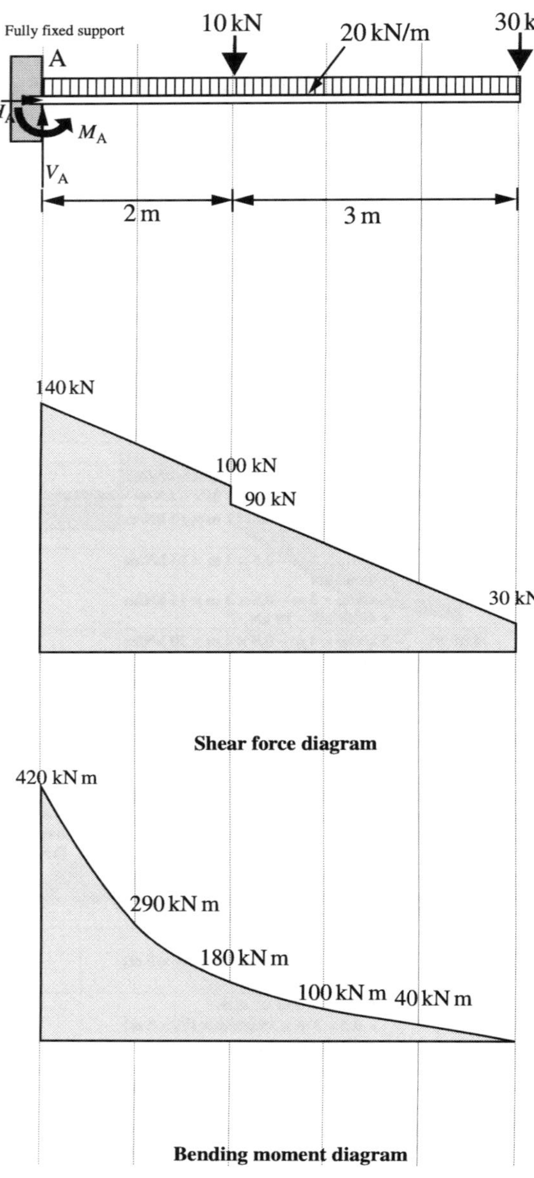

Calculate reactions:

$\Sigma F_x \quad H_A = 0\,\text{kN}$

$\Sigma F_y \quad V_A - (20\,\text{kN/m} \times 5\,\text{m}) - 10\,\text{kN} - 30\,\text{kN} = 0\,\text{kN}$

$V_A - 100 - 10 - 30 = 0\,\text{kN}$

$V_A = 140\,\text{kN}$

Take moments about support A:

$\Sigma M_z = -(20\,\text{kN/m} \times 5\,\text{m} \times 5\,\text{m/2}) - (10\,\text{kN} \times 2\,\text{m}) - (30\,\text{kN} \times 5\,\text{m}) - M_A = 0\,\text{kN m}$

$M_A = 420\,\text{kN m}$

Calculate shear force along beam:

Distance from support		Shear force (kN)
0 m	0 kN 140 kN	0 140
1 m	140 kN – 20 kN/m × 1 m	120
2 m	140 kN – 20 kN/m × 2 m 140 kN – 20 kN/m × 2 m – 10 kN	100 90
3 m	140 kN – 20 kN/m × 3 m – 10 kN	70
4 m	140 kN – 20 kN/m × 5 m – 10 kN	50
5 m	140 kN – 20 kN/m × 5 m – 10 kN 140 kN – 20 kN/m × 5 m – 10 kN – 30 kN	30 0

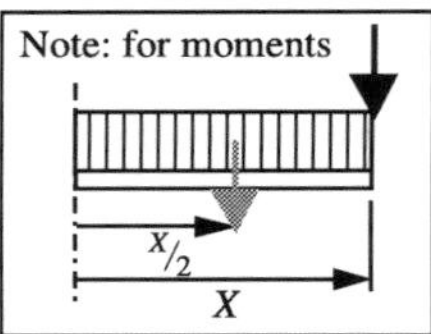

Calculate bending moments along beam:

Distance from *free* end		Bending moment (kN m)
0 m	0	0
1 m	30 kN × 1 m + 20 kN/m × 1 m × 0.5 m	40
2 m	30 kN × 2 m + 20 kN/m × 2 m × 1 m	100
3 m	30 kN × 3 m + 20 kN/m × 3 m × 1.5 m	180
4 m	30 kN × 4 m + 20 kN/m × 4 m × 2 m + 10 kN × 1 m	290
5 m	30 kN × 5 m + 20 kN/m × 5 m × 2.5 m + 10 kN × 2 m	420

Maximum bending moment will occur at support A= 420 kN m

Example 3.37: Cantilevered beam

For the beam of Example 3.37 shown below:
(a) calculate the reactions at the supports V_A, H_A and M_A;
(b) plot the shear force and bending moment diagrams;
(c) determine the position and magnitude of the maximum bending moment.

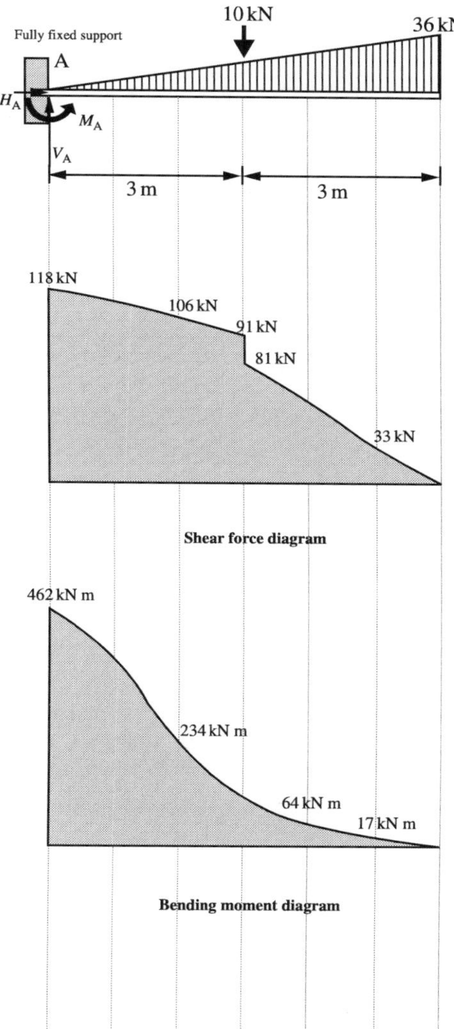

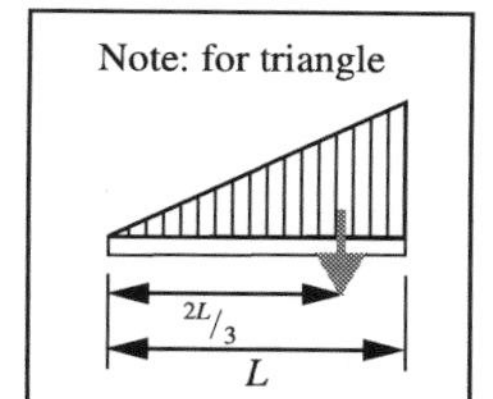

Calculate reactions:

$\Sigma F_x \quad H_A = 0$ kN

$\Sigma F_y \quad V_A - 10\text{ kN} - (0.5 \times 6\text{ m} \times 36\text{ kN/m}) = 0\text{ kN}$

$V_A - 10 - 108 = 0$ kN

$V_A = 118$ kN

Take moments about support A:

$\Sigma M_z = -(10\text{ kN} \times 3\text{ m}) + (0.5 \times 6\text{ m} \times 36\text{ kN/m} \times \tfrac{2}{3} \times 6\text{ m}) - M_A = 0\text{ kN m}$

$M_A = 462$ kN m

We note that the triangular load varies at a rate of 6 kN per metre length (i.e 6 kN/m@1 m, 12 kN/m@2 m, 18 kN/m@3 m etc.)

Calculate shear force along beam:

Distance from support		Shear force (kN)
0 m	0 kN 118 kN	0 118
1 m	118 kN – 0.5 × 1 m × 6 kN/m	115
2 m	118 kN – 0.5 × 2 m × 12 kN/m	106
3 m	118 kN – 0.5 × 3 m × 18 kN/m 118 kN – 0.5 × 3 m × 18 kN/m – 10 kN	91 81
4 m	118 kN – 0.5 × 4 m × 24 kN/m – 10 kN	60
5 m	118 kN – 0.5 × 5 m × 30 kN/m – 10 kN	33
6 m	118 kN – 0.5 × 6 m × 36 kN/m – 10 kN	0

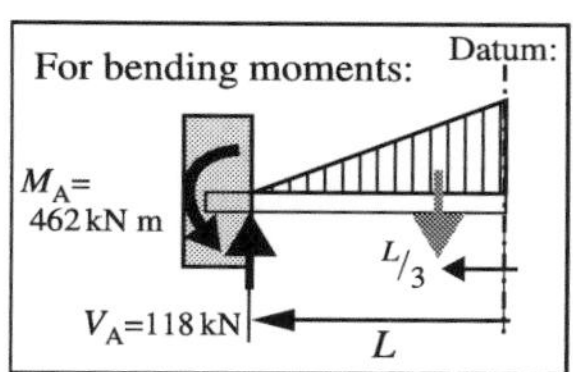

Calculate bending moments along beam:

Distance from support		Bending moment (kN m)
0 m	– 462 kN m	– 462
1 m	– 462 kN m + 0.5 × 1 m × 6 kN/m × (⅓ × 1 m) + 118 kN × 1 m	– 344
2 m	– 462 kN m + 0.5 × 2 m × 12 kN/m × (⅓ × 2 m) + 118 kN × 2	– 234
3 m	– 462 kN m + 0.5 × 3 m × 18 kN/m × (⅓ × 3 m) + 118 kN × 3 m	– 135
4 m	– 462 kN m + 0.5 × 4 m × 24 kN/m × (⅓ × 4 m) + 118 kN × 4 m – 10 kN × 1 m	–64
5 m	– 462 kN m + 0.5 × 5 m × 30 kN/m × (⅓ × 5 m) + 118 kN × 5 m – 10 kN × 2 m	– 17
6 m	– 462 kN m + 0.5 × 6 m × 36 kN/m × (⅓ × 6 m) + 118 kN × 6 m – 10 kN × 3 m	0

Maximum bending moment will occur at the support = 462 kN m

Having completed this chapter you should be able to:

- ***Identify different loading configurations***
- ***Identify different types of support***
- ***Determine if a structure is statically determinate***
- ***Calculate beam reactions for simply supported and cantilever beams***
- ***Draw shear force diagrams for simply supported and cantilever beams***
- ***Draw bending moment diagrams for simply supported and cantilever beams***

Further Problems 3

For each of the beams shown:

(a) calculate the horizontal and vertical reactions at the supports;

(b) plot the shear force and bending moment diagrams;

(c) determine the position and magnitude of the maximum bending moment.

1.

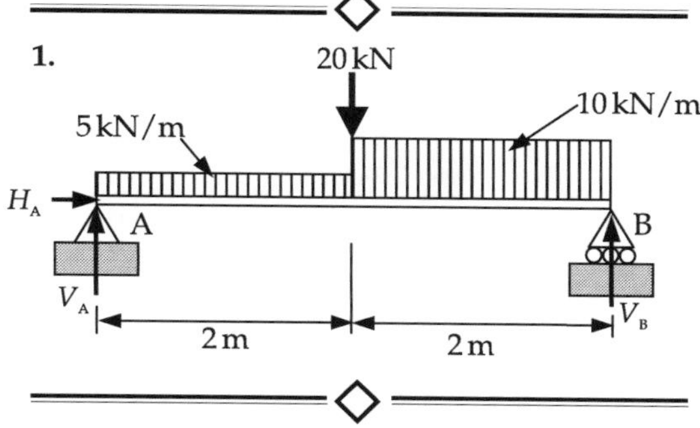

2.

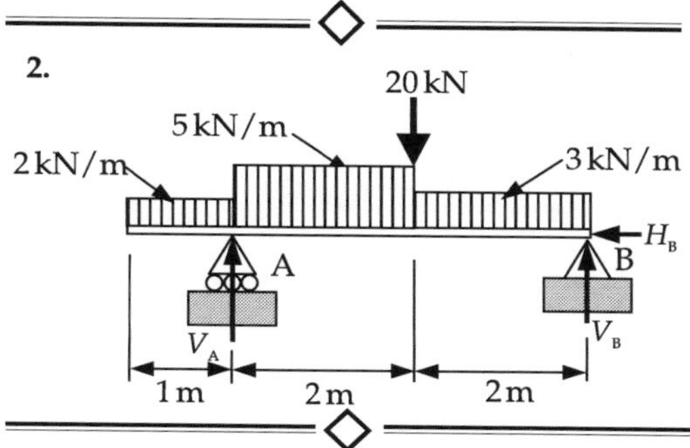

3.

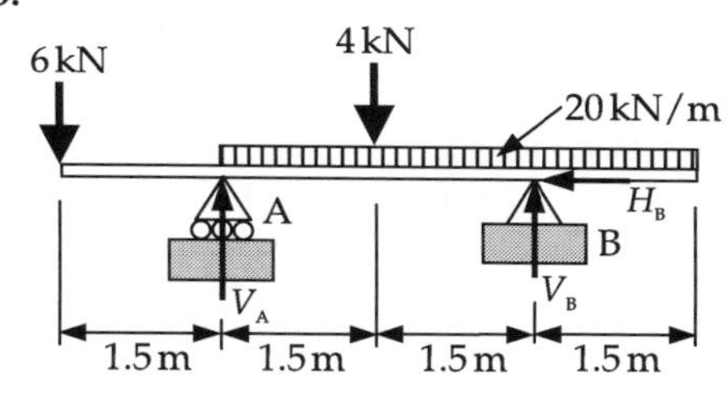

4.

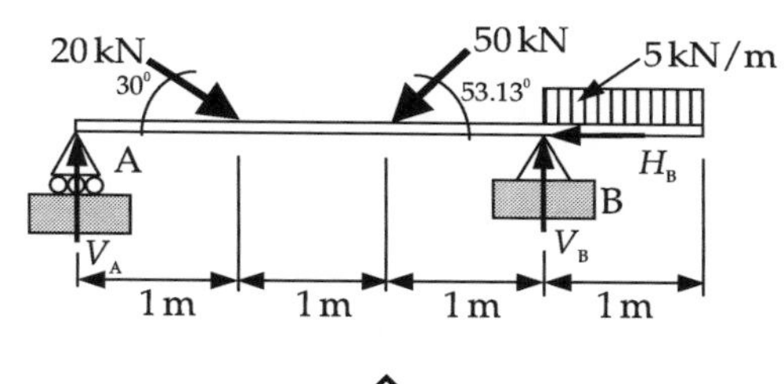

5.

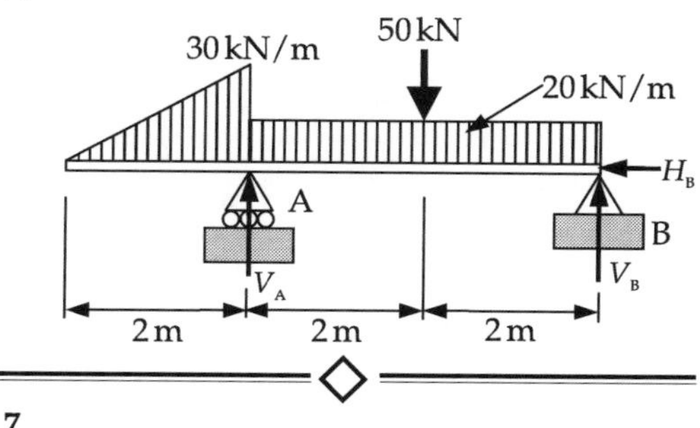

6.

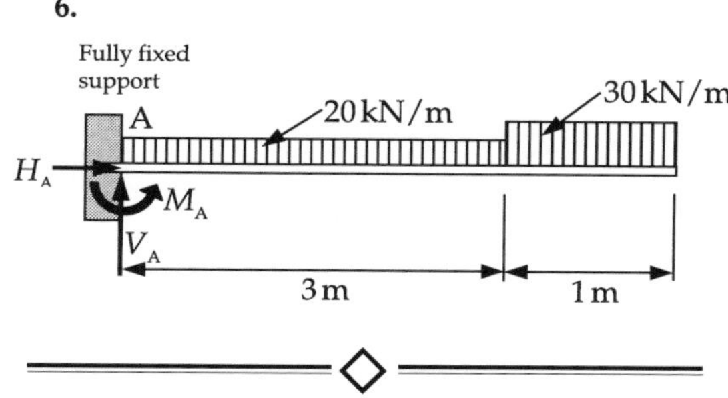

7.

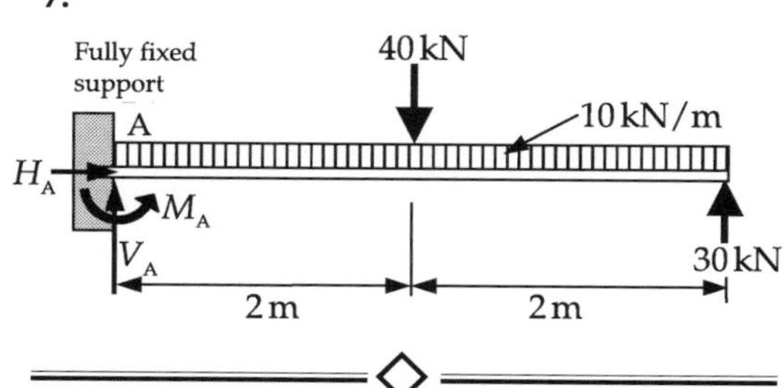

8.

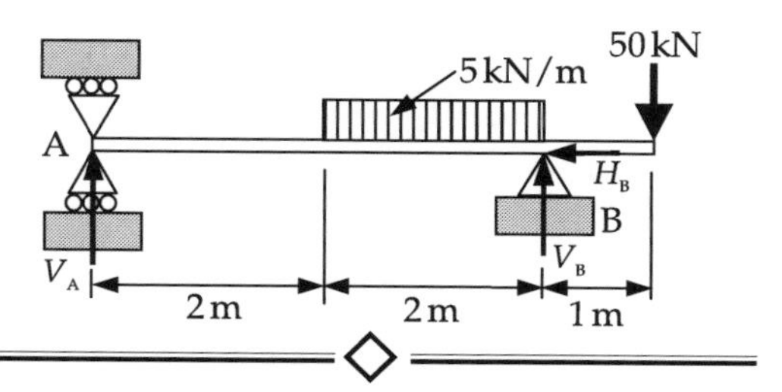

Solutions to Further Problems 3

1.

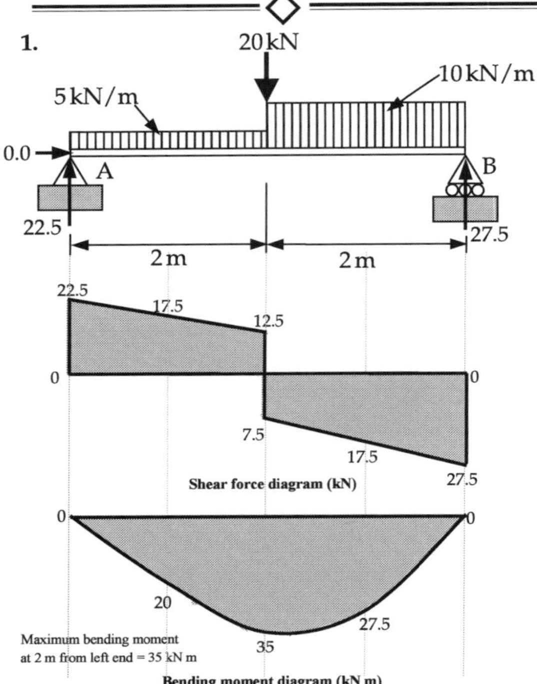

Shear force diagram (kN)

Maximum bending moment at 2 m from left end = 35 kN m

Bending moment diagram (kN m)

2.

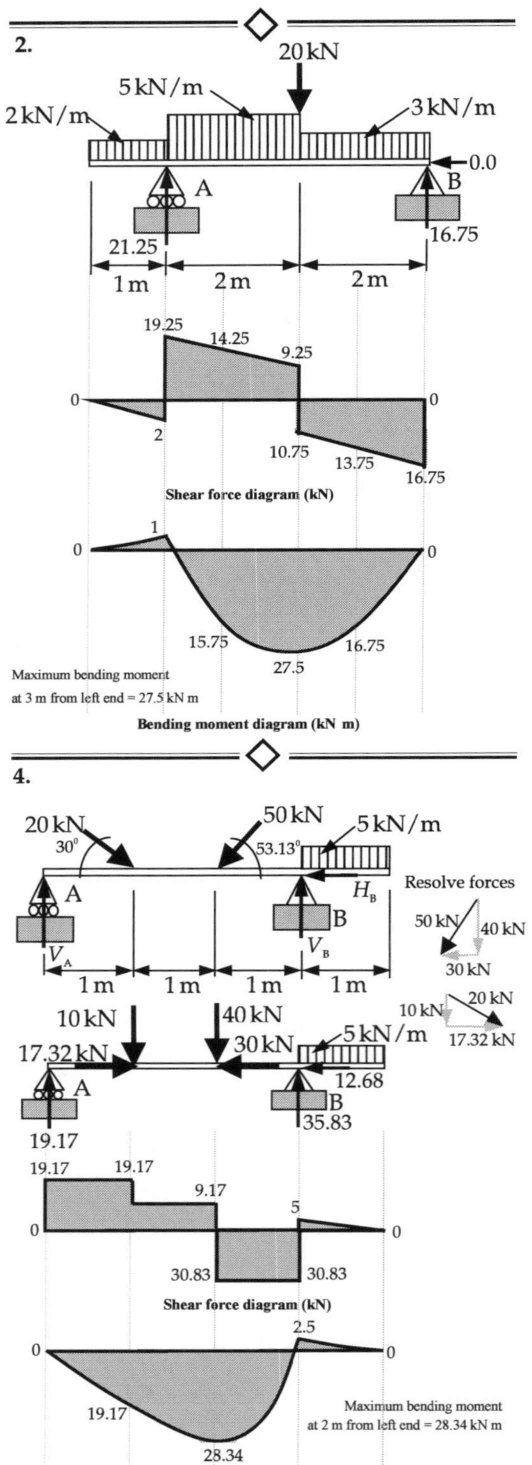

Shear force diagram (kN)

Maximum bending moment at 3 m from left end = 27.5 kN m

Bending moment diagram (kN m)

Shear force diagram (kN)

Maximum bending moment at 2 m from left end = 28.34 kN m

Bending moment diagram (kN m)

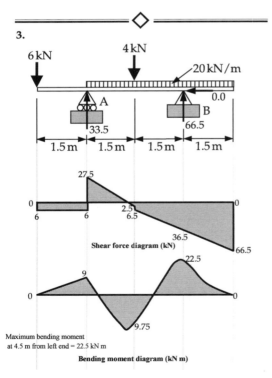

3.

6 kN

4 kN

20 kN/m

A

B

0.0

33.5

66.5

1.5 m 1.5 m 1.5 m 1.5 m

27.5

0 0

6 6 2.5 6.5

36.5

66.5

Shear force diagram (kN)

22.5

9

0 0

9.75

Maximum bending moment at 4.5 m from left end = 22.5 kN m

Bending moment diagram (kN m)

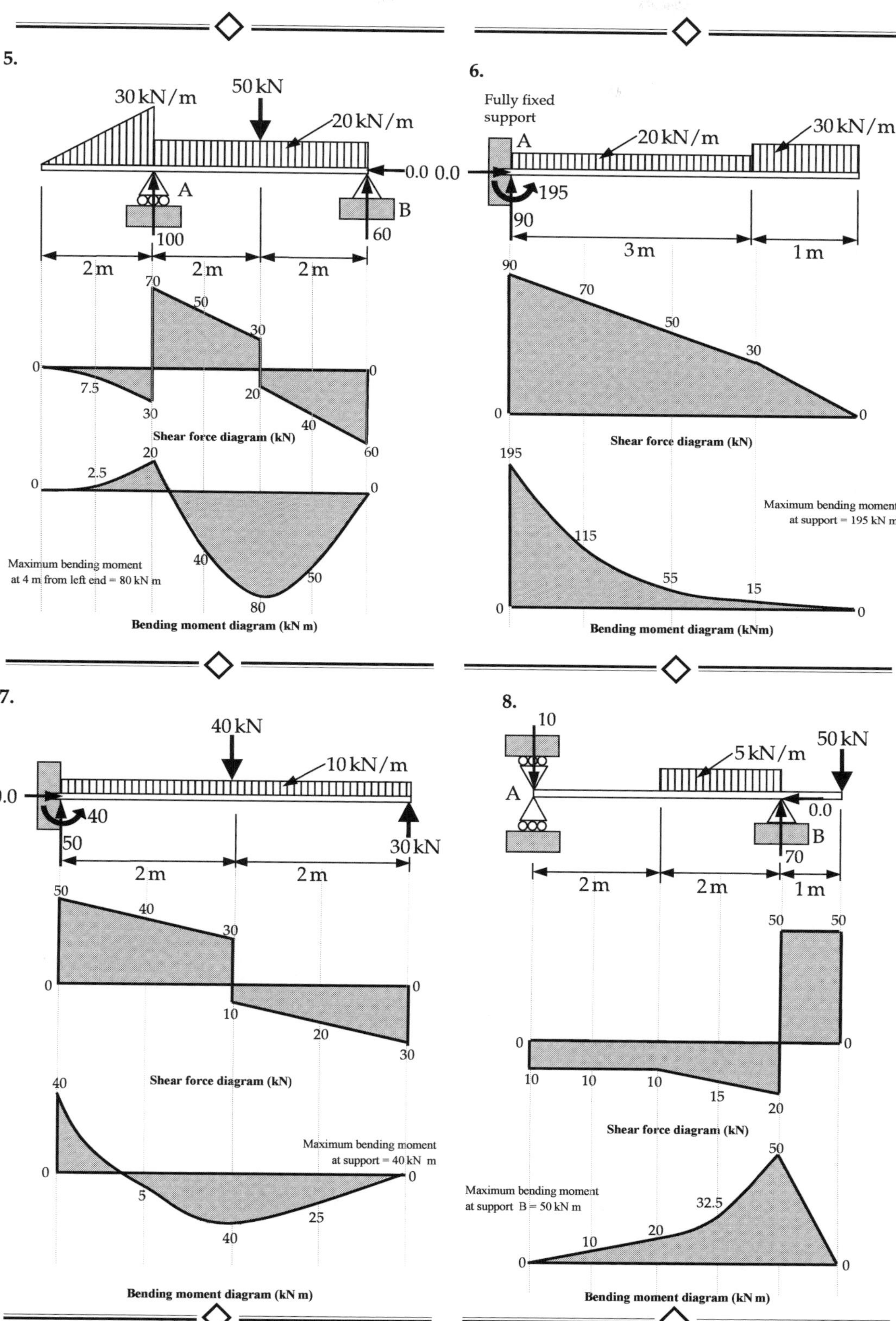
5.
30 kN/m
50 kN
20 kN/m
A
B
0.0
100
60
2 m
2 m
2 m
70
50
30
0
7.5
30
20
40
0
60
Shear force diagram (kN)
20
2.5
0
0
40
50
80
Maximum bending moment
at 4 m from left end = 80 kN m
Bending moment diagram (kN m)
6.
Fully fixed
support
A
20 kN/m
30 kN/m
0.0
195
90
3 m
1 m
90
70
50
30
0
0
Shear force diagram (kN)
195
115
55
15
0
0
Maximum bending moment
at support = 195 kN m
Bending moment diagram (kNm)
7.
40 kN
10 kN/m
0.0
40
50
30 kN
2 m
2 m
50
40
30
0
0
10
20
30
Shear force diagram (kN)
40
0
0
5
40
25
Maximum bending moment
at support = 40 kN m
Bending moment diagram (kN m)
8.
10
5 kN/m
50 kN
A
B
0.0
70
2 m
2 m
1 m
50
50
0
0
10
10
10
15
20
Shear force diagram (kN)
50
32.5
20
10
0
0
Maximum bending moment
at support B = 50 kN m
Bending moment diagram (kN m)

4. Three Pin Frames

> ***In this chapter we will learn about:***
> - ***Different types of portal frames***
> - ***Determination of reactions for frame structures***
> - ***Plotting axial thrust diagrams***
> - ***Plotting bending moment and shear force diagrams for frame structures***

4.1 Introduction

Previously we have considered simple beam structures which are effectively one-dimensional – they have length but no height. We will now extend our studies to consider two-dimensional frames.

Figure 4.1 shows some simple two-dimensional frames. These structures are generally known as 'portal frames' and are most often constructed in steel or reinforced concrete, though some examples formed in timber (wood) do exist. They can be formed in various configurations, with pitched (sloping) or flat tops, and are commonly used to span 25–30 metres.

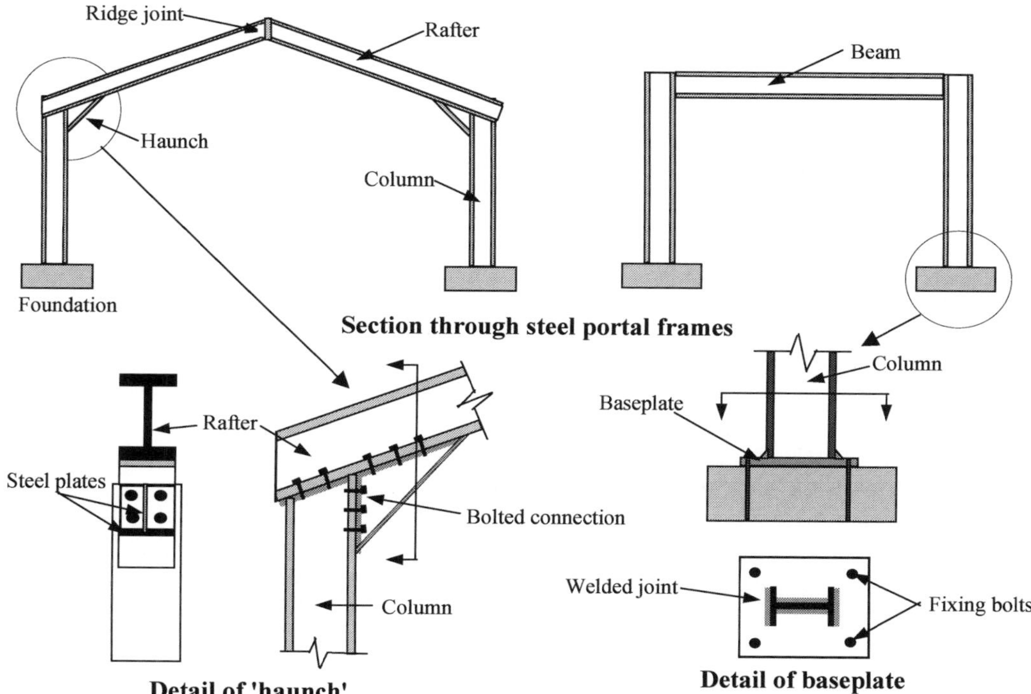

Figure 4.1: Typical portal frame details

The connection of the frame to its foundations is formed using methods that will ensure it is fully fixed, though it is possible to design the connection as a pin. The connection between the upright column members and the beam or rafter of three pin frame structures is normally assumed to be rigid in order to ensure stability of the frame. It is common to strengthen these joints by forming a haunch, as shown in the detail of Figure 4.1, to ensure that they remain stiff.

Portal frames are commonly used in the construction of single-storey light industrial units and warehouses, where large uninterrupted floor areas are essential. These buildings are often enclosed by attaching steel sheets to the frames (cladding).

In analysing such frames we must remember that they are in fact three-dimensional structures. Whilst our analysis will include loads applied in the horizontal X direction and vertical Y direction, no account will be taken of those loads applied in the Z direction. This, in most cases, will be a second-order quantity. However when considering the overall stability of the building, in particular its resistance to lateral movement, due consideration of these effects must be taken.

Consider the frame shown for the building of Figure 4.2.

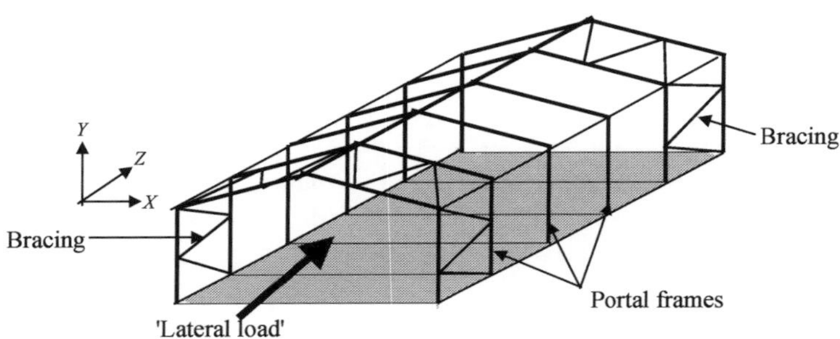

Figure 4.2: Portal framed building

The building will be subjected to lateral loads such as wind pressure. These loads must be resisted by the structure, and we would normally design bracing in the end and/or middle bays of the building to resist these forces. The bracing will transfer loads back to the foundation of the structure. The initial design of the portal frames would not normally include these 'out-of-plane' loads as they will be included in the design of the secondary structures, in this case the bracing.

The simplification involved in designing two-dimensional structures for 'real-world' situations must always be borne in mind, particularly when completing final analysis and 'sizing' of suitable structural members. There is little point in designing an adequate two-dimensional frame which has no lateral stability and therefore falls over!

4.2 Layout of Three Pin Frames

Portal frames can be designed with either pinned or fixed joints at the junction between the frame and foundation. In terms of their structural analysis, they may be drawn as shown in Figure 4.3a, b and c.

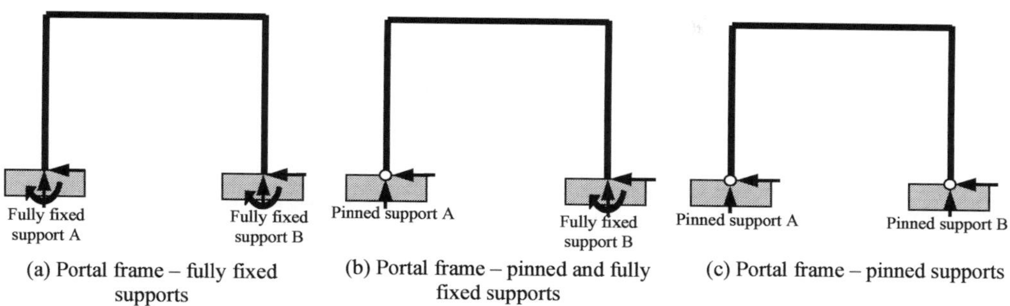

(a) Portal frame – fully fixed supports

(b) Portal frame – pinned and fully fixed supports

(c) Portal frame – pinned supports

Figure 4.3: Types of supports on portal frame structures

In order for us to carry out an analysis of such structures we must first check to see if they are statically determinate. We will consider each of the structures of Figure 4.3 and apply the test for static determinacy detailed in section 3.3 and given as:

$$(m + r - f) \; - 2j = \text{ degree of indeterminacy} \qquad [3.2]$$

For portal frame (a):

Number of members m	=	3		= 3
Number of rigid joints r	=	4		= 4
Number of roller releases f	=	0		= 0
Number of joints j	=	2	$-2j = 2(2)$	= – 4
Static Indeterminacy of portal frame (a)				**= 3°**

For portal frame (b):

Number of members m	=	3		= 3
Number of rigid joints r	=	3		= 4
Number of roller releases f	=	0		= 0
Number of joints j	=	2	$-2j = 2(2)$	= – 4
Static Indeterminacy of portal frame (b)				**= 2°**

For portal frame (c):

Number of members m	=	3		= 3
Number of rigid joints r	=	2		= 4
Number of roller releases f	=	0		= 0
Number of joints j	=	2	$-2j = 2(2)$	= – 4
Static Indeterminacy of portal frame (c)				**= 1°**

So, even though the portal frame structure (c) has pin joints at both the supports, we cannot solve it using the equations of statics alone. We could insert a roller joint at one of the supports as shown in Figure 4.4, however such a structure would tend to be unstable and would not be very economic.

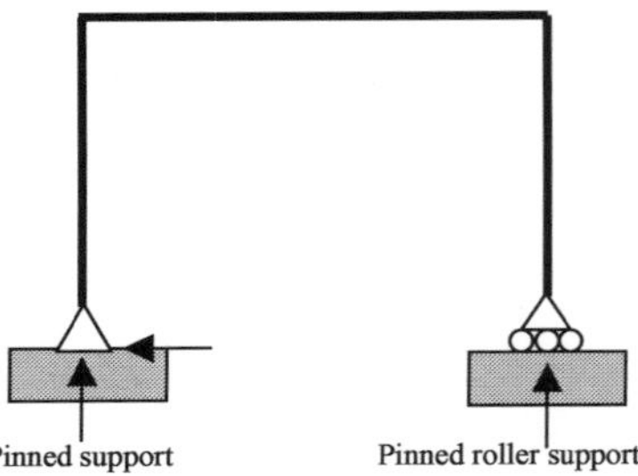

Figure 4.4: Statically determinate portal frame

The analysis of such a structure would follow a similar procedure to that adopted for a beam.

A much more useful structure may be formed by inserting a pin joint into the frame. This joint can be located at any position (apart from at a support!) on the structure, as shown in Figure 4.5.

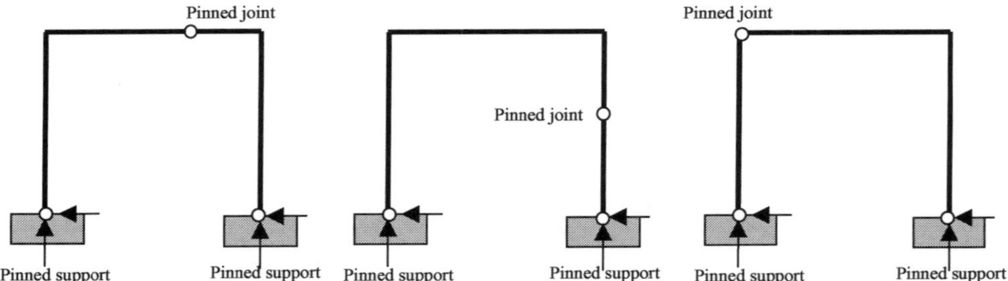

Figure 4.5: Typical 'three pin frames'

The consequence of including an internal pin joint in the frame is to provide us with a further equation in our analysis. The three equations of statics have already been stated as:

$$\sum F_x = 0; \qquad \sum F_y = 0; \qquad \text{and} \qquad \sum M_z = 0;$$

and we have solved various statically determinate structures using only these relationships. The structure to be considered here, ignoring the internal pin, is statically indeterminate to 1° (see Figure 4.3c), however we know that a pin joint will provide no resistance to moment. We can therefore form a further equation based on the assumption that:

$$\sum M_z \text{ (at internal pin)} = 0$$

We now have four equations which can be used to find the four 'unknown' forces at the reactions.

4.3 Determination of Reactions

The determination of reactions for three pin frames follows a similar procedure to that adopted to determine beam reactions.

We must, as before, derive equations for horizontal and vertical forces, and for moments about one of the supports. For equilibrium of the frame, each of these equations must sum to zero. We can derive a further equation because we know the sum of the moments at the internal pin joint must also equal zero. It does not matter whether we sum moments to the left or right of the pin because, as shown in Figure 4.6, both solutions **must** yield the same result (equal to zero).

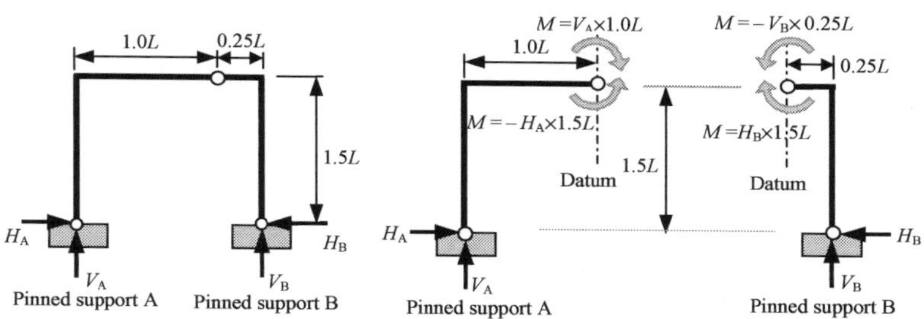

Figure 4.6: Determination of moments about internal pin joint

For the frame of Figure 4.6, the moments about each side of the internal pin are given by:

ΣM_z (to left of internal pin) $= 0 = (V_A \times 1.0L) + (-H_A \times 1.5L)$

ΣM_z (to right of internal pin) $= 0 = (-V_B \times 0.25L) + (H_B \times 1.5L)$

$\therefore \; (V_A \times 1.0L) + (-H_A \times 1.5L) = (-V_B \times 0.25L) + (H_B \times 1.5L)$

It is important to note that in the analysis of frames we must now calculate bending moments for forces applied in both the horizontal and vertical planes. All distances, as before, are measured as the distance from a datum to the point of application of the load in the horizontal and vertical plane. It is also important to note that, with this type of frame, horizontal reactions will still be induced, even if no horizontal loads are applied. Consider the analogy of standing on a small stool. The tendency for the legs of the stool is to try to spread sideways, and this will be resisted by the stiffness of the legs and the friction between the floor and the bottom of the legs. In our frame, this 'spreading' action will be resisted by horizontal reactions at the supports.

The procedure to be followed in order to calculate the reactions at the supports of a frame is therefore:

Sum all horizontal forces and equate to zero:	$\Sigma F_x = 0$
Sum all vertical forces and equate to zero:	$\Sigma F_y = 0$
Sum all moments about one of the reactions and equate to zero:	$\Sigma M_z = 0$
Sum all moments about the internal pin and equate to zero:	$\Sigma M_z = 0$

We can then solve the four equations formed from these relationships simultaneously in order to solve for the unknown forces.

Consider the frame of Example 4.1.

Example 4.1: Three pin frame supporting a single vertical point load

The three pin frame shown in Figure 4.7 is supported at points A and F on pin joints and has an internal pin joint at point C. The frame is loaded with a single point load at point D. Determine the reactions at the supports A and F.

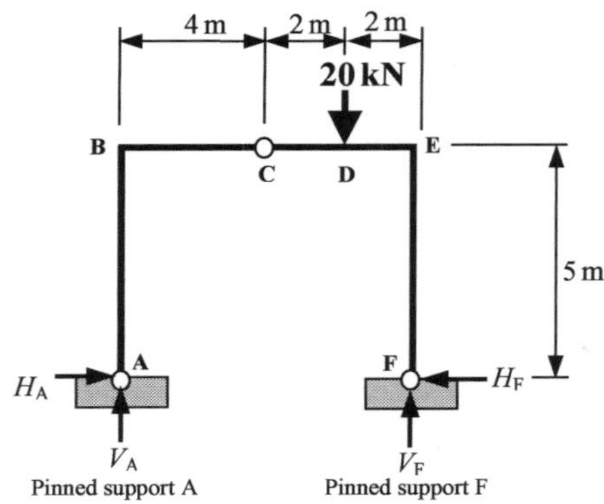

Figure 4.7: Three pin frame – Example 4.1

As with the analysis of beams, we will need to consider how the support reactions will ensure the equilibrium of the structure. In most cases we commence our analysis with the proposition that the vertical reactions are upwards. A negative result will indicate that the direction assumed was incorrect and that the force is in the opposite direction. We can adopt a similar approach for the horizontal reactions, however it is necessary to adopt a 'sensible' solution to the problem and propose reactions based on critical judgement and our understanding of how the supports will resist the loads. The direction of reactions proposed will not always be correct, however an incorrect proposal will again yield a negative answer in the solution which will indicate that the direction was wrong. In the case of Example 4.1, in order to ensure equilibrium of the structure, we assume the horizontal forces act in opposite (inwards) directions. From observation we also note that, for horizontal equilibrium, they must be equal in value. We will now calculate the reactions for the frame of Example 4.1:

(Sign convention: $+Y$ upwards, $+X$ to the right)

Summation of horizontal forces $\quad \Sigma F_x = H_A - H_F = 0$

$$\therefore H_A = H_F \qquad [1]$$

Summation of vertical forces $\quad \Sigma F_y = V_A + V_F - 20\,\text{kN} = 0$

$$\therefore V_A + V_F = 20\,\text{kN} \qquad [2]$$

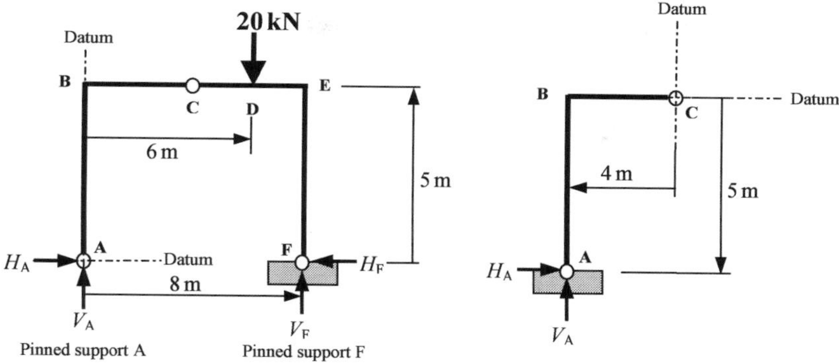

Figure 4.8: Position of datum for calculation of moments – Example 4.1

With reference to Figure 4.8:

Taking moments about support A: $\Sigma M_z = (V_A \times 0\,\text{m}) + (H_A \times 0\,\text{m}) + (20\,\text{kN} \times (4\,\text{m}+2\,\text{m}))$
$- (V_F \times (4\,\text{m} + 2\,\text{m} + 2\,\text{m})) - (H_F \times 0\,\text{m}) = 0$
$= 120 - 8\,V_F = 0$ [3]

We can solve this equation now since it contains only one unknown value such that:

$8\,V_F = 120 \qquad \therefore\ V_F = 120/8 = 15\,\text{kN}$

We can now form our fourth equation by taking moments about the internal pin as shown in Figure 4.8. In this case we take moments about the left side of the pin, as this will be the easiest to resolve. Hence:

Taking moments about internal pin to left side: $\Sigma M_z = (V_A \times 4\,\text{m}) - (H_A \times 5\,\text{m}) = 0$
$= 4\,V_A - 5\,H_A = 0$
$\therefore\ 4\,V_A = 5\,H_A$ [4]

We now have four equations:

$H_A = H_F$ [1]
$V_A + V_F = 20\,\text{kN}$ [2]
$120 - 8V_F = 0$ (from which $V_F = 15\,\text{kN}$) [3]
and $4\,V_A = 5\,H_A$ [4]

Substituting the value for V_F obtained from equation [3] into equation [2] we have:

$V_A + 15 = 20\,\text{kN} \qquad \therefore V_A = 5\,\text{kN}$

Substituting the value for V_A above into equation [4] gives:

$4(5) = 5\,H_A \qquad \therefore H_A = 4\,\text{kN}$

Substituting the value for H_A from above into equation [1] gives:

$4 = H_F \qquad \therefore H_F = 4\,\text{kN}$

The final values for the reactions are therefore as shown in Figure 4.9.

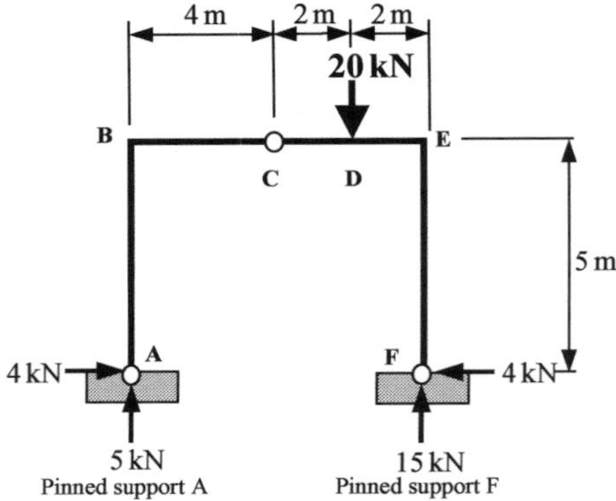

Figure 4.9: Reactions calculated for Example 4.1

Example 4.2: Three pin frame supporting a vertical and horizontal point load

The three pin frame shown in Figure 4.10 is supported at A and F on pin joints and has an internal pin joint at E. The frame supports point loads at B and D. Determine the reactions at the supports A and F.

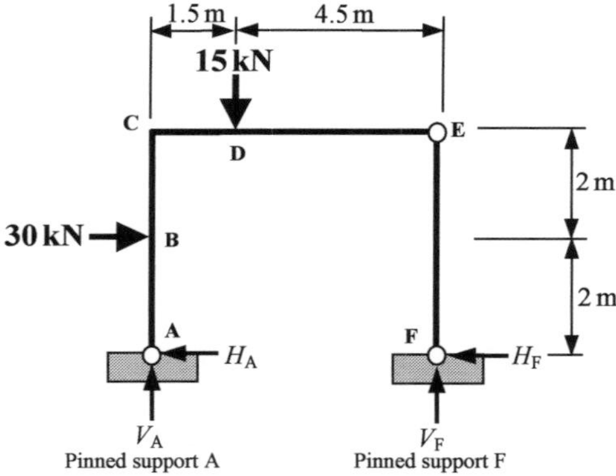

Figure 4.10: Three pin frame – Example 4.2

Here we have a horizontal applied load that must be resisted by the support reactions. We assume that the reactions act in the opposite direction to the load in order to ensure the structure is in equilibrium.

We now calculate the reactions as before:

Summation of horizontal forces $\Sigma F_x = 30\,\text{kN} - H_A - H_F = 0$

$\therefore\ 30\,\text{kN} = H_A + H_F$ [1]

Summation of vertical forces $\Sigma F_y = V_A + V_F - 15\,\text{kN} = 0$

$\therefore\ V_A + V_F = 15\,\text{kN}$ [2]

With reference to Figure 4.11 and omitting reactions at support A (which have zero distance!):

Taking moments about support A: $\Sigma M_z = (30\,\text{kN} \times 2\,\text{m}) + (15\,\text{kN} \times 1.5\,\text{m})$
$- (V_F \times (1.5\,\text{m} + 1.5\,\text{m} + 3\,\text{m})) - (H_F \times 0\,\text{m}) = 0$
$= 60 + 22.5 - 6\,V_F = 0$ [3]

We can solve this equation now since it only contains one unknown value such that:
$6\,V_F = 82.5$ $\quad\therefore\ V_F = 82.5/6 = 13.75\,\text{kN}$

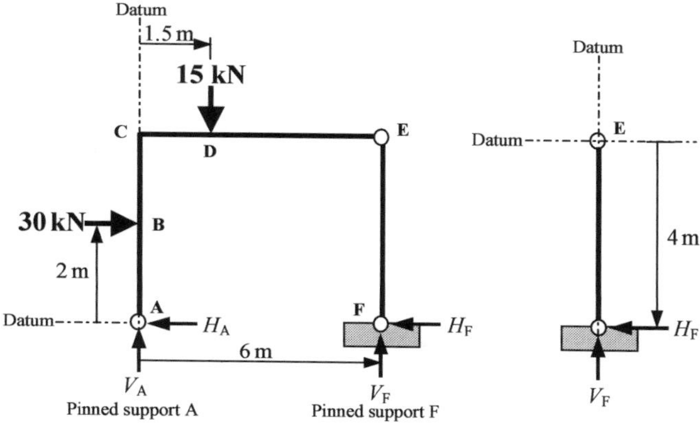

Figure 4.11: Position of datum for calculation of moments – Example 4.2

In this case we will create a fourth equation by taking moments to the right of the internal pin such that:

Taking moments about internal pin to right side: $\Sigma M_z = (V_F \times 0\,\text{m}) + (H_F \times 4\,\text{m}) = 0$
$= 4\,H_F = 0$ [4]

We can solve this equation now since it only contains one unknown value such that:
$H_F = 0$

We now have four equations:

$30\,\text{kN} = H_A + H_F$ [1]
$V_A + V_F = 15\,\text{kN}$ [2]
$60 + 22.5 - 6\,V_F = 0$ (from which $V_F = 13.75\,\text{kN}$) [3]
and $4\,H_F = 0$ (from which $H_F = 0\,\text{kN}$) [4]

Substituting the value for V_F obtained from equation [3] into equation [2] we have:
$V_A + 13.75 = 15\,\text{kN}$ $\quad\therefore V_A = 1.25\,\text{kN}$

Substituting the value for H_F from equation [4] into equation [1] gives:
$30\,\text{kN} = H_A + 0\,\text{kN}$ $\quad\therefore H_A = 30\,\text{kN}$

The final values for the reactions are shown in Figure 4.12.

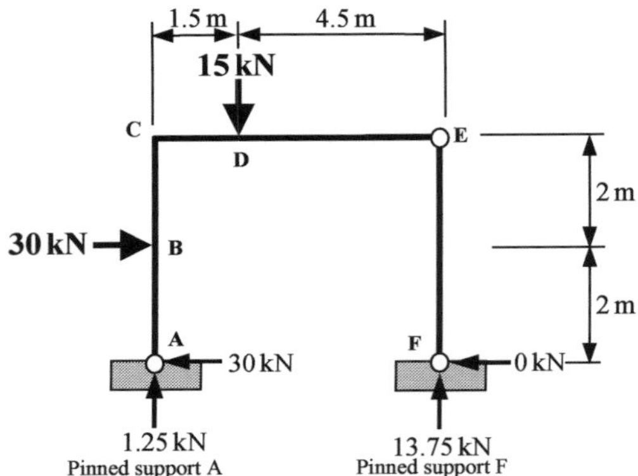

Figure 4.12: Reactions calculated for Example 4.2

Example 4.3: Three pin frame supporting two vertical and one horizontal point load

The three pin portal frame shown in Figure 4.13 is supported at points A and E with pin joints and has an internal pin joint at point C. Determine the reactions at the supports A and E.

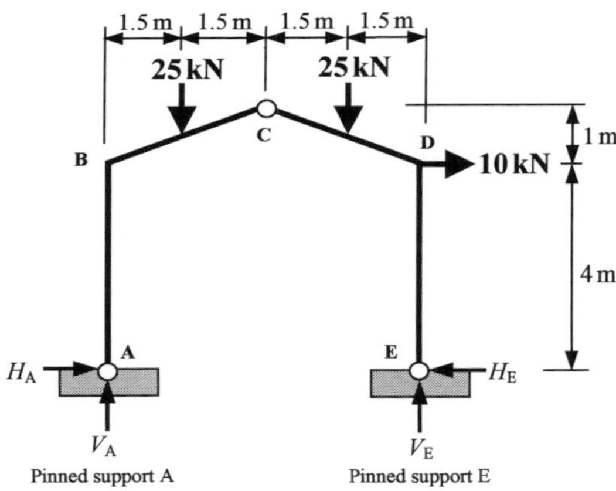

Figure 4.13: Three pin portal frame – Example 4.3

We calculate the reactions as before:

Summation of horizontal forces $\quad \Sigma F_x = 10\,\text{kN} + H_A - H_E = 0$

$$\therefore\ 10\,\text{kN} = -H_A + H_E \qquad [1]$$

Summation of vertical forces $\quad \Sigma F_y = V_A + V_E - 25\,\text{kN} - 25\,\text{kN} = 0$

$$\therefore\ V_A + V_E = 50\,\text{kN} \qquad [2]$$

With reference to Figure 4.14 and omitting reactions at 0 m from datum (at support A):

Taking moments about support A: $\Sigma M_z = (25\,\text{kN} \times 1.5\,\text{m}) + (25\,\text{kN} \times (1.5\,\text{m} + 1.5\,\text{m} + 1.5\,\text{m}))$

$$+(10\,\text{kN} \times 4\,\text{m}) - (V_E \times (1.5\,\text{m} + 1.5\,\text{m} + 1.5\,\text{m} + 1.5\,\text{m})) = 0$$

$$= 37.5 + 112.5 + 40 - 6\,V_E = 0 \qquad [3]$$

$$6\,V_E = 190 \qquad \therefore\ V_E = 190/6 = 31.67\,\text{kN}$$

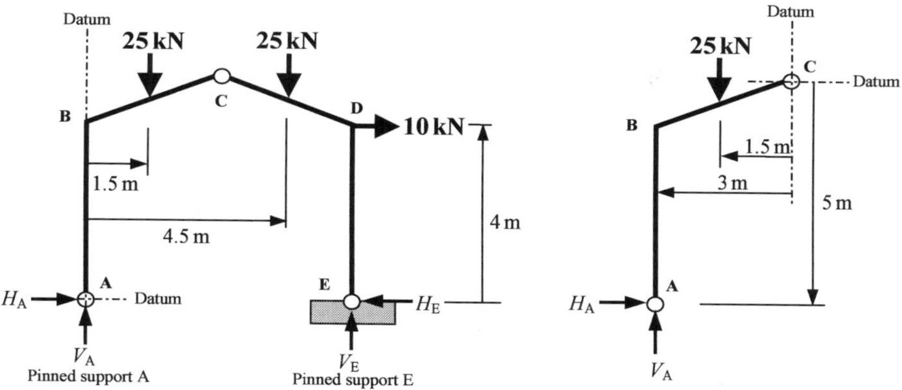

Figure 4.14: Position of datum for calculation of moments – Example 4.3

Taking moments about internal pin to left side: $\Sigma M_z = (V_A \times 3\,\text{m}) - (H_A \times 5\,\text{m}) - (25\,\text{kN} \times 1.5\,\text{m}) = 0$

$$= 3V_A - 5H_A - 37.5 = 0$$

$$\therefore\ 3V_A - 5H_A = 37.5 \qquad [4]$$

We now have four equations:

$$10\,\text{kN} = -H_A + H_E \qquad [1]$$

$$V_A + V_E = 50\,\text{kN} \qquad [2]$$

$$190 - 6V_E = 0 \qquad \text{(from which } V_E = 31.67\,\text{kN)} \qquad [3]$$

and

$$3V_A - 5H_A = 37.5 \qquad [4]$$

Substituting the value for V_E obtained from equation [3] into equation [2] we have:

$$V_A + 31.67 = 50\,\text{kN} \qquad \therefore V_A = 18.33\,\text{kN}$$

Substituting the value for V_A from above into equation [4] gives:

$$3(18.33) - 5H_A = 37.5 \qquad \therefore H_A = -3.5\,\text{kN}$$

Note: the negative sign indicates that we chose the wrong direction for this reaction!

Substituting the answer for H_A obtained from equation [4] into equation [1] will give:

$$10\,\text{kN} = -(-3.5) + H_E \qquad \therefore H_E = 6.5\,\text{kN}$$

The final directions and values for the reactions for Example 4.3 are shown in Figure 4.15.

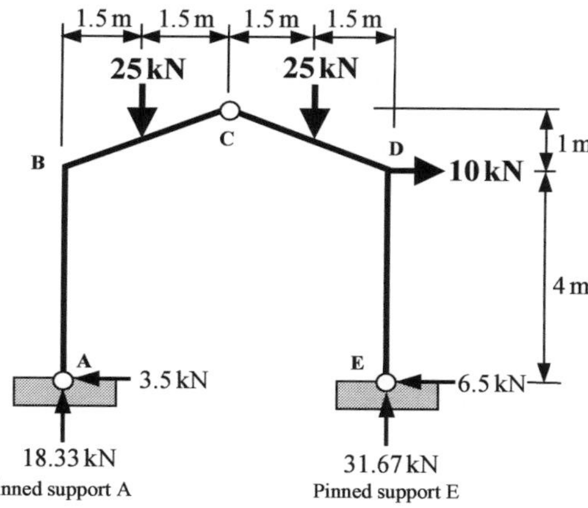

Figure 4.15: Reactions calculated for Example 4.3

We will now consider the analysis of a non-symmetric frame.

Example 4.4: Non-symmetric three pin frame supporting a vertical and horizontal point load

The non-symmetric three pin frame shown in Figure 4.16 is supported at A and E on pin joints and has an internal pin joint at point C. The frame is loaded with a vertical and a horizontal point load at joint D. Determine the reactions at the supports A and E.

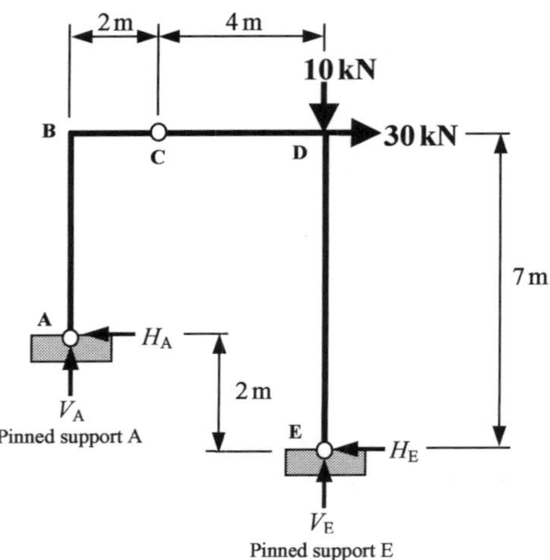

Figure 4.16: Three pin frame – Example 4.4

Summation of horizontal forces $\quad \Sigma F_x = -H_A - H_E + 30\,\text{kN} = 0$

$$\therefore H_A + H_E = 30\,\text{kN} \qquad [1]$$

Summation of vertical forces $\quad \Sigma F_y = V_A + V_E - 10\,\text{kN} = 0$

$$\therefore V_A + V_E = 10\,\text{kN} \qquad [2]$$

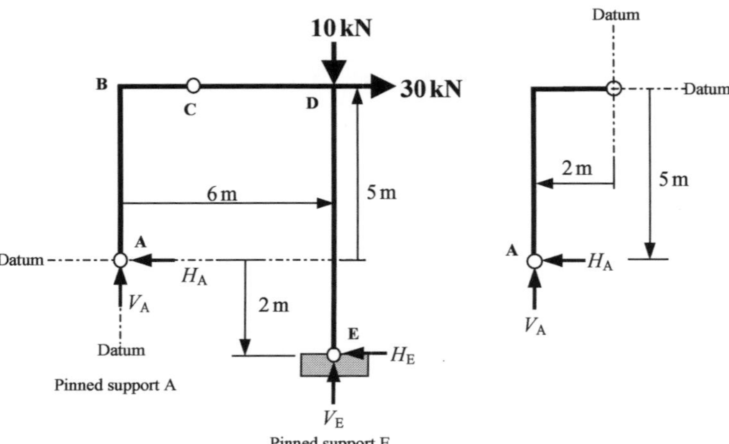

Figure 4.17: Position of datum for calculation of moments – Example 4.4

Taking moments about support A: $\Sigma M_z = (10\,\text{kN} \times (2\,\text{m} + 4\,\text{m})) + (30\,\text{kN} \times 5\,\text{m})$
$- (V_E \times (2\,\text{m} + 4\,\text{m})) + (H_E \times 2\,\text{m}) = 0$
$= 210 - 6V_E + 2H_E = 0$
$\therefore\ 6V_E = 210 + 2H_E$ [3]

Taking moments about internal pin to left side: $\Sigma M_z = (V_A \times 2\,\text{m}) + (H_A \times 5\,\text{m}) = 0$
$= 2V_A + 5H_A = 0$
$\therefore 2V_A = -5H_A$ [4]

We now have four equations:

$H_A + H_E = 30\,\text{kN}$ [1]
$V_A + V_E = 10\,\text{kN}$ [2]
$6V_E = 210 + 2H_E$ [3]
and $2V_A = -5H_A$ [4]

From [4] we have: $2V_A = -5H_A$ $\therefore\ V_A = -2.5H_A$ [5]
And from [3] we have $6V_E = 210 + 2H_E$ or $V_E = 35 + 0.333H_E$ [6]

Substituting [5] and [6] into equation [2] gives: $(-2.5H_A) + (35 + 0.333H_E) = 30\,\text{kN}$
Simplifying: $-2.5H_A + 0.333H_E = -5\,\text{kN}$ [7]

Multiply [1] × 2.5 and solve [1] and [7] simultaneously.

[1] × 2.5 $2.5H_A + 2.5H_E = 75\,\text{kN}$ [8]
[8] + [7] $2.5H_A + 2.5H_E + ((-2.5H_A) + 0.333H_E)$
$= 75\,\text{kN} + (-5\,\text{kN})$
$\therefore\ 2.8333H_E = 70\,\text{kN}$ and $H_E = 24.706\,\text{kN}$

Substituting the value for H_E obtained from above into equation [1] we have:
$H_A + 24.706 = 30\text{kN}$ $\therefore H_A = 5.294\,\text{kN}$

Substituting the value for H_A above into equation [4] gives:
$2V_A = -5(5.294)$ $\therefore V_A = -13.235\,\text{kN}$

Substituting the value for V_A above into equation [2] gives:
$-13.235 + V_E = 10$ $\therefore V_E = 3.235\,\text{kN}$

The final directions and values for the reactions are therefore as shown in Figure 4.18.

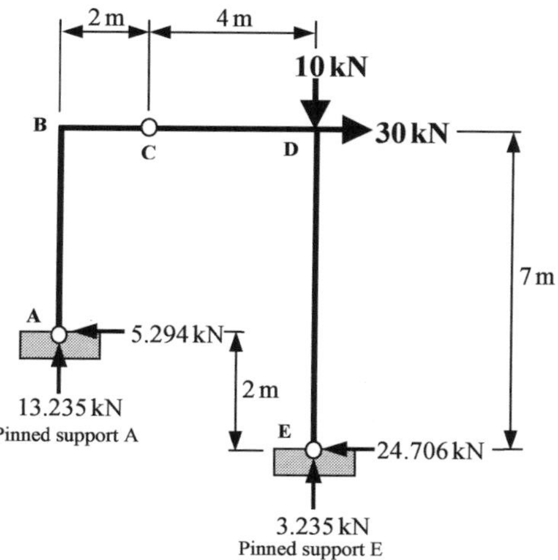

Figure 4.18: Reactions calculated for Example 4.4

The methods proposed here can equally be applied to 'arches' and frames supporting distributed loads. (DLs).

Example 4.5: Three pin arch supporting a vertical distributed load

The three pin arch shown in Figure 4.19 is supported at A and C on pin joints and has an internal pin joint at B. The arch supports a vertically downward UDL of 10 kN/m along its entire length. Determine the reactions at the supports A and C.

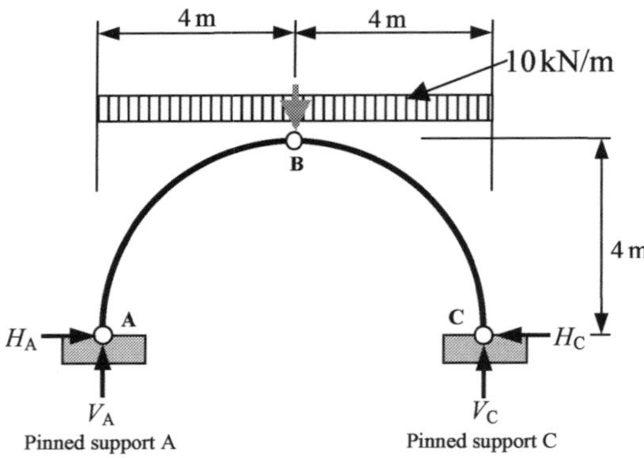

Figure 4.19: Three pin arch – Example 4.5

As before, this uniformly distributed load (UDL) acts through its centroid and distances will be calculated from the datum to this point.

Summation of horizontal forces $\Sigma F_x = H_A - H_C = 0$

$\therefore H_A = H_C$ [1]

Summation of vertical forces $\Sigma F_y = V_A + V_C - (10\,\text{kN} \times (4\,\text{m} + 4\,\text{m})) = 0$

$\therefore\ V_A + V_C = 80\,\text{kN}$ [2]

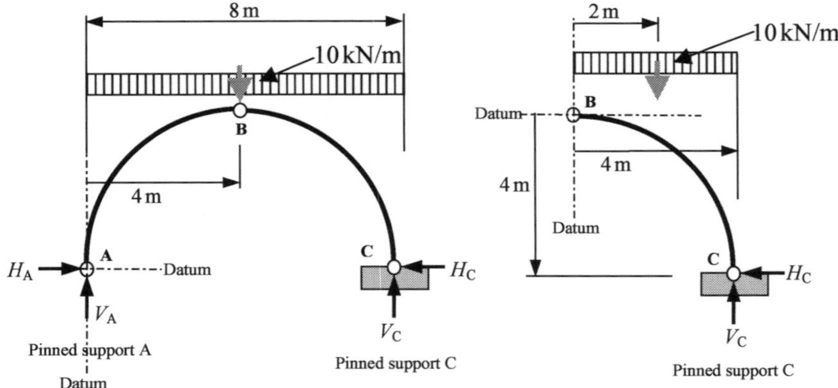

Figure 4.20: Position of datum for calculation of moments – Example 4.5

Taking moments about support A: $\Sigma M_z = (10\,\text{kN/m} \times 8\,\text{m} \times (8\,\text{m}/2)) - (V_C \times 8\,\text{m}) - (H_C \times 0\,\text{m}) = 0$

$= 320 - 8\,V_C = 0$ [3]

$8\,V_C = 320$ $\quad\therefore\ V_C = 320/8 = 40\,\text{kN}$

We now form the fourth equation by taking moments about the internal pin as shown in Figure 4.20. In this case, calculations will be similar for either side of the internal pin. We will therefore calculate moments to the right of the internal pin such that:

Taking moments about internal pin to right side: $\Sigma M_z = (V_C \times 4\,\text{m}) - (H_C \times 4\,\text{m})$

$$- (10\,\text{kN/m} \times 4\,\text{m} \times (4\,\text{m}/2)) = 0$$

$$= 4V_C - 4H_C - 80 = 0$$

$\therefore 4V_C - 4H_C = 80$ or (dividing by 4) $V_C - H_C = 20$ [4]

We now have four equations:

$H_A = H_C$ [1]

$V_A + V_C = 80\,\text{kN}$ [2]

$8V_C = 320$ (from which $V_C = 40\,\text{kN}$) [3]

and $V_C - H_C = 20$ [4]

Substituting the value for V_C obtained from equation [3] into equation [2] we have:

$V_A + 40 = 80\,\text{kN}$ $\quad \therefore V_A = 40\,\text{kN}$

Substituting the value for V_C obtained from equation [3] into equation [4]:

$40 - H_C = 20$ $\quad \therefore H_C = 20\,\text{kN}$

Substituting the value for H_C above into equation [1] gives:

$H_A = 20$ $\quad \therefore H_A = 20\,\text{kN}$

The final values for the reactions are shown in Figure 4.21:

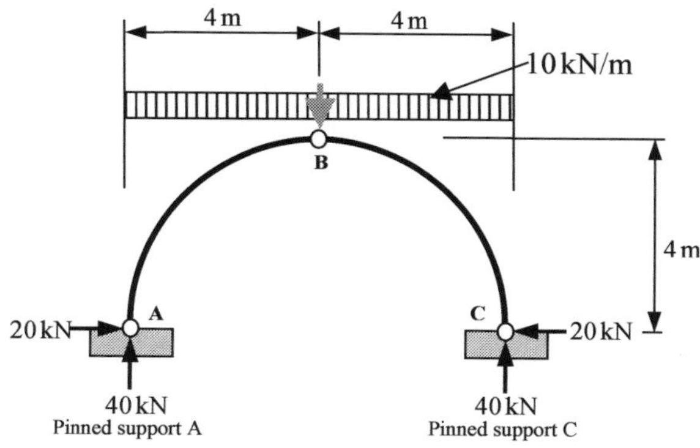

Figure 4.21: Reactions for Example 4.5

Example 4.6: Three pin frame supporting a horizontal distributed and horizontal and vertical point loads

The three pin frame shown in Figure 4.22 is supported at A and E on pin joints and has an internal pin joint at C. The frame supports a vertical point load on member BC and a horizontal point load at joint D. It also supports a horizontal UDL applied along the length of member AB. Determine the reactions at the supports A and E.

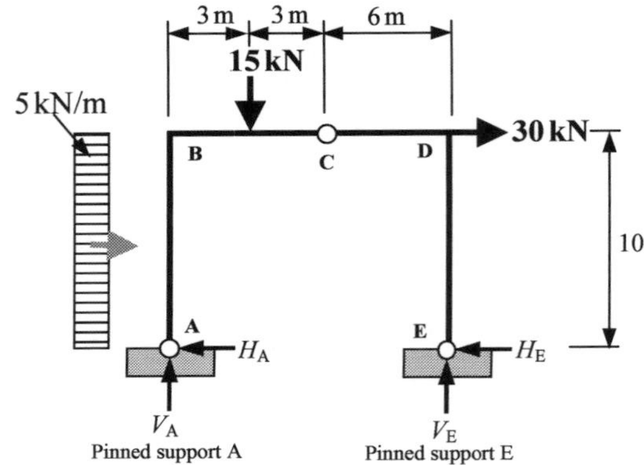

Figure 4.22: Three pin frame – Example 4.6

Calculate reactions:

Summation of horizontal forces $\sum F_x = 30\,\text{kN} + (5\,\text{kN/m} \times 10\,\text{m}) - H_A - H_E = 0$

$\therefore\ 80\,\text{kN} = H_A + H_E$ [1]

Summation of vertical forces $\sum F_y = V_A + V_E - 15\,\text{kN} = 0$

$\therefore\ V_A + V_E = 15\,\text{kN}$ [2]

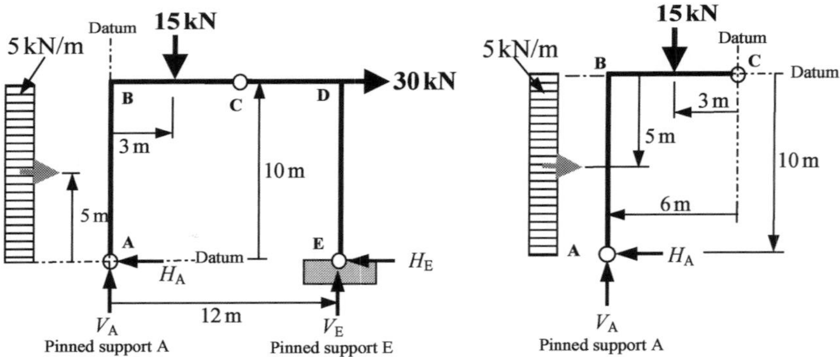

Figure 4.23: Position of datum for calculation of moments – Example 4.6

Taking moments about support A: $\sum M_z = (5\,\text{kN/m} \times 10\,\text{m} \times (10\,\text{m}/2)) + (15\,\text{kN} \times 3\,\text{m}) + (30\,\text{kN} \times 10\,\text{m})$

$- (V_E \times (3\,\text{m} + 3\,\text{m} + 6\,\text{m})) = 0$

$= 250 + 45 + 300 - 12\,V_E = 0$ [3]

$12\,V_E = 595 \qquad \therefore\ V_E = 595/12 = 49.583\,\text{kN}$

Taking moments to left of internal pin : $\sum M_z = -(15\,\text{kN} \times 3\,\text{m}) - (5\,\text{kN/m} \times 10\,\text{m} \times (10\,\text{m}/2))$

$+ (V_A \times 6\,\text{m}) + (H_A \times 10\,\text{m}) = 0$

$= -45 - 250 + 6\,V_A + 10\,H_A = 0$

$6V_A + 10H_A = 295$ [4]

We now have four equations:

$80 = H_A + H_E$ [1]

$V_A + V_E = 15\,\text{kN}$ [2]

$12\,V_E = 595$ (from which $V_E = 49.583\,\text{kN}$) [3]

and $6V_A + 10H_A = 295$ [4]

Substituting the value for V_E obtained from equation [3] into equation [2] we have:

$$V_A + 49.583 = 15\,\text{kN} \qquad \therefore V_A = -34.583\,\text{kN}$$

Substituting the value for V_A from above into equation [4] gives:

$$6(-34.583) + 10H_A = 295 \qquad \therefore H_A = 50.25\,\text{kN}$$

Substituting the value for V_A from above into equation [1] gives:

$$80 = 50.25 + H_E \qquad \therefore H_E = 29.75\,\text{kN}$$

The final values for the reactions are shown in Figure 4.24.

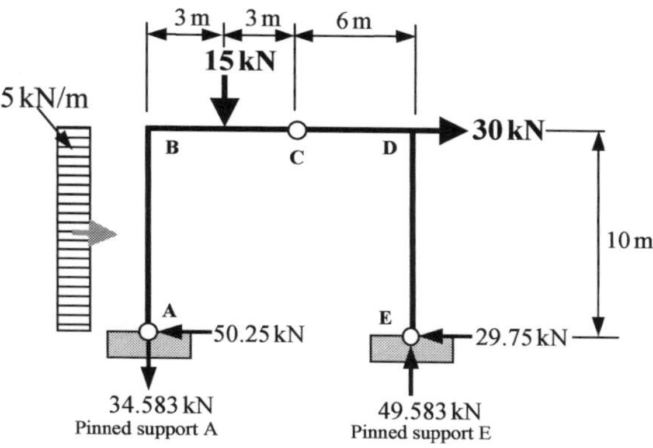

Figure 4.24: Reactions for Example 4.6

Example 4.7: Non-symmetric three pin frame supporting horizontal and vertical distributed loads and a horizontal point load

The non-symmetric three pin frame shown in Figure 4.25 is supported at A and E on pin joints and has an internal pin joint at point C. The frame supports a point load at D, a vertical downward UDL along member BD and a horizontal triangular distributed load along member AB which varies from 0 kN/m at A to a maximum of 18 kN/m at B. Determine the reactions at the supports A and E.

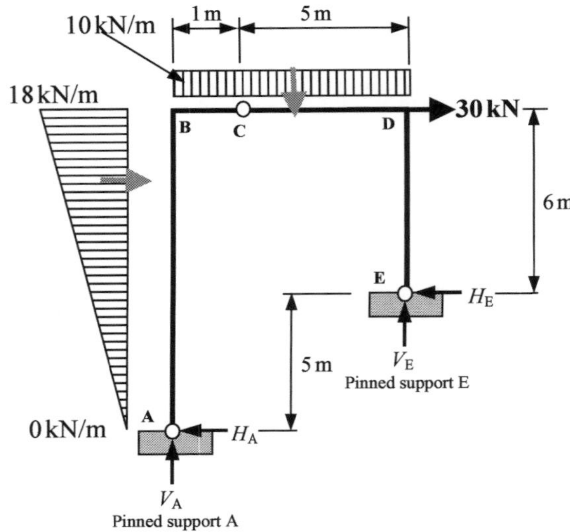

Figure 4.25: Three pin frame – Example 4.7

Here we have a horizontal triangular load applied to member AB. We have already considered such loads applied to beams and a similar method of analysis is adopted.

Calculate the reactions:

Summation of horizontal forces $\qquad \sum F_x = 30\,\text{kN} + (0.5 \times 18\,\text{kN/m} \times 11\,\text{m}) - H_A - H_E = 0$

$$30\,\text{kN} + 99\,\text{kN} = H_A + H_E$$

$$\therefore\ H_A + H_E = 129\,\text{kN} \qquad [1]$$

Summation of vertical forces $\qquad \sum F_y = V_A + V_E - (10\,\text{kN/m} \times (1\,\text{m} + 5\,\text{m})) = 0$

$$\therefore\ V_A + V_E = 60\,\text{kN} \qquad [2]$$

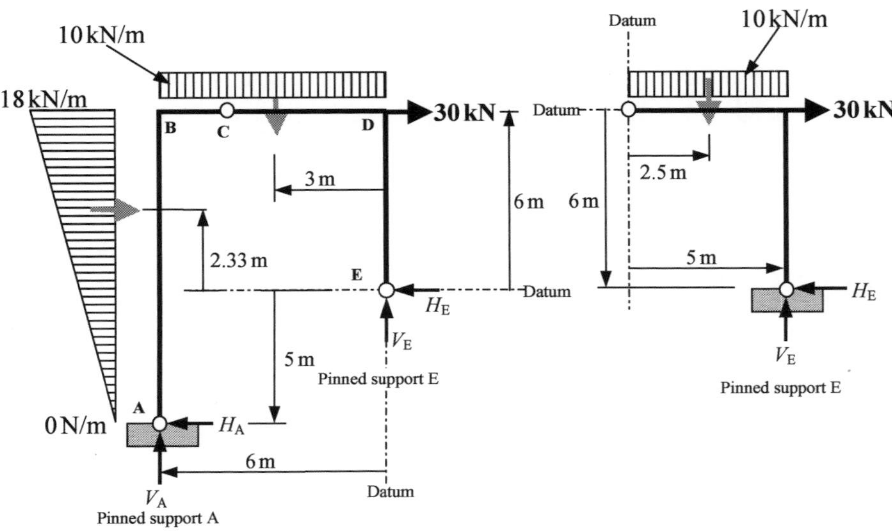

Figure 4.26: Position of datum for calculation of moments – Example 4.7

Taking moments about support E: $\Sigma M_z = (30\,\text{kN} \times 6\,\text{m}) - (10\,\text{kN/m} \times 6\,\text{m} \times (6\,\text{m}/2))$
$+ (0.5 \times 18\,\text{kN/m} \times 11\,\text{m} \times ((^2/_3 \times 11\,\text{m}) - 5\,\text{m}))$
$+ (V_A \times 6\,\text{m}) + (H_A \times 5\,\text{m}) = 0$
$= 180 - 180 + 231 + 6V_A + 5H_A = 0$

$\therefore\ 6V_A + 5H_A = -231$ [3]

Taking moments to right of internal pin: $\Sigma M_z = (10\,\text{kN/m} \times 5\,\text{m} \times 5\,\text{m}/2) - (30\,\text{kN} \times 0\,\text{m})$
$- (V_E \times 5\,\text{m}) + (H_E \times 6\,\text{m}) = 0$
$= 125 - 5V_E + 6H_E = 0$

$\therefore\ 5V_E - 6H_E = 125$ [4]

Our four equations are:

$H_A + H_E = 129$ [1]
$V_A + V_E = 60$ [2]
$6V_A + 5H_A = -231$ [3]
and $5V_E - 6H_E = 125$ [4]

Solving these equations simultaneously we find:

$V_A = -71.82\,\text{kN}$
$H_A = 39.99\,\text{kN}$
$V_E = 131.82\,\text{kN}$
$H_E = 89.01\,\text{kN}$

The reactions for the structure are shown in Figure 4.27.

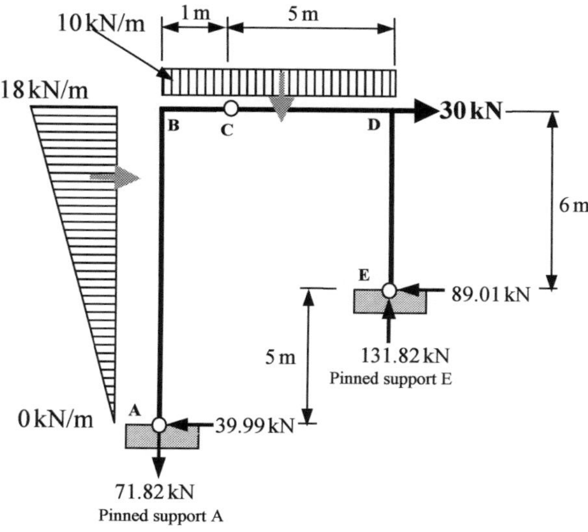

Figure 4.27: Reactions calculated for Example 4.7

Example 4.8: Three pin frame with an inclined member supporting a vertical distributed load and horizontal point loads

The three pin frame shown in Figure 4.28 is supported at A and D on pin joints and has an internal pin joint at C. The frame supports point loads on members AB and at joint C, and has a vertical downward UDL applied along the length of member BC. Determine the reactions at the supports A and D.

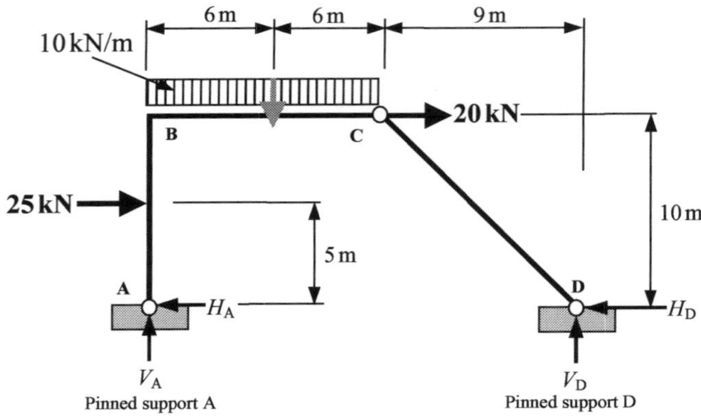

Figure 4.28: Three pin frame – Example 4.8

Calculate reactions:

Summation of horizontal forces $\quad \Sigma F_x = 25\,\text{kN} + 20\,\text{kN} - H_A - H_D = 0$

$\therefore\ 45\,\text{kN} = H_A + H_D$ [1]

Summation of vertical forces $\quad \Sigma F_y = V_A + V_D - (10\,\text{kN/m} \times (6\,\text{m} + 6\,\text{m})) = 0$

$\therefore V_A + V_D = 120\,\text{kN}$ [2]

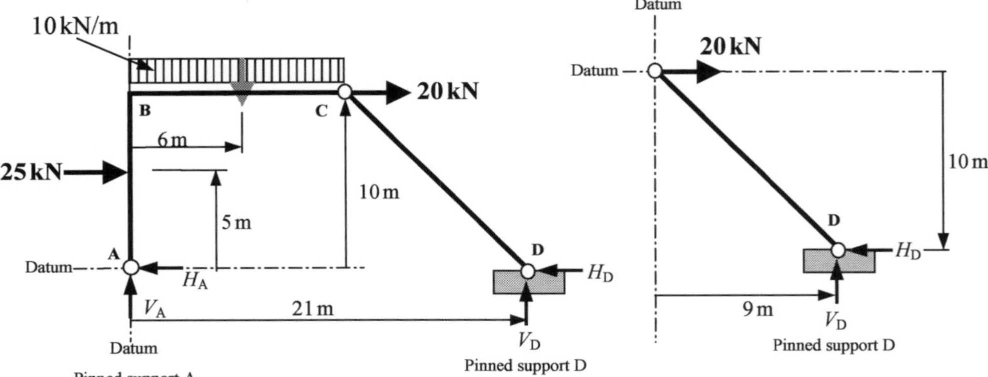

Figure 4.29: Position of datum for calculation of moments – Example 4.8

Taking moments about support A: $\Sigma M_z = (25\,\text{kN} \times 5\,\text{m}) + (10\,\text{kN/m} \times 12\,\text{m} \times (12\,\text{m}/2))$

$+\ (20\,\text{kN} \times 10\,\text{m}) - (V_D \times (6\,\text{m} + 6\,\text{m} + 9\,\text{m})) = 0$

$= 125 + 720 + 200 - 21\,V_D = 0$ [3]

$21\,V_D = 1045 \qquad \therefore\ V_D = 1045/21 = 49.762\,\text{kN}$

Taking moments to left of internal pin: $\Sigma M_z = (20\,\text{kN} \times 0\,\text{m}) - (V_D \times 9\,\text{m}) + (H_D \times 10\,\text{m}) = 0$

$$-9V_D + 10H_D = 0 \qquad [4]$$

We now have four equations:

$$H_A + H_D = 45\,\text{kN} \qquad [1]$$
$$V_A + V_D = 120\,\text{kN} \qquad [2]$$
$$21\,V_D = 1045 \quad \text{(from which } V_D = 49.762\,\text{kN)} \qquad [3]$$
and
$$-9V_D + 10H_D = 0 \qquad [4]$$

Solving these equations simultaneously we find that:

$V_A = 70.238\,\text{kN}$ $\qquad V_D = 49.762\,\text{kN}$
$H_A = 0.2142\,\text{kN}$ $\qquad H_E = 44.7858\,\text{kN}$

The final values for the reactions are as shown in Figure 4.30.

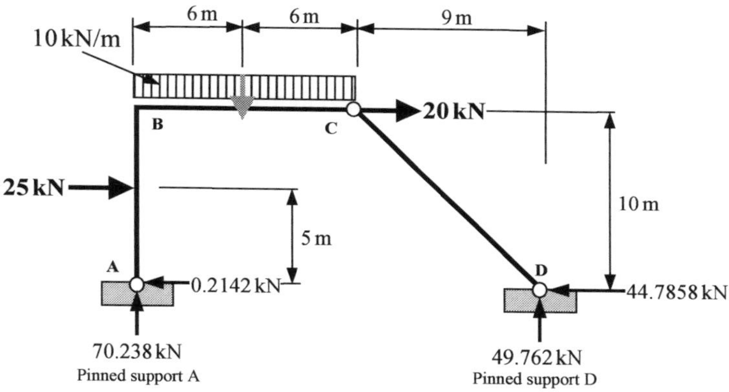

Figure 4.30: Reactions calculated for Example 4.8

4.4 Axial Thrust Diagrams

We have already learnt how to draw both shear force and bending moment diagrams for beams, and the methods applied earlier can be used for three pin frame structures. However, in the design of many structures we also need to consider the forces applied along the line of the member as well as the shear forces, which are perpendicular to it, and the bending moments. Axial thrust diagrams plot the magnitude and direction of forces that act along the line of the member, as shown in Figure 4.31. These diagrams are simple to construct, since changes in the shape of the diagram can only occur at the point of application of a load.

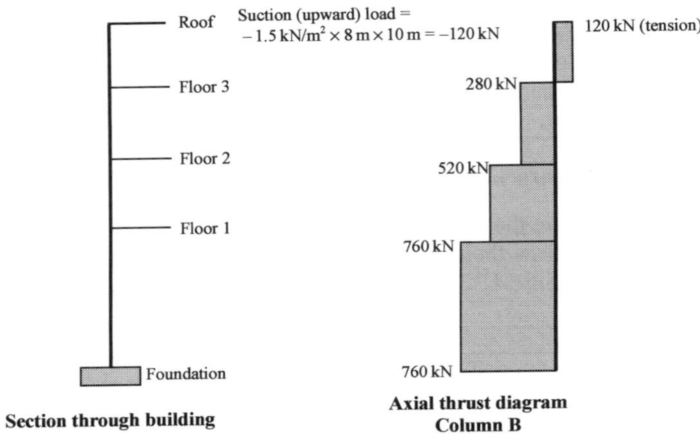

Figure 4.36: Modified axial thrust diagram for column B

From Figure 4.36 we note that the suction load from the roof induces a tensile resistance in column B between the roof and floor 3. The effect on our axial thrust diagram is shown, with the tension in the top part of the column plotted to the right of the vertical axis. Values below are calculated by subtracting the tensile force from the applied compressive loads.

Axial thrust diagrams are useful for the design of many frame structures, particularly in the design of columns, struts and ties which are members loaded predominantly alone their length.

We will now complete the analysis of some three pin frames and draw the Axial thrust, Shear force and Bending moment diagrams for each example.

4.5 Analysis of Three Pin Frames

We are now able to complete the analysis of three pin frames. In order to do this we will:

(1) calculate the reactions at the supports;
(2) draw the axial thrust diagram;
(3) draw the shear force diagram;
and (4) draw the bending moment diagram.

We will also calculate the position and magnitude of the maximum bending moment, noting that this will occur at the point where the shear force passes through zero.

The analysis can be undertaken, as with beams, by dividing the frame into sections and calculating the total shear force and bending moments about a datum. We must, however, bear in mind that it is necessary to calculate moments for both horizontal and vertical loads. This fact adds some complexity to the problem but, in all other respects, follows the same procedures adopted for the analysis of beams.

We will, initially, calculate shear forces and bending moments at specific points on the frame. However, as you become more proficient with this analysis technique, it may be possible to reduce calculation time by selecting only key points on the structure.

Example 4.9: Analysis of a three pin frame supporting a single vertical point load and a horizontal distributed load

The three pin frame shown in Figure 4.37 is supported at points A and F on pin joints and has an internal pin joint at point C. The frame is loaded with a point load at point D and a horizontal UDL along member AB. Determine the reactions at the supports A and F and plot the axial thrust, shear force and bending moment diagrams for the frame.

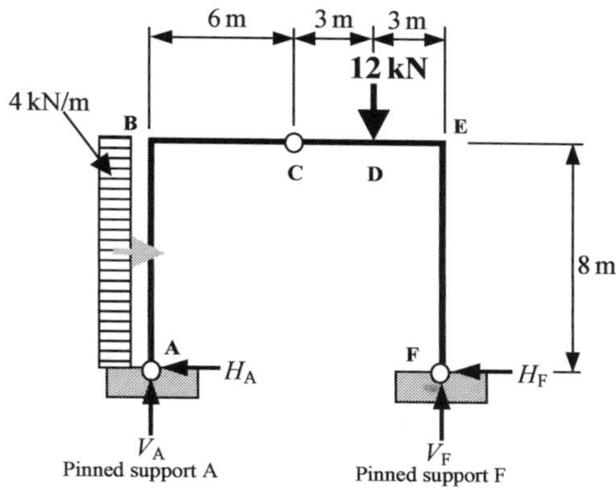

Figure 4.37: Three pin frame – Example 4.9

Summation of horizontal forces $\quad \Sigma F_x = (4\,\text{kN/m} \times 8\text{m}) - H_A - H_F = 0$

$\therefore H_A + H_F = 32\,\text{kN}$ [1]

Summation of vertical forces $\quad \Sigma F_y = V_A + V_F - 12\,\text{kN} = 0$

$\therefore V_A + V_F = 12\,\text{kN}$ [2]

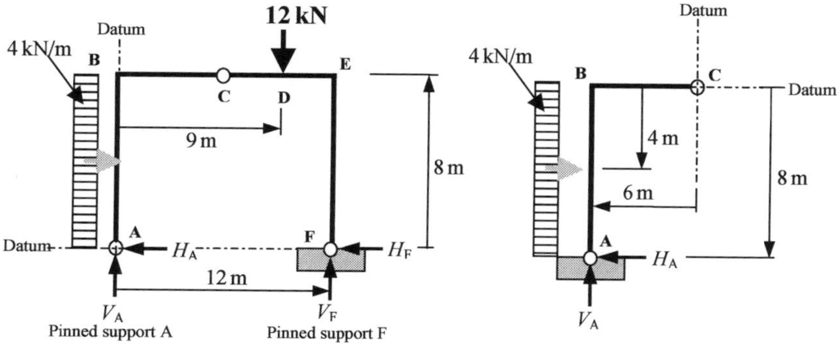

Figure 4.38: Position of datum for calculation of moments – Example 4.9

Taking moments about support A: $\Sigma M_z = (4\,\text{kN/m} \times 8\,\text{m} \times (8\,\text{m}/2)) + (12\,\text{kN} \times (6\text{m} + 3\,\text{m}))$

$- (V_F \times (6\text{m} + 3\,\text{m} + 3\,\text{m})) = 0$

$= 128 + 108 - 12\,V_F = 0$

$= 236/12 = V_F \qquad \therefore V_F = 19.667\,\text{kN}$

Taking moments about internal pin to left side: $\Sigma M_z = -(4\,\text{kN/m} \times 8\,\text{m} \times (8\,\text{m}/2)) + (V_A \times 6\,\text{m}) + (H_A \times 8\,\text{m}) = 0$

$$6V_A + 8H_A = 128\,\text{kN}$$

Solving equations simultaneously we find:

$V_A = -7.667\,\text{kN}$ $\quad V_F = 19.667\,\text{kN}$

$H_A = 21.75\,\text{kN}$ $\quad H_F = 10.25\,\text{kN}$

The final values for the reactions are therefore as shown in Figure 4.39.

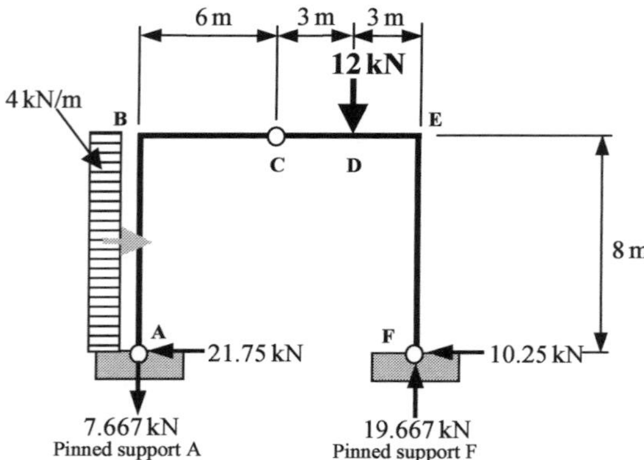

Figure 4.39: Reactions calculated for Example 4.9

We can draw the axial thrust diagram using the results indicated in Figure 4.39 as shown in Figure 4.40.

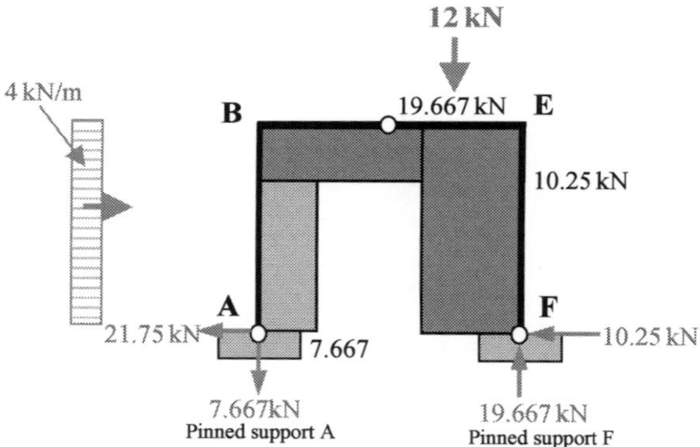

Figure 4.40: Axial thrust diagram for Example 4.9

From Figure 4.40 we note that the axial thrust along member AB is equal to the reaction V_A and is 7.667 kN (tension). The axial thrust along member EF is equal to the reaction V_F and is 19.667 kN (compression). The axial thrust in member BE is equal to the horizontal reaction at H_F. This is equal and opposite to the horizontal load applied at joint B which is:

$$\text{Horizontal load at joint B} = -21.75\,\text{kN} + (4\,\text{kN/m} \times 8\,\text{m}) = 10.25\,\text{kN}$$

To construct the shear force and bending moment diagrams we divide the frame into sections. In this example we will calculate values at 2-m intervals commencing at support A.

Shear force and bending moment at support A:

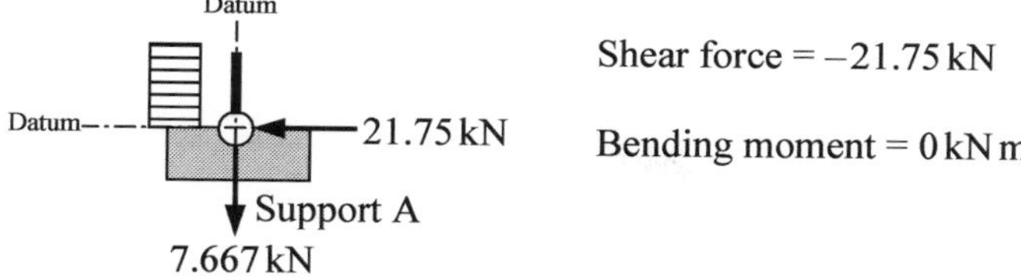

Shear force = –21.75 kN

Bending moment = 0 kN m

Shear force and bending moment at 2 m from support A:

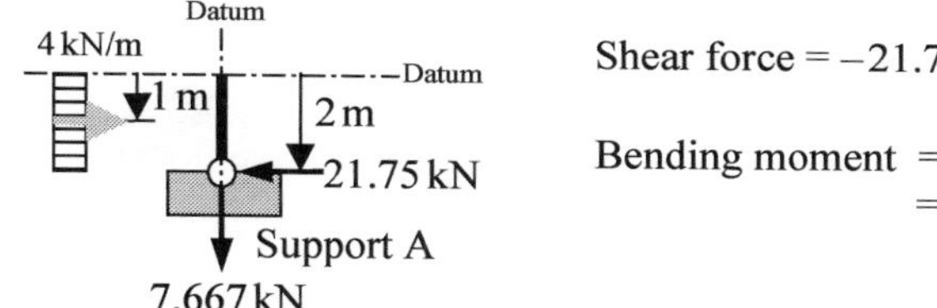

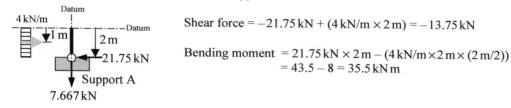

Shear force = –21.75 kN + (4 kN/m × 2 m) = – 13.75 kN

Bending moment = 21.75 kN × 2 m – (4 kN/m × 2 m × (2 m/2))
= 43.5 – 8 = 35.5 kN m

Similarly:

Shear force and bending moment at 4 m from support A:

Shear force = –21.75 kN + (4 kN/m × 4 m) = –5.75 kN

Bending moment = 21.75 kN × 4 m – (4 kN/m × 4 m × (4 m/2)) = 87 – 32 = 55 kN m

Shear force and bending moment at 6 m from support A:

Shear force = –21.75 kN + (4 kN/m × 6 m) = 2.25 kN

Bending moment = 21.75 kN × 6 m – (4 kN/m × 6 m × (6 m/2)) = 130.5 – 72 = 58.5 kN m

Shear force and bending moment at 8 m from support A:

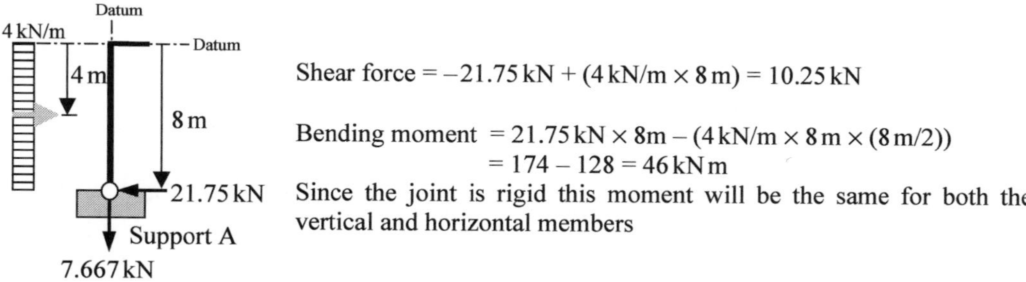

Shear force = –21.75 kN + (4 kN/m × 8 m) = 10.25 kN

Bending moment = 21.75 kN × 8m – (4 kN/m × 8 m × (8 m/2))
= 174 – 128 = 46 kN m

Since the joint is rigid this moment will be the same for both the vertical and horizontal members

Note at this point we must move horizontally along member BE. The shear force will therefore be that perpendicular to the member, i.e. – 7.667 kN at joint B. The calculation of moments will however continue as before.

Shear force and bending moment at 10 m from support A (2 m along member BE):

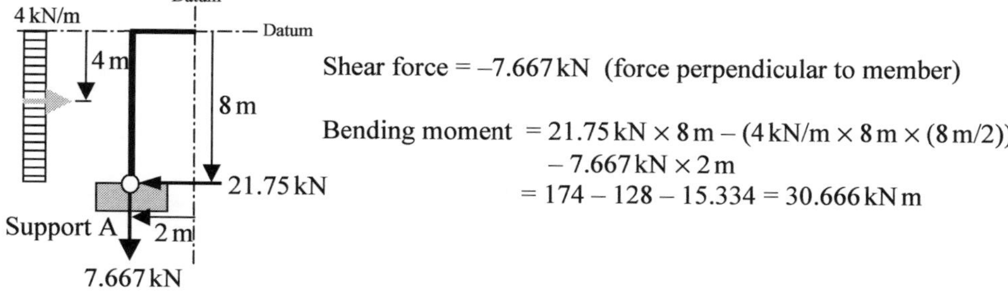

Shear force = –7.667 kN (force perpendicular to member)

Bending moment = 21.75 kN × 8 m – (4 kN/m × 8 m × (8 m/2))
– 7.667 kN × 2 m
= 174 – 128 – 15.334 = 30.666 kN m

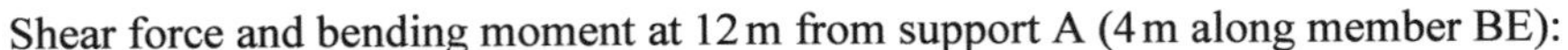

Shear force and bending moment at 12 m from support A (4 m along member BE):

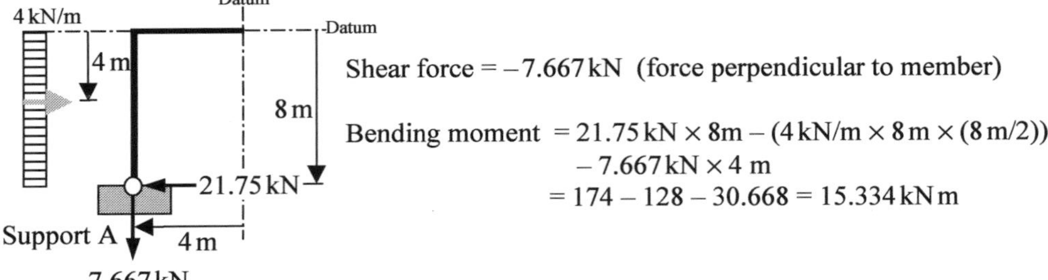

Shear force = –7.667 kN (force perpendicular to member)

Bending moment $= 21.75\,\text{kN} \times 8\text{m} - (4\,\text{kN/m} \times 8\,\text{m} \times (8\,\text{m}/2))$
$- 7.667\,\text{kN} \times 4\ \text{m}$
$= 174 - 128 - 30.668 = 15.334\,\text{kN m}$

We note that at 14 m from support A, i.e. 6 m along member BE, we have the internal pin. The shear force and bending moment at this point are:

Shear force = –7.667 kN

Bending moment $= 21.75\,\text{kN} \times 8\,\text{m} - (4\,\text{kN/m} \times 8\,\text{m} \times (8\,\text{m}/2)) - 7.667 \times 6\,\text{m}$
$= 174 - 128 - 46 = 0\,\text{kN m}$

Note that we have shear force, even at the pin!

Similarly:

Shear force and bending moment at 16 m from support A (8 m along member BE):

Shear force = –7.667 kN

Bending moment $= 21.75\,\text{kN} \times 8\text{m} - (4\,\text{kN/m} \times 8\,\text{m} \times (8\,\text{m}/2)) - 7.667\,\text{kN} \times 8\,\text{m}$
$= 174 - 128 - 61.336 = -15.336\,\text{kN m}$

Shear force and bending moment at 17 m from support A (9 m along member BE):

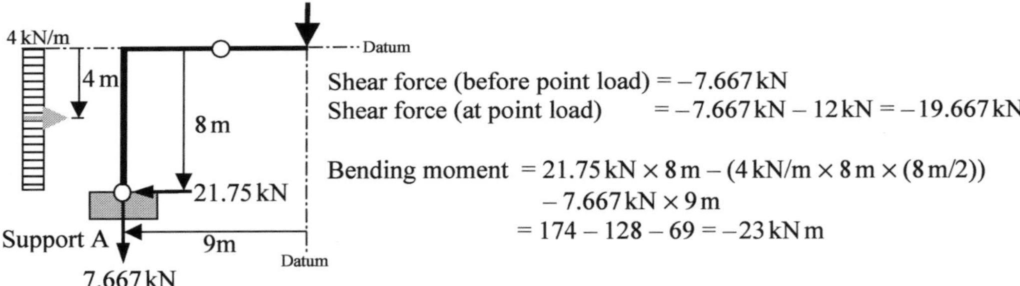

Shear force (before point load) = –7.667 kN

Shear force (at point load) = –7.667 kN – 12 kN = –19.667 kN

Bending moment $= 21.75\,\text{kN} \times 8\,\text{m} - (4\,\text{kN/m} \times 8\,\text{m} \times (8\,\text{m}/2))$
$- 7.667\,\text{kN} \times 9\,\text{m}$
$= 174 - 128 - 69 = -23\,\text{kN m}$

We can continue to calculate moments around the frame, but at this point it might be easier to switch to calculating moments from support F. We will find changing our point of reference quite useful in the analysis of many complex structures.

Shear force and bending moment at 11 m from support F (3 m along member EB):

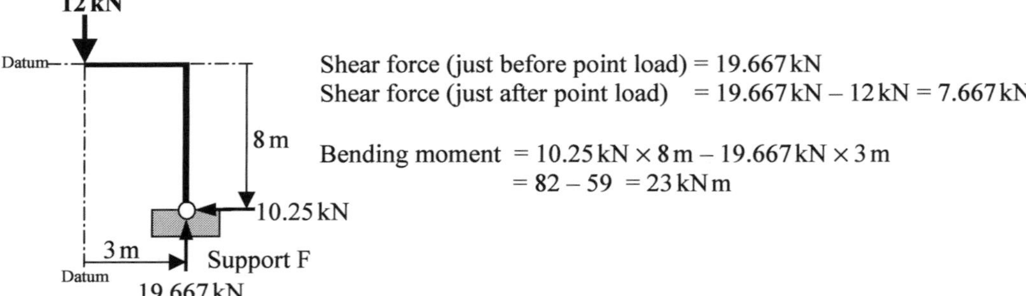

Shear force (just before point load) = 19.667 kN

Shear force (just after point load) = 19.667 kN – 12 kN = 7.667 kN

Bending moment $= 10.25\,\text{kN} \times 8\,\text{m} - 19.667\,\text{kN} \times 3\,\text{m}$
$= 82 - 59 = 23\,\text{kN m}$

Similarly:
Shear force and bending moment at 10 m from support F (2 m along member EB):

Shear force = 19.667 kN

Bending moment $= 10.25\,\text{kN} \times 8\,\text{m} - 19.667\,\text{kN} \times 2\,\text{m}$
$= 82 - 39.334 = 42.666\,\text{kN m}$

Shear force and bending moment at 8 m from support F (at joint E):

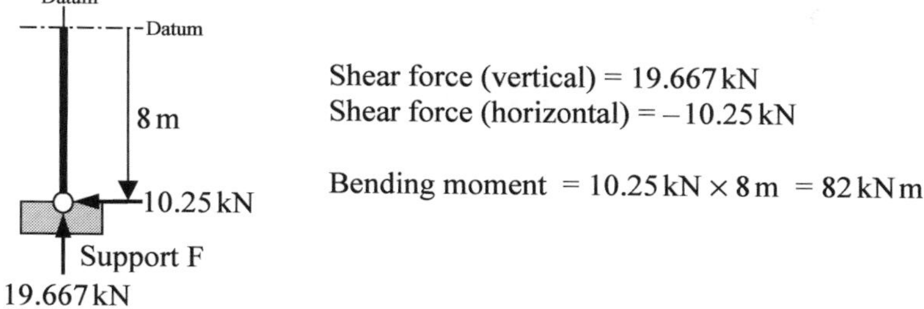

Shear force (vertical) = 19.667 kN
Shear force (horizontal) = – 10.25 kN

Bending moment $= 10.25\,\text{kN} \times 8\,\text{m} = 82\,\text{kN m}$

We can now calculate the moments back to support F such that:
Shear force and bending moment at 6 m from support F:

Shear force = – 10.25 kN

Bending moment $= 10.25\,\text{kN} \times 6\,\text{m} = 61.5\,\text{kN m}$

Shear force and bending moment at 4 m from support F:

Shear force = – 10.25 kN

Bending moment $= 10.25\,\text{kN} \times 4\,\text{m} = 41\,\text{kN m}$

Shear force and bending moment at 2 m from support F:

Shear force = – 10.25 kN

Bending moment $= 10.25\,\text{kN} \times 2\,\text{m} = 20.5\,\text{kN m}$

Shear force and bending moment at 0 m from support F:

Shear force = – 10.25 kN

Bending moment $= 0\,\text{kN m}$

The shear force and bending moment diagrams are shown in Figure 4.41.

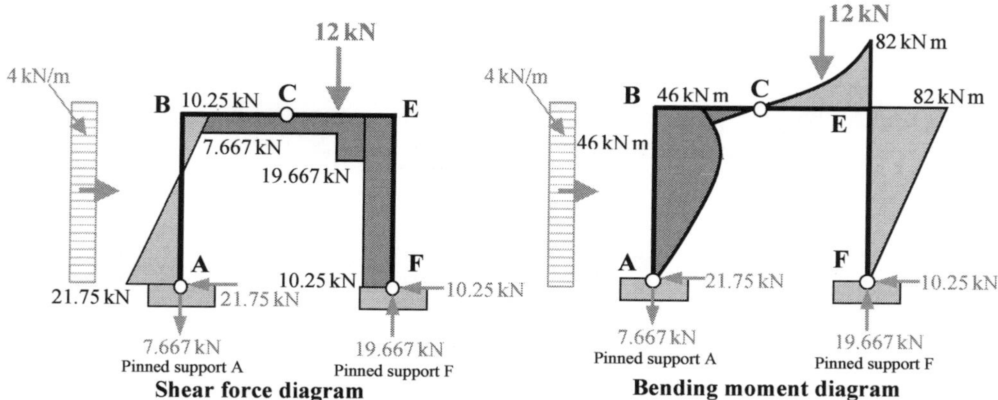

Figure 4.41: Shear force and bending moment diagrams for frame – Example 4.9

Example 4.10: Analysis of a non-symmetric three pin frame supporting a single horizontal and vertical point load and a vertical distributed load

The three pin frame shown in Figure 4.42 is supported at A and D on pin joints and has an internal pin at joint C. The frame is loaded with point loads at C and a vertical downward UDL along member BC. Determine the reactions at supports A and D and plot the axial thrust, shear force and bending moment diagrams for the frame.

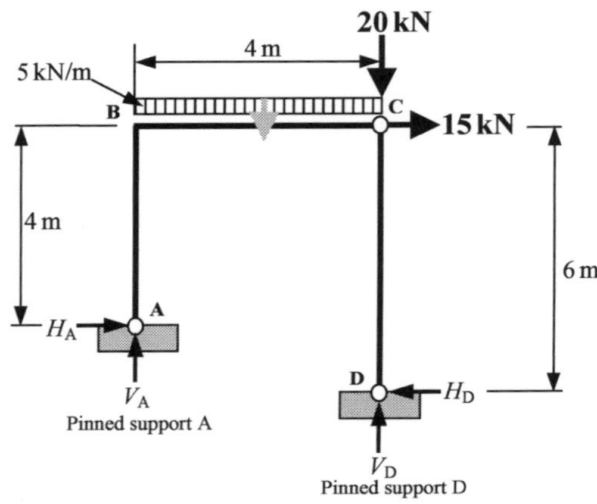

Figure 4.42: Three pin frame – Example 4.10

Calculate reactions:

Summation of horizontal forces $\quad \Sigma F_x = 15\,\text{kN} + H_A - H_D = 0$

$\therefore H_A - H_D = -15\,\text{kN}$ [1]

Summation of vertical forces $\quad \Sigma F_y = V_A + V_D - 20\,\text{kN} - (5\,\text{kN/m} \times 4\,\text{m}) = 0$

$\therefore V_A + V_D = 40\,\text{kN}$ [2]

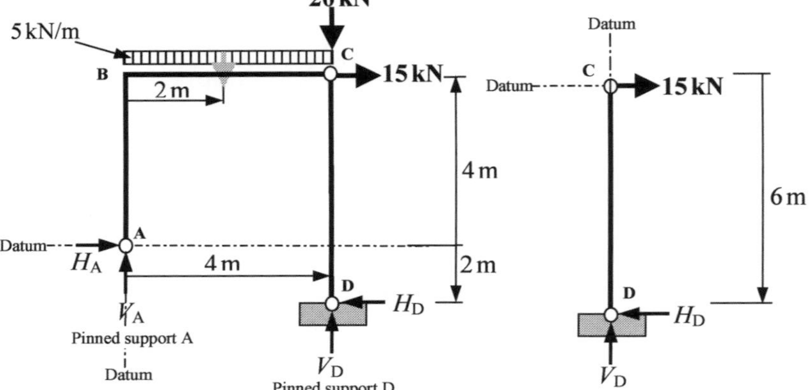

Figure 4.43: Position of datum for calculation of moments – Example 4.10

Taking moments about support A: $\Sigma M_z = (5\,\text{kN/m} \times 4\,\text{m} \times 4\,\text{m}/2) + (20\,\text{kN} \times 4\,\text{m}) + (15\,\text{kN} \times 4\,\text{m})$
$+ (H_D \times 2\,\text{m}) - (V_D \times 4\,\text{m}) = 0$
$= 40 + 80 + 60 + 2H_D - 4V_D = 0$
$= 180 + 2H_D = 4V_D$ [3]

Taking moments about internal pin to right side: $\Sigma M_z = H_D \times 6\,\text{m} = 0$
$6H_D = 0\,\text{kN}$

Solving equations simultaneously we find:

$V_A = -5\,\text{kN}$ $\quad V_D = 45\,\text{kN}$

$H_A = -15\,\text{kN}$ $\quad H_D = 0\,\text{kN}$

The final values for the reactions are therefore as shown in Figure 4.44.

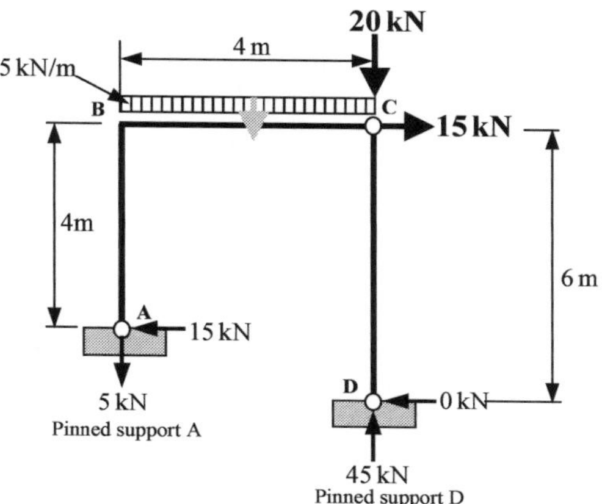

Figure 4.44: Reactions calculated for Example 4.10

We can now construct the axial thrust diagram, as shown in Figure 4.45.

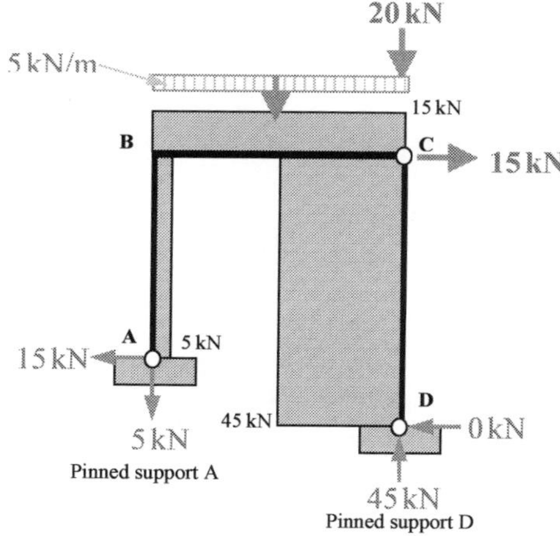

Figure 4.45: Axial thrust diagram for Example 4.10

In this case, to construct the shear force and bending moment diagrams, it is only necessary to calculate the values at principal points on the frame. We can fill in the missing information based on our knowledge of structural analysis.

Shear force and bending moment at support A:

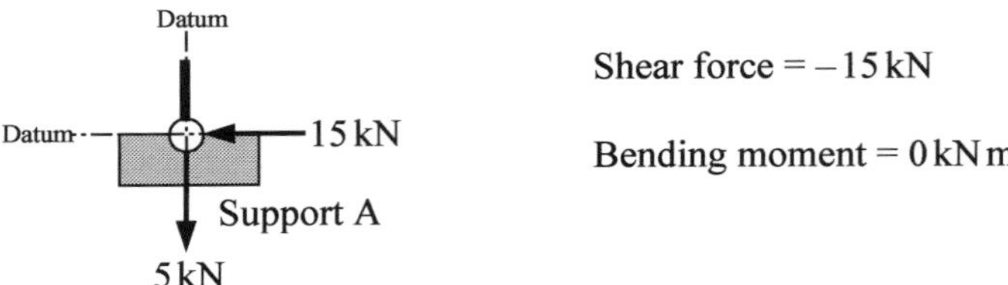

Shear force = – 15 kN

Bending moment = 0 kN m

Shear force and bending moment at 2 m from support A:

Shear force = – 15 kN

Bending moment = 15 kN × 2 m = 30 kN m

Shear force and bending moment at 4 m from support A:

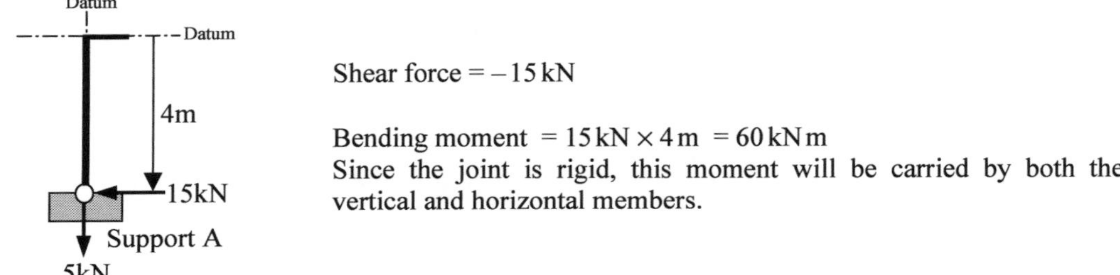

Shear force = – 15 kN

Bending moment = 15 kN × 4 m = 60 kN m

Since the joint is rigid, this moment will be carried by both the vertical and horizontal members.

To construct the shear force and bending moment diagram for the horizontal member BC, we note that the moment at joint B is 60 kN m and the moment at joint C is zero (internal pin). We therefore only require one intermediate point in order to determine the shape of the diagram, noting that the shear force will vary linearly from joint B to C whilst the distribution of bending moments will be of a parabolic form.

Shear force and bending moment at 2 m from joint B along member BC:

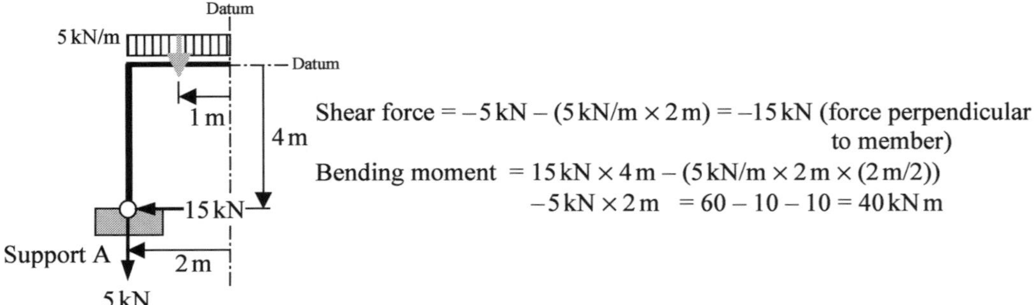

Shear force = – 5 kN – (5 kN/m × 2 m) = –15 kN (force perpendicular to member)

Bending moment = 15 kN × 4 m – (5 kN/m × 2 m × (2 m/2)) – 5 kN × 2 m = 60 – 10 – 10 = 40 kN m

We now have enough information to enable us to sketch the shear force and bending moment diagrams. These are shown in Figure 4.46.

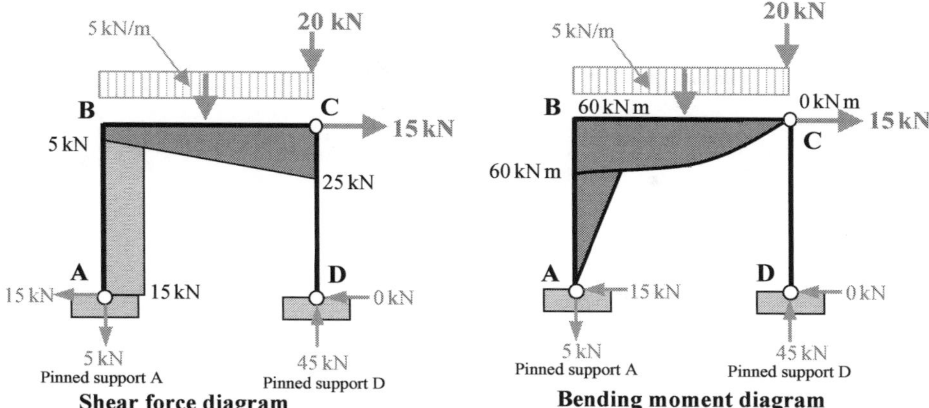

Figure 4.46: Shear force and bending moment diagrams for frame – Example 4.10

◇

Example 4.11: Analysis of a non-symmetric three pin frame supporting a single horizontal point load and a vertical distributed load

The three pin frame shown in Figure 4.47 is supported at A and D on pin joints and has an internal pin at joint B. The frame is loaded with a horizontal point load at joint C and a vertical UDL along member BC. Determine the reactions at the supports A and D, and plot the axial thrust, shear force and bending moment diagrams for the frame.

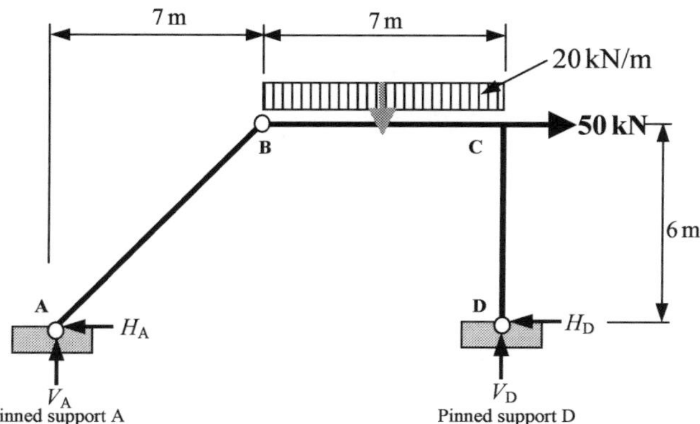

Figure 4.47: Three pin frame – Example 4.11

Calculate reactions:

Summation of horizontal forces $\quad \Sigma F_x = 50\,\text{kN} - H_A - H_D = 0$

$\therefore H_A + H_D = 50\,\text{kN}$ [1]

Summation of vertical forces $\quad \Sigma F_y = V_A + V_D - (20\,\text{kN/m} \times 7\,\text{m}) = 0$

$\therefore V_A + V_D = 140\,\text{kN}$ [2]

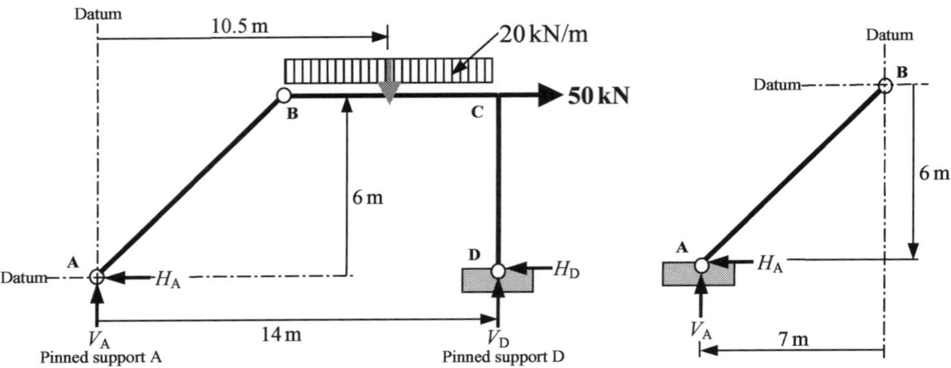

Figure 4.48: Position of datum for calculation of moments – Example 4.11

Taking moments about support A: $\Sigma M_z = (20\,\text{kN/m} \times 7\,\text{m} \times 10.5\,\text{m}) + (50\,\text{kN} \times 6\,\text{m}) - (V_D \times 14\,\text{m}) = 0\,\text{m}$

$= 1470 + 300 - 14\,V_D = 0$

$1770 = 14\,V_D \qquad \therefore\ V_D = 126.43\,\text{kN}$ [3]

Taking moments about internal pin to left side: $\Sigma M_z = V_A \times 7\,\text{m} + H_A \times 6\,\text{m} = 0$

$7\,V_A + 6\,H_A = 0\,\text{kN}$ [4]

Solving equations simultaneously we find:

$V_A = 13.57\,\text{kN} \qquad V_D = 126.43\,\text{kN}$

$H_A = -15.833\,\text{kN} \qquad H_D = 65.833\,\text{kN}$

The final values for the reactions are therefore as shown in Figure 4.49.

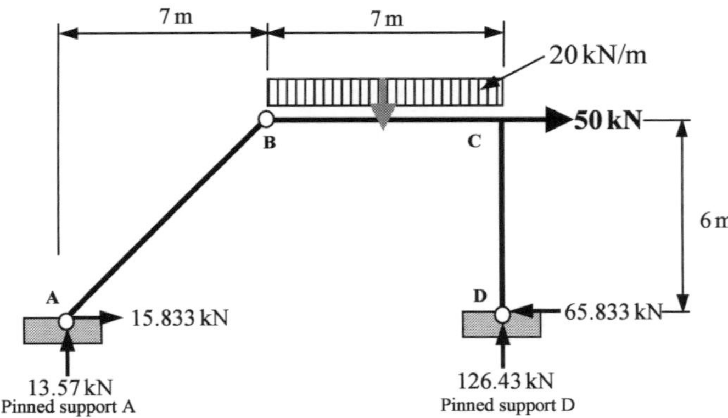

Figure 4.49: Reactions calculated for Example 4.11

Here we note that the reactions at support A are 13.57 kN and 15.833 kN in the vertical and horizontal directions. By definition, the axial thrust is the force along the line of the member and the shear force is the force perpendicular to it. We must therefore derive the force components for the reactions resolved along the line of, and perpendicular to, member AB. Figure 4.50 shows the force diagrams for the resolution of the forces at the reactions.

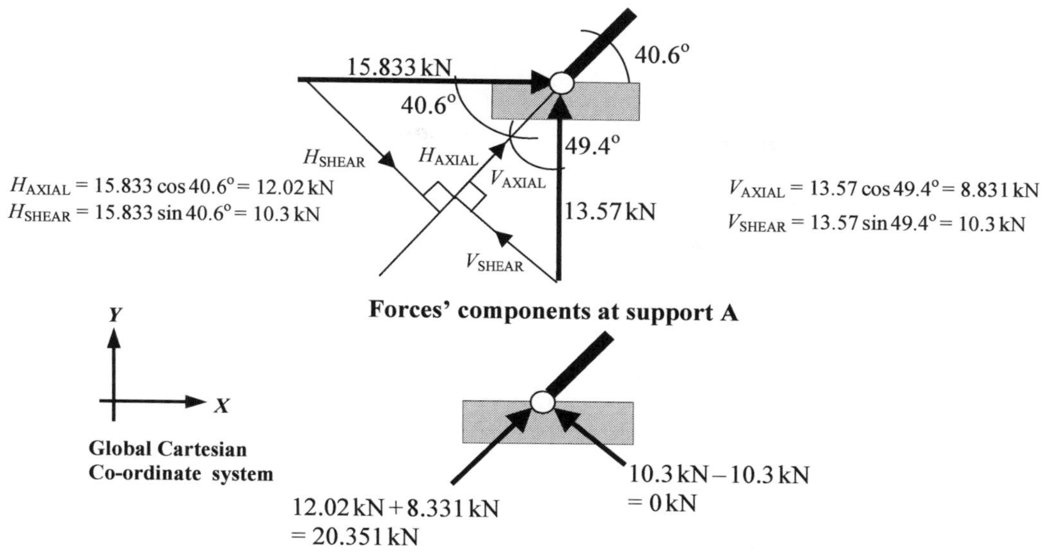

Figure 4.50: Resolution of reactions at support A

We can now plot the axial thrust diagram for the structure, as shown in Figure 4.51.

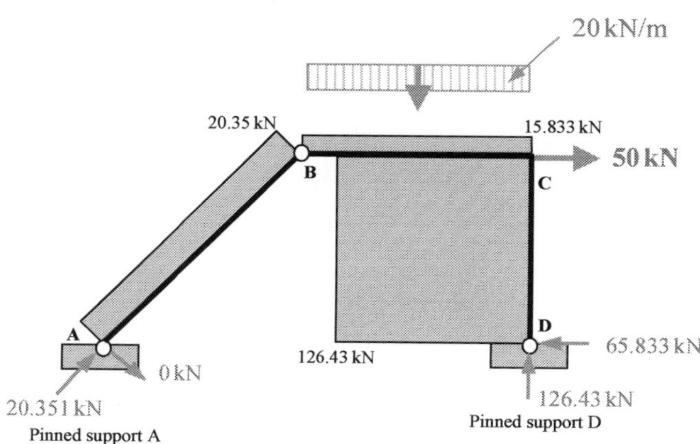

Figure 4.51: Axial thrust diagram – Example 4.11

We will now calculate the principal values necessary to sketch the shear force and bending moment diagrams.

Shear force and bending moment at support A:

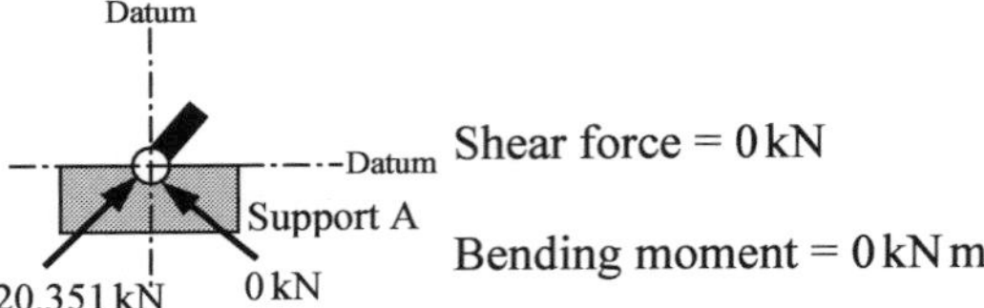

Shear force = 0 kN

Bending moment = 0 kN m

Shear force and bending moment 3.5 m horizontally from support A:

Noting that the shear force along the member AB is zero we will revert to our original horizontal and vertical reactions to calculate bending moments along the member.

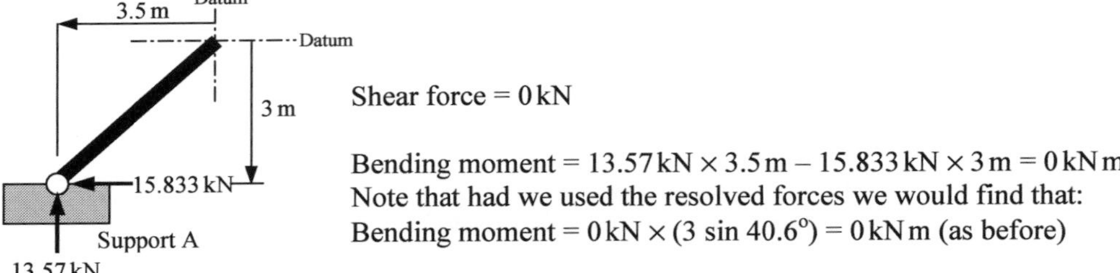

Shear force = 0 kN

Bending moment = 13.57 kN × 3.5 m – 15.833 kN × 3 m = 0 kN m
Note that had we used the resolved forces we would find that:
Bending moment = 0 kN × (3 sin 40.6°) = 0 kN m (as before)

Since the bending moment at joint B is zero then the moments along member AB will be 0 kN m throughout its length.

We now calculate the moments to the right of joint C:

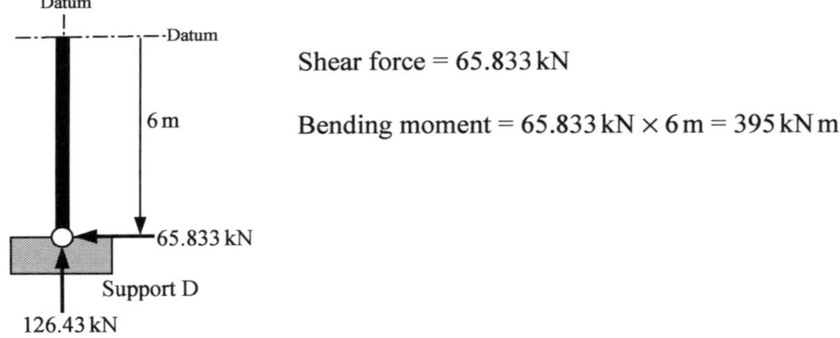

Shear force = 65.833 kN

Bending moment = 65.833 kN × 6 m = 395 kN m

We must now calculate the bending moments along the member BC. We will do this for every 1 m length commencing at joint C and moving toward joint B. This is required because the distribution of bending moments along the member are not easily determined by observation.

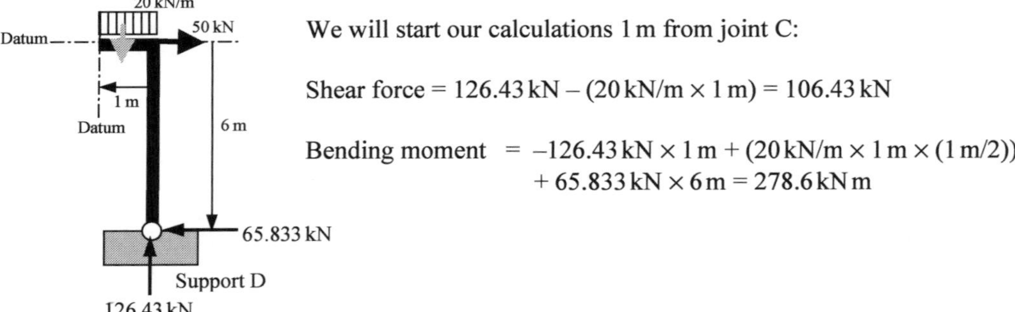

We will start our calculations 1 m from joint C:

Shear force = 126.43 kN – (20 kN/m × 1 m) = 106.43 kN

Bending moment = –126.43 kN × 1 m + (20 kN/m × 1 m × (1 m/2)) + 65.833 kN × 6 m = 278.6 kN m

Moving along the beam BC we find:

2 m from joint C

Shear force = 126.43 kN – (20 kN/m × 2 m) = 86.43 kN

Bending moment = – 126.43 kN × 2 m + (20 kN/m × 2 m × (2m/2)) + 65.833 kN × 6 m = 182 kN m

3 m from joint C

Shear force = 126.43 kN – (20 kN/m × 3 m) = 66.43 kN

Bending moment = –126.43 kN × 3 m + (20 kN/m × 3 m × (3 m/2)) + 65.833 kN × 6 m = 108 kN m

4 m from joint C

Shear force = 126.43 kN – (20 kN/m × 4 m) = 46.43 kN

Bending moment = –126.43 kN × 4 m + (20 kN/m × 4 m × (4 m/2)) + 65.833 kN × 6 m
= 49.28 kN m

5 m from joint C

Shear force = 126.43 kN – (20 kN/m × 5 m) = 26.43 kN

Bending moment = –126.43 kN × 5 m + (20 kN/m × 5 m × (5 m/2)) + 65.833 kN × 6 m
= 12.88 kN m

6 m from joint C

Shear force = 126.43 kN – (20 kN/m × 6 m) = 6.43 kN

Bending moment = –126.43 kN × 6 m + (20 kN/m × 6 m × (6 m/2)) + 65.833 kN × 6 m
= –3.58 kN m

7 m from joint C (at joint B = pin)

Shear force = 126.43 kN – (20 kN/m × 7 m) = –13.57 kN

Bending moment = –126.43 kN × 7 m + (20 kN/m × 7 m × (7 m/2)) + 65.833 kN × 6 m = 0 kN m

We can now draw the shear force and bending moment diagrams as shown in Figure 4.52.

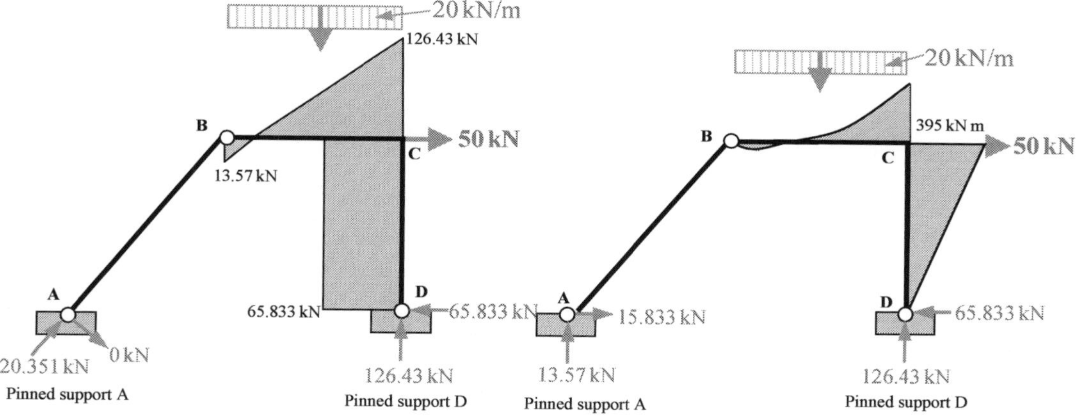

Figure 4.52: Shear force and bending moment diagram for frame – Example 4.11

◇

We will now complete some further examples. The layout used for these worked examples should be noted and adopted for future calculations. However, before we proceed to complete these examples, it is worth reviewing the construction of each of the force diagrams. Firstly, in the axial thrust diagram we note that changes only occur at positions where load is applied. We also note that it may be necessary to resolve vertical and horizontal reactions in order to plot the axial thrust for inclined members. This also applies to shear force diagram. With shear forces we note that changes take place at the point of application of a point load and vary linearly for UDLs. These will prove to be useful checks when constructing diagrams and will also enable us to complete the drawings with minimal calculation.

For bending moment diagrams we note that:

(a) bending moment at pin supports = 0
(b) bending moment at internal pin = 0
(c) bending moments are transferred in rigid joints (same in column and beam)
(d) bending moment diagrams for point loads vary linearly (trianglar distribution)
(e) bending moment diagrams for UDLs vary in a parabolic form.

These facts not only allow use to check our calculations, but will also enable us to produce a preliminary sketch of the bending moment diagram and determine which principal values are required and which elements of the structure warrant detailed examination.

Consider the frame of Figure 4.53.

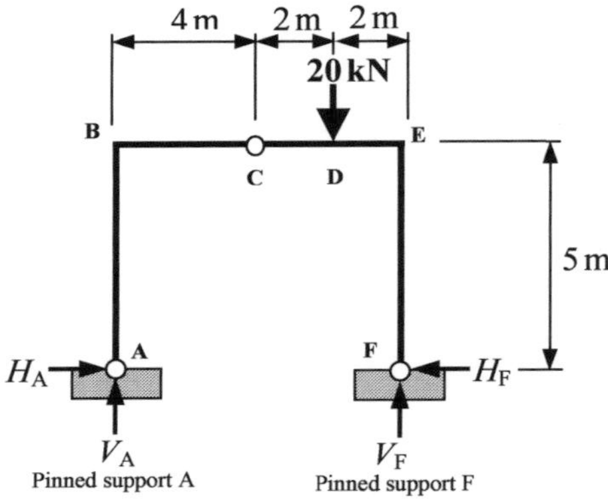

Figure 4.53: Three pin frame

We can, without calculation, sketch the 'known' elements of the bending moment diagram, as shown in Figure 4.54.

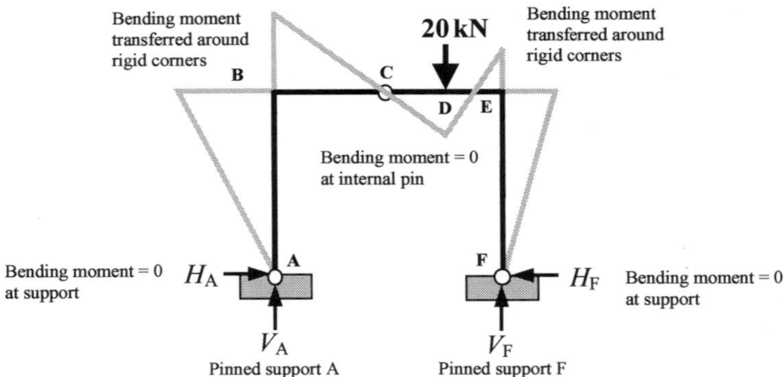

Figure 4.54: Preliminary sketch of bending moment diagram

Whilst some calculations will be necessary to confirm the values, and also to determine the actual distribution of the bending moments along members BC and CE, it is useful to be able to visualise the diagram before we commence. Errors in calculation can then be easily identified and corrected.

Example 4.12: Three pin frame carrying horizontal and vertical loads

The frame shown as Example 4.12 below is supported on pin joints at A and E and has an internal pin at joint C:

(a) calculate the reactions at the supports;

(b) plot the axial thrust, shear force and bending moment diagrams.

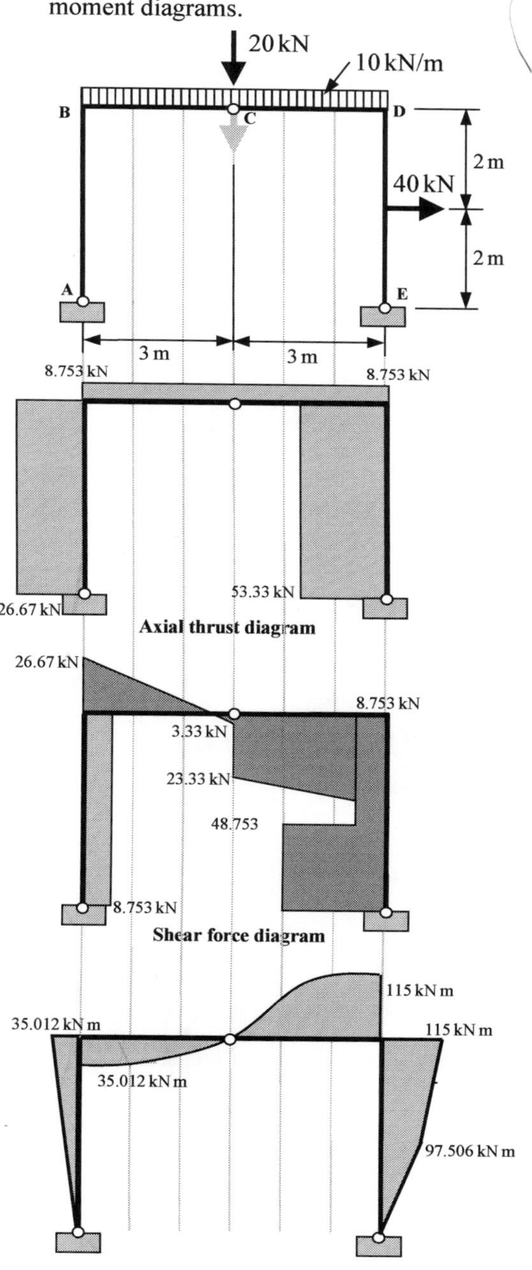

Solution:

Calculate reactions:

Assume reactions applied as:

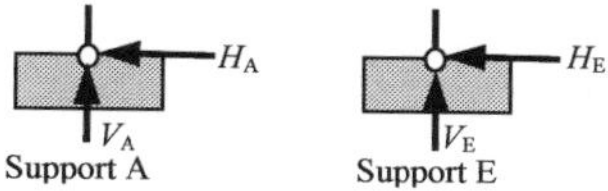

$$\Sigma F_x \quad H_A + H_E = 40\,\text{kN} \qquad [1]$$

$$\Sigma F_y \quad V_A + V_E - 20\,\text{kN} - (10\,\text{kN/m} \times 6\,\text{m}) = 0\,\text{kN}$$

$$V_A + V_E = 80\,\text{kN} \qquad [2]$$

Take moments about support A:

$$\Sigma M_z = 20\,\text{kN} \times 3\,\text{m} + (10\,\text{kN/m} \times 6\,\text{m} \times (6\,\text{m}/2)) + (40\,\text{kN} \times 2\,\text{m}) - V_E \times 6\,\text{m} = 0\,\text{kN m}$$

$$6\,V_E = 320\,\text{kN m} \qquad [3]$$

Take moments to left of pin C:

$$\Sigma M_z = V_A \times 3\,\text{m} + H_A \times 4\,\text{m} - (10\,\text{kN/m} \times 3\,\text{m} \times (3\,\text{m}/2)) = 0\,\text{kN m}$$

$$3\,V_A + 4\,H_A = 45\,\text{kN m} \qquad [4]$$

Solving equations simultaneously gives:

$V_A = 26.67\,\text{kN}$ $\quad V_E = 53.33\,\text{kN}$

$H_A = -8.753\,\text{kN}$ $\quad H_E = 48.753\,\text{kN}$

Axial thrust for frame:

Member		Axial thrust (kN)
AB	26.67 kN (vertical reaction at A)	26.67
BC	8.753 kN (horizontal reaction at A)	8.753
CD	8.753 kN (horizontal reaction at A) = 48.753 kN (horizontal reaction at E) – 40 kN (load)	8.753 8.753
DE	53.33 kN (vertical reaction at E)	53.33

Principal shear forces on frame:

Position		Shear force (kN)
Support A	8.753 kN (horizontal reaction at A)	8.753
Joint B (member AB)	8.753 kN	8.753
Joint B (member BC)	26.67 kN	26.67
Joint C	26.67 kN — 10 kN/m × 3 m 26.67 kN — 10 kN/m × 3 m – 20 kN	–3.33 –23.33
Joint D	26.67 kN – 10 kN/m × 6 m – 20 kN	–53.33
Support E	– 48.753 kN (horizontal reaction at E)	–48.753
2 m above support E	– 48.753 kN + 40 kN	8.753

Principal bending moments for frame:

Position		Bending moment (kN m)
Joint B	8.753 kN × 4 m	35.012
2 m along BD	– 8.753 × 4 m + 26.67 kN × 2 m – 10 kN/m × 2 m × 1 m	1.672
4 m along BD	– 8.753 × 4 m + 26.67 kN × 4 m – 10 kN/m × 4 m × 2 m – 20 kN × 1 m	–28.332
6 m along BD	– 8.753 × 4 m + 26.67 kN × 6 m – 10 kN/m × 6 m × 3 m – 20 kN × 3 m	–115
Joint D	48.753 kN × 4 m – 40 kN × 2 m	115
2 m below D	48.753 kN × 2 m	97.506

Example 4.13: Symmetric three pin frame carrying a vertical load

The frame shown as Example 4.13 below is supported on pin joints at A and D and has an internal pin at joint C:

(a) calculate the reactions at the supports;

(b) plot the axial thrust, shear force and bending moment diagrams.

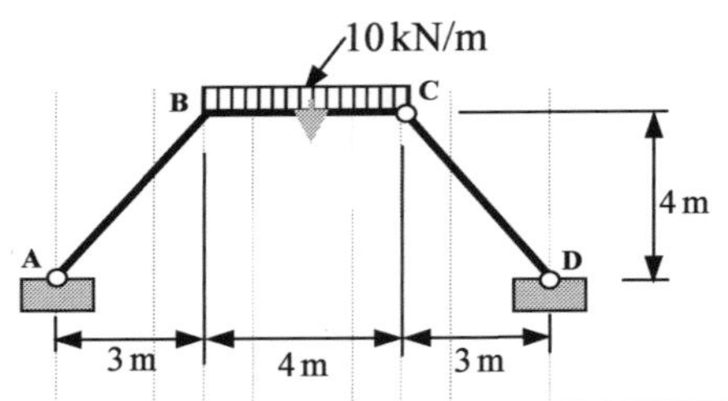

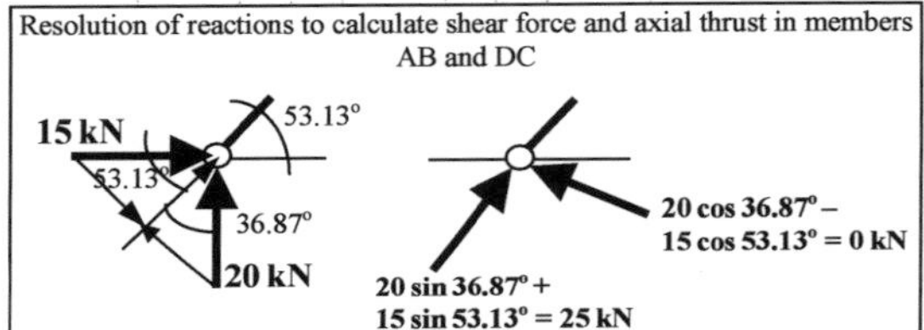

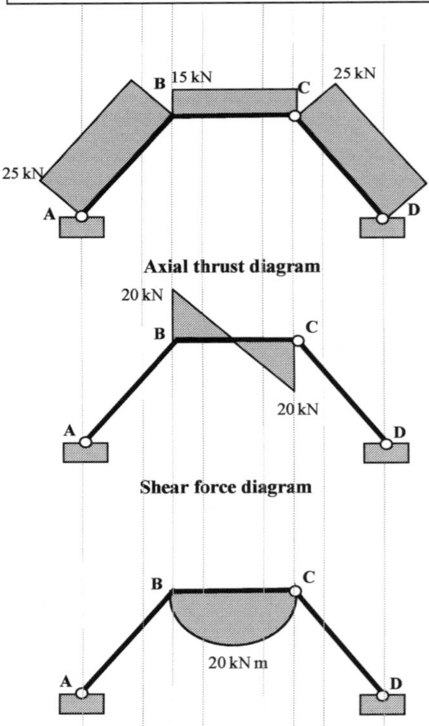

Bending moment diagram

Solution:

Calculate reactions:

Assume reactions applied as:

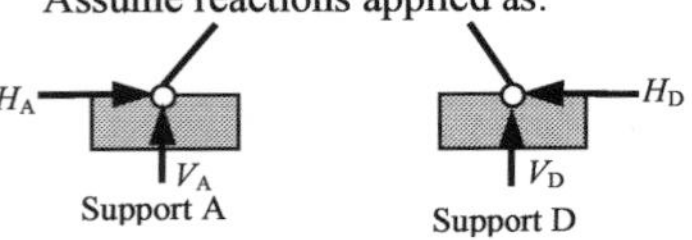

$\Sigma F_x \quad H_A - H_D = 0\,\text{kN}$ [1]

$\Sigma F_y \quad V_A + V_D - (10\,\text{kN/m} \times 4\,\text{m}) = 0\,\text{kN}$

$V_A + V_D = 40\,\text{kN}$ [2]

Take moments about support A:

$\Sigma M_z = (10\,\text{kN/m} \times 4\,\text{m} \times (3\,\text{m} + (4\,\text{m}/2))) - V_D \times 10\,\text{m} = 0\,\text{kN m}$

$10\,V_D = 200\,\text{kN m}$ [3]

Take moments to right of pin C:

$\Sigma M_z = H_D \times 4\,\text{m} - V_D \times 3\,\text{m} = 0\,\text{kN m}$

$4\,H_D = 3\,V_D$ [4]

Solving equations simultaneously gives:

$V_A = 20\,\text{kN}$ $\quad V_D = 20\,\text{kN}$

$H_A = 15\,\text{kN}$ $\quad H_E = 15\,\text{kN}$

Axial thrust for frame:

Member		Axial thrust (kN)
AB	25 kN (resultant of reactions at A)	25
BC	15 kN (horizontal reactions at A and D)	15
CD	25 kN (resultant of reactions at D)	25

Note: Axial compression in member CD drawn on outside of frame in this case (for clarity).

Principal shear forces on frame:

Position		Shear force (kN)
Support A	0 kN (resultant of reactions at A)	0
Joint B (member BC)	20 kN (vertical reaction at A)	20
Midway along BC	20 kN – 10 kN/m × 2 m	0
Joint C (member BC)	20 kN – 10 kN/m × 4 m 20 kN – 10 kN/m × 4 m + 20 kN	–20 0
Support D	0 kN (resultant of reactions at D)	0

Principal bending moments for frame:

Position		Bending moment (kN m)
Joint B	15 kN × 4 m – 20 kN × 3 m	0
1 m along BC	15 kN × 4 m – 20 kN × 4 m + 10 kN/m × 1m × 0.5 m	–15
2 m along BC	15 kN × 4 m – 20 kN × 5 m + 10 kN/m × 2 m × 1 m	–20
3 m along BC	15 kN × 4 m – 20 kN × 6 m + 10 kN/m × 3 m × 1.5 m	–15
Joint C	–15 kN × 4 m + 20 kN × 1.5 m	0

Example 4.14: Three pin frame carrying a horizontal and a vertical load

The frame shown as Example 4.14 below is supported on pin joints at A and E and has an internal pin at joint C:

(a) calculate the reactions at the supports;

(b) plot the axial thrust, shear force and bending moment diagrams.

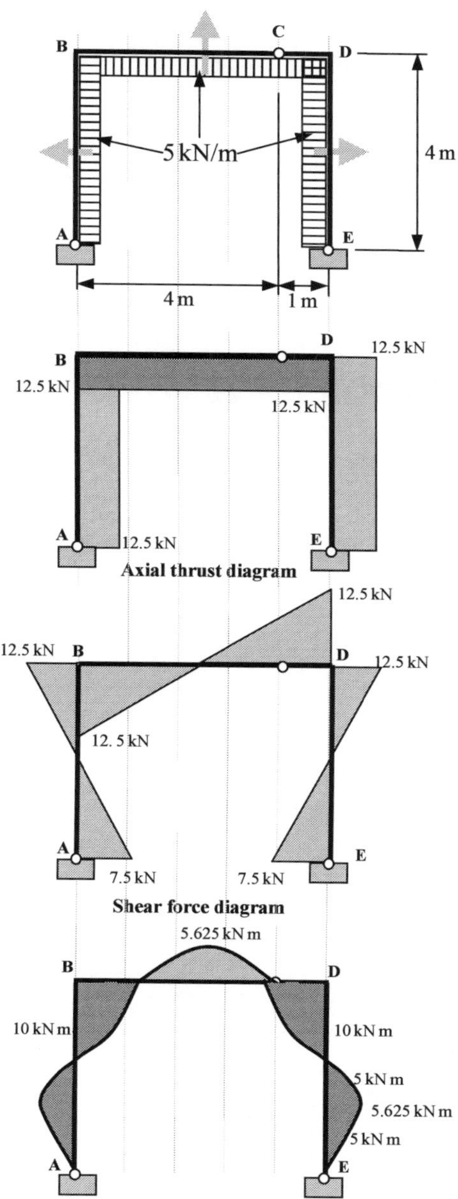

Solution:

Calculate reactions:

Assume reactions applied as:

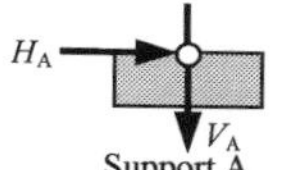

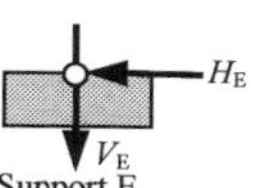

ΣF_x $H_A - 5\text{ kN/m}\times 4\text{ m} + 5\text{ kN/m}\times 4\text{ m} - H_E = 0\text{ kN}$

$H_A - H_E = 0\text{ kN}$ [1]

ΣF_y $-V_A - V_E + (5\text{ kN/m}\times 5\text{ m}) = 0\text{ kN}$

$V_A + V_E = 25\text{ kN}$ [2]

Take moments about support A:

$\Sigma M_z = -(5\text{ kN/m}\times 4\text{ m}\times(4\text{ m}/2))$
$-(5\text{ kN/m}\times 4\text{ m}\times(4\text{ m}/2))$
$+(5\text{ kN/m}\times 4\text{ m}\times(4\text{ m}/2)) + V_E\times 5\text{ m} = 0\text{ kNm}$

$5V_E = 62.5\text{ kNm}$ [3]

Take moments to right of pin C:

$\Sigma M_z = -(5\text{ kN/m}\times 4\text{ m}\times(4\text{ m}/2)) + V_E\times 1\text{ m} + H_E\times 4\text{ m}$
$-(5\text{ kN/m}\times 1\text{ m}\times(1\text{ m}/2)) = 0\text{ kNm}$

$V_E + 4H_E = 42.5\text{ kNm}$ [4]

Solving equations simultaneously gives:

$V_A = 12.5\text{ kN}$ $V_E = 12.5\text{ kN}$

$H_A = 7.5\text{ kN}$ $H_E = 7.5\text{ kN}$

Axial thrust for frame:

Member		Axial thrust (kN)
AB	12.5 kN (vertical reaction at A)	12.5
BD	7.5 kN – 5 kN/m × 4 m	– 12.5
DE	12.5 kN (vertical reaction at E)	12.5

Principal shear forces on frame:

Position		Shear force (kN)
Support A	7.5 kN (horizontal reaction at A)	7.5
Joint B (member AB)	7.5 kN – 5 kN/m × 4 m	– 12.5
Joint B (member BD)	– 12.5 kN (vertical reaction at A)	– 12.5
Joint D (member BD)	– 12.5 kN + 5 kN/m × 4 m – 12.5 kN + 5 kN/m × 4 m – 12.5 kN	12.5 0
Joint D (member DE)	– 7.5 kN + 5 kN/m × 4 m	12.5
Support E	– 7.5 kN (horizontal reaction at E)	7.5

Principal bending moments for frame:

Position		Bending moment (kN m)
1 m above joint A	– 7.5 kN × 1 m + 5 kN/m × 1 m × (1 m/2)	– 5
1.5 m above joint A	– 7.5 kN × 1.5 m + 5 kN/m × 1.5 m × (1.5 m/2)	– 5.625
2 m above joint A	– 7.5 kN × 2 m + 5 kN/m × 2 m × (2 m/2)	– 5
3 m above joint A	– 7.5 kN × 3 m + 5 kN/m × 3 m × (3 m/2)	0
Joint B	– 7.5 kN × 4 m + 5 kN/m × 4 m × (4 m/2)	10
Midway member BD	– 7.5 kN × 4 m + 5 kN/m × 4 m × (4 m/2) + 5 kN/m × 2.5 m × 2.5 m/2 – 12.5 kN × 2.5 m	– 5.625
Joint C	7.5 kN × 4 m – 5 kN/m × 4 m × (4 m/2) – 5 kN/m × 1 m × 1 m/2 + 12.5 kN × 1 m	0
Joint D	7.5 kN × 4 m – 5 kN/m × 4 m × (4 m/2)	– 10

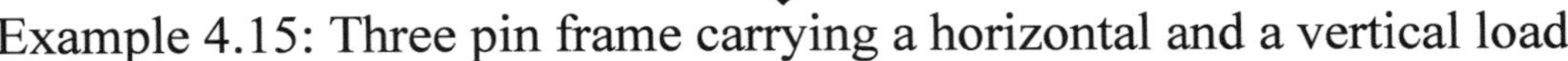

Example 4.15: Three pin frame carrying a horizontal and a vertical load

The frame shown as Example 4.15 below is supported on pin joints at A and D and has an internal pin at joint B:

(a) calculate the reactions at the supports;

(b) plot the axial thrust, shear force and bending moment diagrams.

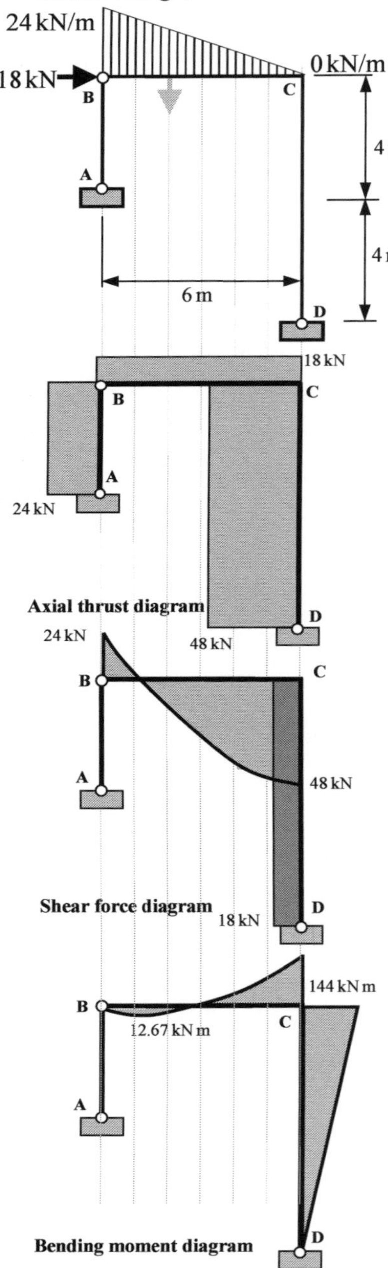

Solution:

Calculate reactions:

Assume reactions applied as:

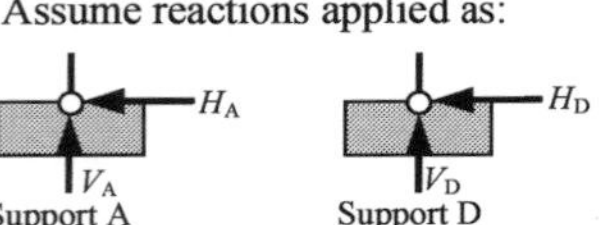

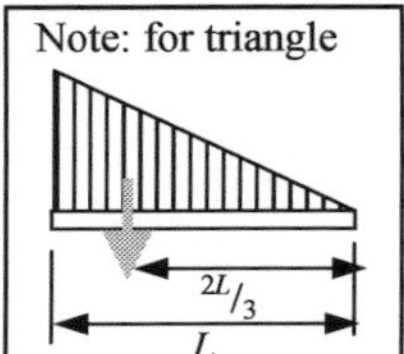

$\Sigma F_x \quad -H_A - H_D + 18\text{ kN} = 0\text{ kN}$

$H_A + H_D = 18\text{ kN}$ [1]

$\Sigma F_y \quad V_A + V_D - (0.5 \times 24\text{ kN/m} \times 6\text{ m}) = 0\text{ kN}$

$V_A + V_D = 72\text{ kN}$ [2]

Take moments about support D:

$\Sigma M_z = -(0.5 \times 24\text{ kN/m} \times 6\text{ m} \times (^2/_3 \times 6\text{ m}))$

$+ (18\text{ kN} \times 8\text{ m}) + V_A \times 6\text{ m} - H_A \times 4\text{ m} = 0\text{ kN m}$

$6V_A - 4H_A = 144\text{ kN m}$ [3]

Take moments to left of pin B:

$\Sigma M_z = H_A \times 4\text{ m} - (10\text{ kN/m} \times 3\text{ m} \times (3\text{ m}/2)) = 0\text{ kN m}$

$4H_A = 0\text{ kN m}$ [4]

Solving equations simultaneously gives:

$V_A = 24\text{ kN}$ $\quad V_D = 48\text{ kN}$

$H_A = 0\text{ kN}$ $\quad H_D = 18\text{ kN}$

Axial thrust for frame:

Member		Axial thrust (kN)
AB	24 kN (vertical reaction at A)	24
BC	18 kN (horizontal load at B)	18
CD	24 kN (vertical reaction at D)	24

Principal shear forces on frame:

Position		Shear force (kN)
Support A	0 kN (horizontal reaction at A)	0
Joint B (member AB)	0 kN (note: 18 kN load transferred along member BC ∴ no shear in AB)	0
Joint B (member BD)	24 kN (vertical reaction at A)	24
Joint C (member BC)	24 kN – (0.5 × 24 kN/m × 6 m) 24 kN – (0.5 × 24 kN/m × 6 m) + 48 kN	– 48 0
Joint D (member CD)	18 kN	18
Support D	0 kN (horizontal reaction at D)	0

Principal bending moments for frame:

Position		Bending moment (kN m)
Joint B	0 kN m	0
6 m from joint C	18 kN × 8 m – 48 kN × 6 m + (0.5 × 24 kN/m × 6 m × ($^1/_3$ × 6 m))	0
5 m from joint C	18 kN × 8 m – 48 kN × 5 m + (0.5 × 20 kN/m × 5 m × ($^1/_3$ × 5 m))	– 12.67
4 m from joint C	18 kN × 8 m – 48 kN × 4 m + (0.5 × 16 kN/m × 4 m × ($^1/_3$ × 4 m))	– 5.33
2 m from joint C	18 kN × 8 m – 48 kN × 2 m + (0.5 × 8 kN/m × 2 m × ($^1/_3$ × 2 m))	53.33
Joint C	18 kN × 8 m	144
Midway on member CD	18 kN × 4 m	72

Example 4.16: Three pin frame carrying horizontal loads

The frame shown as Example 4.16 below is supported on pin joints at A and D and has an internal pin at joint B:

(a) calculate the reactions at the supports;

(b) plot the axial thrust, shear force and bending moment diagrams.

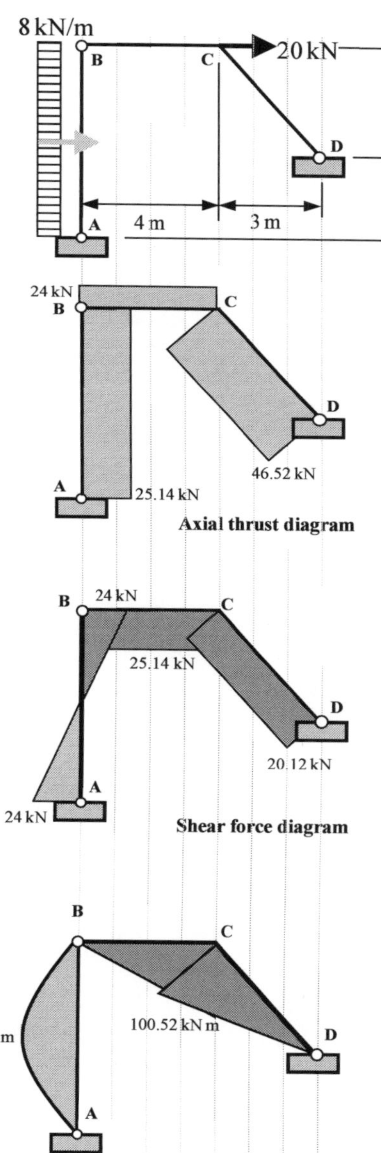

Axial thrust diagram

Shear force diagram

Bending moment diagram

Solution:

Calculate reactions:

Assume reactions applied as:

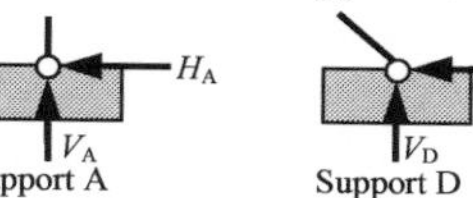

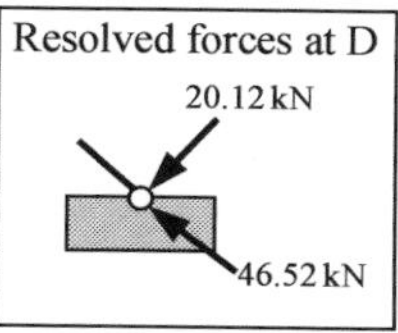

$\Sigma F_x \quad -H_A - H_D + 8\text{ kN/m} \times 6\text{ m} + 20\text{ kN} = 0\text{ kN}$

$H_A + H_D = 68\text{ kN}$ [1]

$\Sigma F_y \quad V_A + V_D = 0\text{ kN}$ [2]

Take moments about support A:

$\Sigma M_z = (8\text{ kN/m} \times 6\text{ m} \times (6\text{ m}/2)) - V_D \times 7\text{ m} - H_A \times 2\text{ m}$

$= 0\text{ kN m}$

$7V_D - 2H_D = 264\text{ kN m}$ [3]

Take moments to left of pin B:

$\Sigma M_z = -H_A \times 6\text{ m} + (8\text{ kN/m} \times 6\text{ m} \times (6\text{ m}/2)) = 0\text{ kN m}$

$6H_A = 144\text{ kN m}$ [4]

Solving equations simultaneously gives:

$V_A = -25.14\text{ kN}$ $\quad V_D = 25.14\text{ kN}$

$H_A = 24\text{ kN}$ $\quad H_D = 44\text{ kN}$

Axial thrust for frame:

Member		Axial thrust (kN)
AB	−25.14 kN (vertical reaction at A)	−25.14
BC	8 kN/m × 6 m − 24 kN	24
CD	46.52 kN (resolved vertical reaction at D)	46.52

Principal shear forces on frame:

Position		Shear force (kN)
Support A	24 kN (horizontal reaction at A)	24
Joint B (member AB)	24 kN − 8 kN/m × 6 m	−24
Joint B (member BD)	−25.14 kN (vertical reaction at A)	−25.14
Joint C (member BC)	−25.14 kN −25.14 kN + 25.14 kN	−25.14 0
Joint D (member CD)	20.12 kN (resolved component at support D)	20.12

Principal bending moments for frame:

Position		Bending moment (kN m)
Support A	0 kN m	0
2 m up from support A	24 kN × 2 m − 8 kN/m × 2 m × (2 m/2)	32
3 m up from support A	24 kN × 3 m − 8 kN/m × 3 m × (3 m/2)	36
4 m up from support A	24 kN × 4 m − 8 kN/m × 4 m × (4 m/2)	32
Joint B	44 kN × 4 m − 25.15 kN × 7 m	0
Midway of member BC	44 kN × 4 m − 25.15 kN × 5 m	50.25
Joint C	44 kN × 4 m − 25.15 kN × 3 m	100.55

Having completed this chapter you should now be able to:

- ***Calculate the reactions for various forms of three pin frames***
- ***Plot for various frame structures***
  - ***Axial thrust diagrams***
  - ***Shear force diagrams***
  - ***Bending moment diagrams***
- ***Determine the position and magnitude of the maximum bending moment***

Further Problems 4

For the three pin frames shown:
(a) calculate the reactions at the supports;
(b) plot the axial thrust, shear force and bending moment diagrams.

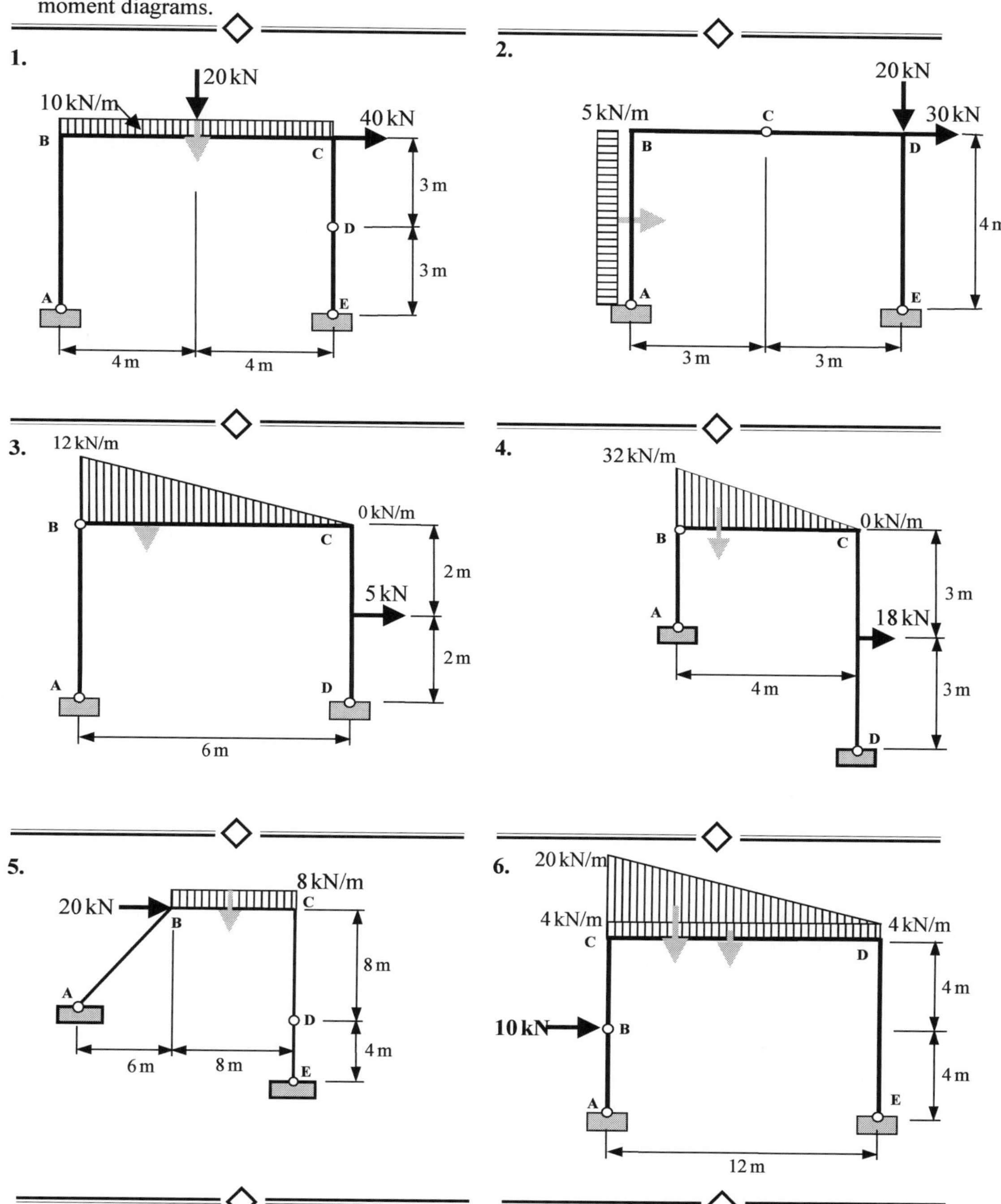

Solutions to Further Problems 4

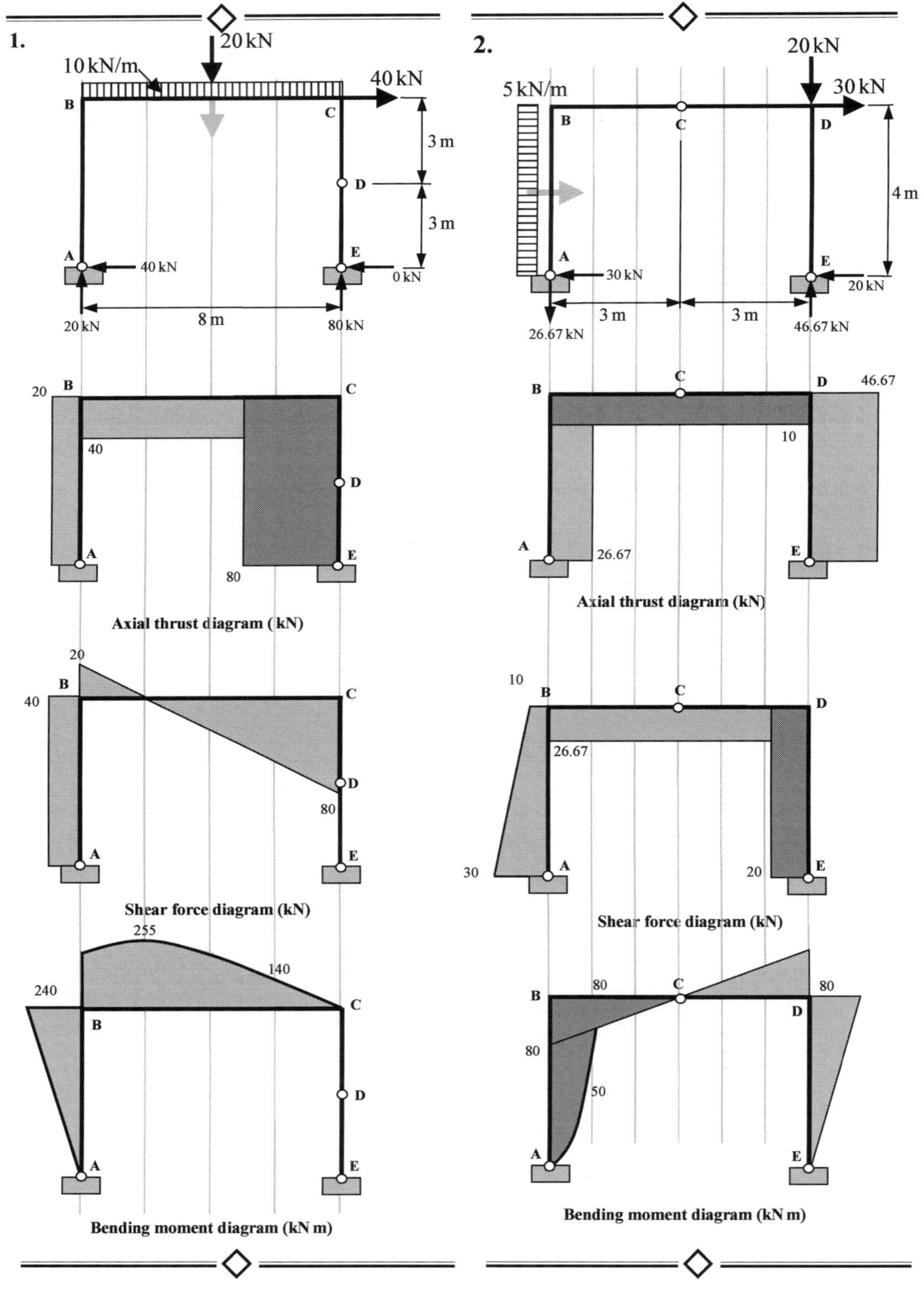

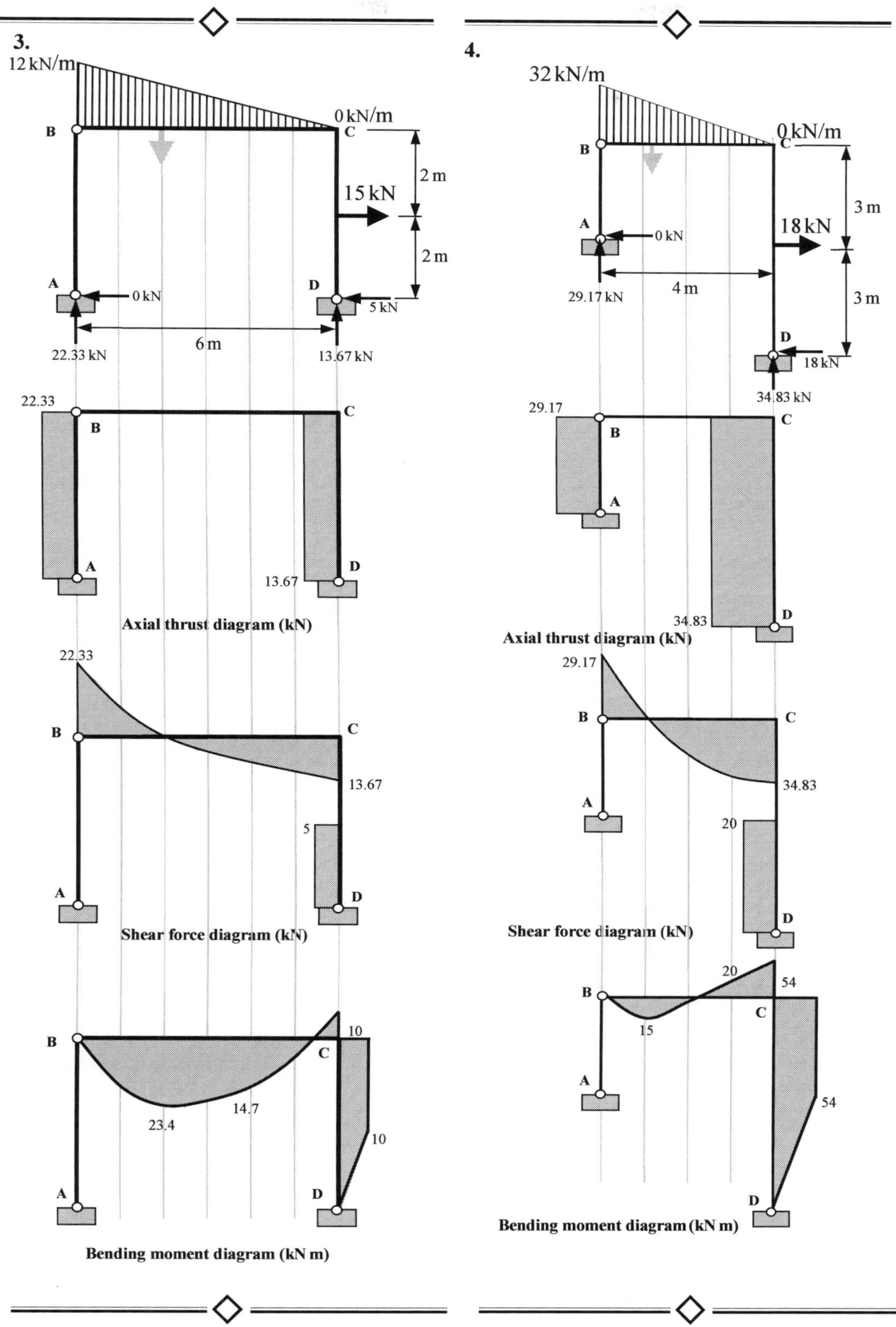
3.
12 kN/m
0 kN/m
B
C
2 m
15 kN
2 m
A
0 kN
D
5 kN
22.33 kN
6 m
13.67 kN
22.33
13.67
Axial thrust diagram (kN)
22.33
13.67
5
Shear force diagram (kN)
10
23.4
14.7
10
Bending moment diagram (kN m)
4.
32 kN/m
0 kN/m
3 m
0 kN
18 kN
29.17 kN
4 m
3 m
18 kN
34.83 kN
29.17
34.83
Axial thrust diagram (kN)
29.17
34.83
20
Shear force diagram (kN)
20
54
15
54
Bending moment diagram (kN m)

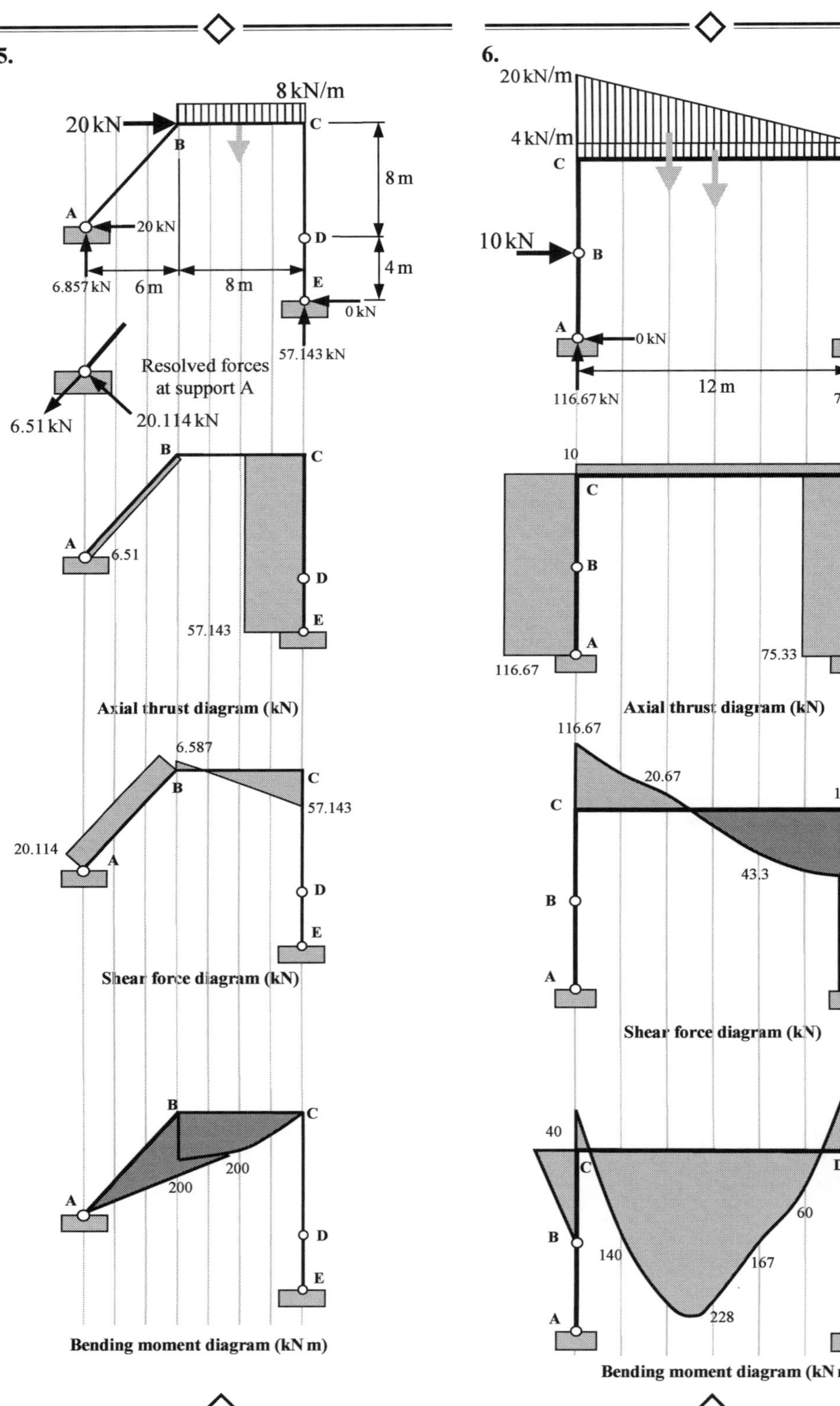
5.
8 kN/m
20 kN
B
C
8 m
A
20 kN
D
4 m
6.857 kN
6 m
8 m
E
0 kN
57.143 kN
Resolved forces at support A
6.51 kN
20.114 kN
6.51
57.143
Axial thrust diagram (kN)
6.587
57.143
20.114
Shear force diagram (kN)
200
200
Bending moment diagram (kN m)
6.
20 kN/m
4 kN/m
4 kN/m
C
D
4 m
10 kN
B
4 m
E
A
0 kN
10 kN
12 m
116.67 kN
75.33 kN
10
116.67
75.33
Axial thrust diagram (kN)
116.67
20.67
10
43.3
75.33
Shear force diagram (kN)
80
40
80
60
140
167
228
Bending moment diagram (kN m)

5. Deflection of Beams and Frames

In this chapter we will learn how:
- ***Load patterns and support restraints affect deflection***
- ***To sketch deflected forms of structures***
- ***To derive equations to determine the magnitude of deflection for a given structure and load pattern***

5.1 Introduction

Deflection can be defined as: Lateral bending or turn, deviation (literally or figuratively); (Physics) movement of instrument's pointer from its zero position (from Late Latin *deflexio)*.

When a structure is placed under load it will bend, deflect or displace. This deflection will depend on a number of factors including:

(1) the geometry of the structure, including shape and flexural rigidity of member;
(2) the flexibility (or rigidity) of the material used;
(3) the restraint placed on the system by its support;
(4) the load pattern.

These factors will determine the magnitude, direction and amount of deflection.

Whilst the amount that a structure deflects can be determined by experimentation or calculation, the approximate shape of its deflected form may be predicted. Knowing the deflected form of the structure prior to carrying out any detailed analysis will provide a useful 'check' to ensure the validity of results. We now consider the factors that affect the deflection of a structure.

5.2 Effects of Supports and Load Patterns

All the factors listed above have an effect on the deflection of a structure. The geometry of a structure will greatly determine its deflected form. A triangulated structure has an inherent rigidity due to its shape and the interaction of its members, whilst a rectangular structure relies heavily on the rigidity of its members and its joints. Under a given load, a flexible structure will deflect more than a rigid structure and the properties of the material have a great influence on this. Consider a structure constructed of steel compared to a similar structure constructed of rubber! These factors will be considered by the designer to ensure the structure is adequate for its purpose, safe, and that it meets all 'in service' requirements in its constructed form.

At the boundary of the structure will be the supports. We have already discussed three different types of support in Chapter 1. These were defined as:

(1) fully fixed (or encastré);
(2) simple support, pin joint or knife edge;
(3) simple support and roller.

It is important to note that these are NOT the only types of support that exist. Many other types can be postulated, including both vertical and rotational spring supports as shown in Figure 5.1. This type of support is used extensively in the analysis of foundations where consideration must be given to the reaction of the soil under load and its deformation (compaction or heave) over time. The analysis of spring systems is the basis of the 'stiffness' method, which is an advanced structural analysis technique and has been adapted for many commercial computer packages.

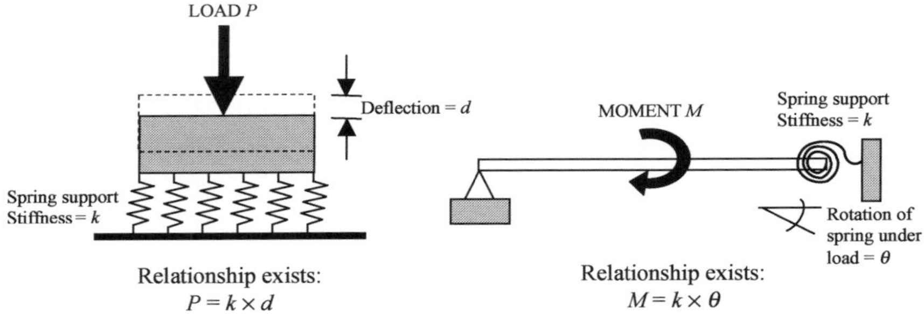

Figure 5.1: Spring support systems

The analysis of such systems is based on the assumption that the displacement of the structure, and hence the springs, is related to the magnitude of the applied load and the stiffness of the spring support. The stiffer the spring, the less it will displace under a constant load. This relationship has also been developed and extended to provide the basis for the Finite Element Method (FEM) of analysis.

It is useful to note that many more support systems can be developed, however we will concentrate here on the three basic types detailed earlier, looking specifically at the deflections which take place at the support when the structure is loaded.

5.2.1 Fully Fixed Support

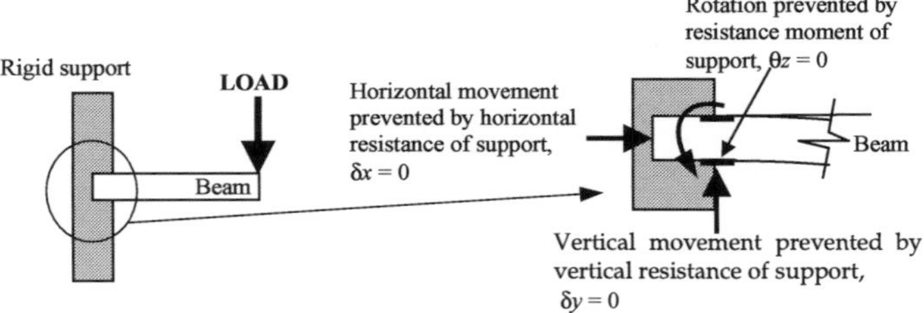

Figure 5.2: Fully fixed support

As shown in Figure 5.2 and discussed in Chapter 3, at a fully fixed support there is no vertical or horizontal displacement and no rotation. Along the beam, displacement will occur under load, however the support itself will not yield. The member is therefore restrained and cannot move vertically or horizontally. Also, the structure cannot rotate at the support. This means that at the point where the member joins the support, the member must remain at a constant angle (in the case of Figure 5.2, horizontal) whatever the load, and this angle cannot alter unless the support fails.

In terms of applied forces, the support will resist vertical displacement with a vertical reaction which is equal and opposite to the vertical applied forces. It will resist horizontal displacement with a horizontal reaction which is equal and opposite to the horizontal applied forces, and it will resist rotation with a fixed end moment (FEM) at the support which is equal and opposite to the total applied moment.

5.2.2 Simple Support

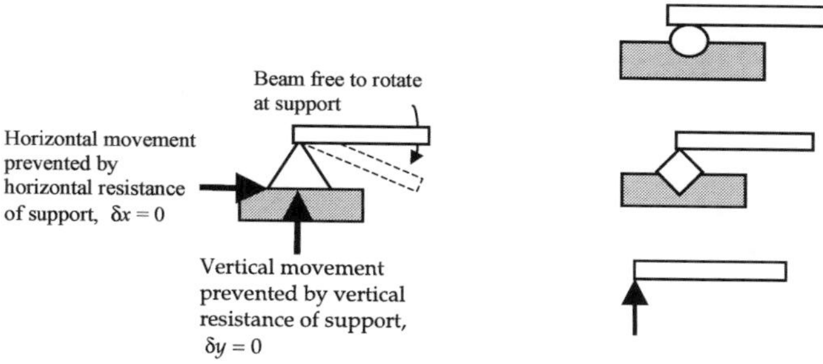

Figure 5.3: Simple support

Typical simple supports and pin joints are discussed in Chapter 3, and shown in Figure 5.3. They provide no resistance to rotation but will resist both vertical and horizontal movement. Therefore, the resisting forces which occur at a simple support are a vertical reaction and a horizontal reaction. The support will not provide a resisting moment. The structure is therefore free to rotate at the support.

5.2.3 Roller Joints

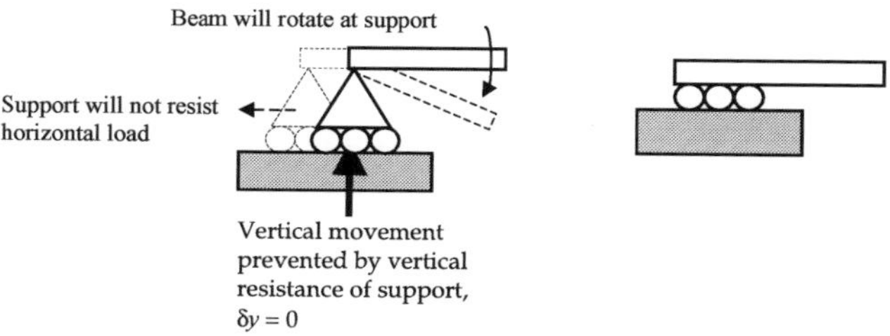

Figure 5.4: Roller support

Typical roller supports are discussed in Chapter 3, and shown in Figure 5.4. They have no resistance to rotation or horizontal movement and can only resist vertical movement. Therefore, the only resisting force that occurs at a roller support is a vertical reaction. The support will not provide a horizontal resisting force or a resisting moment. The structure is therefore free to deflect horizontally or to rotate at the support.

The different types of support therefore have a different effect on the deflection of the structure. They provide restraints at the boundary of our structural systems. Understanding the effect of each type of support on a structure will enable us to predict deflections at the boundary.

5.2.4 Load Patterns

The deflected form of the structure will also be affected by the magnitude, direction and pattern of load applied to it. Members will deflect under applied loads. The movement will normally be in the direction of the load but consideration must be given to the magnitude and type of loading if an accurate prediction of the deflected form is to be derived. When considering these factors, some ‘engineering judgement’ will be required. Not all solutions will be obvious and in some circumstances a variety of feasible options may exist. In such cases we must opt for the most logical solution, always bearing in mind that we may have to modify our prediction at a later stage of the analysis.

Patterns of load have a major influence on the deflected form of a structure. At this early stage in your studies it is often very useful to carry out some simple experiments using a flexible material to form a beam supported at its boundary and loaded in various combinations and patterns. The deflected form can then be observed and plotted.

5.3 Sketching Displacements

When sketching the deflected form of a structure we often exaggerate the deflection scale. This is useful as it will enable us to easily visualise what is happening to the structure since, in real structures, deflections are normally small in comparison to the structure's overall dimensions. The deflected form is drawn using the original, ‘undeflected’ structure as the origin and the deflection is shown by the amount that the structure deviates from this origin, as illustrated in Figure 5.5.

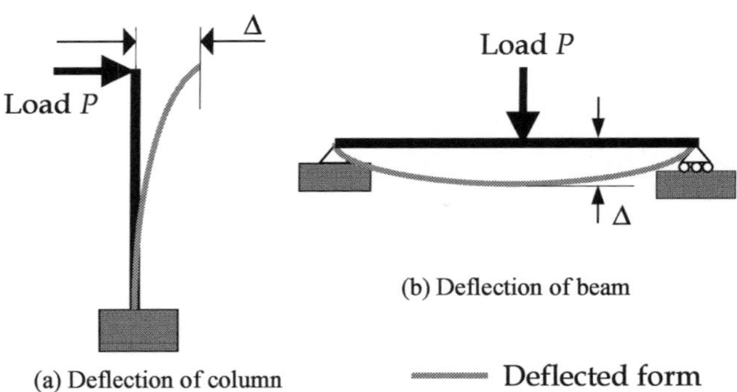

Figure 5.5: Predicted deflection of a column and a beam

When predicting the deflected form of any structure we must also ensure that compatibility is maintained. This requires that the deflected shape be compatible with any external constraints and that any joint movements which take place are applied to all members connected at that point. Compatibility demands that any movements at a joint be ‘compatible’ in all members, as shown in Figure 5.6.

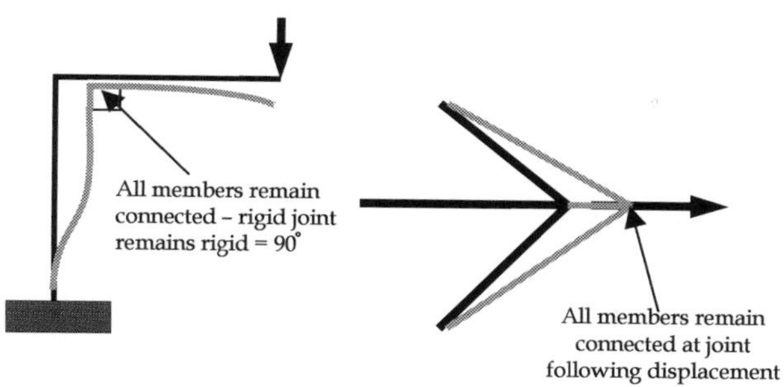

Figure 5.6: Compatibility at joints

Therefore, in order to predict the deflected form of any structure we will consider:

(1) support (boundary) conditions;
(2) position and magnitude of all applied loads;
(3) joint compatibility.

We will now apply our knowledge of these factors and consider a number of examples to demonstrate a process which can be adopted to sketch the deflected forms of a number of simple structures.

Example 5.1: Sketching the deflected form of a beam supporting a single vertical point load

Consider the beam system of Example 5.1, shown in Figure 5.7. Sketch the deflected form of the structure.

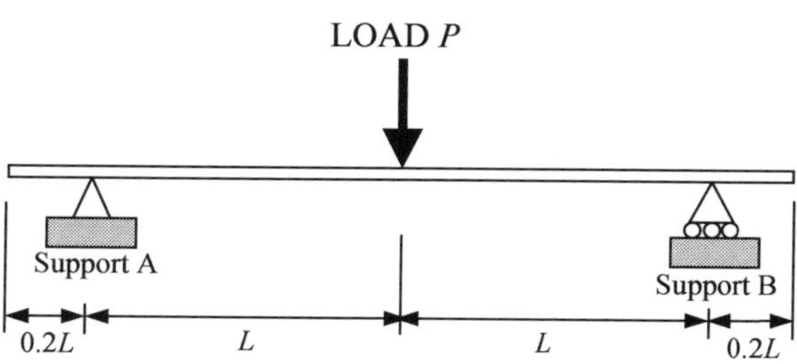

Figure 5.7: Beam – Example 5.1

We commence the analysis of this beam by listing the factors that will influence the deflected form of the structure. Such that:

(a) The load P will cause a vertical downward deflection of the beam.
(b) There will be no vertical displacement of the beam at supports A and B.
(c) The overhanging section of the beam to the left of support A and to the right of support B do not carry load and therefore will not deflect downward.
(d) Support B **may** displace horizontally.

We can now plot these 'known' factors on our diagram, as shown in Figure 5.8.

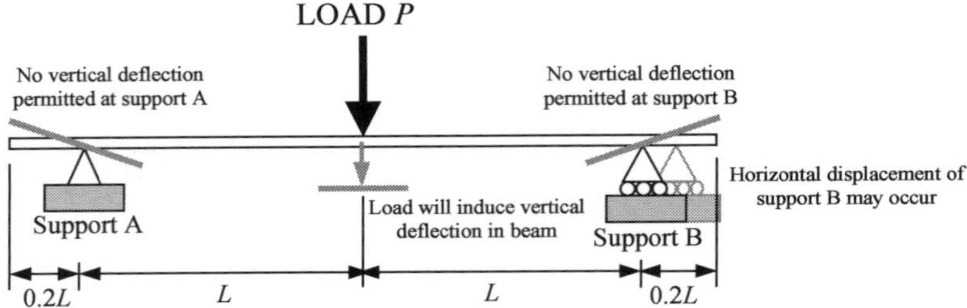

Figure 5.8: Plot of 'known' factors for beam – Example 5.1

We must now use our 'engineering judgement' to interpolate the missing elements in our sketch and predict the deflected form as shown in Figure 5.9.

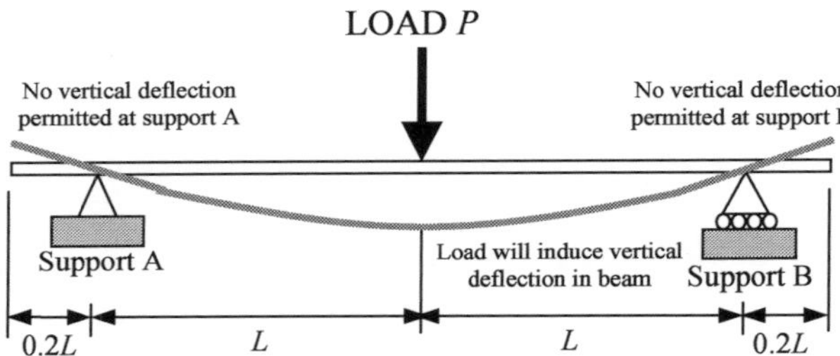

Figure 5.9: Sketch of the predicted deflected form for beam – Example 5.1

We note from Figure 5.9 that the ends of the beam deflect in an upward direction. This occurs because there is no applied or restraining force at or after the supports, and the beam is therefore free to rotate. Following this simple procedure, we shall attempt to sketch the deflected forms of further structural systems.

◇

Example 5.2: Sketching the displacement of a simply supported beam carrying three vertical point loads

Consider the beam system of Example 5.2 shown in Figure 5.10, and sketch the deflected form of the structure.

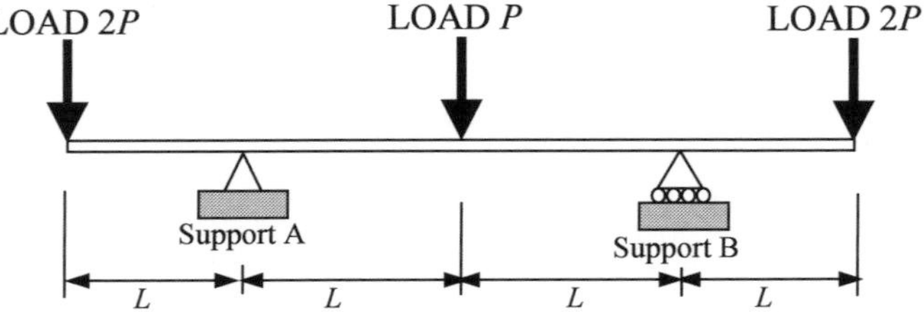

Figure 5.10: Beam – Example 5.2

We again commence our analysis by listing the factors that will affect the deflected form of the structure:

(a) The load $2P$ at each of the cantilevered ends will cause a vertical downward deflection of the beam at this point. This will cause the centre of the beam, between the supports, to deflect upwards (rotation at the supports).
(b) The load P will cause a downward displacement at mid-span.
(c) There will be no vertical displacement at support A or at support B.
(d) Support B **may** displace horizontally.

We can now plot these 'known' facts on our diagram as shown in Figure 5.11.

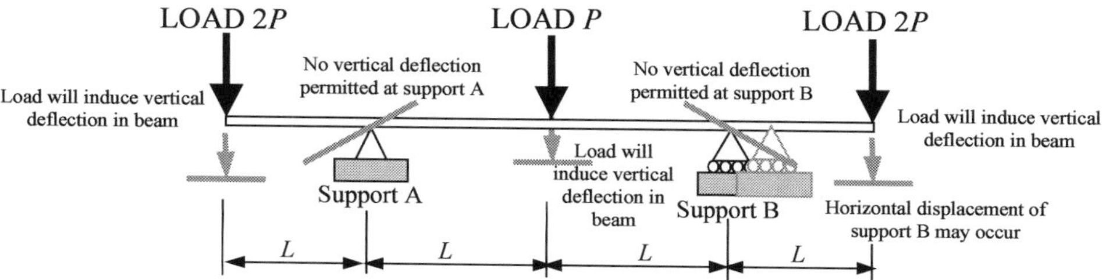

Figure 5.11: Plot of 'known' factors for beam – Example 5.2

We note that we have an incompatibility problem, since the loads at the ends of the beam will cause an upward displacement between supports A and B as the beam rotates whilst the load at mid-span will also cause a downward movement. We also note that the magnitude of the loads at the ends of the beam are twice that at the centre. We can therefore assume that the deflection induced by the loads at the ends of the beam would be greater than that induced at mid-span. The load at mid-span will therefore reduce the overall upward deflection of the beam, but is not large enough to diminish it completely. The sketch of the deflected form is shown in Figure 5.12.

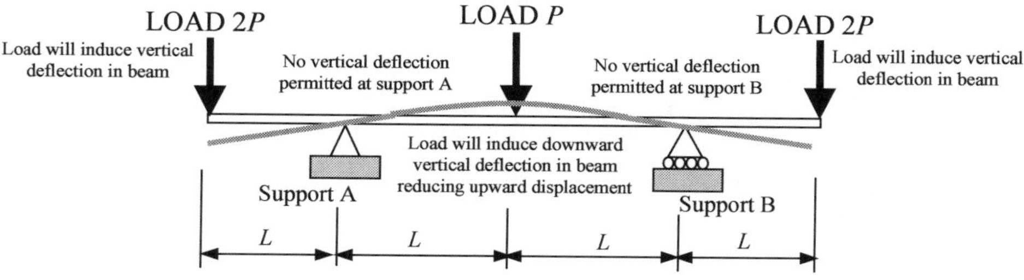

Figure 5.12: Sketch of the predicted deflected form for beam – Example 5.2

Example 5.3: Sketching the deflected form of a cantilever beam supporting a single vertical point load

Consider the cantilever beam system of Example 5.3, shown in Figure 5.13. Sketch the deflected form of the structure.

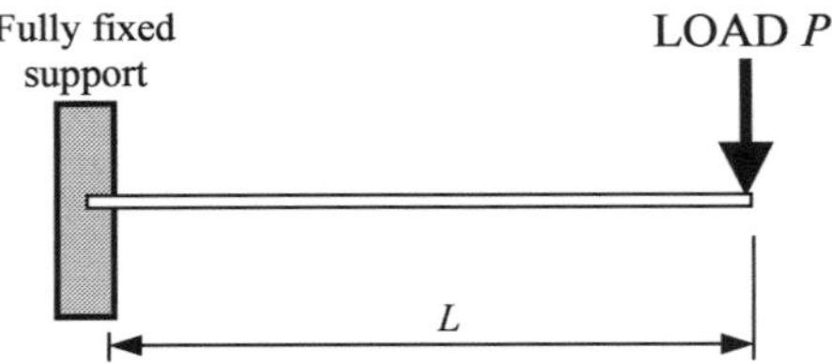

Figure 5.13: Cantilever beam – Example 5.3

Factors which will affect the deflected form of this structure are:

(a) The load P at the cantilevered end will cause a vertical downward deflection of the beam at this point.
(b) There will be no vertical or horizontal displacement at the support.
(c) There will be no rotation at the support

We can now plot these 'known' factors on our diagram, as shown in Figure 5.14.

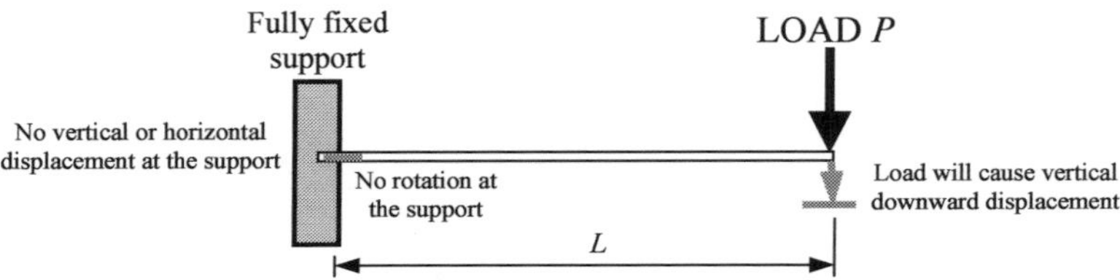

Figure 5.14: Details of 'known' factors for cantilevered beam – Example 5.3

Based on these known factors we can sketch the deflected form of the cantilevered beam as shown in Figure 5.15.

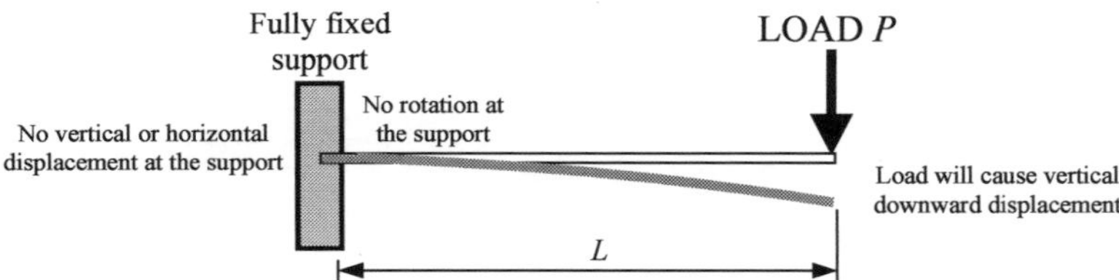

Figure 5.15: Sketch of the predicted deflected form for beam – Example 5.3

We note that at the junction of the beam and its support, no rotation occurs and the beam therefore remains in its original horizontal position at this point.

◇

Example 5.4: Sketching the deflected form of a cantilever beam supporting a vertical point and a distributed load

Consider the cantilever beam system of Example 5.4, shown in Figure 5.16. Sketch the deflected form of the structure.

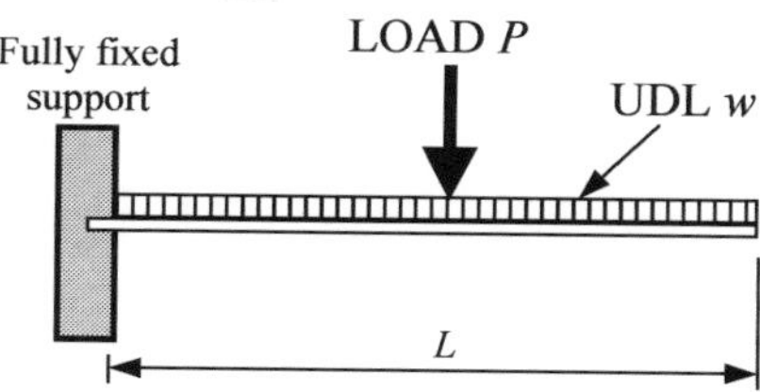

Figure 5.16: Cantilever beam – Example 5.4

In this case the structure supports a uniformly distributed load. The effect of this load is to increase the deflection in the downward direction. We complete the analysis as before.

Factors which will affect the deflected form of the structure are:

(a) The load P will cause a vertical downward deflection of the beam.
(b) The UDL w will cause a vertical downward deflection of the beam.
(c) There will be no vertical or horizontal displacement at the support.
(d) There will be no rotation at the support.

We can now plot these 'known' factors on our diagram, as shown in Figure 5.17.

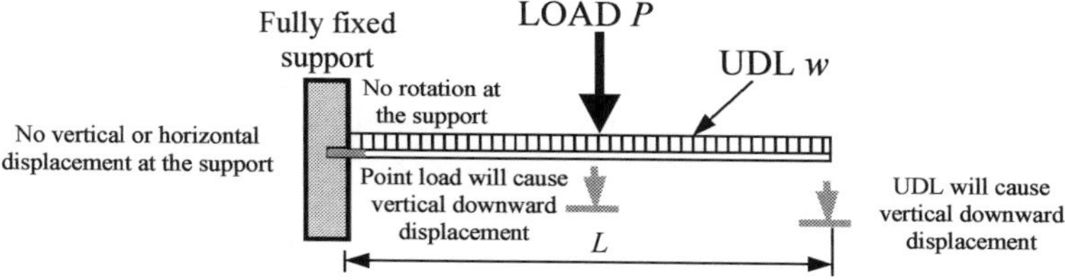

Figure 5.17: Details of known factors for cantilevered beam – Example 5.4

The effect of both loads is to cause a downward, vertical, displacement. The combined effect is additive, however as our sketch contains no 'values' it will be plotted as shown in Figure 5.18.

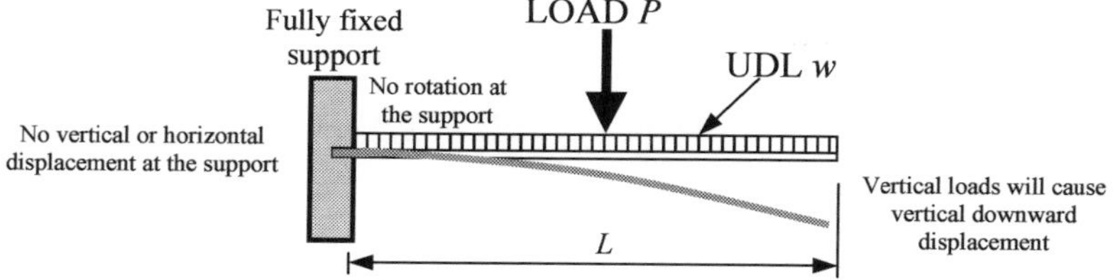

Figure 5.18: Sketch of the predicted deflected form for beam – Example 5.4

We can extend our predictions to statically indeterminate beams and frames as demonstrated in Examples 5.5, 5.6 and 5.7.

Example 5.5: Sketching the deflected form of a 'propped' cantilever beam supporting a vertical point and a distributed load

Consider the beam system of Example 5.5, shown in Figure 5.19. Sketch the deflected form of the structure.

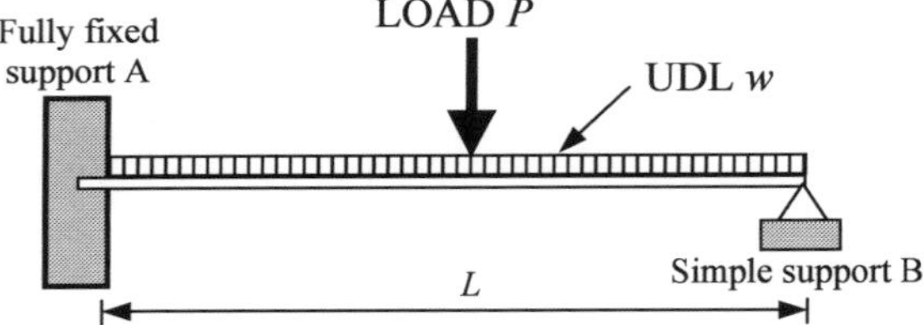

Figure 5.19: 'Propped' cantilever beam – Example 5.5

The beam of Figure 5.19 is generally known as a 'propped' cantilever, as the free end of the beam is provided with a simple support which 'props' it in position. Factors which will affect the deflection of this structure are:

(a) The load *P* will cause a vertical downward deflection of the beam.
(b) The UDL *w* will cause a vertical downward deflection of the beam.
(c) There will be no vertical or horizontal displacement at the support A.
(d) There will be no rotation at the support A.
(e) There will be no vertical or horizontal displacement at the support B (however rotation of the beam at support B is permissible).

We can now plot these 'known' factors on our diagram, as shown in Figure 5.20.

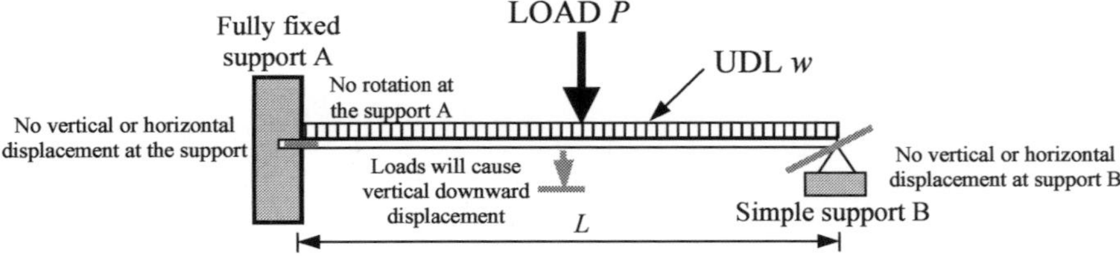

Figure 5.20: Plot of known factors for propped cantilever beam – Example 5.5

Having plotted our 'known' factors we can interpolate the missing data. The predicted deflected form of the propped cantilever beam is shown in Figure 5.21.

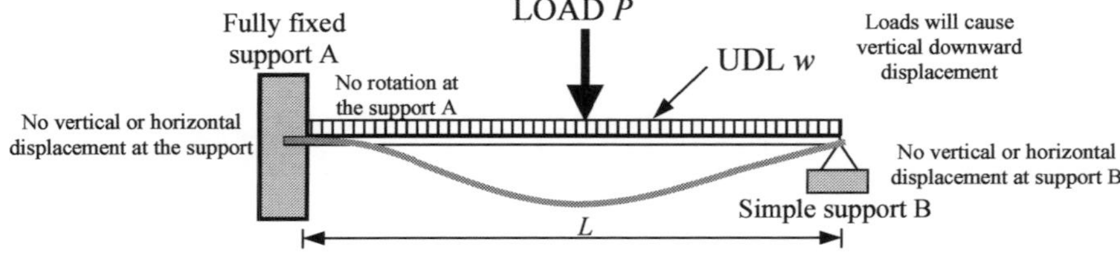

Figure 5.21: Sketch of the predicted deflected form for beam – Example 5.5

Example 5.6: Sketching the deflected form of a portal frame supporting a vertical point and a distributed load

Consider the portal frame structure of Example 5.6, which has a full fixed support at A and a pin support at B as shown in Figure 5.22. The frame supports a vertical downward UDL and a point load. Sketch the deflected form of the structure. Note that all internal joints on the frame are to be considered as rigid.

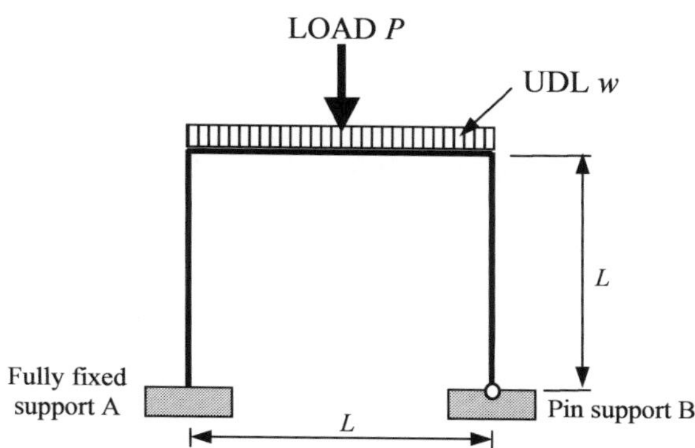

Figure 5.22: Portal frame structure – Example 5.6

For this structure we note that the joints between the members of the frame are rigid, and therefore the angle between the upright column and horizontal beam member will remain constant (90° in this case).

Factors which will affect the deflection of this structure are therefore:

(a) The load *P* will cause a vertical downward deflection of the beam.
(b) The UDL *w* will cause a vertical downward deflection of the beam.
(c) There will be no vertical or horizontal displacement at the support A.
(d) There will be no rotation at the support A.
(e) There will be no vertical or horizontal displacement at the support B (however rotation of the column at support B is permissible).
(f) The internal joints are rigid (therefore remain at 90°).

We can now plot these 'known' factors on our diagram, as shown in Figure 5.23.

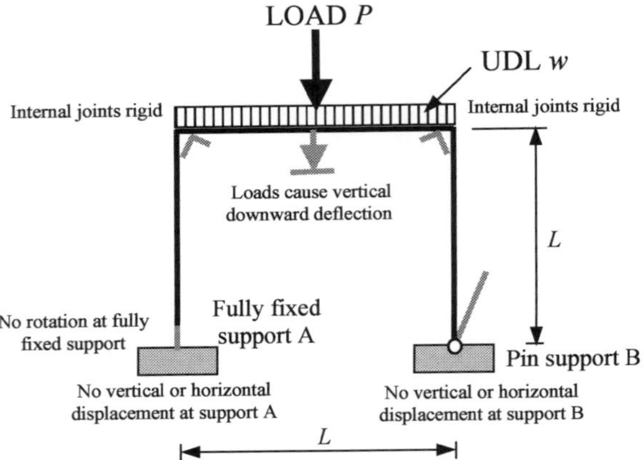

Figure 5.23: Plot of known factors for portal frame – Example 5.6

From Figure 5.23, we can sketch our prediction of the deflected form, as shown in Figure 5.24.

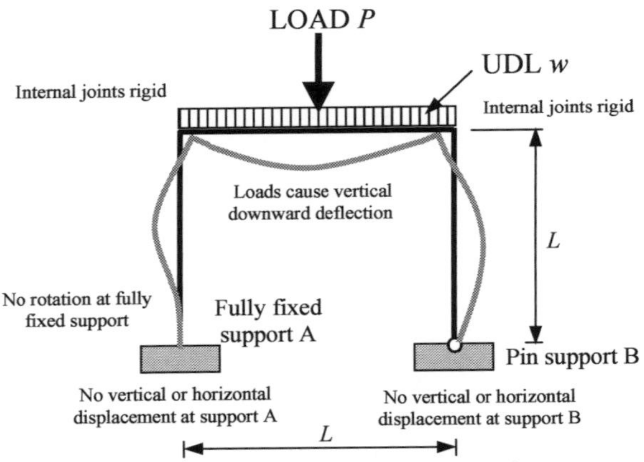

Figure 5.24: Sketch of the predicted deflected form for portal frame – Example 5.6

Note that our predicted form allows rotation at support B whilst no rotation is permitted at support A. Note also that internal joints remain at 90°.

Having consider examples with vertical loads, we will now apply the same procedure to a frame supporting both horizontal and vertical loads.

◇

Example 5.7: Sketching the deflected form of a portal frame supporting a vertical point load and a horizontal distributed load

Consider the portal frame structure of Example 5.7, which has a pin support at A and B, as shown in Figure 5.25. The frame supports a vertical downward point load at mid-span of the beam member and a horizontal UDL along the left column. Sketch the deflected form of the structure. Note that all internal joints on the frame are to be considered as rigid and the point load is $^{1}/_{10}$ th the magnitude of the UDL.

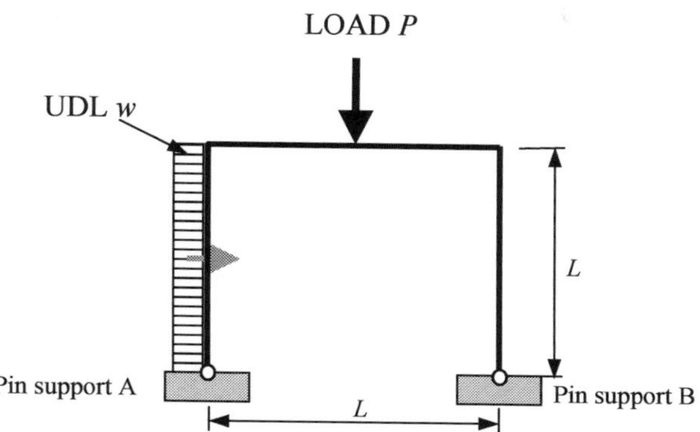

Figure 5.25: Portal frame structure – Example 5.7

Factors which will affect the deflection of this structure are:

(a) The load P will cause a vertical downward deflection of the beam
(b) The UDL w will cause a horizontal deflection of the column.
(c) There will be no vertical or horizontal displacement at the support A.
(d) There will be no vertical or horizontal displacement at the support A or B (however rotation of the columns at supports A and B is permissible).
(e) The internal joints are rigid (therefore they will remain at 90°).

We also note that the point load is only $^{1}/_{10}$ th of the UDL therefore the major displacement will be in the horizontal plane.

We can now plot these 'known' factors on our diagram, as shown in Figure 5.26.

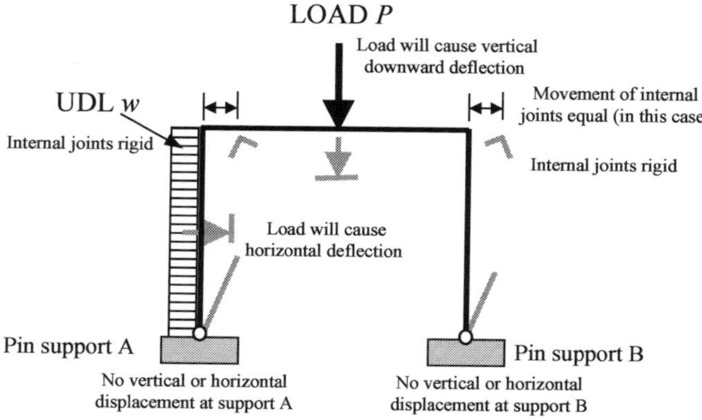

Figure 5.26: Plot of known factors for portal frame – Example 5.7

From Figure 5.26 we can plot a prediction of the deflected form, however it is important to note that in this case a number of solutions exist. We should always be aware that this is only a prediction; we may have to modify our solution following detailed calculations. A sketch of one prediction of the deflected form of the frame is shown in Figure 5.27.

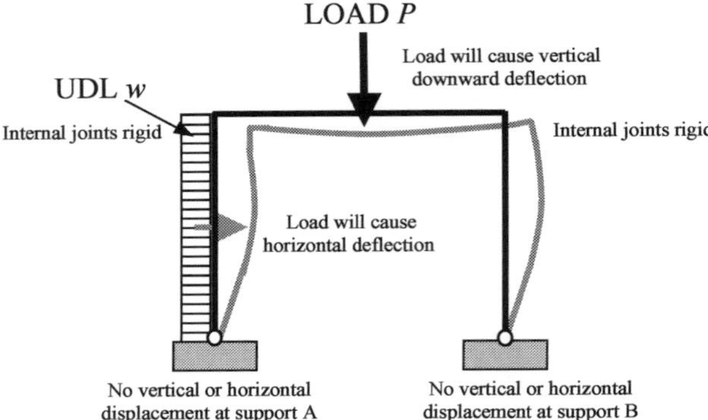

Figure 5.27: Sketch of predicted deflected form for portal frame – Example 5.7

As previously stated, a number of alternative predictions for the deflected form could be proposed. It is possible that, after plotting the known factors, the deflected form could have been predicted as that shown in Figure 5.28, however experience will show this to be a less likely approximation.

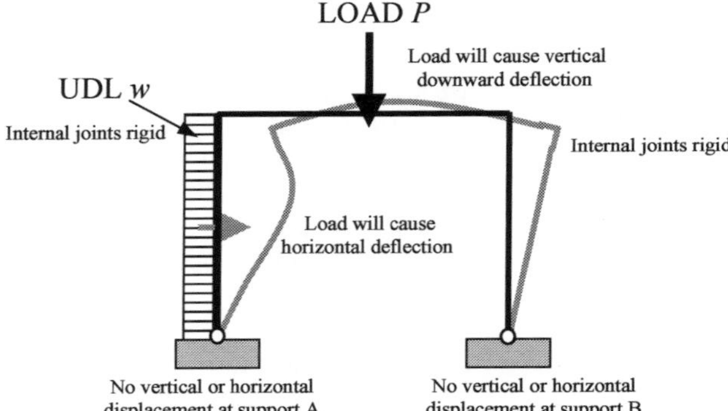

Figure 5.28: Sketch of alternative deflected form for portal frame – Example 5.7

◇

It is important to bear in mind that sketching the deflected form of structures, both simple and complex, requires practice. It would be impossible to cover every eventuality in this text, and as the loading and geometry of structures increase in complexity it may not be possible to predict the deflected form using observations and hand calculations alone. However, even when a computer solution is adopted, we can still use the procedures developed here to check the validity of the results. The deflected form of the structure must still obey the rules given for support restraints, loads, joint conditions and compatibility.

Knowing the deflected form of a structure can provide a very useful check of our analysis. This prediction is however of little use to the designer unless we are able to predict the magnitude of the deflection that will take place. As stated in Chapter 1, various Standards and Codes of Practice limit the allowable deflection of structural members in order to ensure the overall stability of the structure and to ensure movements do not cause distress or danger to the occupants. This distress can take various forms and might include discomfort due to excessive movements or damage caused to other elements of the building. A common failure caused by excessive deflection is the formation of cracks in brickwork and plaster. The amount of allowable deflection is governed by a number of factors, and many Codes specify limits for various types of structural member. A typical example of such limits is shown in the abstract of British Standard 5950 – Structural use of steelwork in building[1] of Figure 5.29.

Table 5: Deflection limits other than for pitched roof portal frames	
(a) Deflection of beams due to unfactored imposed loads	
Cantilevers	Length/180
Beams carrying plaster or other brittle finish	Length/360
All other beams	Length/200
Purlins and sheeting rails	See code
(b) Horizontal deflection of columns other than portal frames due to unfactored imposed and wind loads	
Tops of columns in single-storey buildings	Height/300
In each storey of a building with more than one storey	Height of storey under consideration/300
(c) Crane gantry girders	
Vertical deflection due to static wheel loads	Span/600
Horizontal deflection (calculated on the top flange properties alone) due to crane surge.	Span/500
NOTE 1: On low pitched and flat roofs the possibility of ponding needs consideration. NOTE 2: For limiting deflections in runway beams refer to BS 2853.	

Figure 5.29: Abstract from BS5950[1] of limiting deflections in steel members

Codes of practice therefore specify maximum values for deflections that must not be exceeded. Various methods exist for calculating deflections for specified loading situations and formula for regularly encountered situations are given in a number of references,[2,3] examples of which are shown in the table of Figure 5.30.

Details of beam and loading	Location of maximum deflection	Maximum deflection
Cantilever, length L, point load P kN at free end	Free end	$\delta = \left[\dfrac{PL^3}{3EI}\right]$
Simply supported, P kN at midspan ($L/2$, $L/2$)	Length/2	$\delta = \left[\dfrac{PL^3}{48EI}\right]$
Simply supported, length L, uniform load w kN/m	Length/2	$\delta = \left[\dfrac{5wL^4}{384EI}\right]$

Figure 5.30: Formula to calculate the maximum deflection of various structural members

We therefore need to develop methods to enable us to determine the amount of deflection for a structure under various load conditions. Such a method, based on an integration technique, has been proposed by Macaulay and is commonly used to derive equations that can be used to predict the magnitude of deflection of beams subjected to various loading configurations.

1. British Standards Institution, BS 5950: 1985: Structural use of steelwork in buildings: Part 1: Code of practice for the design of simple and continuous construction: hot rolled sections.
2. *Steel Designers Manual*, The Steel Construction Institute, London, 1995.
3. Reynolds and Steedman, *Reinforced Concrete Designers Handbook*, 10th edn, E&F Spon: New York, 1988.

5.4 Macaulay's Method

Various methods exist to determine the amount of deflection produced by loading on a structural member. These include:

(1) a mathematical solution based on what is commonly called the Method of slope–deflection;

(2) the area moment method as proposed by Mohr and commonly called Mohr's Theorem;

(3) graphical solutions (not generally used).

These methods are really variations on the same root principles, and the method adopted here will be based on a mathematical solution developed from the slope–deflection method. Many texts exist which explain these methods in much greater detail if you wish to research them further.

We have already discussed the Engineers' bending equation in Chapter 2, which is commonly written as:

$$\frac{M}{I}=\frac{\sigma}{y}=\frac{E}{R}$$

For the analysis of deflection problems this can be simplified to:

$$\frac{M}{I}=\frac{E}{R}$$

and:

$$\frac{M}{IE}=\frac{1}{R} \qquad [5.1]$$

We have previously defined these terms as:

M = Bending moment
E = Elastic (or Young's) Modulus
I = Second Moment of Area
R = Radius of curvature.

The term R, the radius of curvature, is directly related to the deflection of the member. Consider the small section of a beam shown in Figure 5.31.

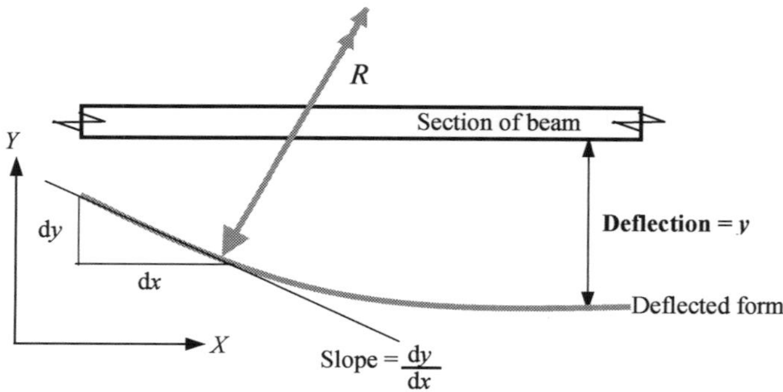

Figure 5.31: Section of deflected beam

The radius of curvature (R) for deflected beams is very large since the deformations are small in comparison to the overall length. The radius can be found with reference to the analysis of a catenary curve. In differential terms the radius of curvature may thus be derived as:

$$\frac{1}{R}=\frac{{}^{\mathrm{d}^2y}\!\big/_{\mathrm{d}x^2}}{\left(1+\left({}^{\mathrm{d}y}\!\big/_{\mathrm{d}x}\right)^2\right)^{\frac{3}{2}}} \qquad [5.2]$$

We already know that displacements are very small and so the slope term, ${}^{\mathrm{d}y}/_{\mathrm{d}x}$ must also be very small. The square of this term, contained in equation [5.2], will therefore have a very small influence on the equation, in fact the denominator will tend towards 1 and may therefore be considered as a secondary function and ignored. Equation [5.2] can therefore be simplified to:

$$\frac{1}{R}=\frac{\mathrm{d}^2y}{\mathrm{d}x^2} \qquad [5.3]$$

Substituting equation [5.3] into [5.1] we have:

$$\frac{M}{EI}=\frac{\mathrm{d}^2y}{\mathrm{d}x^2} \qquad [5.4]$$

Equation [5.4] can also be expressed as

$$M=EI\frac{\mathrm{d}^2y}{\mathrm{d}x^2} \qquad [5.5]$$

Therefore, if we can derive an equation for the bending moments of our beam, and then integrate this equation twice, we will be able to determine the deflection y.

We will now proceed to derive the deflection equations for a number of examples remembering that for integration:

$$\int x^n\mathrm{d}x=\frac{1}{n+1}x^{n+1}+C \qquad [5.6]$$

where C = constant of integration.

Example 5.8: Deflection of a cantilever beam supporting a distributed load (1)

Consider the cantilever beam of Figure 5.32. The beam supports a uniformly distributed load of w kN/m along its length. Derive a general expression to calculate the deflection along the beam, and determine the maximum deflection.

Assume the flexural rigidity (EI) for the beam is constant along its length.

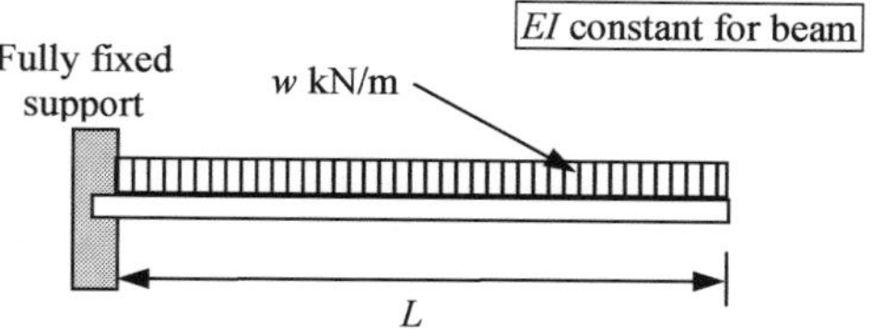

Figure 5.32: Cantilever beam – Example 5.8

In order to derive an equation for the bending moments along the beam, we must first isolate a section of the beam and derive an expression for the moments about the datum as shown in Figure 5.33. The section should be selected such that it includes all the loads applied to the beam. This will provide us with a deflection equation that can be applied along the entire beam length.

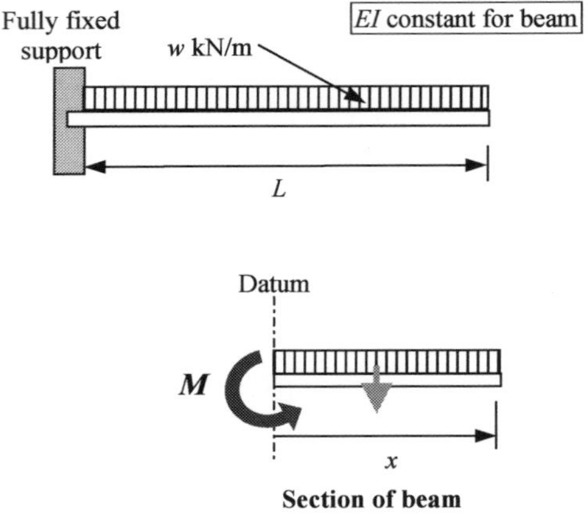

Figure 5.33: Details of section of beam – Example 5.8

The section is taken at some distance x along the beam. In the first instance we will measure our distances from the datum to the free end, for ease of calculation. We could similarly derive an equation by measuring the distance back to the support, as will be demonstrated in Example 5.9.

Taking moments about the datum and noting that the balancing moment M is anticlockwise, we have:

$$-M = \frac{wx^2}{2} \qquad [5.7]$$

Substituting this expression for M into equation [5.5]:

$$-M = -EI\frac{\mathrm{d}^2y}{\mathrm{d}x^2} = \frac{wx^2}{2} \qquad [5.8]$$

Integrating equation [5.8] we have:

$$-EI\frac{\mathrm{d}y}{\mathrm{d}x} = \frac{wx^3}{6} + A \qquad [5.9]$$

where A = constant of integration.

Integrating equation [5.9] will give:

$$-EI\ y = \frac{wx^4}{24} + Ax + B \qquad [5.10]$$

where B is the second constant of integration.

In order to solve for the constants of integration, we must consider what happens at the boundaries of our system, in this case at the support. At the support we know there is no vertical or horizontal displacement and that there is no rotation. These known facts will form the basis of our **boundary conditions,** which we will now apply to equations [5.9] and [5.10].

Our first boundary condition is derived from the fact that there is no rotation at the support. Therefore the slope at the support is zero. The support is located at a distance L from the free end such that:

$$\text{at } x = L \qquad {}^{dy}/_{dx}\,(\text{slope}) = 0$$

We can substitute these values into equation [5.9] and:

$$-EI\frac{dy}{dx} = -EI \times 0 = \frac{wL^3}{6} + A$$

$$\text{Therefore:} \quad A = -\frac{wL^3}{6}$$

Our second boundary condition is derived from the fact that there is no vertical displacement at the support. Therefore y is zero at the support. The support is located at a distance of $x = L$ such that:

$$\text{at } x = L \qquad y\,(\text{deflection}) = 0$$

We can substitute these values into equation [5.10] along with that for the constant A such that:

$$-EI\ y = -EI \times 0 = \frac{wL^4}{24} - \frac{wL^3}{6}L + B$$

$$\text{From which we find:} \quad B = \frac{wL^4}{8}$$

Substituting the values for the constants of integration A and B into equation [5.10] we derive the **general solution** for the structure:

$$-EI\ y = \frac{wx^4}{24} - \frac{wL^3}{6}x + \frac{wL^4}{8} \qquad [5.11]$$

This general solution can be used to determine the deflection at any distance from the free end by substituting the distance for x in equation [5.11].

For instance, the maximum deflection will occur at the free end of the cantilever beam, that is the point when $x = 0$. By substituting this value into the general solution of [5.11] we can find the maximum deflection such that:

at $x = 0$ from equation [5.11]:

$$-EI\ y = \frac{w(0)^4}{24} - \frac{wL^3}{6}(0) + \frac{wL^4}{8}$$

which will give:

$$y_{\max} = -\frac{wL^4}{8EI}$$

In this example, we derived our equations with reference to the free end. It is also possible to derive a similar equation commencing from the support. However, when deriving the equation from the support we must ensure that we include any moments (commonly known as Fixed End Moments (FEM)) or forces developed. We must also include any vertical or horizontal reactions. The effect of these must be included in our derivation of the general expression for bending moments.

◇

Example 5.9: Deflection of a cantilever beam supporting a distributed load (2)

Consider the cantilever beam of Figure 5.34. The beam supports a uniformly distributed load of w kN/m along its length. Derive an expression to calculate the deflection along the beam. What will the maximum deflection be?

Assume the flexural rigidity (EI) for the beam is constant along its length.

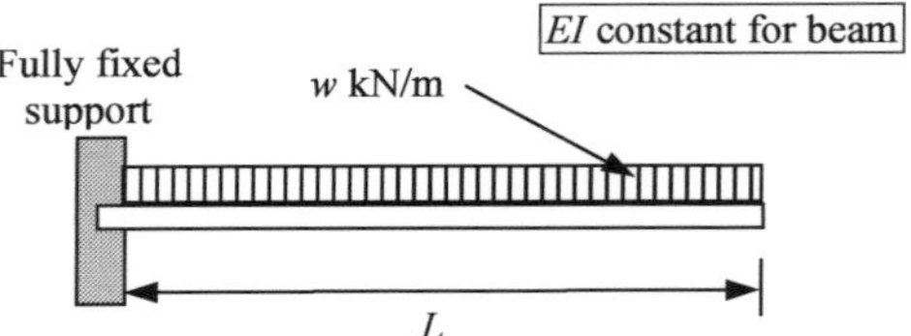

Figure 5.34: cantilever beam – Example 5.9

We will again derive an equation for the bending moments of the beam by isolating a section and deriving an expression for the moments about the datum as shown in Figure 5.35. However, in this case we will calculate our distances to the support. We must therefore include values for the moments induced by the vertical reaction and the fixed end moment (FEM) in our bending moment expression.

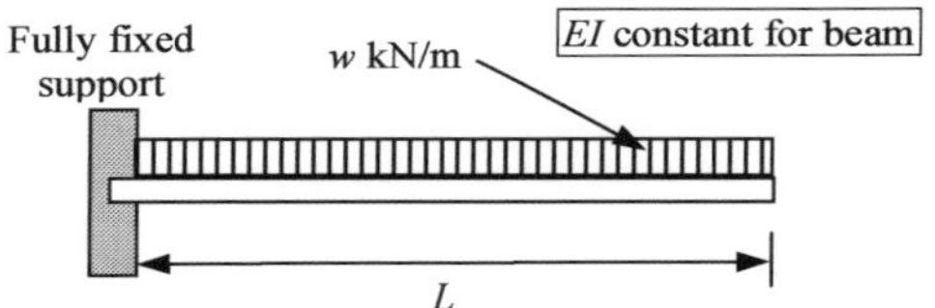

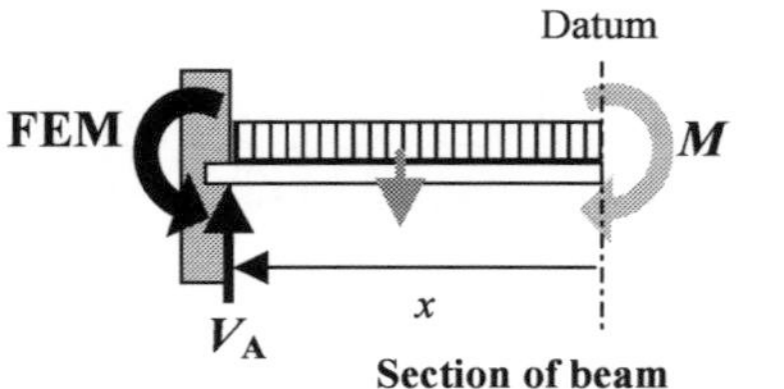

Figure 5.35: Details of section of beam – Example 5.9

The values of the fixed end moment (FEM) and vertical reaction at the support for a cantilever beam supporting a UDL are (see ref. 2 or 3 listed on page 231):

$$\text{FEM} = \frac{wL^2}{2} \qquad\qquad V_A = wL$$

We now take moments about the datum, noting that M is, in this case, assumed to be clockwise:

$$M = \underset{\text{(reaction)}}{wLx} - \underset{\text{(FEM)}}{\frac{wL^2}{2}} - \underset{\text{(applied load)}}{\frac{wx^2}{2}}$$

Substituting this expression for bending moments M into equation [5.5] we have:

$$EI\frac{d^2y}{dx^2} = wLx - \frac{wL^2}{2} - \frac{wx^2}{2} \qquad [5.12]$$

Integrating equation [5.12] we have:

$$EI\frac{dy}{dx} = \frac{wLx^2}{2} - \frac{wL^2}{2}x - \frac{wx^3}{6} + A \qquad [5.13]$$

where A = constant of integration.

Integrating equation [5.13] will give:

$$EI\ y = \frac{wLx^3}{6} - \frac{wL^2}{4}x^2 - \frac{wx^4}{24} + Ax + B \qquad [5.14]$$

where B is the second constant of integration.

The **boundary conditions** will again be derived based on our knowledge of what happens at the support. The first boundary condition is again derived from the fact that there is no rotation at the support, therefore the slope at the support is zero. The support has zero distance in the x direction and hence:

at $x = 0$ $\quad {}^{dy}/_{dx}$ (slope) = 0

We can substitute these values into equation [5.13] such that:

$$EI(0) = \frac{wL(0)^2}{2} - \frac{wL^2}{2}(0) - \frac{w(0)^3}{6} + A$$

therefore: $A = 0$

Our second boundary condition is derived from the fact that there is no vertical displacement at the support, therefore y is zero at the support. The support has zero distance and hence:

at $x = 0$ $\quad y$ (deflection) = 0

We can substitute these values into equation [5.14] along with the previously calculated value for the constant of integration A to find:

$$EI(0) = \frac{wL(0)^3}{6} - \frac{wL^2}{4}(0)^2 - \frac{w(0)^4}{24} + A(0) + B$$

from which we have: $B = 0$

Substituting the values for the constants of integration A and B into equation [5.14], we derive the **general solution** for the structure:

$$EI\ y = \frac{wLx^3}{6} - \frac{wL^2}{4}x^2 - \frac{wx^4}{24} \qquad [5.15]$$

We can now use this expression to determine the maximum deflection which will occur at the free end of the cantilever beam, that is the point when $x = L$. Substituting this value into the general solution of [5.15] we find the maximum deflection such that:

At $x = L$ from equation [5.15]:

$$EI\ y = \frac{wL(L)^3}{6} - \frac{wL^2}{4}(L)^2 - \frac{w(L)^4}{24}$$

which will give $$y_{max} = -\frac{wL^4}{8EI}$$

As should be expected, this is the same as the result derived from the analysis commenced at the free end.

This integration technique can be applied to various types of beams where the load is a continuous function (load constant). However, this technique is not applicable to beams where a stepped function (e.g. point or varying loads) is to be considered. Macaulay's method is a modification to the procedure that allows us to analyse beams with stepped functions. Because it is a modification, it is sometimes referred to as Macaulay's variation.

Consider the analysis of the simply supported beam of Example 5.10.

Example 5.10: Deflection of a simply supported beam carrying a single vertical point load at mid-span

The simply supported beam of Figure 5.36 supports a point load of P kN at mid-span. Derive an expression to calculate the deflection along the beam. What will the maximum deflection be?

Assume the flexural rigidity (EI) for the beam is constant along its length.

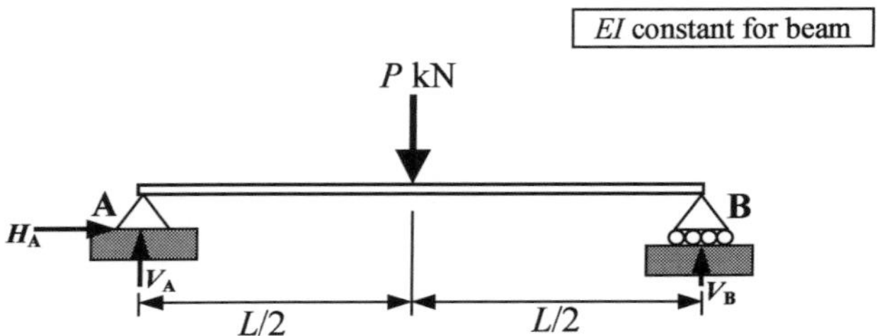

Figure 5.36: Simply supported beam – Example 5.10

First, we need to calculate the reactions such that:

Summation of all horizontal forces $\Sigma F_x = H_A + 0 = 0$ [1]

$\therefore H_A = 0$ kN

Summation of all vertical forces $\Sigma F_y = V_A + V_B - P\text{ kN} = 0$

$V_A + V_B = P$ kN [2]

Summation of all moments about support A:

$\Sigma M_z = (P\text{ kN} \times L/2) - (V_B \times L) = 0$ [3]

$= (PL/2) - V_BL = 0$

$V_B = P/2$

Back-substituting V_B into equation [2] gives $V_A = P/2$

We must now derive an expression for the bending moments about a datum, as shown in Figure 5.37.

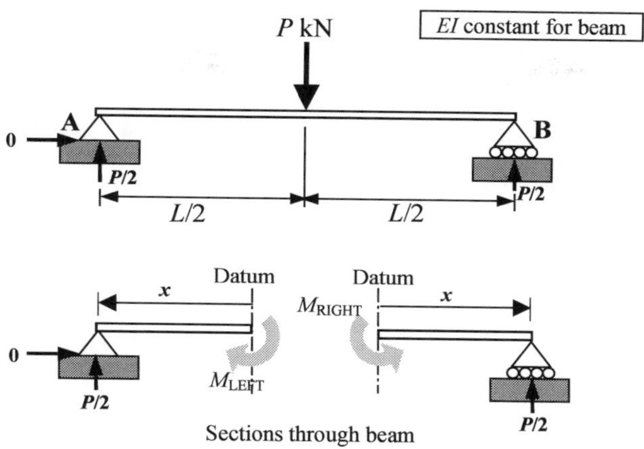

Figure 5.37: Detail of sections to left and right of load *P* – Example 5.10

It can be seen that the moment equations derived from either of the sections indicated in Figure 5.37 will be equal and opposite:

$$M_{LEFT} = -M_{RIGHT} = (P/2)\,x$$

The expression for deflection derived from these moment equations can, in this case, be applied to any part of the beam. This is because the beam is symmetrical and symmetrically loaded. Maximum deflection will occur at mid-span and from symmetry we know that the displacement at $^{L}/_{4}$ will be equal to that at $^{3L}/_{4}$. It may be expedient to derive an expression for the moments that will be applicable to any part of the structure. In order to do this we must locate our datum beyond the point load, as illustrated in Figure 5.38.

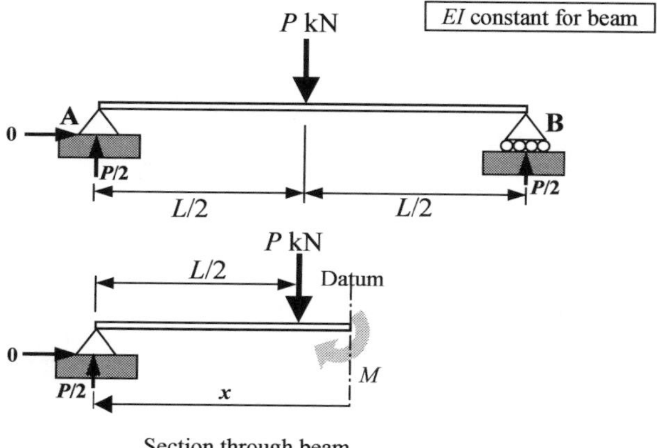

Figure 5.38: Section of beam with datum located beyond the point load – Example 5.10

We now derive our moment equation about the datum such that:

$$M = \frac{P}{2}x - P\left[x - \frac{L}{2}\right]$$

It is important to note here the use of the square brackets for the term $[x - {}^{L}/_{2}]$. These brackets constitute a convention which is used in Macaulay's method. When we integrate a term enclosed within these square brackets we are required to carry out the integration of the brackets and **NOT** of its contents. Therefore, in this method, $[F]$ when integrated becomes ${}^{1}/_{2}[F]^2$. This convention must be followed in order to derive the deflection equation. We now continue with the solution:

Substituting the expression for bending moments M into equation [5.5]:

$$M = EI\frac{d^2y}{dx^2} = \frac{P}{2}x - P\left[x - \frac{L}{2}\right]$$

Integrating this equation (and remembering to integrate the brackets [ ] and NOT their contents) we have:

$$EI\frac{dy}{dx} = \frac{P}{4}x^2 - \frac{P}{2}\left[x - \frac{L}{2}\right]^2 + A$$

where A = constant of integration.

And integrating the above equation again will give:

$$EI\ y = \frac{P}{12}x^3 - \frac{P}{6}\left[x - \frac{L}{2}\right]^3 + Ax + B$$

where B is the second constant of integration.

We must now define some boundary conditions based on our knowledge of what happens as the structure deflects and from the constraints placed at the support. The first boundary condition is derived from our understanding of how the structure deflects. We know that, in this case, maximum deflection will occur at mid-span and that the deflection either side of the mid-span is symmetrical. The deflected form of the beam is shown in Figure 5.39.

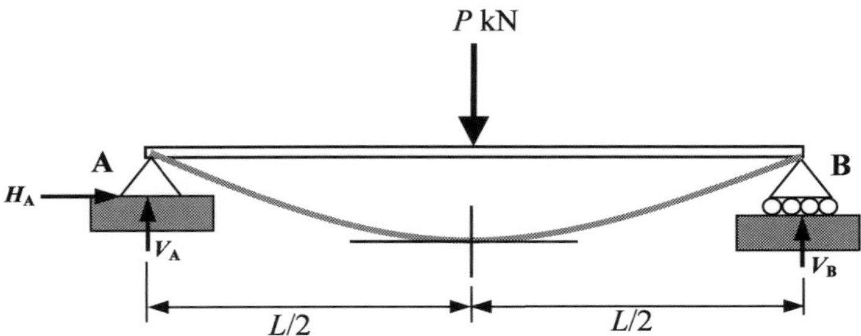

Figure 5.39: Deflected form of beam – Example 5.10

We note that, from support A to mid-span, the deflection is increasing. From mid-span to support B, the deflection is reducing. At mid-span, the deflection is a maximum and the slope (${}^{dy}/_{dx}$) at this point is zero. It is important to remember that the location of this point will depend on a number of factors, including the beams support and its load configuration – **it will not always be located at mid-span.**

From this we can derive our first boundary condition since:

at $x = {}^{L}/_{2}$ ${}^{dy}/_{dx}$ (slope) = 0

We can substitute this boundary condition into the expression for ${}^{dy}/_{dx}$ such that:

$$EI(0) = \frac{P}{4}\left(\frac{L}{2}\right)^2 - \frac{P}{2}\left[\left(\frac{L}{2}\right) - \frac{L}{2}\right]^2 + A$$

We must remember here to ensure that we apply the operators to the entire contents of the brackets such that:

$$\left(\frac{L}{2}\right)^2 = \left(\frac{L^2}{2^2}\right) = \left(\frac{L^2}{4}\right)$$

Hence for the constant of integration A:

$$A = -\frac{PL^2}{16} + \frac{P}{2}[0]^2 = -\frac{PL^2}{16}$$

Our second boundary condition is derived from the fact that there is no vertical displacement at the supports. Therefore y is zero at support A. The support has zero distance and hence:

$$\text{at } x = 0 \qquad y\,(\text{deflection}) = 0$$

We can substitute these values into the equation for y such that:

$$EI(0) = \frac{P}{12}(0)^3 - \frac{P}{6}\left[(0) - \frac{L}{2}\right]^3 + A(0) + B$$

from which we find that: $B = 0$.

It is important to note that the contents of the square brackets, that is $[(0) - {}^{L}/_{2}]$ will yield a negative answer. This is meaningless, since the expression refers to a physical length measured on the beam. It can therefore be considered as zero. We will find that we can adopt this approach in all instances that yield a negative result. Having solved for the constants of integration, we can now derive our **general solution**:

$$EI\;y = \frac{P}{12}x^3 - \frac{P}{6}\left[x - \frac{L}{2}\right]^3 - \frac{PL^2}{16}x$$

This equation can now be used to determine the deflection at any part of the beam.

To find maximum deflection we must put $x = L/2$ into the equation:

$$EI\;y = \frac{P}{12}\left(\frac{L}{2}\right)^3 - \frac{P}{6}\left[\frac{L}{2} - \frac{L}{2}\right]^3 - \frac{PL^2}{16}\left(\frac{L}{2}\right)$$

$$\text{this will give } y_{max} = -\frac{PL^3}{48EI}$$

We can also calculate the deflection at ${}^{L}/_{4}$ and ${}^{3L}/_{4}$ thus:

at $x = {}^{L}/_{4}$ (negative value = 0)

$$EI\;y = \frac{P}{12}\left(\frac{L}{4}\right)^3 - \frac{P}{6}\left[\frac{L}{4} - \frac{L}{2}\right]^3 - \frac{PL^2}{16}\left(\frac{L}{4}\right)$$

This will give y at ${}^{L}/_{4} = -\dfrac{11PL^3}{768EI}$

at $x = {}^{3L}/_{4}$

$$EI\;y = \frac{P}{12}\left(\frac{3L}{4}\right)^3 - \frac{P}{6}\left[\frac{3L}{4} - \frac{L}{2}\right]^3 - \frac{PL^2}{16}\left(\frac{3L}{4}\right)$$

and y at ${}^{3L}/_{4} = -\dfrac{11PL^3}{768EI}$

As expected, the deflection at the 'quarter' points are equal. We can follow the same procedure, substituting any value for L into our general expression, to determine the deflection at that point.

We will now solve some further examples using the same procedure.

Example 5.11: Deflection of a simply supported beam carrying three vertical point loads

The simply supported beam of Figure 5.40 supports point loads of P kN as shown. Derive a general expression to calculate the deflection at any point along the beam. What will the maximum deflection be? What will the deflection be at a distance of $^{L}/_{12}$ from support A? Assume the flexural rigidity (EI) for the beam is constant along its length.

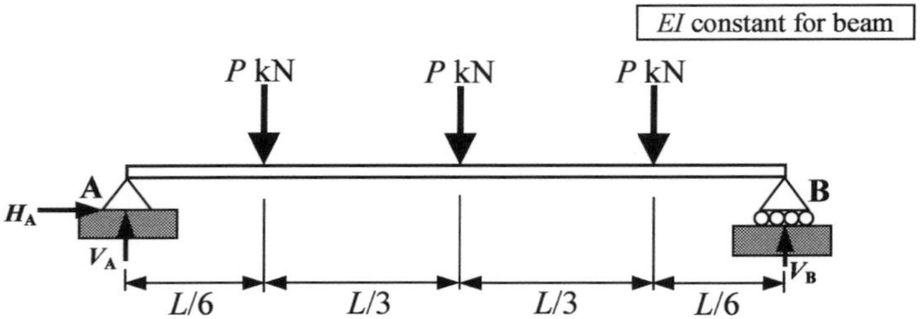

Figure 5.40: Simply supported beam – Example 5.11

First, we need to calculate the reactions such that:

Summation of all horizontal forces $\quad \sum F_x = H_A + 0 = 0 \quad$ [1]

$\therefore H_A = 0$ kN

Summation of all vertical forces $\quad \sum F_y = V_A + V_B - P\text{ kN} - P\text{ kN} - P\text{ kN} = 0$

$V_A + V_B = 3P$ kN $\quad$ [2]

Summation of all moments about support A:

$$\sum M_z = (P\text{ kN} \times L/6) + (P\text{ kN} \times L/2) + (P\text{ kN} \times 5L/6) - (V_B \times L) = 0 \quad [3]$$

$$= (3PL/2) - V_BL = 0$$

$$V_B = 3P/2$$

Back-substituting V_B into equation [2] gives $\quad V_A = 3P/2$

We must now derive an expression for bending moments as shown in Figure 5.41.

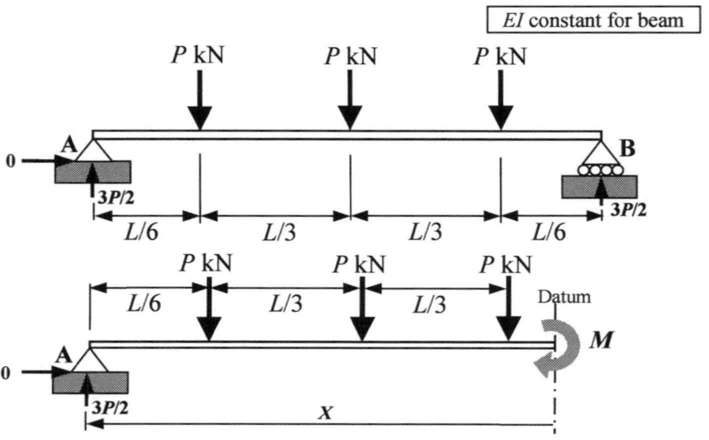

Figure 5.41: Section of beam – Example 5.11

We now derive our moment equation about the datum such that:

$$M = \frac{3P}{2}x - P\left[x - \frac{L}{6}\right] - P\left[x - \frac{L}{2}\right] - P\left[x - \frac{5L}{6}\right]$$

Substituting this expression into equation [5.5]:

$$M = EI\frac{d^2y}{dx^2} = \frac{3P}{2}x - P\left[x - \frac{L}{6}\right] - P\left[x - \frac{L}{2}\right] - P\left[x - \frac{5L}{6}\right]$$

Integrating this equation (and remembering to integrate brackets [ ], NOT contents) we have:

$$EI\frac{dy}{dx} = \frac{3P}{4}x^2 - \frac{P}{2}\left[x - \frac{L}{6}\right]^2 - \frac{P}{2}\left[x - \frac{L}{2}\right]^2 - \frac{P}{2}\left[x - \frac{5L}{6}\right]^2 + A$$

where A = constant of integration.

And integrating this equation once again will give:

$$EI\ y = \frac{3P}{12}x^3 - \frac{P}{6}\left[x - \frac{L}{6}\right]^3 - \frac{P}{6}\left[x - \frac{L}{2}\right]^3 - \frac{P}{6}\left[x - \frac{5L}{6}\right]^3 + Ax + B$$

where B is the second constant of integration.

We must now define our boundary conditions. The first boundary condition is again derived from our understanding of how the structure deflects. We know that, in this case, maximum deflection will occur at mid-span and that the deflection either side of the mid-span is symmetrical.

From this we can derive our first boundary condition since:

at $x = {}^{L}/_{2}$ $\quad {}^{dy}/_{dx}$ (slope) = 0

We can substitute this boundary condition into our ${}^{dy}/_{dx}$ expression such that:

$$EI(0) = \frac{3P}{4}\left(\frac{L}{2}\right)^2 - \frac{P}{2}\left[\frac{L}{2} - \frac{L}{6}\right]^2 - \frac{P}{2}\left[\frac{L}{2} - \frac{L}{2}\right]^2 - \frac{P}{2}\left[\frac{L}{2} - \frac{5L}{6}\right]^2 + A$$

Hence for the constant of integration A:

$$A = -\frac{3P}{4}\left(\frac{L^2}{4}\right) - \frac{P}{2}\left[\frac{L^2}{9}\right] - \frac{P}{2}[0]^2 - \frac{P}{2}[0]^2 = -\frac{19PL^2}{144EI}$$

Our second boundary condition is derived from the fact that there is no vertical displacement at support A. Therefore y is zero at the support. The support has zero distance and hence:

at $x = 0 \quad y$ (deflection) = 0

We can substitute these values into the equation for y such that:

$$EI(0) = \frac{3P}{12}(0)^3 - \frac{P}{6}\left[(0) - \frac{L}{6}\right]^3 - \frac{P}{6}\left[(0) - \frac{L}{2}\right]^3 - \frac{P}{6}\left[(0) - \frac{5L}{6}\right]^3 + A(0) + B$$

from which we find that: $B = 0$

Having solved for the constants of integration, we can now derive our **general solution** which is:

$$EI\ y = \frac{3P}{12}x^3 - \frac{P}{6}\left[x - \frac{L}{6}\right]^3 - \frac{P}{6}\left[x - \frac{L}{2}\right]^3 - \frac{P}{6}\left[x - \frac{5L}{6}\right]^3 - \frac{19PL^2}{144EI}x$$

This equation can now be used to determine the deflection at any part of the beam.

To find maximum deflection we must put $x = L/2$ into the equation:

$$EI\ y = \frac{3P}{12}\left(\frac{L}{2}\right)^3 - \frac{P}{6}\left[\frac{L}{2}-\frac{L}{6}\right]^3 - \frac{P}{6}\left[\frac{L}{2}-\frac{L}{2}\right]^3 - \frac{P}{6}\left[\frac{L}{2}-\frac{5L}{6}\right]^3 - \frac{19PL^2}{144EI}\left(\frac{L}{2}\right)$$

Hence:

$$EI\ y = \frac{3P}{12}\left(\frac{L^3}{8}\right) - \frac{P}{6}\left[\frac{L^3}{27}\right] - \frac{P}{6}[0]^3 - \frac{P}{6}[0]^3 - \frac{19PL^2}{144}\left(\frac{L}{2}\right) = -\frac{53PL^3}{1296}$$

This will give $y_{max} = -\dfrac{53PL^3}{1296EI}$

We can also calculate the deflection at $^L/_{12}$ thus:
at $x = {}^L/_{12}$

$$EI\ y = \frac{3P}{12}\left(\frac{L}{12}\right)^3 - \frac{P}{6}\left[\frac{L}{12}-\frac{L}{6}\right]^3 - \frac{P}{6}\left[\frac{L}{12}-\frac{L}{2}\right]^3 - \frac{P}{6}\left[\frac{L}{12}-\frac{5L}{6}\right]^3 - \frac{19PL^2}{144EI}\left(\frac{L}{12}\right)$$

and

$$EI\ y = \frac{3P}{12}\left(\frac{L^3}{1728}\right) - \frac{P}{6}[0]^3 - \frac{P}{6}[0]^3 - \frac{P}{6}[0]^3 - \frac{19PL^2}{144EI}\left(\frac{L}{12}\right)$$

This will give y at $^L/_{12} = -\dfrac{25PL^3}{2304EI}$

◇

Example 5.12: Deflection of a simply supported beam carrying three vertical point loads

The simply supported beam of Figure 5.42 supports three point loads as shown:

(a) derive an expression to calculate the deflection along the beam;

(b) determine the position and magnitude of the maximum deflection.

Assume the flexural rigidity (EI) for the beam = 1.4×10^4 kN m^2.

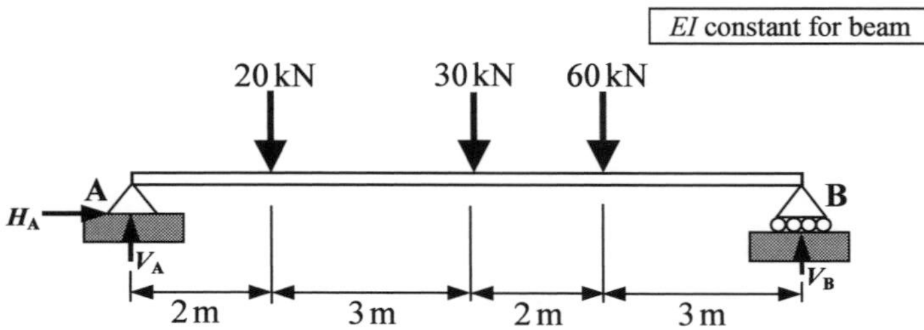

Figure 5.42: Simply supported beam – Example 5.12

First, we need to calculate the reactions such that:

Summation of all horizontal forces $\Sigma F_x = H_A + 0 = 0$ [1]

$\therefore H_A = 0$ kN

Summation of all vertical forces $\Sigma F_y = V_A + V_B - 20\,\text{kN} - 30\,\text{kN} - 60\,\text{kN} = 0$

$V_A + V_B = 110\,\text{kN}$ [2]

Summation of all moments about support A:

$$\Sigma M_z = (20\text{ kN} \times 2\text{ m}) + (30\text{ kN} \times 5\text{ m}) + (60\text{ kN} \times 7\text{ m}) - (V_B \times 10\text{ m}) = 0 \quad [3]$$

$$= (610) - 10V_B = 0$$

$$V_B = 61\text{ kN}$$

Back-substituting V_B into equation [2] gives $V_A = 49\text{ kN}$

We must now derive an expression for bending moments about the datum shown in Figure 5.43.

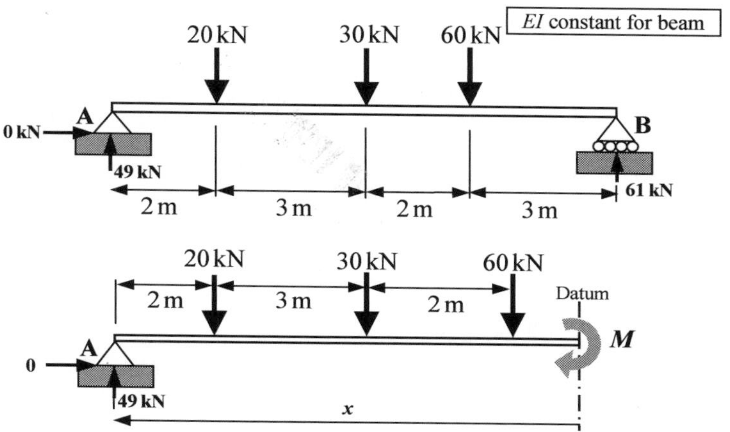

Figure 5.43: Section of beam – Example 5.12

We now derive our moment equation about the datum such that:

$$M = 49x - 20[x-2] - 30[x-5] - 60[x-7]$$

Substituting this expression into equation [5.5]:

$$M = EI\frac{d^2y}{dx^2} = 49x - 20[x-2] - 30[x-5] - 60[x-7]$$

Integrating the equation we have:

$$EI\frac{dy}{dx} = \frac{49}{2}x^2 - \frac{20}{2}[x-2]^2 - \frac{30}{2}[x-5]^2 - \frac{60}{2}[x-7]^2 + A$$

And integrating this equation once again will give:

$$EI\ y = \frac{49}{3}x^3 - \frac{20}{3}[x-2]^3 - \frac{30}{3}[x-5]^3 - \frac{60}{3}[x-7]^3 + Ax + B$$

and

$$EI\ y = 16.333x^3 - 6.667[x-2]^3 - 10[x-5]^3 - 20[x-7]^3 + Ax + B$$

In this case we do not know where the maximum deflection will occur and therefore we will not be able to locate the point at which the slope is a minimum. We must therefore derive both of our boundary condition equations from the supports such that:

at $x = 0$, $y = 0$ (no vertical deflection at support A)

and at $x = 10$, $y = 0$ (no vertical deflection at support B)

We can substitute these boundary condition into the expression for y such that:

at $x = 0,\ y = 0$

$$EI\ (0) = 16.333(0)^3 - 6.667[(0)-2]^3 - 10[(0)-5]^3 - 20[(0)-7]^3 + A(0) + B$$

gives $B = 0$

at $x = 10,\ y = 0$

$$EI\ (0) = 16.333(10)^3 - 6.667[(10)-2]^3 - 10[(10)-5]^3 - 20[(10)-7]^3 + A(10) + (0)$$

gives $A = -1112.98$

The **general solution** for this example will therefore be:

$$EI\ y = 16.333x^3 - 6.667[x-2]^3 - 10[x-5]^3 - 20[x-7]^3 - 1112.98x$$

We do not know where the maximum deflection will occur, however we do know that at the position of maximum deflection the slope will be zero. We can therefore use our slope equation to find the position of zero slope. Consider the derived equation for slope, modified to include the value for the constant of integration A:

$$EI\frac{dy}{dx} = \frac{49}{2}x^2 - \frac{20}{2}[x-2]^2 - \frac{30}{2}[x-5]^2 - \frac{60}{2}[x-7]^2 - 1112.98$$

A complex mathematical solution to this equation can be derived by minimising the function, however it is often more expedient to use an iterative technique. We therefore use our judgement to determine where the maximum deflection might take place. We then substitute this value for x into the slope equation and check to see if it equates to zero. If it does not equal zero we modify our distance until it is within an acceptable limit. It is usually only necessary to calculate the position to three decimal places. Thus, for this example we will commence our calculations at 5 m from support A. The slope equation at 5 m will therefore be:

Try $x = 5\,\text{m}$
$$EI\frac{dy}{dx} = \frac{49}{2}(5)^2 - \frac{20}{2}[5-2]^2 - \frac{30}{2}[0]^2 - \frac{60}{2}[0]^2 - 1112.98$$

which will yield $EI\,{}^{dy}/_{dx} = -409.52$ (not zero, therefore try new distance)

Try $x = 8\,\text{m}$
$$EI\frac{dy}{dx} = \frac{49}{2}(8)^2 - \frac{20}{2}[8-2]^2 - \frac{30}{2}[8-5]^2 - \frac{60}{2}[8-7]^2 - 1112.98$$

which will yield $EI\,{}^{dy}/_{dx} = -69.98$ (not zero, but less than that for 5 m)

From these results, we see that we are converging on the solution. Further modification to the length will yield the results shown in the table of Figure 5.44.

Distance x from support A	$EI\,{}^{dy}/_{dx}$
8.5 m	−16.605
8.6 m	−7.76
8.65 m	−3.566
8.7 m	0.475
8.695 m	0.08
8.694 m	0.002

Figure 5.44: Results of iterations to find position of zero slope – Example 5.12

Maximum deflection will therefore occur at approximately 8.694 m from support A. Substituting this value, and that given for the flexural rigidity EI into the general solution will yield the maximum deflection, such that:

$$1.4\times10^{4}\ y = 16.333(8.694)^3 - 6.667[8.694-2]^3 - 10[8.694-5]^3 - 20[8.694-7]^3 - 1112.98(8.694)$$

which gives: $y_{max} = -0.1103\,\text{m}$.

Should we require to calculate the deflection at any specified point on the beam, we can insert the appropriate x value into the general solution to obtain an answer.

Example 5.13: Deflection of a simply supported beam carrying a trapezoidal distributed load

The simply supported beam of Figure 5.45 supports a trapezoidal load;

(a) derive an expression to calculate the deflection along the beam;

(b) determine the position and magnitude of the maximum deflection.

Assume the flexural rigidity (EI) for the beam = $12\times10^4\,\text{kNm}^2$.

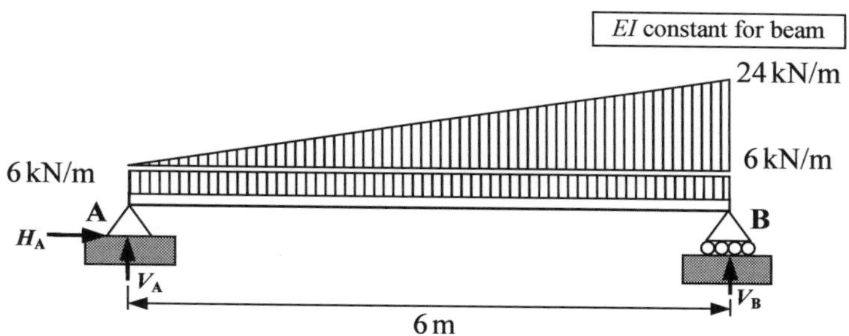

Figure 5.45: Simply supported beam – Example 5.13

For ease of calculation, we can break down the trapezoidal load into a UDL and a triangular load. The UDL will have a value of 6 kN/m and the triangular load will vary from 0 kN/m at support A to a maximum of 24 kN/m at support B.

Calculate reactions:

Summation of all horizontal forces $\quad \Sigma F_x = H_A + 0 = 0 \quad$ [1]

$\therefore H_A = 0\,\text{kN}$

Summation of all vertical forces $\quad \Sigma F_y = V_A + V_B - (6\,\text{kN/m}\times 6\,\text{m}) - (0.5\times 24\,\text{kN}\times 6\,\text{m}) = 0\,\text{kN}$

$V_A + V_B = 108\,\text{kN} \quad$ [2]

Summation of all moments about support A

$$\Sigma M_z = (6\,\text{kN/m}\times 6\,\text{m}\times(6\,\text{m}/2)) + (0.5\times 24\,\text{kN/m}\times 6\,\text{m}\times\{{}^2/_3\times 6\,\text{m}\}) - (V_B\times 6\,\text{m}) = 0 \quad [3]$$

$$= (396) - 6V_B = 0$$

$$V_B = 66\,\text{kN}$$

Back-substituting V_B into equation [2] gives $\quad V_A = 42\,\text{kN}$

Before we derive an expression for the bending moments, we note that the triangular load will vary in intensity along the beam. We must therefore derive a general expression for this change in intensity. We note that the beam is 6 m in length and that the intensity varies from 0 to 24 kN/m. The load therefore varies at a rate of (24 kN/m)/(6 m) = 4 kN/m per metre span of beam.

We must now derive an expression for bending moments as shown in Figure 5.46.

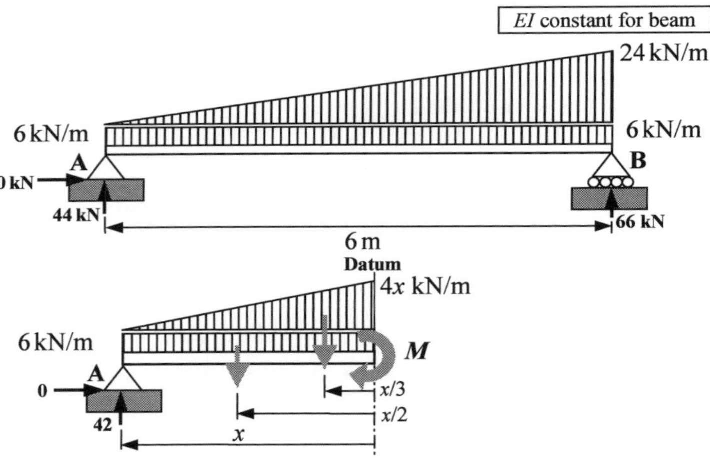

Figure 5.46: Section of beam – Example 5.13

We now derive our moment equation about the datum such that:

$$M = 42x - \frac{6x^2}{2} - \left(4x \times \frac{x}{2} \times \frac{x}{3}\right) = 42x - \frac{6x^2}{2} - \frac{4x^3}{6}$$

Substituting this expression into equation [5.5] gives:

$$M = EI\frac{\mathrm{d}^2y}{\mathrm{d}x^2} = 42x - \frac{6x^2}{2} - \frac{4x^3}{6}$$

Integrating this equation we have:

$$EI\frac{\mathrm{d}y}{\mathrm{d}x} = \frac{42}{2}x^2 - \frac{6x^3}{6} - \frac{4x^4}{24} + A$$

And integrating this equation once again will give:

$$EI\ y = \frac{42}{6}x^3 - \frac{6x^4}{24} - \frac{4x^5}{120} + Ax + B$$

and

$$EI\ y = 7x^3 - \frac{x^4}{4} - \frac{x^5}{30} + Ax + B$$

We again do not know where the maximum deflection will occur and thus we cannot locate the point of zero slope. We therefore derive both of our boundary conditions from the supports such that:

at $x = 0$, $y = 0$ (no vertical deflection at support A)

and at $x = 6$, $y = 0$ (no vertical deflection at support B)

We can substitute these boundary condition into the expression for y:

at $x = 0,\ y = 0$

$$EI\ (0) = 7(0)^3 - \frac{(0)^4}{4} - \frac{(0)^5}{30} + A(0) + B$$

gives $B = 0$

at $x = 6,\ y = 0$

$$EI\ (0) = 7(6)^3 - \frac{(6)^4}{4} - \frac{(6)^5}{30} + A(6) + (0)$$

gives $A = -154.8$

The **general solution** to this example will therefore be:

$$EI\ y = 7x^3 - \frac{x^4}{4} - \frac{x^5}{30} - 154.8x$$

In order to determine the maximum deflection we must again find the position of zero slope using the slope equation modified to include the value for the constant of integration A such that:

$$EI\frac{dy}{dx} = \frac{42}{2}x^2 - \frac{6x^3}{6} - \frac{4x^4}{24} - 154.8$$

In this case we commence our calculations at 3.5 m from support A and iterate values using the slope equation as before. The results are shown in the table of Figure 5.47.

Distance x from support A	$EI\ ^{dy}/_{dx}$
3.5 m	34.57
3 m	−6.3
3.2 m	10
3.1 m	1.827
3.07 m	−0.616
3.077 m	−0.047

Figure 5.47: Results of iterations to find position of zero slope – Example 5.13

Maximum deflection will therefore occur at approximately 3.077 m from support A.

Substituting this value, and that given for the flexural rigidity EI into the general solution will yield the maximum deflection, such that:

$$12\times10^4\ y = 7(3.077)^3 - \frac{(3.077)^4}{4} - \frac{(3.077)^5}{30} - 154.8(3.077)$$

which gives: $y_{max} = -0.0025$ m.

Example 5.14: Deflection of a simply supported beam carrying a point and a distributed load

The simply supported beam of Figure 5.48 supports a UDL and a point load as shown:

(a) derive an expression to calculate the deflection along the beam;

(b) determine the position and magnitude of the maximum deflection.

Assume that the beam has a Modulus of Elasticity (E) = 200 kN/mm^2 and a Second Moment of Area (I_{xx}) = 150×10^6 mm^4.

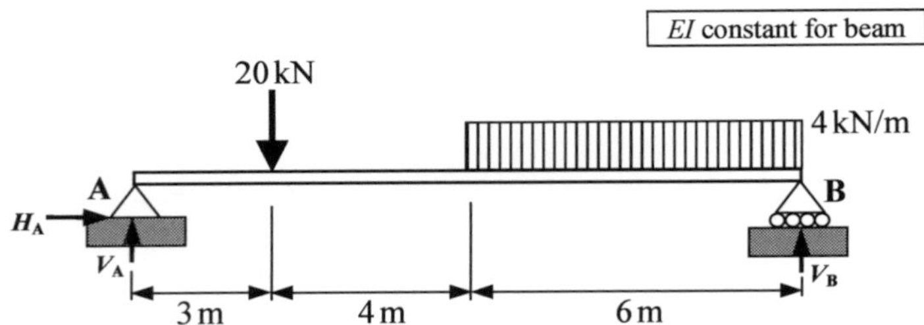

Figure 5.48: Simply supported beam – Example 5.14

In this example we have a discontinuous UDL and we cannot apply the procedure directly to the problem. In order to derive a deflection equation for this beam, we must insert two 'imaginary' loads in order to create a beam with a continuous UDL. Thus our analysis will be based on the modified beam of Figure 5.49.

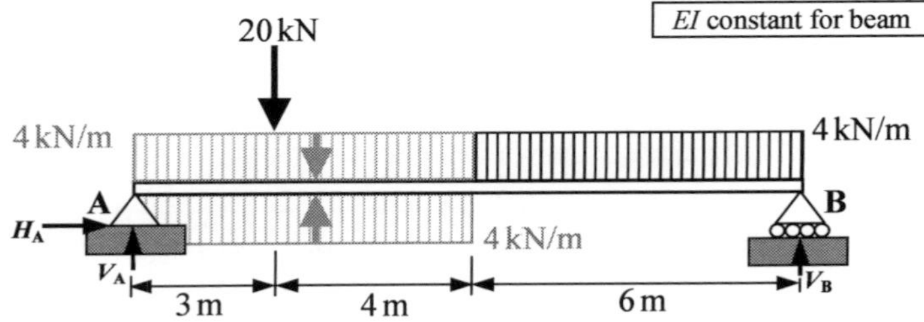

Figure 5.49: Modified beam – Example 5.14

It should be noted that the additional loads have a zero net effect on the overall loading of the beam.

Calculate reactions for modified system:

Summation of all horizontal forces $\quad \Sigma F_x = H_A + 0 = 0$ [1]

$\therefore H_A = 0\,\text{kN}$

Summation of all vertical forces $\quad \Sigma F_y = V_A + V_B - (4\,\text{kN/m} \times 13\,\text{m}) - 20\,\text{kN} + (4\,\text{kN} \times 7\,\text{m}) = 0\,\text{kN}$

$V_A + V_B = 44\,\text{kN}$ [2]

Summation of all moments about support A

$$\Sigma M_z = (4\,\text{kN/m} \times 13\,\text{m} \times (13\,\text{m}/2)) + (20\,\text{kN} \times 3\,\text{m}) - (4\,\text{kN/m} \times 7\,\text{m} \times (7\,\text{m}/2)) - (V_B \times 13\,\text{m}) = 0 \quad [3]$$

$$= 300 - 13V_B = 0$$

$$V_B = 23.08\,\text{kN}$$

Back-substituting V_B into equation [2] gives $\quad V_A = 20.92\,\text{kN}$

We will now derive an expression for bending moments, as shown in Figure 5.50.

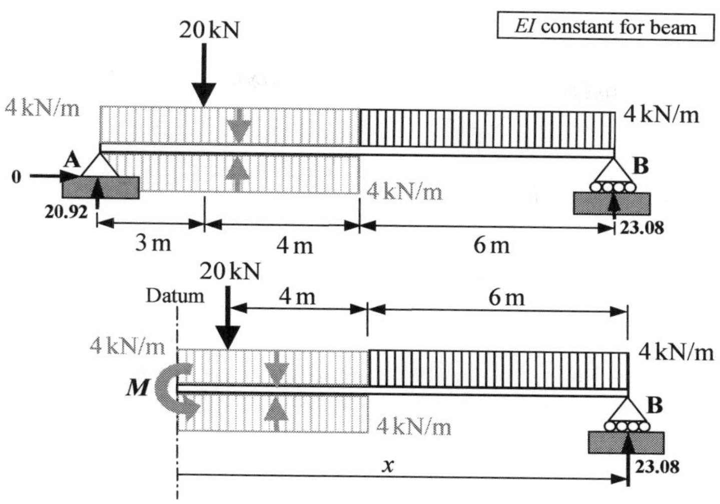

Figure 5.50: Section of beam – Example 5.14

We now derive our moment equation about the datum such that:

$$-M = -\frac{4}{2}[x-6]^2 + \frac{4x^2}{2} + 20[x-10] - 23.08x$$

Substituting this expression into equation [5.5]:

$$-M = -EI\frac{\mathrm{d}^2 y}{\mathrm{d}x^2} = -2[x-6]^2 + \frac{4x^2}{2} + 20[x-10] - 23.08x$$

Integrating this equation we have:

$$-EI\frac{\mathrm{d}y}{\mathrm{d}x} = -\frac{2}{3}[x-6]^3 + \frac{4x^3}{6} + \frac{20}{2}[x-10]^2 - \frac{23.08}{2}x^2 + A$$

And integrating this equation once again will give:

$$-EI\ y = -\frac{2}{12}[x-6]^4 + \frac{4x^4}{24} + \frac{20}{6}[x-10]^3 - \frac{23.08}{6}x^3 + Ax + B$$

and

$$-EI\ y = -0.1667[x-6]^4 + 0.1667x^4 + 3.33[x-10]^3 - 3.847x^3 + Ax + B$$

We again do not know where the maximum deflection will occur and therefore we cannot easily locate the point where the slope will be zero. We therefore derive both of our boundary conditions from consideration of the supports such that:

	at $x = 0$,	$y = 0$	(no vertical deflection at support A)
and	at $x = 13$,	$y = 0$	(no vertical deflection at support B)

We can substitute these boundary condition into the expression for y such that:

at $x = 0,\ y = 0$

$$-EI(0) = -0.1667[(0)-6]^4 + 0.1667(0)^4 + 3.33[0-10]^3 - 3.847(0)^3 + A(0) + B$$

gives $B = 0$

at $x = 13,\ y = 0$

$$-EI(0) = -0.1667[13-6]^4 + 0.1667(13)^4 + 3.33[13-10]^3 - 3.847(13)^3 + A(13) + (0)$$

gives $A = 307.78$

The **general solution** for this example will therefore be:

$$-EI\ y = -0.1667[x-6]^4 + 0.1667x^4 + 3.33[x-10]^3 - 3.847x^3 + 307.78x$$

We will now determine the point at which the slope is zero using the slope equation ($^{dy}/_{dx}$) modified to include the value for the constant of integration A such that:

$$-EI\frac{dy}{dx} = -\frac{2}{3}[x-6]^3 + \frac{4x^3}{6} + \frac{20}{2}[x-10] - \frac{23.08}{2}x^2 + 307.78$$

For this example we commence calculations at 7 m from support B and iterate values using the slope equation. The results are shown in the table of Figure 5.51.

Distance x from support B	$EI\ ^{dy}/_{dx}$
7 m	−29.66
6.75 m	−13.25
6.5 m	3.224
6.55 m	−0.075
6.549 m	−0.0095
6.548 m	0.0504

Figure 5.51: Results of iterations to find position of zero slope – Example 5.14

We can assume that maximum deflection will therefore occur at approximately 6.549 m from support B.

Substituting this value, and those given for the Modulus of Elasticity (E) and Second Moment of Area (I_{xx}) into the general solution will yield the maximum deflection (ignoring the 20 kN load which will yield a negative result):

$$-(200\,000\times 15\times 10^7)/10^9\ y = -0.1667[6.549-6]^4 + 0.1667(6.549)^4 - 3.847(6.549)^3 + 307.78(6.549)$$

which gives: $y_{max} = 0.041$ m.

◇

Example 5.15: Deflection of a fully fixed beam carrying a distributed load

The ‘built in’ beam of Figure 5.52 supports a UDL along its length as shown:

(a) derive an expression to calculate the deflection along the beam;

(b) determine the position and magnitude of the maximum deflection in terms of the beams flexural rigidity (EI).

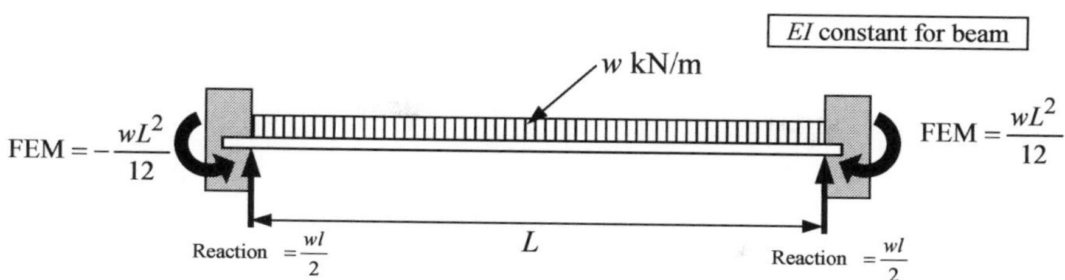

Figure 5.52: Fully fixed ('built in') beam – Example 5.15

We note the reactions are give as:

Vertical reactions = $wL/2$ Fixed end moments = $wL^2/12$

We now derive an expression for bending moments as shown in Figure 5.53.

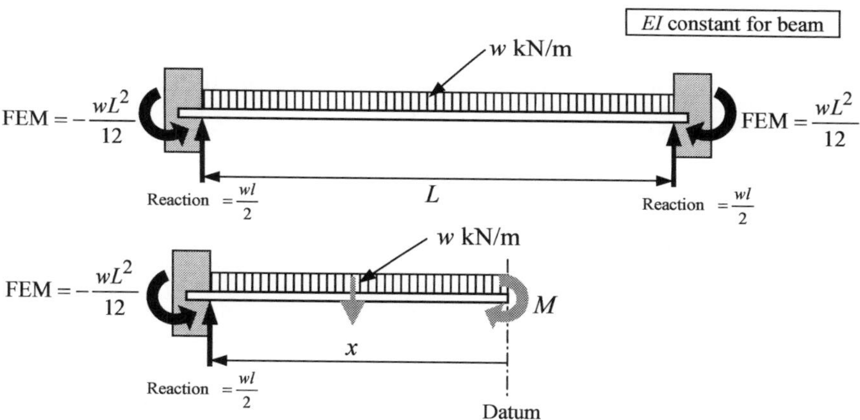

Figure 5.53: Section of beam – Example 5.15

The moment equation about the datum will be:

$$M = -\frac{wL^2}{12} + \frac{wL}{2}x - \frac{wx^2}{2}$$

Substituting this expression for bending moments M into equation [5.5]:

$$M = EI\frac{d^2y}{dx^2} = -\frac{wL^2}{12} + \frac{wL}{2}x - \frac{wx^2}{2}$$

Integrating the equation we have:

$$EI\frac{dy}{dx} = -\frac{wL^2}{12}x + \frac{wL}{4}x^2 - \frac{wx^3}{6} + A$$

And integrating this equation once again will give:

$$EI\ y = -\frac{wL^2}{24}x^2 + \frac{wL}{12}x^3 - \frac{wx^4}{24} + Ax + B$$

The boundary conditions can be derived from the supports. At support A, the rotation and hence the slope will be zero. Also the vertical displacement will be zero hence:

at $x = 0$, $y = 0$ (no vertical deflection at support A)

and at $x = 0$, ${}^{dy}/_{dx} = 0$ (no rotation at support A)

We can substitute these boundary condition into expressions for ${}^{dy}/_{dx}$ and y such that:
at $x = 0$, ${}^{dy}/_{dx} = 0$

$$EI\,(0) = -\frac{wL^2}{12}(0) + \frac{wL}{4}(0)^2 - \frac{w(0)^3}{6} + A$$

gives $A = 0$

at $x = 0$, $y = 0$

$$EI\ (0) = -\frac{wL^2}{24}(0)^2 + \frac{wL}{12}(0)^3 - \frac{wx^4}{24} + A(0) + B$$

gives $B = 0$

The **general solution** to this example will therefore be:

$$EI\ y = -\frac{wL^2}{24}x^2 + \frac{wL}{12}x^3 - \frac{wx^4}{24}$$

Maximum deflection will, in this case, take place at mid-span. In terms of its flexural rigidity the maximum deflection at mid-span will therefore be:
at $x = L/2$

$$EI\ y = -\frac{wL^2}{24}\left(\frac{L}{2}\right)^2 + \frac{wL}{12}\left(\frac{L}{2}\right)^3 - \frac{w\left(\frac{L}{2}\right)^4}{24}$$

The solution of this equation will yield: $y_{\max} = \dfrac{wL^4}{384EI}$

Example 5.16: Deflection of a cantilever beam of varying flexural rigidity (*EI*) carrying a point load

The cantilever beam of Figure 5.54 supports a point load at its free end. The flexural rigidity (*EI*) of the beam is 2*EI* for section A–B and *EI* for section B–C, as shown. Calculate the maximum deflection at the free end.

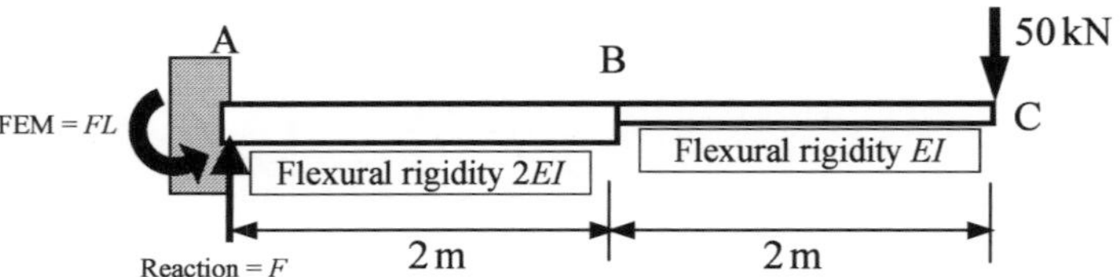

Figure 5.54: Cantilever beam – Example 5.16

We note the reactions can be determined from Figure 5.54 as:

Vertical reaction = 50 kN Fixed end moment = 200 kN m

To complete this problem we will use the Law of linear superposition detailed in Chapter 3. In order to analyse this beam we must split the problem at point B, where we will apply imaginary 'balancing' loads and moments in order to ensure the system remains in equilibrium. These loads and moments, shown in Figure 5.55, will have a zero net effect on the overall structure. We will also derive expressions for the bending moments for each section in order to develop a solution for the complete structure.

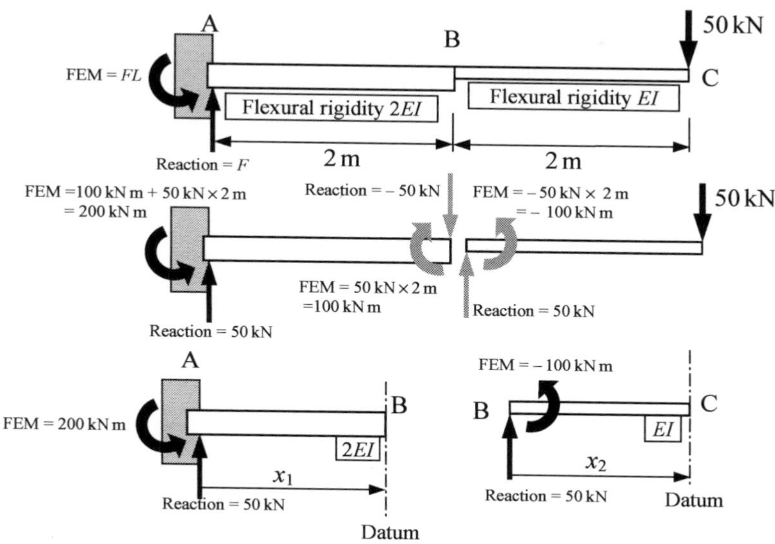

Figure 5.55: Sections of cantilever beam – Example 5.16

We now derive moment equations for each section in turn about its datum such that:

For section A–B

$$M = 200 - 50x_1$$

For section B–C

$$M = 100 - 50x_2$$

Substituting these expressions for bending moments M into equation [5.5]:

$$M = 2EI\frac{\mathrm{d}^2y}{\mathrm{d}x^2} = 200 - 50x_1 \qquad M = EI\frac{\mathrm{d}^2y}{\mathrm{d}x^2} = 100 - 50x_2$$

Integrating these equations we have:

$$2EI\frac{\mathrm{d}y}{\mathrm{d}x} = 200x_1 - \frac{50}{2}x_1^2 + A \qquad EI\frac{\mathrm{d}y}{\mathrm{d}x} = 100x_2 - \frac{50}{2}x_2^2 + A$$

And integrating these equations once again will give:

$$2EI\ y = \frac{200}{2}x_1^2 - \frac{50}{6}x_1^3 + Ax_1 + B \qquad EI\ y = \frac{100}{2}x_2^2 - \frac{50}{6}x_2^3 + Ax_2 + B$$

The boundary conditions can be derived from the support for section A–B and from the loads and moment applied at point B for section B–C. At support A, the rotation and hence the slope will be zero. Also the vertical displacement will be zero. At point B, the slope in member BC will be the same as that already determined for member AB, and the deflection can be determined using the deflection equation for AB above. Hence for member AB, the boundary conditions will be:

at $x_1 = 0$, $y = 0$ gives $B = 0$
at $x_1 = 0$, ${}^{dy}/_{dx} = 0$ gives $A = 0$

For member BC:

Boundary condition (1) at $x_2 = 0$, ${}^{dy}/_{dx}$ = slope at point B in member AB
Hence:
At $x_2 = 0$
$$\frac{dy}{dx} = \frac{1}{2EI}\left(200x_1 - \frac{50}{2}x_1^2\right)$$
And since at point B $x_1 = 2$ m, then ($A = 0$, **B** $= 0$):
$$\frac{dy}{dx} = \frac{1}{2EI}\left(200 \times 2 - \frac{50}{2} \times 2^2\right) = \frac{150}{EI}$$

Substituting this value, and the value for $x_2 = 0$, in the slope equation for member AB we have:

$$EI\frac{dy}{dx} = 100x_2 - \frac{50}{2}x_2^2 + A$$
$$EI\frac{150}{EI} = 100(0) - \frac{50}{2}(0) + A$$

which gives $A = 150$

Boundary condition (2) at $x_2 = 0$, y = deflection at point B in member AB
Hence:
At $x_2 = 0$
$$2EI\ y = \frac{200}{2}x_1^2 - \frac{50}{6}x_1^3 + Ax_1 + B$$
And since at point B $x_1 = 2$ m, then ($A = 0$, $B = 0$):
$$2EI\ y = \frac{200}{2}(2)^2 - \frac{50}{6}(2)^3 = \frac{166.66}{EI}$$

Substituting this value, and the value for $x_2 = 0$, in the deflection equation for member AB and the value for constant A previously calculated as –150 we have:

$$EI\ \frac{166.66}{EI} = \frac{100}{2}(0) - \frac{50}{6}(0) - 150(0) + B$$
which gives $B = 166.66$

The **general solution** for each section will therefore be:

For $x = 0$ to 2 m from support A:
$$2EI\ y = 100x_1^2 - 8.333x_1^3$$

For $x = 2$ to 4 m from support A:
$$EI\ y = 50x_2^2 - 8.333x_2^3 + 150x_2 + 166.66$$

Maximum deflection will occur at the free end. In terms of its flexural rigidity the maximum deflection will therefore be:
$x_2 = 2$ m
$$y_{max} = \left(\frac{50x_2^2 - 8.333x_2^3 + 150x_2 + 166.66}{EI}\right)$$
$$y_{max} = \left(\frac{50(2)^2 - 8.333(2)^3 + 150(2) + 166.66}{EI}\right) = \frac{600}{EI}$$

Example 5.17: Deflection of a simply supported beam with a cantilever end carrying a point and a distributed load

The simply supported beam of Figure 5.56 supports a point load and a UDL as shown:

(a) derive an expression to calculate the deflection along the beam;

(b) determine the position and magnitude of the maximum deflection in terms of the beams flexural rigidity (*EI*).

Assume the flexural rigidity is constant for the length of the beam.

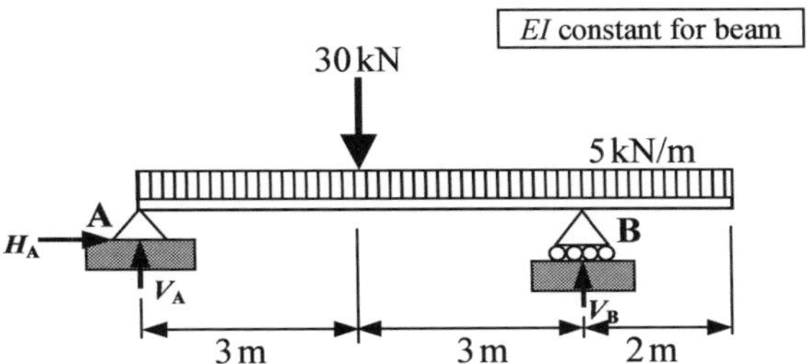

Figure 5.56: Simply supported beam – Example 5.17

First, we calculate the reactions such that:

Summation of all horizontal forces $\Sigma F_x = H_A + 0 = 0$ [1]

$\therefore H_A = 0\text{ kN}$

Summation of all vertical forces $\Sigma F_y = V_A + V_B - 30\text{ kN} - (5\text{ kN} \times 8\text{ m}) = 0$

$V_A + V_B = 55\text{ kN}$ [2]

Summation of all moments about support A

$$\Sigma M_z = (30\text{ kN} \times 3\text{ m}) + (5\text{ kN/m} \times 8\text{ m} \times (8\text{ m}/2)) - (V_B \times 6\text{ m}) = 0 \quad [3]$$

$$250 - 6\,V_B = 0$$

$$V_B = 41.667\text{ kN}$$

Back-substituting V_B into equation [2] gives $V_A = 28.333\text{ kN}$

We must now derive an expression for bending moments, as shown in Figure 5.57, noting that the section includes both supports.

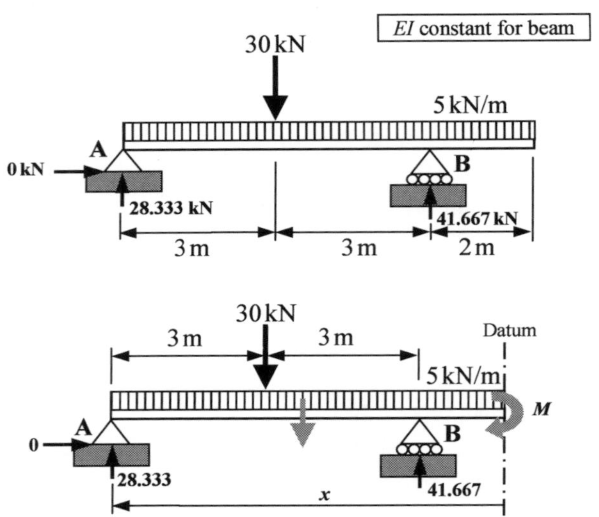

Figure 5.57: Section of beam – Example 5.17

Moment equation about datum:

$$M = 28.333x - 30[x-3] - \frac{5}{2}x^2 + 41.667[x-6]$$

Substituting the expression into equation [5.5]:

$$M = EI\frac{\mathrm{d}^2 y}{\mathrm{d}x^2} = 28.333x - 30[x-3] - \frac{5}{2}x^2 + 41.667[x-6]$$

Integrating this equation gives:

$$EI\frac{\mathrm{d}y}{\mathrm{d}x} = \frac{28.333}{2}x^2 - \frac{30}{2}[x-3]^2 - \frac{5}{6}x^3 + \frac{41.667}{2}[x-6]^2 + A$$

And integrating this equation once again we have:

$$EI\ y = \frac{28.333}{6}x^3 - \frac{30}{6}[x-3]^3 - \frac{5}{24}x^4 + \frac{41.667}{6}[x-6]^3 + Ax + B$$

and

$$EI\ y = 4.722x^3 - 5[x-3]^3 - 0.1042x^4 + 6.9445[x-6]^3 + Ax + B$$

Boundary conditions:

at $x = 0$, $y = 0$ (no vertical deflection at support A)

and at $x = 6$, $y = 0$ (no vertical deflection at support B)

$x = 0$, $y = 0$ gives $B = 0$

at $x = 6$, $y = 0$

$$EI\ y = 4.772(6)^3 - 5[6-3]^3 - 0.1042[6-3]^4 + 6.9445[6-6]^3 + A(6) + (0)$$

which gives $A = -126.79$

The **general solution** will therefore be:

$$EI\ y = 4.722x^3 - 5[x-3]^3 - 0.1042x^4 + 6.9445[x-6]^3 - 126.76x$$

We note that one point at which the slope angle will change is at support B, however the slope will not necessarily change from a positive to a negative direction! We will commence our calculations to find the maximum deflection, assuming that it occurs between the supports, at 4 m from support A, using the slope equation:

$$EI\frac{dy}{dx} = \frac{28.333}{2}x^2 - \frac{30}{2}[x-3]^2 - \frac{5}{6}x^3 + \frac{41.667}{2}[x-6]^2 - 126.76$$

The results of successive iterations are shown in the table of Figure 5.58.

Distance x from support A	$EI\ ^{dy}/_{dx}$
4.0 m	31.573
3.5 m	7.301
3.1 m	−15.594
3.3 m	− 3.783
3.36 m	− 0.379
3.367 m	0.0132

Figure 5.58: Results of iterations to find position of zero slope – Example 5.17

Maximum deflection between supports A and B will therefore occur at approximately 3.367 m from support A.

Substituting this value into the general solution will yield the maximum deflection, such that:

$$EI\ y = 4.772(3.367)^3 - 5[3.367-3]^3 - 0.1042(3.367)^4 + 6.9445[3.367-6]^3 - 126.76(3.367)$$

which gives: $y = -\dfrac{258.29}{EI}$

In this case we should also check the deflection at the end of the cantilever, to the left of support B.

Deflection at free end, i.e. at $x = 8$ m:

$$EI\ y = 4.772(8)^3 - 5[8-3]^3 - 0.1042(8)^4 + 6.9445[8-6]^3 - 126.76(8)$$

which gives: $y = -\dfrac{432.94}{EI}$

Therefore, in Example 5.17, maximum deflection will occur at the end of the cantilever.

◇

Macaulay's method, coupled with the procedures developed here, can be adapted to derive deflection equations for various structures and loading configurations. However, it may be necessary, in certain cases, to break down the problem into separate elements. Having derived the solution for each element, the deflections can be calculated. Using the Law of Linear Superposition proposed in Chapter 3, the results may then be added to find the final solution to the problem.

It is also possible to approximate the deflections of complex structures using the simple formulae listed in Figure 5.30. Having calculated the bending moments for a beam, we can define an equivalent UDL or point load and, applying a suitable safety factor, we can then calculate the deflection at mid-span. This will normally yield a result that is an over-estimate, however in many instances we only require an indication of the deflection, not an exact value.

For example, consider a combination of loading on a beam of span 10 m which produces a maximum moment of 60 kN m (assume $E = 200\ 000\ \text{N/mm}^2$ and $I = 10\,000\ \text{cm}^4$).

For a continuous UDL on a beam, maximum bending moment $= \dfrac{wL^2}{8}$

Therefore, for a bending moment of 60 kN m:

Equivalent UDL $= [(60 \times 8)/(10^2)] \times 1.25$ (safety factor) $= 6$ kN/m

$$\text{Maximum deflection} = y_{max} = \frac{5wL^4}{384EI} = \frac{5 \times 6 \times 10\ 000^4}{384 \times 200 \times 10^3 \times 10\ 000 \times 10^4} = 39.1\ \text{mm}$$

Whilst this simplification may be used to give some indication of the approximate magnitude of deflection to which the beam will be subjected under load, **extreme caution** should be adopted when attempting to utilise such methods for detailed design. The method is a very useful aid to assist in checking results obtained from a more sophisticated analysis, such as those used in computer programs, but would not normally be employed for design purposes.

Having completed this chapter you should now be able to:

- ***Sketch the deflected forms of various simple beam and frame structures***
- ***Derive equations to determine the magnitude of deflection for a given structure and loading***
- ***Determine the position of maximum deflection and its magnitude***

Further Problems 5

1. For each of the structures (a), (b) and (c) sketch the deflected form (all internal joints are to be considered as rigid).

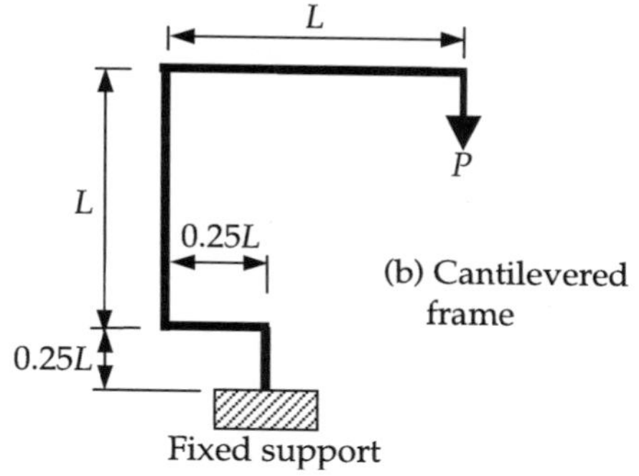

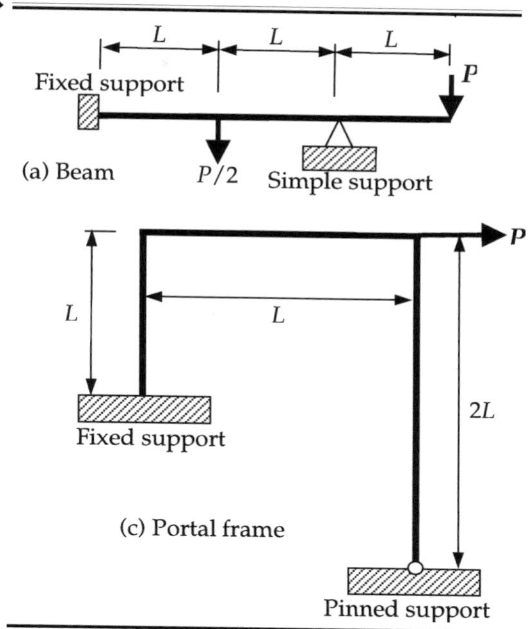

2. For the beam shown in Figure Question 2:
   (a) Derive a general expression for the deflection of the beam.
   (b) For a similar beam of span 10 m the deflection is limited to 30 mm; assuming the beam has a Modulus of Elasticity (E) = 200 kN/mm^2 and a Second Moment of Area (I) = 70 000 cm^4 calculate the maximum allowable value for W.

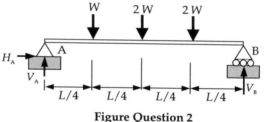

Figure Question 2

3. For the propped cantilever of Figure Question 3:
   (a) derive a general expression for the deflection of the beam;
   (b) determine the position of maximum deflection and its magnitude (E and I are given).

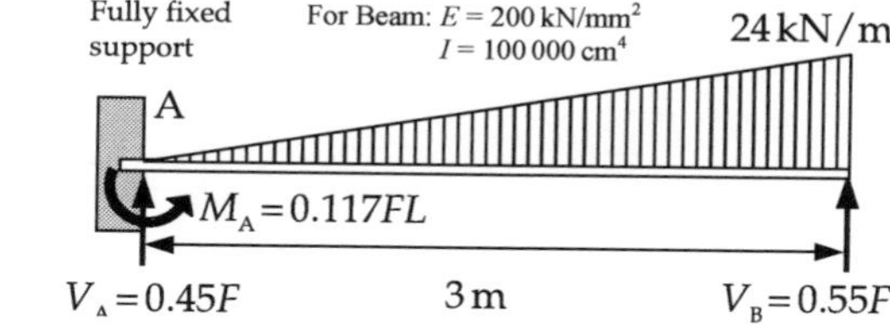

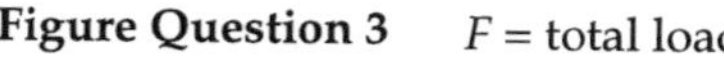
Figure Question 3 F = total load

4. For the simply supported beam shown in Figure Question 4:
   (a) derive a general expression for the deflection of the beam;
   (b) determine the maximum deflection at $L/2$ in terms of the beams flexural rigidity (EI).

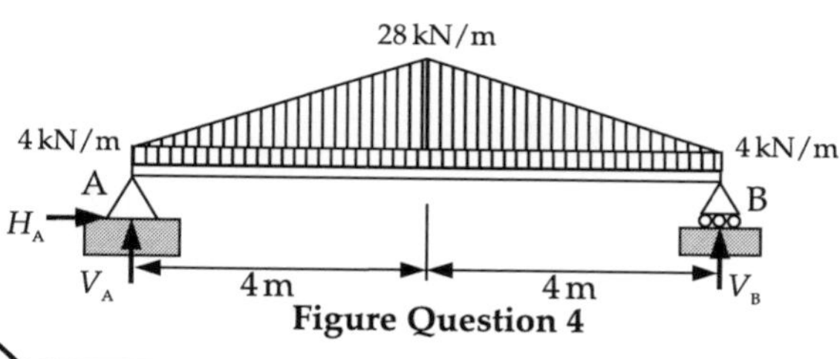

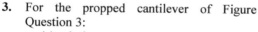

Figure Question 4

Solutions to Further Problems 5

1. Possible solutions.

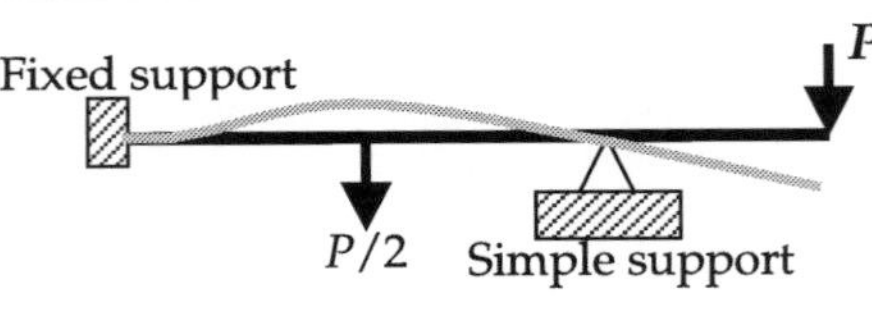

(a) Beam

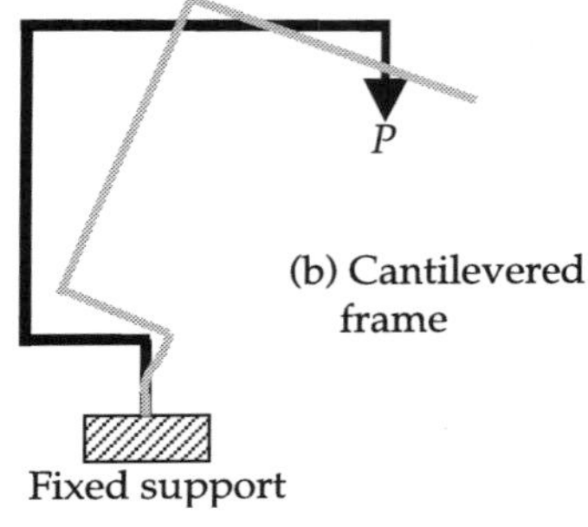

(b) Cantilevered frame

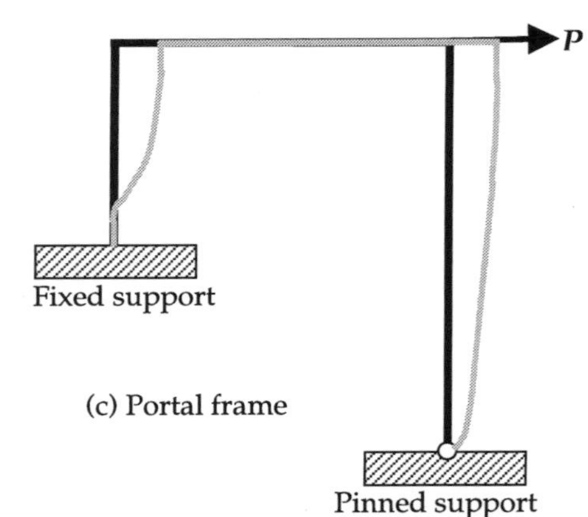

(c) Portal frame

---

2. (a) $V_A = 2.25\,W$ $\quad V_B = 2.75\,W$

$$M = 2.25Wx - 2W\left[x - \frac{3L}{4}\right] - 2W\left[x - \frac{L}{2}\right] - W\left[x - \frac{L}{4}\right]$$

Boundary conditions

$x = 0,\ y = 0$ gives $B = 0$

$x = L,\ y = 0$ gives $A = -0.26WL^2$

General solution

Maximum deflection at 5.543 m from support A (found by substituting various values for length x into slope equation and solving resultant quadratic)

$$EI\,y = 0.375\,Wx^3 - \frac{W}{3}\left[x - \frac{3L}{4}\right]^3 - \frac{W}{3}\left[x - \frac{L}{2}\right]^3 - \frac{W}{6}\left[x - \frac{L}{4}\right]^3 - 0.26\,WL^2x$$

(b) $E = 200\,\text{kN/mm}^2$ $\quad I = 70\,000\,\text{cm}^4$

$L = 10\,000\,\text{mm}$ $\quad x = 5.115\,\text{mm}$

which gives $W = 48.955\,\text{kN}$

---

3. (a) $F = 36\,\text{kN}$ $\quad M_A = 12.6\,\text{kN m}$

$V_A = 16.2\,\text{kN}$ $\quad V_B = 19.8\,\text{kN}$

$$M = 16.2x - 12.6 - \frac{4}{3}x^3$$

Boundary conditions

$x = 0,\ y = 0$ gives $B = 0$

$x = 0,\ {}^{dy}/_{dx} = 0$ gives $A = 0$

General solution

$$EI\,y = 2.7x^3 - 6.3x^2 - 0.067x^5$$

(b) Maximum deflection at 1.795 m from support A = 6.24 mm

4. (a) $V_A = 64\,\text{kN}$ $V_B = 64\,\text{kN}$

Derive a solution for half of beam (symmetry):

$$M = 64x - 2x^2 - x^3$$

Boundary conditions

$x = 0,\ y = 0$ gives $B = 0$

$x = 4,\ {}^{dy}/_{dx} = 0$ gives $A = 405.3$

General solution

$$EI\,y = 10.667x^3 - 0.167x^4 - 0.05x^5 + 405.3$$

(b) Maximum deflection at 4 m from support A

$$y_{max} = -\frac{993.46}{EI}$$

6. Pin Jointed Frames

In this chapter we will learn how to:

- ***Determine the external and internal forces on pin jointed frames using***
  - ***A graphical technique***
  - ***The method of sections***
  - ***Joint resolution***
- ***Plot frame force diagrams***

6.1 Introduction

Pin jointed frames and trusses are an important class of structure with a variety of applications. They are commonly used to span long distances and have been used in the construction of roof structures, road and rail bridges, sporting stadia and many other applications where clear unobstructed spaces are required. Spans of 50–100 m are not uncommon. The structure's strength lies in its geometric form, being constructed of triangles, which is an inherently strong shape. They also offer an economic solution since the size of the members of the frame are small, in comparison to the span of the structure. This is possible because frame members are designed to carry tensile and compressive loads only. These structures are formed with pin (hinged) joints to all internal members and we know from our previous studies that a pin joint cannot carry bending moment (see Chapter 3). In our analysis, the members forming a pin jointed frame or truss will not carry loads perpendicular to their length and, therefore, no bending moment will be induced in the members. All loads will be applied at the pin joints, which again offer no resistance to bending moment, therefore the members will only support axial tensile and compressive loads. Some examples of the application of typical pin jointed frames and trusses are shown in Figure 6.1.

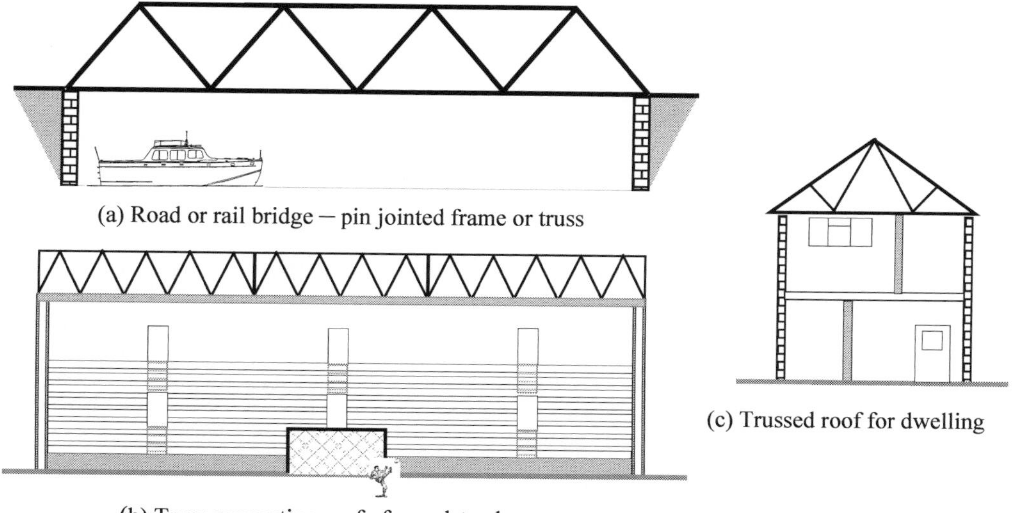

(a) Road or rail bridge – pin jointed frame or truss

(b) Truss supporting roof of grandstand

(c) Trussed roof for dwelling

Figure 6.1: Typical applications for pin jointed frames and trusses

It should be noted that, in some circumstances such as the trusses forming the roof of a building, the members might be subjected to loads along their length which will induce bending moments. However, in the design of the members for the truss we will be required to simplify the pin jointed system, by resolving all loads to the joints. In the design of the frame members we would then consider the moments induced by the 'real' loads. This will require a more complex member design since it will involve both moments and axial loads; however this type of design is accommodated in most current Codes of Practice. Typical stages of the analysis and design of a roof truss are illustrated in Figure 6.2.

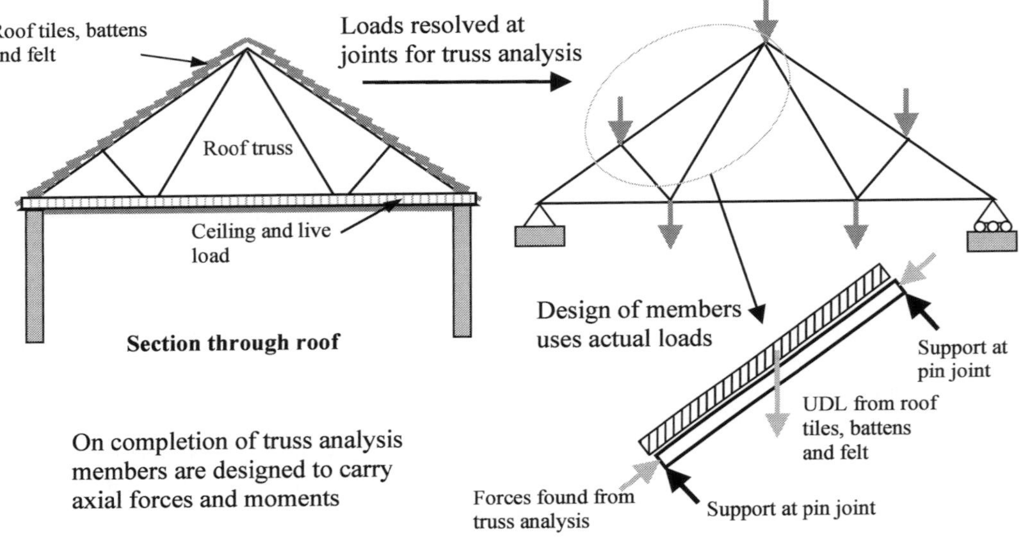

Figure 6.2: Typical stages in simplified analysis and design of a roof truss

The typical stages for the design of a roof truss might therefore be summarised as:

(1) Calculate all dead and live loads supported by the roof structure.
(2) Resolve all UDLs to point loads applied at the joint positions.
(3) Analyse truss.
(4) Design the individual members of the truss to carry loads determined from (3) and original calculated UDL.

The analysis of pin jointed frames and trusses will therefore be based on systems which support point loads at joint positions. All uniformly distributed loads along a member must be resolved and applied as point loads. Analysing members supporting UDLs as simply supported beams and calculating the reactions will allow us to determine appropriate values. The reaction forces can then be reversed and applied as point loads at the joint. This is shown diagrammatically in Figure 6.3.

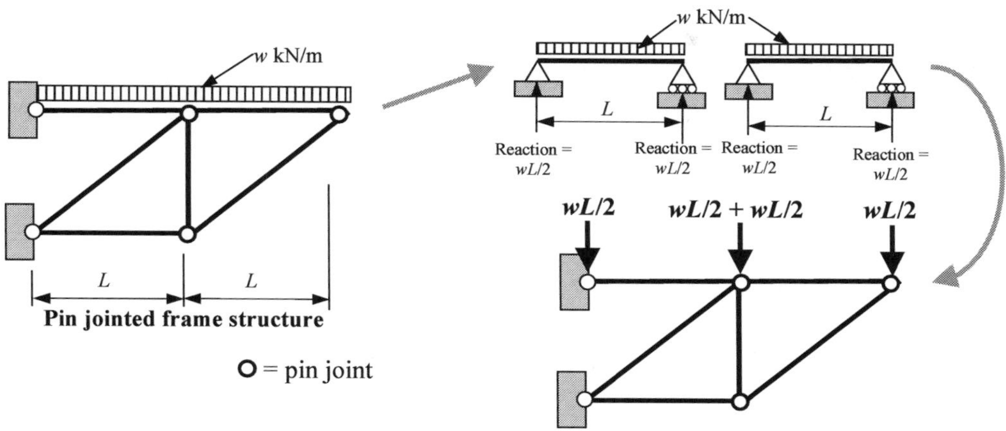

Figure 6.3: Resolution of a uniformly distributed load on a simple frame

In the analysis of pin jointed frames we consider the joints between members as being formed with frictionless pins. This allows members to rotate freely at their ends. Though it may not always be possible to form these frictionless pin joints in practice, for the purpose of our analysis we will ignore any forces induced by friction as rotation of the members takes place.

Having considered some typical applications for this type of structure, we will now proceed to look at some of the many methods of analysis available for pin jointed frames and trusses.

6.2 Methods of Analysis

The purpose of any analysis is to determine the magnitude and type of forces that the members of the structure will have to resist under a specified load configuration and also the reactions at the supports. The overall structure must be in equilibrium. In some cases we will be able to calculate the reactions using a similar procedure to that adopted for the analysis of beams, the frame being considered as a solid structure and the forces and moments calculated. However, this may not always be the easiest solution. In the analysis of some frames it will be difficult to predict the forces at the supports. In such cases we can often commence our analysis by locating a joint with one or more 'known' and a maximum of two 'unknown' forces. We will then derive equations to solve for the 'unknowns'. Following an analysis of the entire frame, the design of the individual members can be completed.

Many methods of analysis exist. A number of these techniques have been developed into commercially available computer programs that provide fast, efficient and accurate solutions to problems. However, it is still important that we have a basic understanding of the analysis in order that we are able to analyse simple frame structures and to check the analysis of more complex frames completed by computers. The accuracy of any computer solution relies on the accuracy of the input data. Always remember 'Rubbish In = Rubbish Out'! The methods developed here can later be extended to more advanced analysis techniques such as virtual work and flexibility.

We will now consider three of the many analysis techniques available. These are:

(a) a graphical solution;
(b) the method of sections;
(c) the method of resolution of forces at joints.

In the following sections we will introduce both a graphical solution and the method of sections. A more detailed explanation of the method of resolution of forces at joints will be given in section 6.3.

6.2.1 Graphical Solution

We have already discussed, in Chapter 1, the resolution of a force and the construction of forces diagrams. We know that any two force vectors acting at a point can be resolved into a single force, and that a single force acting at some angle to the horizontal or vertical can be replaced by its horizontal and vertical components, as shown in Figure 6.4. We can calculate the forces mathematically or using a force diagram, that is, a scale drawing that plots each force, as a vector, to a suitable scale. Unknown forces can be read directly from the scale drawing. In order to solve for the forces in a frame structure, we must extend this procedure to construct a composite force diagram (sometimes referred to as a Maxwell diagram) for the entire frame.

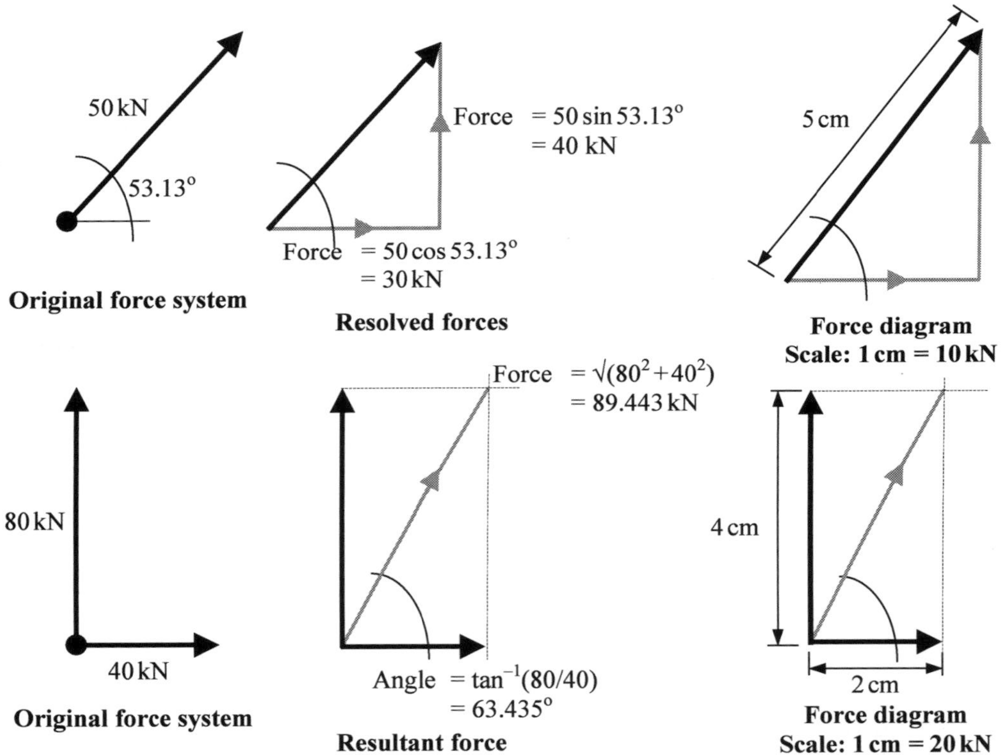

Figure 6.4: Resolution of forces and force diagrams

The construction of the composite force diagram for a frame structure can be complex and, in order to avoid confusion, a very specific procedure and convention is adopted known as Bow's notation. Before commencing the construction of the composite force diagram for a simple structure, it is important to note that the magnitude of the forces in any member of a pin jointed frame is directly related to the structure's geometry. A relationship exists between the ratio of the lengths of the members of the frame and the magnitude of the forces it carries, as illustrated in Figure 6.5.

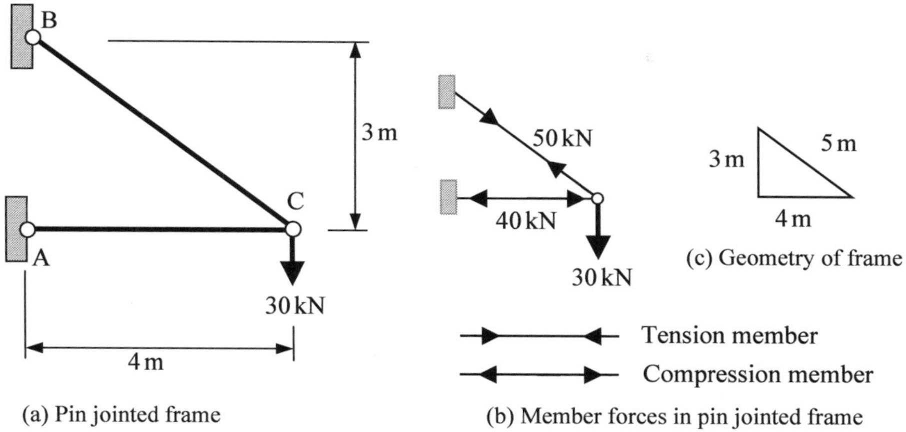

Figure 6.5: Pin jointed frame structure

Analysis of the pin jointed frame ABC in Figure 6.5 will show that the applied load of 30 kN at joint C will induce a compressive force of 40 kN in member AC and a tensile force of 50 kN in member BC. We note that there exists a ratio between the force triangle and the geometry of the structure such that, if we select any force and calculate its ratio with the structure's geometry, we will find it to be 1 m = 10 kN. This is an important relationship that will prove useful in the analysis of pin jointed frame structures using the method of resolution of forces at a joint.

Example 6.1: Analysis of a pin jointed frame (1)

We will now proceed to analyse the frame of Example 6.1, shown in Figure 6.6, using a graphical method.

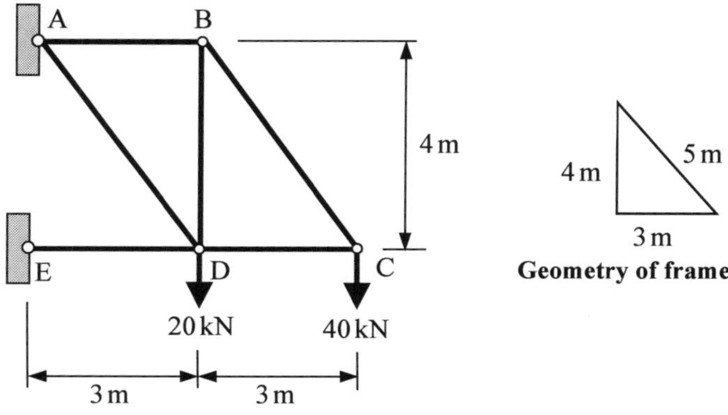

Figure 6.6: Pin jointed frame – Example 6.1

In order to commence our analysis, we must first use Bow's notation and number the spaces on the structure as shown in Figure 6.7. We note that space (1) covers the area between the two reactions. Area (2) covers the space from reaction A to the load at C, and area (3) is located between the loads. Area (4) extends from the load of 20 kN at D to support E. We will also plot the position and directions of the reactions. It does not matter at this stage if the direction indicated is wrong. The analysis will correct any errors, however it is important to maintain equilibrium of the structure and our prediction of the reactions should always bear this in mind. In this case the vertical reaction at A is considered to be in the opposite direction to the applied loads. There cannot be a vertical reaction at E as the member DE cannot induce a vertical load (resolution of a force through 90° = 0).

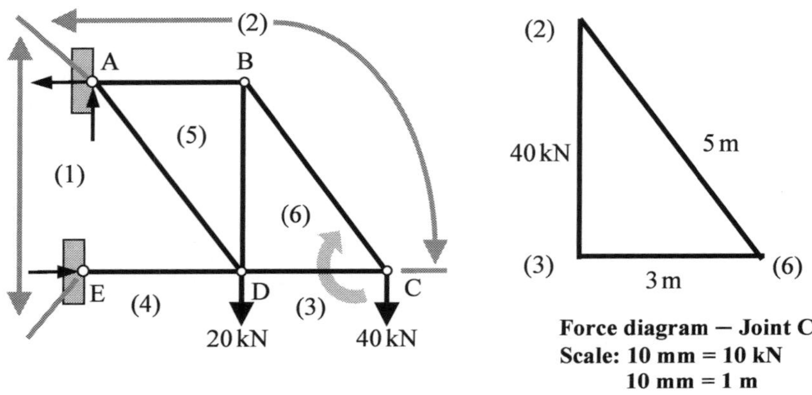

Figure 6.7: Bow's notation and force triangle at joint C – Example 6.1

We commence our analysis at joint C by drawing the force polygon (triangle in this case) at the joint. We **must always** start at a joint with only **two unknowns** in order to be able to complete the analysis. In some cases we will need to calculate the reactions. However, in this instance, joint C has one 'known' applied load and two 'unknown' member forces. We commence by drawing a vertical line, downward, from (2) to (3) to represent the 40 kN load. The numbering is determined directly from Bow's notation. We then move clockwise around the joint. The next force will be that induced in member DC. We note that the member is 3 m long, horizontal and goes from space (3) to (6). We therefore draw a line, at a suitable scale, to represent this member on our diagram. We also note that the line in Figure 6.7 runs from joint (3) on the left to joint (6) on the right. This is necessary, because we would be unable to draw the line for member BC if DC was in any other direction. The direction of the line (3)–(6) is also important as it indicates the direction of force $\overrightarrow{DC}$. We will adopt the convention for indicating compression and tension members indicated in Figure 6.5 throughout this book. Therefore the member DC is in compression as the member force pushes toward joint C. We do not mark the direction of forces on our force diagram because, whilst the force in DC is left-to-right at joint C, it will be right-to-left at joint D. Thus, marking arrows on the force diagram can lead to confusion!

Having constructed the force diagram for joint C, we now look for the next joint with two or less 'unknowns', in this case joint B.

In order to analyse joint B, we first construct the force diagram, noting that we have already plotted member BC at joint C. We commence the diagram at this member and move clockwise around the joint as before. The force diagram is plotted in Figure 6.8a. These force vectors are added to those plotted at joint C to create the composite diagram for the two joints shown in Figure 6.8b.

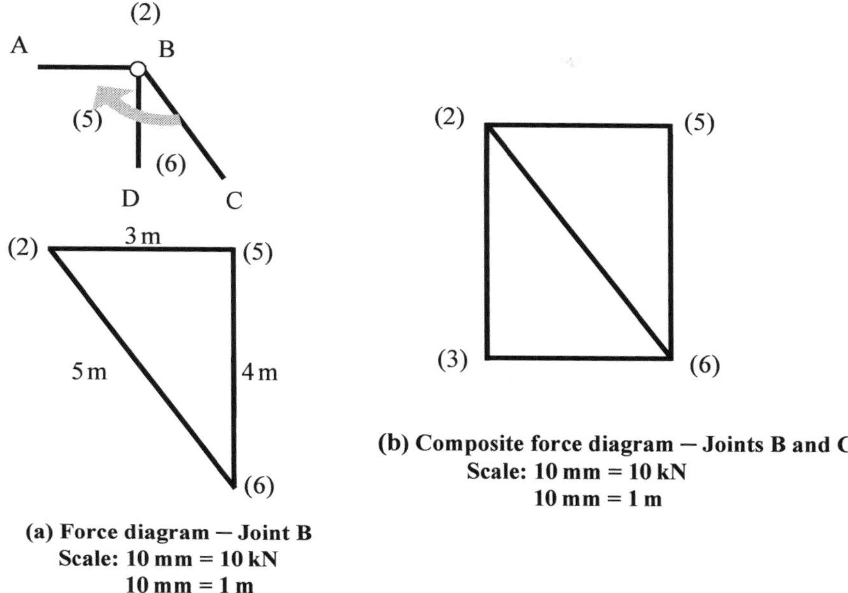

(a) Force diagram — Joint B
Scale: 10 mm = 10 kN
10 mm = 1 m

(b) Composite force diagram — Joints B and C
Scale: 10 mm = 10 kN
10 mm = 1 m

Figure 6.8: Force diagram for joint B and composite diagram for joints B and C

Note that the member BC {(2)(6)} appears at both joints and therefore connects the two force diagrams. In practice we would construct the composite force diagram of Figure 6.8b directly, and would not normally construct the force diagrams for each individual joint. We can now analyse joint D, since only two 'unknowns' remain at this joint, and construct the composite force diagram for joints B, C and D. The partial composite force diagram is shown in Figure 6.9b.

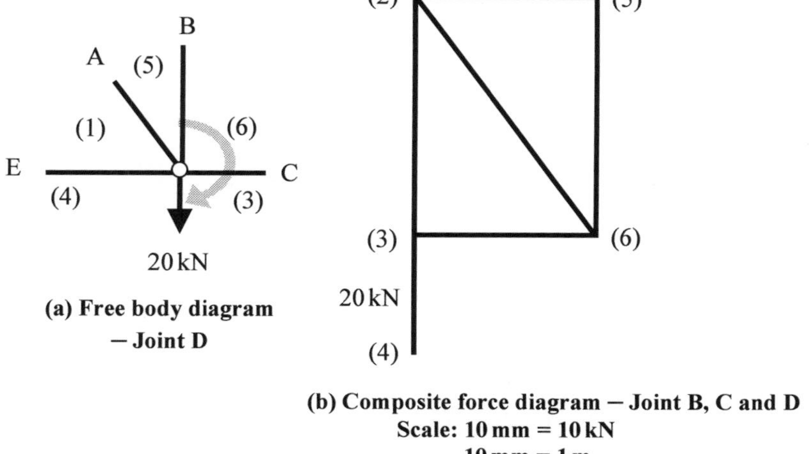

(a) Free body diagram — Joint D

(b) Composite force diagram — Joint B, C and D
Scale: 10 mm = 10 kN
10 mm = 1 m

Figure 6.9: Partial composite force diagram – Example 6.1

Commencing with member BD, we note that the force for this member has already been plotted as (5)(6). Moving clockwise, we find member CD has already been plotted as (6)(3). Next we add the applied load at joint D of 20 kN to produce line (3)(4) on the composite diagram. To complete the force diagram, we must include members DE and AD, however we do not know the magnitude of the forces that they must resist. We do however know their orientation and that they should intersect at point (1) since they are designated (4)(1) and (5)(1) respectively. In order to determine their magnitude we must project a horizontal line from point (4) to represent DE, and project a line from (5) parallel to the line of the member AD until they intersect as shown in Figure 6.10.

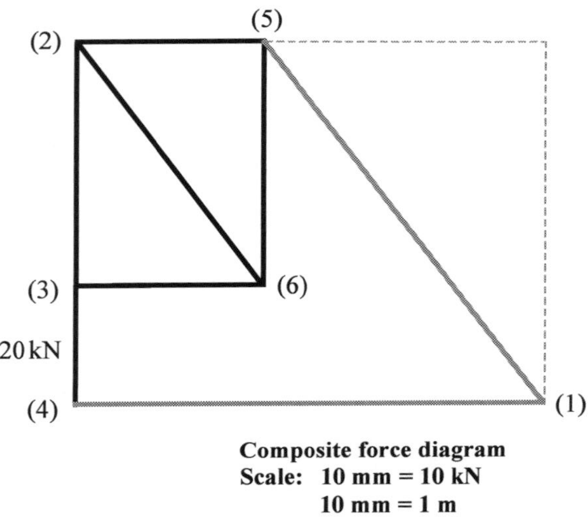

Figure 6.10: Composite force diagram – Example 6.1

We can now measure from this diagram the forces in the members and the reactions, noting that the horizontal reaction at A will be equal to line (1)(4), the vertical reaction will be equal to the projected vertical line (1) to (2), and the horizontal reaction at E will be the line (4)(1). The final forces measured from the composite force diagram are shown in the table of Figure 6.11, the positive and negative signs indicating compression and tension respectively.

Member	Force
AB	−30 kN
BC	−50 kN
BD	+40 kN
CD	+30 kN
DA	−75 kN
DE	+75 kN
Vertical reaction at A	+60kN
Horizontal reaction at A	−75 kN
Horizontal reaction at E	+75 kN

Figure 6.11: Member forces and reactions – Example 6.1

A simple check on the validity of these results is to add all external forces, that is all applied forces and reactions. If the structure is in equilibrium, the sum of all horizontal forces will equal zero as will the sum of all vertical forces. Hence:

Sum of all horizontal forces $\Sigma F_x = 75\,\text{kN} - 75\,\text{kN} = 0\,\text{kN}$
Sum of all vertical forces $\Sigma F_y = -20\,\text{kN} - 40\,\text{kN} + 60\,\text{kN} = 0\,\text{kN}$

$\therefore$ **Structure is in equilibrium**

Though presenting the results in the form of a table is quite acceptable, we would normally complete the problem by marking the magnitude and direction of the member forces and reactions on a force diagram for the structure, as shown in Figure 6.12.

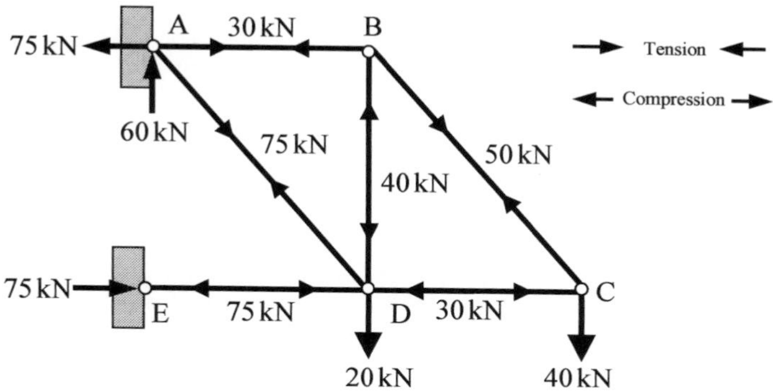

Figure 6.12: Final force diagram – Example 6.1

As the structures we are required to analyse become more complicated, the use of graphical solutions can become quite difficult. As can be seen, it is essential when using these methods to following the correct procedure and conventions. This method is no longer in common use, having been superseded by mathematical and computer solutions, however it can still prove to be useful in the analysis of some simple structures.

6.2.2 Method of Sections

This method of analysis is particularly useful if we only require to calculate the forces in specific members along a frame, for instance, if we can identify the critical members in a complex truss we can use this method to solve for the forces in those members alone. We can also use the method of sections to calculate all member forces, however the solutions can become very complex.

The method of sections requires us to take a 'cut' through the frame and through the members to be analysed. We then use the three equations of statics to solve for the 'unknown' forces. The method is therefore limited in that no more than three members can be 'cut' in any one section, however by varying our cuts we can include all the members required.

Example 6.2: Analysis of a pin jointed frame (2)

Consider the frame of Example 6.2, shown in Figure 6.13, in which we are required to determine the forces in members AB, AD and ED.

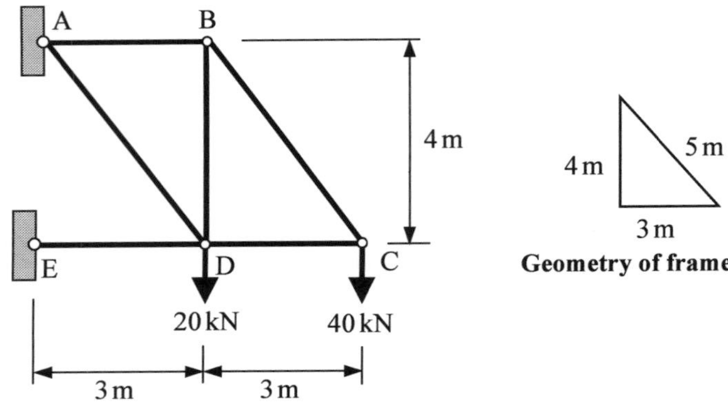

Figure 6.13: Pin jointed frame – Example 6.2

We can see that a single cut through the frame will allow us to include all the members required in the analysis, as shown in Figure 6.14.

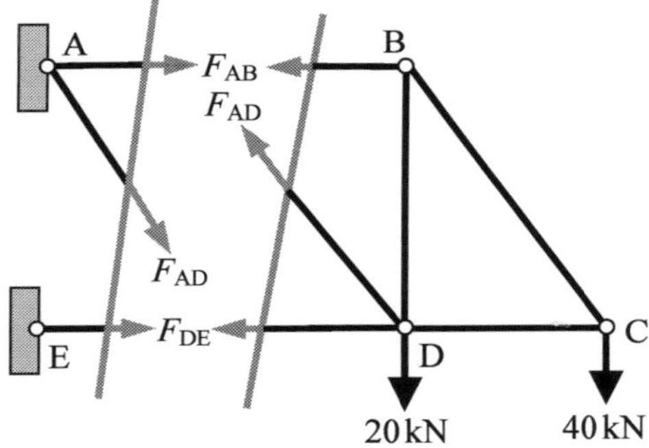

Figure 6.14: Sections of pin jointed frame – Example 6.2

We now consider force and moment equilibrium for the right side of the frame, noting that the force in member AD can be resolved into its horizontal and vertical components and these can be used in the member force equations. We can define our equations for vertical and horizontal forces in exactly the same way as we did when analysing three pin frames. In order to calculate the bending moments, we must first determine where to locate the datum. The datum should be positioned at a point that will reduce the number of 'unknowns' for which we need to solve. This can be achieved by locating it at the intersection point of the line of action of two of the forces in the cut members, as shown for members AB and AD in Figure 6.15. Note that we identify forces in members using F for force followed by the subscript representing the member. We will follow this convention throughout the solution of the following examples.

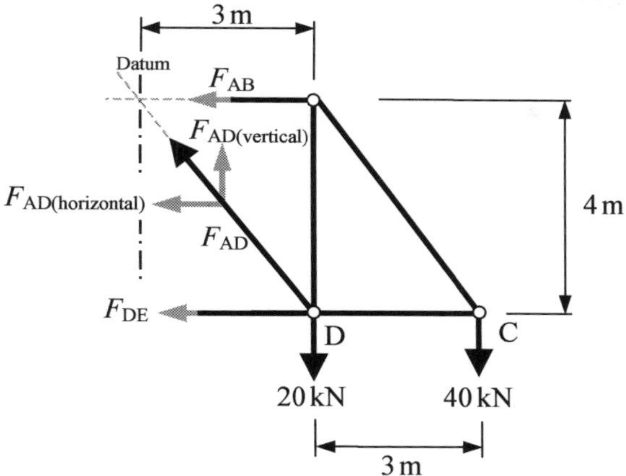

Figure 6.15: Location of datum for calculation of moments – Example 6.2

Applying the three equations of statics, we have:

Summation of horizontal forces $\Sigma F_x = F_{AB} + F_{AD(horizontal)} + F_{DE} = 0\,kN$ [1]

Summation of vertical forces $\Sigma F_y = -20\,kN - 40\,kN + F_{AD(vertical)} = 0\,kN$ [2]

$\therefore F_{AD(vertical)} = 60\,kN$

Summation of moments about datum $F_{AB} \times 0\,m + F_{AD} \times 0\,m + F_{DE} \times 4\,m$

$+ 20\,kN \times 3\,m + 40\,kN \times 6\,m = 0\,kN$ [3]

$\therefore 4F_{DE} + 60\,kNm + 240\,kNm = 0\,kN$

$\therefore F_{DE} = -75\,kN$

We can now calculate the force F_{AD} since we know the vertical component of the force = 60 kN hence:

Ratio of force and geometry triangle = 60 kN:4 m = 15 kN:1 m

Horizontal component of force in member AD (length 3 m) = 15 kN × 3 = 45 kN

Force in member AD (length 5 m) = F_{AD} = 15 kN × 5 m = 75 kN

We can now back-substitute the results for $F_{AD(horizontal)}$ and F_{DE} into equation [1] to determine force F_{AB} such that:

$F_{AB} + 45\,kN - 75\,kN = 0\,kN$

$F_{AB} = 30\,kN$

These results are the same as those calculated in our previous analysis of the same structure.

◇

This method of analysis can easily be applied to any pin jointed frame to quickly determine the forces in specific members. However, in order to calculate the forces in all members of a structure, a large number of 'cuts' (and therefore equations) would be required.

6.3 Method of Resolution of Forces at Joints

This method is based on isolating each joint of the frame, in turn, and ensuring all applied and member forces at the joint are in equilibrium. Having analysed all joints of the structure and ensured that they are each in equilibrium, we can then resolve the forces at the supports to ensure that they balance the forces in the members. If all joints and reactions are in equilibrium, then the whole structure must be in equilibrium.

Again, we must commence our analysis at a joint with only two 'unknowns'. We then move from joint to joint using the results from the previous calculation to reduce the number of 'unknown' member forces. We will adopt a number of simple procedures to assist in using this method of analysis.

First, we will always draw a 'free body diagram' for each joint. This is a simple line drawing, as shown in Figure 6.16, on which we mark all applied forces, member forces and components. We also indicate the geometry of the structure. This diagram will help us to determine forces applied at the joint.

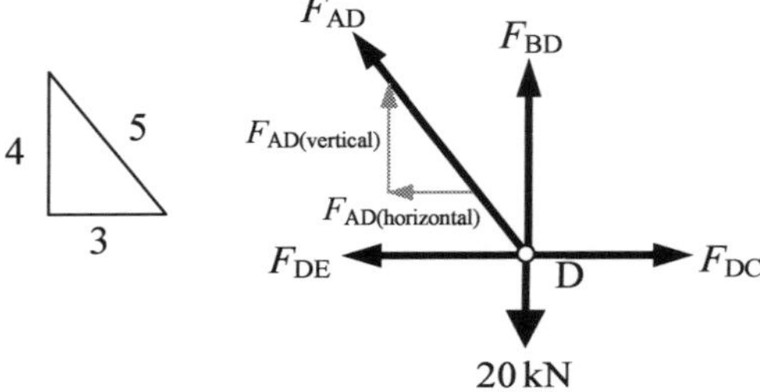

Free body diagram for joint D

Figure 6.16: Free body diagram

Next, we will always commence our analysis by considering the members connected at a joint to be in tension. We will sum all forces using the Cartesian co-ordinate system developed earlier, and we will find that some of the results will yield a negative answer for member forces. This negative sign will indicate that the member is in compression, not tension; however we will not adjust the resultant force in the member until we plot the final force diagram for the structure. We will however ensure that the magnitude and sign associated with a member force are maintained throughout the analysis.

Finally, we will use the concept of Force : Geometry ratio in our calculations. This will help to avoid the errors induced by unnecessary calculation and the confusion that can often result when using sine, cosine and tangent. In fact, the ratios used are only a different way of presenting these common trigonometrical functions.

The method is therefore a logical, almost mechanical process which can follow a very precise set of procedures as follows:

(1) Select a joint with two 'unknown' forces.

(2) Draw a 'free body diagram' for the joint.

(3) Mark on the 'free body diagram' all applied forces, member forces and components (note: members **always** considered to be in **tension**).

(4) Commencing at any member, proceed clockwise around the joint summing all vertical forces or components of forces.

(5) Commencing at any member, proceed clockwise around the joint summing all horizontal forces or components of forces.

(6) Solve the equations formed from steps (4) and (5).

(7) Proceed to the next joint with two 'unknown' forces and repeat steps (2) to (7) until all joints on the frame have been analysed.
(8) Resolve the member forces at the supports and calculate the reactions.

Example 6.3: Analysis of a pin jointed frame (3)

We will now proceed to apply this procedure to the analysis of the frame of Figure 6.17.

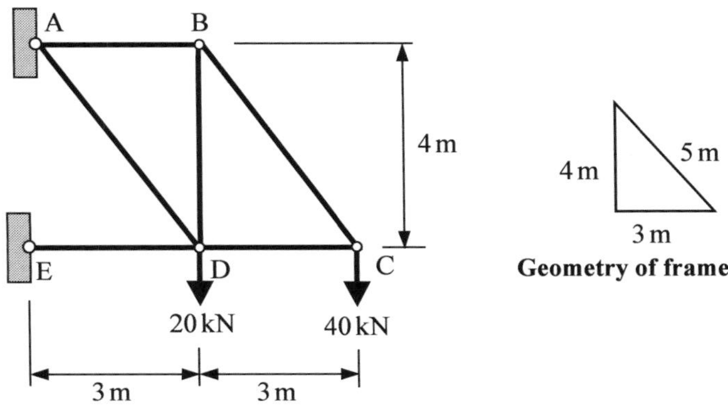

Figure 6.17: Pin jointed frame – Example 6.3

We commence our analysis at joint C with the free body diagram shown in Figure 6.18.

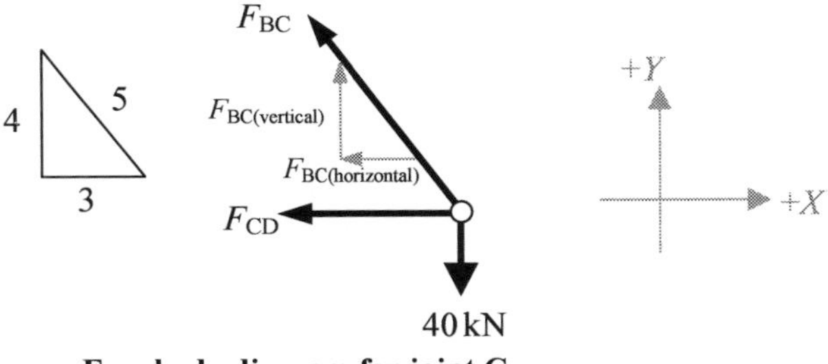

Figure 6.18: Free body diagram for joint C – Example 6.3

Commencing at the applied load and moving clockwise around the joint, we sum all vertical forces and components (note: positive and negative signs for forces are derived from the Cartesian co-ordinate system indicated in Figure 6.18):

Sum of all vertical forces $\quad -40\,\text{kN} + F_{BC(vertical)} = 0\,\text{kN}$

$$\therefore\ F_{BC(vertical)} = 40\,\text{kN}$$

and (from ratio of Force : Geometry = 40 kN : 4 m);

$$F_{BC} = \frac{F_{BC(vertical)}}{4\ m} \times 5\ m = \frac{40\ kN}{4\ m} \times 5\ m = 50\ kN\ (tension)$$

Sum of all horizontal forces $\quad -F_{CD} - F_{BC(horizontal)} = 0\,kN$

Now, since $F_{BC} = 50\,kN$, then:

$$F_{BC(horizontal)} = \frac{F_{BC}}{5\ m} \times 3\,m = \frac{50\ kN}{5\ m} \times 3\ m = 30\ kN$$

Therefore from equation for horizontal forces, $F_{CD} = -30kN$ (compression).

We can now analyse joint B which, having solved for F_{BC}, has two remaining unknown member forces. The free body diagram for joint B is shown in Figure 6.19.

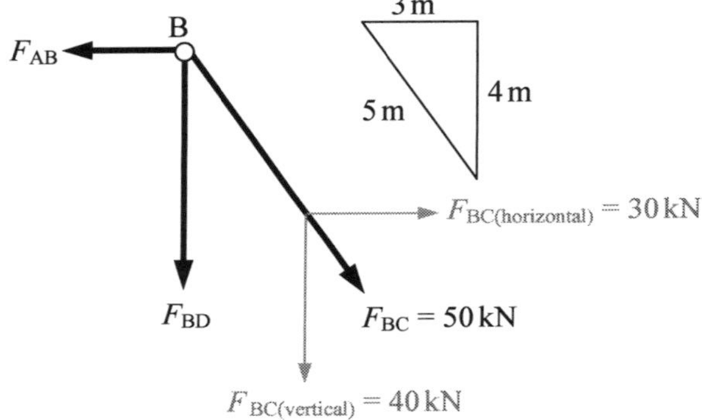

Figure 6.19: Free body diagram for joint B – Example 6.3

Commencing at member BC and moving clockwise around the joint, we sum all vertical forces, and using the components of the force F_{BC} calculated previously we find:

Sum of all vertical forces

$$-F_{BC(vertical)} - F_{BD} = 0\,kN$$
$$\therefore -40\,kN - F_{BD} = 0\,kN$$
$$\therefore F_{BD} = -40\,kN\ (compression)$$

Sum of all horizontal forces

$$-F_{AB} + F_{BC(horizontal)} = 0\,kN$$
$$\therefore -F_{AB} + 30\,kN = 0\,kN$$
$$\therefore F_{AB} = 30\,kN\ (tension)$$

Having solved for members BD and CD, we can now move to joint D where we have two remaining 'unknowns'. The free body diagram for joint D is shown in Figure 6.20.

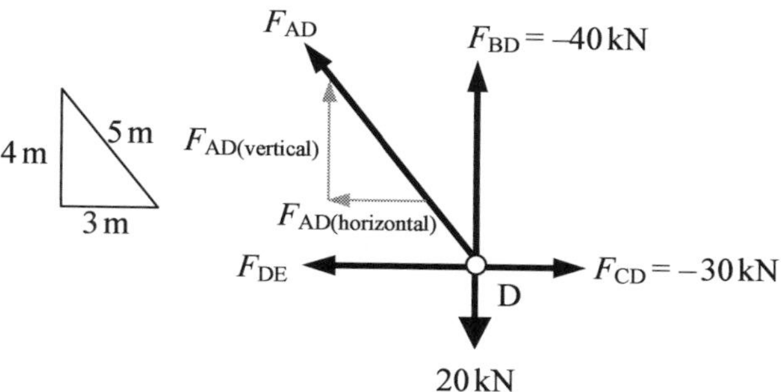

Figure 6.20: Free body diagram for joint D – Example 6.3

Commencing at member BD and moving clockwise around the joint we sum all vertical forces:

Sum of all vertical forces

$$F_{BD} - 20\,kN + F_{AD(vertical)} = 0\,kN$$
$$\therefore\ -40\,kN - 20\,kN + F_{AD(vertical)} = 0\,kN$$
$$\therefore\ F_{AD(vertical)} = 60\,kN$$

Note: force F_{BD} is still considered to be in tension on the free body diagram, however its magnitude $= -40\,kN$.

Also (from ratio of Force : Geometry) = 60 kN : 4 m:

$$F_{AD} = \frac{F_{AD(vertical)}}{4\text{ m}} \times 5\text{ m} = \frac{60\text{ kN}}{4\text{ m}} \times 5\text{ m} = 75\text{ kN (tension)}$$

Sum of all horizontal forces

$$F_{CD} - F_{DE} - F_{AD(horizontal)} = 0\,kN$$

$$F_{AD(horizontal)} = \frac{F_{AD}}{5\text{ m}} \times 3\text{ m} = \frac{75\text{ kN}}{5\text{ m}} \times 3\text{ m} = 45\text{ kN}$$

$\therefore$ Sum of all horizontal forces

$$-30\,kN - F_{DE} - 45\,kN = 0\,kN$$
$$\therefore\ F_{DE} = -75\,kN\text{ (compression)}$$

Having solved for all members on the frame, we must now solve for the reactions. We again draw a free body diagram for each support and indicate the assumed direction of the reaction. Again, any error in our assumption will yield a negative result for the magnitude.

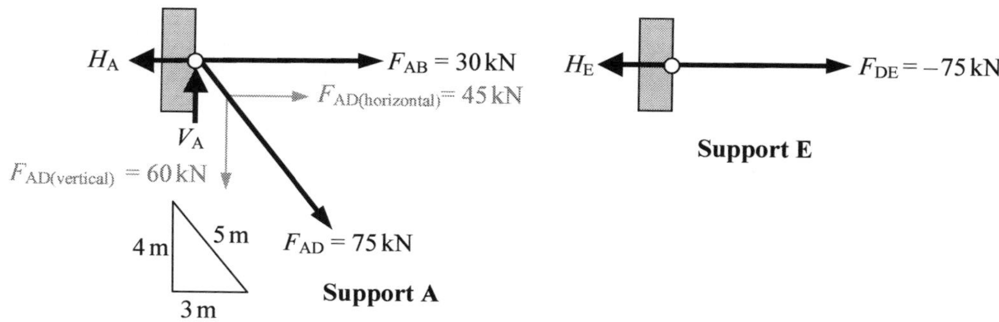

Figure 6.21: Free body diagrams for supports A and E

In order to ensure equilibrium at the supports, we must ensure the forces at the reactions balance the forces exerted by the members. We will define our equations for each support as before, however we note that there cannot be a vertical reaction at support E as no vertical force or component of a force is resolved at that joint.

For support A: Sum of all vertical forces

$$V_A - F_{AD\,(vertical)} = 0\,kN$$
$$\therefore\; V_A - 60\,kN = 0\,kN$$
$$\therefore\; V_A = 60\,kN \text{ (in direction indicated)}$$

Sum of all horizontal forces

$$-H_A + F_{AB} + F_{AD\,(horizontal)} = 0\,kN$$
$$\therefore\; -H_A - 45\,kN - 30\,kN = 0\,kN$$
$$\therefore\; H_A = 75\,kN \text{ (in direction indicated)}$$

For support E: Sum of vertical forces = 0 kN

Sum of all horizontal forces

$$-H_E + F_{DE} = 0\,kN$$
$$\therefore\; -H_E - 75\,kN = 0\,kN$$
$$\therefore\; H_E = -75\,kN \text{ (in opposite direction to that indicated)}$$

The results of the analysis of the pin jointed frame of Example 6.3 are shown Figure 6.22.

Member	Force
AB	−30 kN
BC	−50 kN
BD	+40 kN
CD	+30 kN
DA	−75 kN
DE	+75 kN
Vertical reaction at A	+60 kN
Horizontal reaction at A	−75 kN
Horizontal reaction at E	+75 kN

Figure 6.22: Member forces and reactions – Example 6.3

Checking all external forces, we find:

Sum of all horizontal forces $\Sigma F_x = 75\,kN - 75\,kN = 0\,kN$

Sum of all vertical forces $\Sigma F_y = -20\,kN - 40\,kN + 60\,kN = 0\,kN$

∴ Structure is in equilibrium

Again, we present the results of our analysis in the form of a force diagram, taking due account of the signs associated with each force in order to indicate tension and compression members.

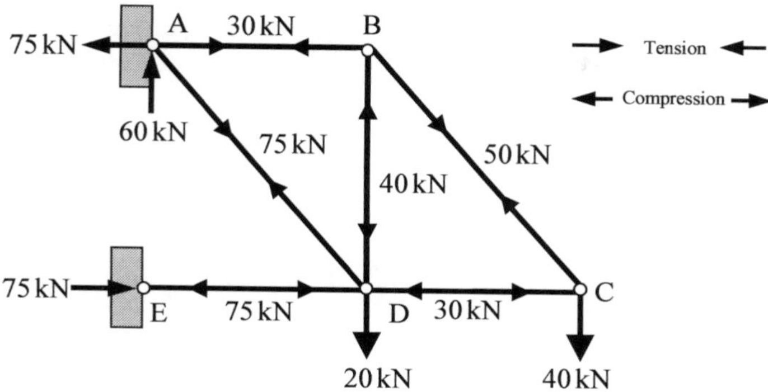

Figure 6.23: Final force diagram – Example 6.3

With reference to Chapter 1, we note that the compressive members in a pin jointed frame such as members BD, CD and DE are often referred to as struts whilst tensile members such as AB, BC and AD are called ties.

◇

A comparison of the results from the three methods of analysis used here shows that each will yield the same results, however the method of resolution of forces in joints offers a systematic and logical approach which can be applied to all forms of pin jointed frames and is easy to use. We will therefore complete a number of examples to demonstrate the ease and versatility of this method.

◇

Example 6.4: Analysis of a pin jointed truss

For the pin jointed truss of Figure 6.24, calculate the magnitude and direction of all member forces and the reactions at the supports A and C.

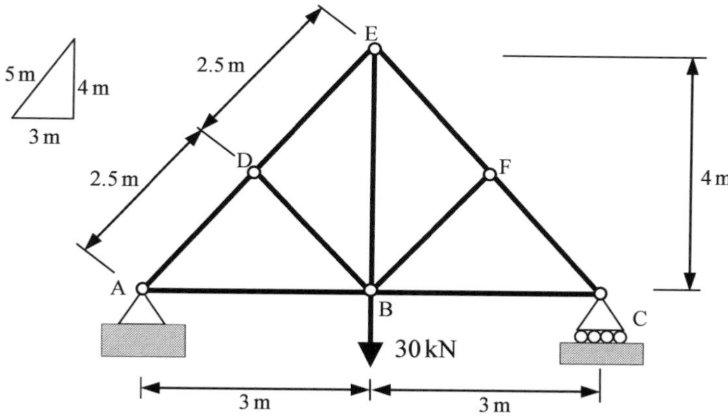

Figure 6.24: Pin jointed truss – Example 6.4

In this example, we will find that we cannot commence our analysis at any of the internal joints of the truss since D, E and F have three 'unknowns' and joint B has four. We note however that the frame is symmetrical and symmetrically loaded, such that at supports A and C we can calculate the reactions:

Sum of all horizontal forces $\Sigma F_x = H_A + H_C = 0\,\text{kN}$ however $H_C = 0\,\text{kN}$ (roller)
$\therefore\ H_A = 0\,\text{kN}$

Sum of all vertical forces $\Sigma F_y = V_A + V_C - 30\,\text{kN} = 0\,\text{kN}$
$\therefore\ V_A + V_C = 30\,\text{kN}$ [1]

Taking moments about A we have:

$$\Sigma M_z = (30\,\text{kN} \times 3\,\text{m}) - (V_C \times 6\,\text{m}) = 0\,\text{kN}$$
$$6\,V_C = 90\,\text{kN}$$
$$\therefore\ V_C = 15\,\text{kN}$$

Back-substituting for V_C into equation [1] gives $V_A = 15\,\text{kN}$

(We could probably have determined these forces by observation!)

We can now commence our analysis at either of the supports, noting that, because of the symmetrical geometry of the structure and its loading, we only need to calculate the member forces for one-half of the truss since both sides will carry equal loads.

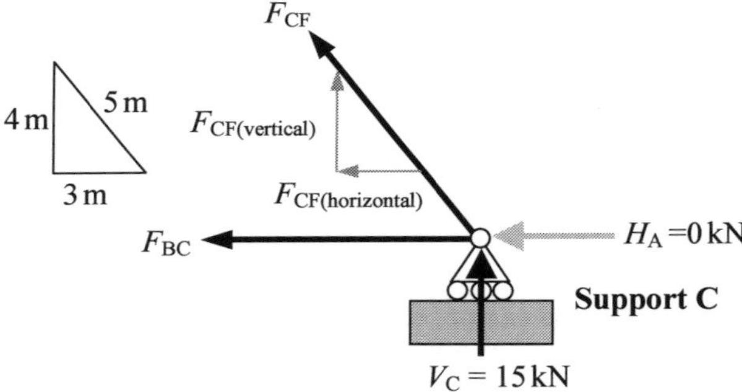

Figure 6.25: Free body diagram for joint C – Example 6.4

Sum of all vertical forces $V_C + F_{CF\,(\text{vertical})} = 0\,\text{kN}$
$\therefore\ 15\,\text{kN} + F_{CF(\text{vertical})} = 0\,\text{kN}$
$\therefore\ F_{CF(\text{vertical})} = -15\,\text{kN}$

and (from ratio of Force : Geometry =15 kN : 4 m):

$$F_{CF} = \frac{F_{CF(\text{vertical})}}{4\ \text{m}} \times 5\ \text{m} = \frac{-15\ \text{kN}}{4\ \text{m}} \times 5\ \text{m} = -18.75\ \text{kN (compression)}$$

Sum of all horizontal forces $-F_{BC} - F_{CF(\text{horizontal})} = 0\,\text{kN}$

and (from ratio of Force : Geometry = 18.75 kN : 5 m):

$$F_{CF(\text{horizontal})} = \frac{F_{CF}}{5\ \text{m}} \times 3\ \text{m} = \frac{-18.75\ \text{kN}}{5\ \text{m}} \times 3\ \text{m} = -11.25\ \text{kN}$$

$\therefore$ Sum of all horizontal forces $-F_{BC} - (-11.25\,\text{kN}) = 0\,\text{kN}$
$\therefore\ F_{BC} = 11.25\,\text{kN}$ (tension)

We now move to joint F which, having solved for member FC, has only two unknowns.

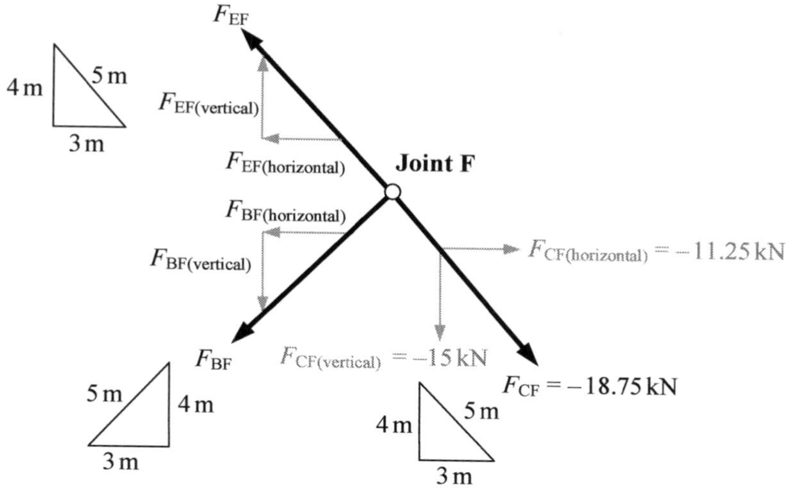

Figure 6.26: Free body diagram for joint F – Example 6.4

Commencing at member CF and moving clockwise around the joint we have:

Sum of all vertical forces
$$-F_{CF\,(vertical)} - F_{BF\,(vertical)} + F_{EF\,(vertical)} = 0\,\text{kN}$$
$$\therefore\ -(-15\text{kN}) - F_{BF\,(vertical)} + F_{EF\,(vertical)} = 0\,\text{kN}$$
$$\therefore\ F_{BF\,(vertical)} + F_{EF\,(vertical)} = -15\,\text{kN} \qquad [1]$$

We cannot solve this equation as it contains two unknowns, however we can, using the structures geometry, relate the components of the forces to the actual member forces such that:

$$F_{BF(vertical)} = \frac{F_{BF}}{5\text{ m}} \times 4\text{ m} = 0.8F_{BF} \qquad \text{and} \qquad F_{EF(vertical)} = \frac{F_{EF}}{5\text{ m}} \times 4\text{ m} = 0.8F_{EF}$$

and substituting these values into equation [1] we have:

$$-0.8F_{BF} + 0.8F_{EF} = -15\,\text{kN} \qquad [2]$$

In order to solve this equation, we need a further relationship for the member forces. This we can derive by considering the sum of the horizontal forces such that:

Sum of all horizontal forces
$$F_{CF\,(horizontal)} - F_{BF\,(horizontal)} - F_{EF\,(horizontal)} = 0\,\text{kN}$$
$$\therefore\ (-11.25\,\text{kN}) - F_{BF\,(horizontal)} - F_{EF\,(horizontal)} = 0\,\text{kN}$$
$$\therefore\ F_{BF\,(horizontal)} - F_{EF\,(horizontal)} = 11.25\,\text{kN} \qquad [3]$$

and from the structure geometry, we find that:

$$F_{BF(horizontal)} = \frac{F_{BF}}{5\text{ m}} \times 3\text{ m} = 0.6F_{BF} \qquad \text{and} \qquad F_{EF(horizontal)} = \frac{F_{EF}}{5\text{ m}} \times 3\text{ m} = 0.6F_{EF}$$

Substituting these values into equation [3] will give:

$$-0.6F_{BF} - 0.6F_{EF} = 11.25\,\text{kN} \qquad [4]$$

We can now solve equations [2] and [4] simultaneously to find that:

$$F_{BF} = 0\,\text{kN}$$
and
$$F_{EF} = -18.75\,\text{kN (compression)}$$

With experience, we would have been able to resolve this joint by observation since the angle $C\hat{F}B = 90^\circ$ and forces cannot resolve through a 90° angle. Also, since the force in member BF is zero, the force in member EF must be equal and opposite to that in member CF in order to ensure equilibrium at the joint. Consider the joint of Figure 6.27.

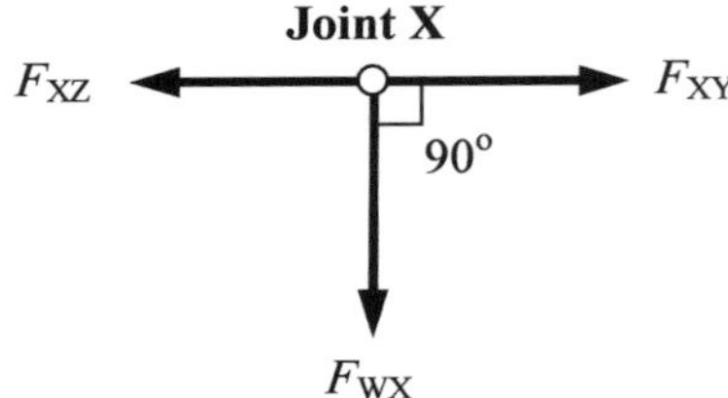

Figure 6.27: Pin joint X

The member WX has zero length in the horizontal plane and the members XY and XZ have zero length in the vertical plane. Therefore:

$$F_{\text{WX(horizontal)}} = \frac{F_{\text{WX}}}{\text{Length WX}} \times 0 \text{ m} = 0 \text{ kN}$$

Similarly, both $F_{\text{XY(vertical)}}$ and $F_{\text{XZ(vertical)}}$ will be zero.

We now proceed with Example 6.4, and analyse joint E.

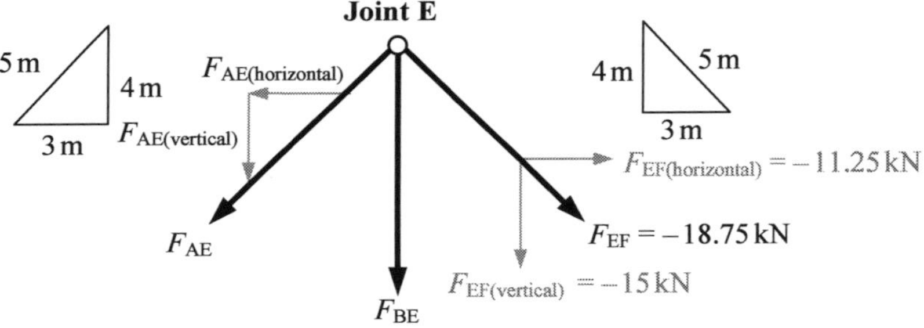

Figure 6.28: Free body diagram for joint E – Example 6.4

Commencing at member EF and moving clockwise around the joint we have:

Sum of all vertical forces

$$-F_{\text{EF (vertical)}} - F_{\text{BE}} - F_{\text{AE (vertical)}} = 0\text{ kN}$$
$$\therefore -(-15\text{ kN}) - F_{\text{BE}} - F_{\text{AE (vertical)}} = 0\text{ kN}$$
$$\therefore F_{\text{BE}} + F_{\text{AE (vertical)}} = 15\text{ kN} \quad [1]$$

We cannot solve this equation as it contains two unknowns but, again, we can relate the components of the force to the actual forces in the members such that:

$$F_{\text{AE(vertical)}} = \frac{F_{\text{AE}}}{5 \text{ m}} \times 4 \text{ m} = 0.8F_{\text{AE}}$$

and substituting these values into equation [1] we have:

$$F_{BE} + 0.8F_{AE} = 15\,\text{kN} \qquad [2]$$

Sum of all horizontal forces

$$F_{EF\,(horizontal)} - F_{AE\,(horizontal)} = 0\,\text{kN}$$
$$\therefore (-11.25\,\text{kN}) \;\; - F_{AE\,(horizontal)} = 0\,\text{kN}$$
$$\therefore F_{AE\,(horizontal)} = -11.25\,\text{kN} \qquad [3]$$

and from consideration of the member geometry, we find that:

$$F_{AE} = \frac{F_{AE(horizontal)}}{3\text{ m}} \times 5\text{ m} = \frac{-11.25\text{ kN}}{3\text{ m}} \times 5\text{ m} = -18.75\text{ kN (compression)}$$

In a similar manner we can find $F_{AE(vertical)} = -15$ kN and substituting this value into equation [3] will give:

$$F_{BE} - 15\,\text{kN} = 15\,\text{kN}$$
$$\therefore F_{BE} = 30\,\text{kN (tension)}$$

Checking all external forces, we find:

Sum of all horizontal forces $\quad \Sigma F_x = 0\,\text{kN} - 0\,\text{kN} = 0\,\text{kN}$

Sum of all vertical forces $\quad \Sigma F_y = -30\,\text{kN} + 15\,\text{kN} + 15\,\text{kN} = 0\,\text{kN}$

We have now solved for all member forces and can draw our final force diagram as shown in Figure 6.29.

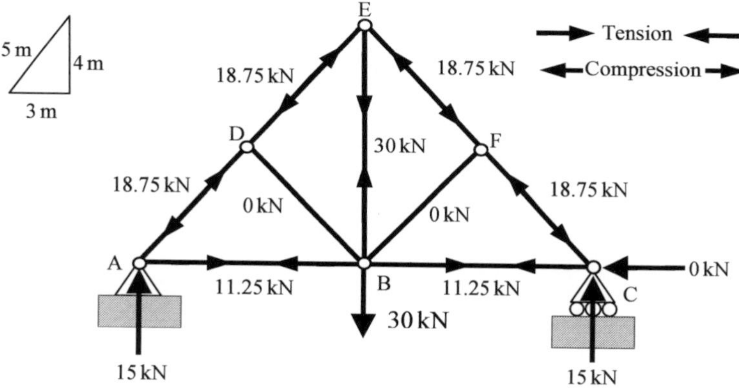

Figure 6.29: Final frame forces and reactions – Example 6.4

From this example we see that it is sometimes necessary to solve the equations for vertical and horizontal forces simultaneously. We also note that, in order to solve these equations, we must substitute the relationship between the force component, vertical or horizontal, and the member force. In the following examples we will mark the components of the member force on the free body diagram, having already taken into account the geometry of the structure.

◇

Example 6.5: Analysis of a pin jointed frame supporting an inclined load

For the pin jointed frame of Figure 6.30, calculate the magnitude and direction of all member forces and the reactions at the supports A and G.

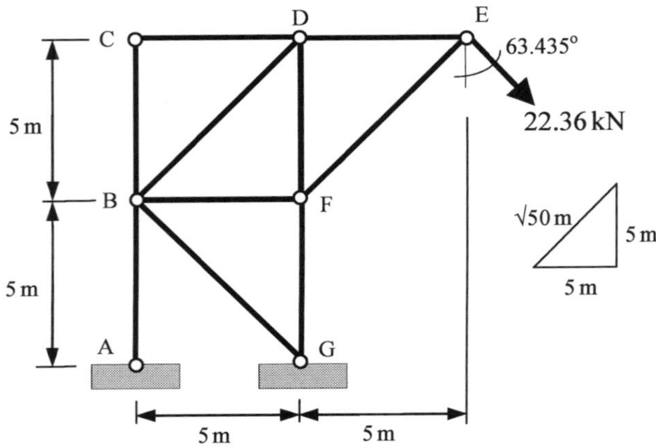

Figure 6.30: Pin jointed frame – Example 6.5

First we note that the force applied at joint E is at an angle of 63.435° to the vertical. For ease of analysis we will resolve this into its horizontal and vertical components such that:

At joint E:

Vertical applied force $= 22.36\,\text{kN} \times \cos 63.435° = 10\,\text{kN}$
Horizontal applied force $= 22.36\,\text{kN} \times \sin 63.435° = 20\,\text{kN}$

We also note that, at joint C, the two members meet at an angle of 90° to each other and there are no other applied forces at that point. In order for this joint to be in equilibrium, we must conclude that the forces in members BC and CD are zero. Our structure can therefore be simplified to that shown in Figure 6.31.

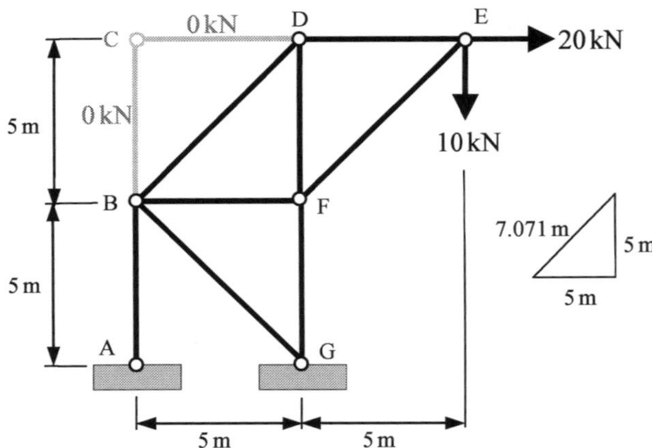

Figure 6.31: Simplified pin jointed frame – Example 6.5

In this case we will commence our analysis at joint E, however we note that whilst both support A and E can provide resistance to horizontal loads, support A can only be resolved if the horizontal force is zero since only a vertical force can be induced by member AB. This will provide us with a preliminary check on our analysis since, for horizontal equilibrium of the complete structure, the horizontal force H_E must equal the total applied horizontal load of 20 kN.

At joint E we have:

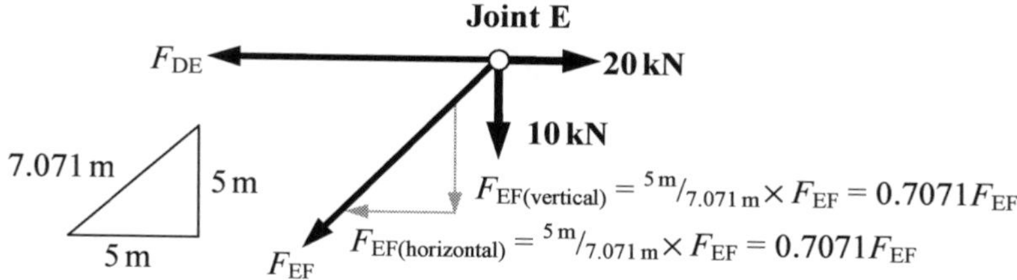

Figure 6.32: Free body diagram for joint E – Example 6.5

Sum of all vertical forces

$$-10\,kN - F_{EF\,(vertical)} = 0\,kN$$
$$\therefore -10\,kN - 0.7071 F_{EF} = 0\,kN$$
$$\therefore F_{EF} = -10\,kN/0.7071 = -14.143\,kN \text{ (compression)}$$

Sum of all horizontal forces

$$-F_{EF\,(horizontal)} - F_{DE} + 20\,kN = 0\,kN$$
$$\therefore -0.7071(-14.143\,kN) - F_{DE} = -20\,kN$$
$$\therefore F_{DE} = 20\,kN + 10\,kN = 30\,kN \text{ (tension)}$$

We now move to joint D noting that we have already determined the force in member CD as 0 kN.

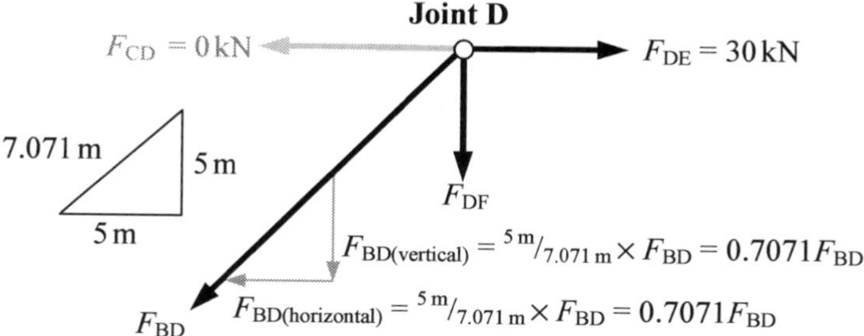

Figure 6.33: Free body diagram for joint D – Example 6.5

Sum of all vertical forces

$$-F_{DF} - F_{BD\,(vertical)} = 0\,kN$$
$$\therefore -F_{DF} - 0.7071 F_{BD} = 0\,kN \qquad [1]$$

Sum of all horizontal forces

$$30\,kN - F_{BD\,(horizontal)} - F_{CD} = 0\,kN$$
$$\therefore 30\,kN - 0.7071 F_{BD} - 0\,kN = 0\,kN$$
$$\therefore F_{BD} = 30\,kN/0.7071 = 42.467\,kN \text{ (tension)}$$

Back-substituting the result for F_{BD} into equation [1] gives: $-F_{DF} - 0.7071(42.467) = 0\,kN$

$$\therefore F_{DF} = -30\,kN \text{ (compression)}$$

We will now solve for the forces at joint F.

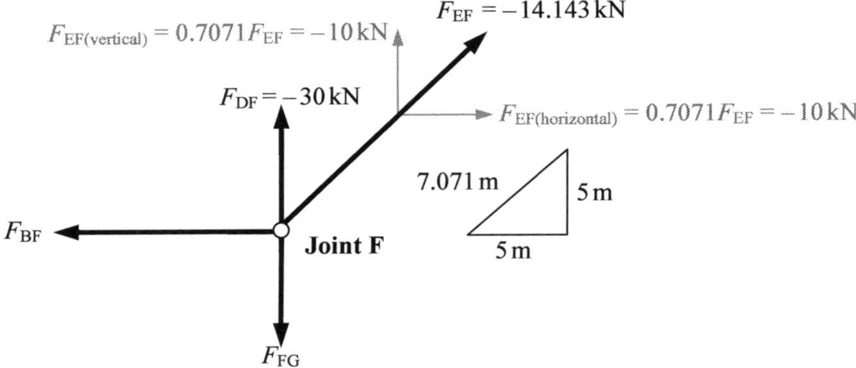

Figure 6.34: Free body diagram for joint F – Example 6.5

Sum of all vertical forces

$$F_{EF\,(vertical)} - F_{FG} + F_{DF} = 0\,\text{kN}$$
$$\therefore -10\,\text{kN} - F_{FG} - 30\,\text{kN} = 0\,\text{kN}$$
$$\therefore F_{FG} = -40\,\text{kN (compression)}$$

Sum of all horizontal forces

$$F_{EF\,(horizontal)} - F_{BF} = 0\,\text{kN}$$
$$\therefore -10\,\text{kN} - F_{BF} = 0\,\text{kN}$$
$$\therefore F_{BF} = -10\,\text{kN (compression)}$$

Next we resolve joint B.

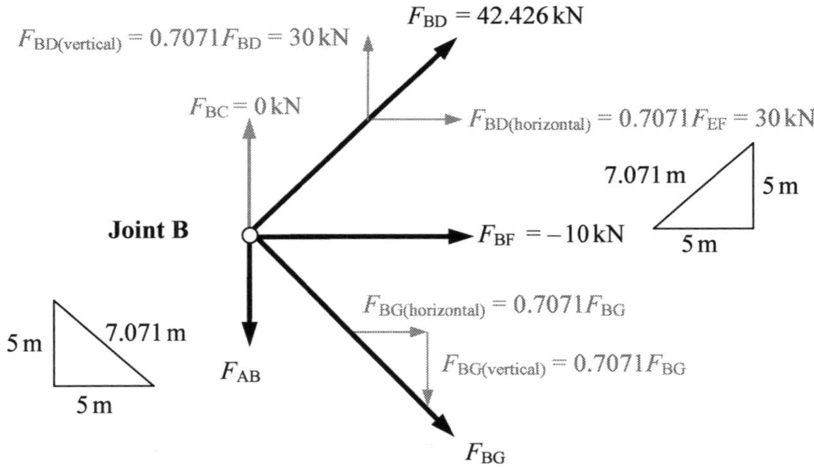

Figure 6.35: Free body diagram for joint B – Example 6.5

Sum of all vertical forces

$$F_{BD\,(vertical)} - F_{BG(vertical)} - F_{AB} + F_{BC} = 0\,\text{kN}$$
$$\therefore 30\,\text{kN} - 0.7071F_{BG} - F_{AB} - 0\,\text{kN} = 0\,\text{kN}$$
$$\therefore 0.7071F_{BG} + F_{AB} = 30\,\text{kN} \qquad [1]$$

Sum of all horizontal forces $F_{BD\,(horizontal)} + F_{BF} + F_{BG\,(horizontal)} = 0\,kN$

$\therefore\ 30\,kN - 10\,kN + 0.7071F_{BG} = 0\,kN$

$\therefore\ 0.7071F_{BG} = -20\,kN$ [2]

Solving equations [1] and [2] simultaneously we find: $F_{AB} = 50\,kN$ (tension)

and: $F_{BG} = -28.286\,kN$ (compression)

All that remains is to solve for the forces at the supports A and G.

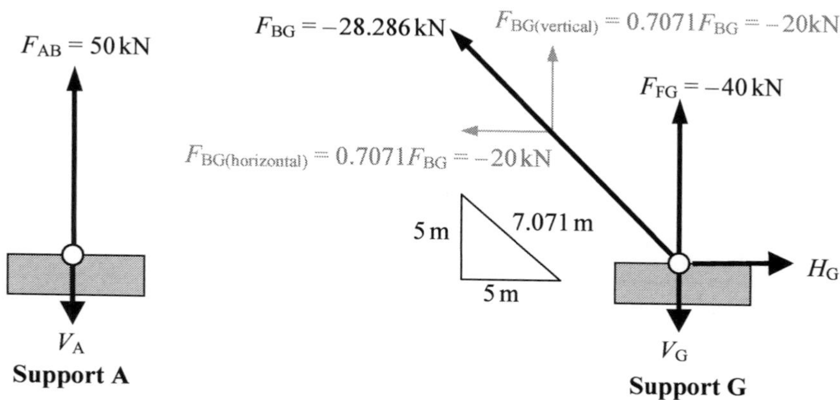

Figure 6.36: Free body diagram at supports A and G – Example 6.5

For support A: Sum of all vertical forces $-V_A + F_{AB} = 0\,kN$

$\therefore\ -V_A + 50\,kN = 0\,kN$

$\therefore\ V_A = 50\,kN$ (in direction indicated)

Sum of all horizontal forces $= 0\,kN$

For support G: Sum of vertical forces $F_{FG} - V_G + F_{BG\,(vertical)} = 0\,kN$

$\therefore\ -40\,kN - V_G + (-20\,kN) = 0\,kN$

$\therefore\ V_G = -60\,kN$ (in the opposite direction to that indicated)

Sum of all horizontal forces $H_G - F_{BG(horizontal)} = 0\,kN$

$\therefore\ H_G - (-20\,kN) = 0\,kN$

$\therefore\ H_G = -20\,kN$ (in the opposite direction to that indicated)

We will now check our results, calculating the sum of all external forces such that:

Sum of all horizontal forces $\Sigma F_x = 20\,kN - 20\,kN = 0\,kN$

Sum of all vertical forces $\Sigma F_y = -10\,kN - 50\,kN + 60\,kN = 0\,kN$

We can now draw a diagram of the final frame forces.

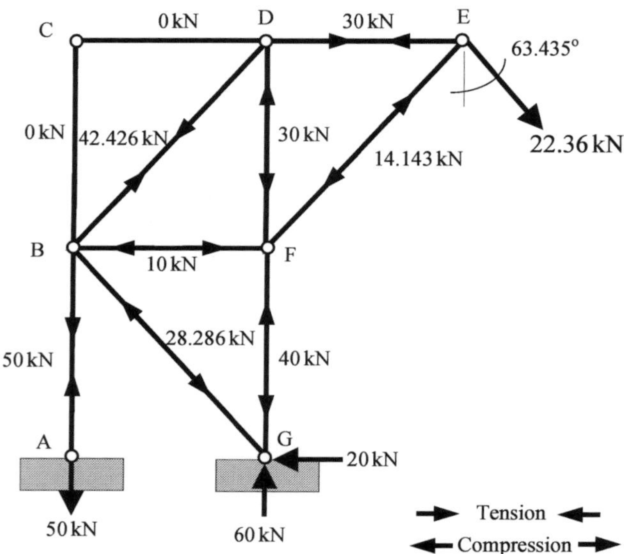

Figure 6.37: Final frame forces – Example 6.5

We will now proceed to use the method demonstrated here to analyse a number of pin jointed frame structures, solving for all member forces and forces at the reactions.

However, before proceeding further, we should note that in some cases it will be necessary to solve for the reactions in order to analyse a structure. We will also make use of symmetry, where applicable, and the resolution of joints that involve two perpendicular member forces to simplify the analysis. Care must be taken when analysing these 'right angle' joints, as illustrated in Figure 6.38. The loading arrangement of the left frame will not induce a load in member CG, whilst that on the right will require the force F_{CG} to be P kN (tension) in order for the joint to remain in equilibrium.

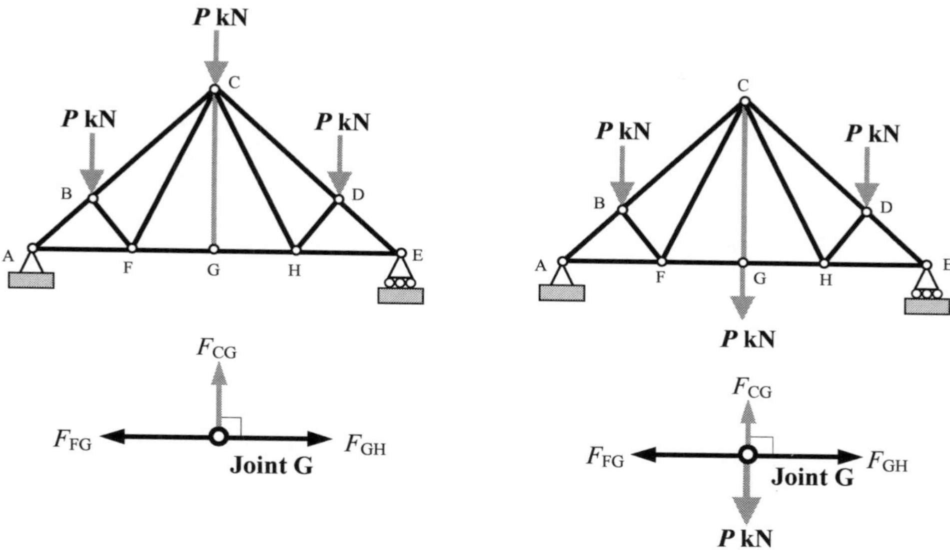

Figure 6.38: Analysis of 'right angle' (90°) joints

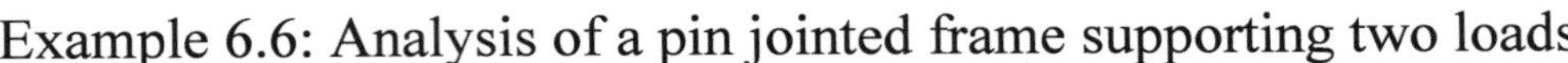

Example 6.6: Analysis of a pin jointed frame supporting two loads

For the pin jointed frame of Example 6.6, determine the magnitude and type of forces in all members and the forces at the supports A and E.

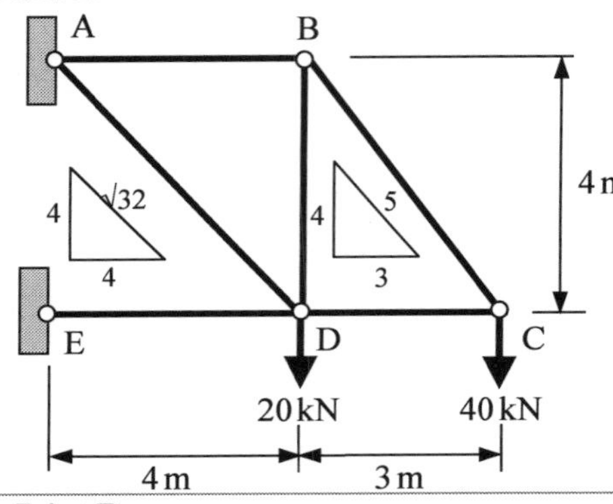

1. Joint C

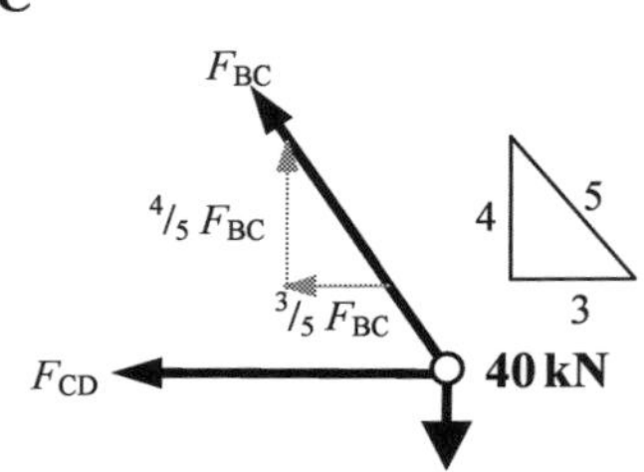

Sum vertical forces:

$-40 + F_{BC(vert)} = 0$

$-40 + 0.8F_{BC} = 0$

$F_{BC} = 50\,\text{kN(tens.)}$

Sum horizontal forces:

$-F_{CD} - F_{BC(horiz)} = 0$

$-F_{CD} - 0.6F_{BC} = 0$

$-F_{CD} - 0.6(50) = 0$

$F_{CD} = -30\,\text{kN(comp.)}$

2. Joint B

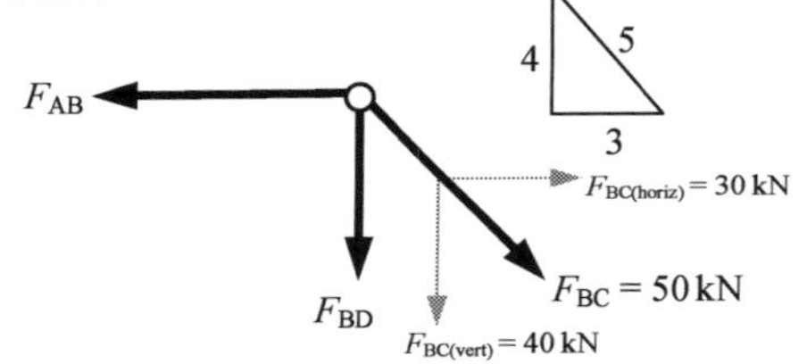

Sum vertical forces:

$-F_{BD} - F_{BC(vert)} = 0$

$-F_{BD} - 40 = 0$

$F_{BD} = -40\,\text{kN(comp.)}$

Sum horizontal forces:

$-F_{AB} + F_{BC(horiz)} = 0$

$-F_{AB} + 30 = 0$

$F_{AB} = 30\,\text{kN(tens.)}$

3. Joint D

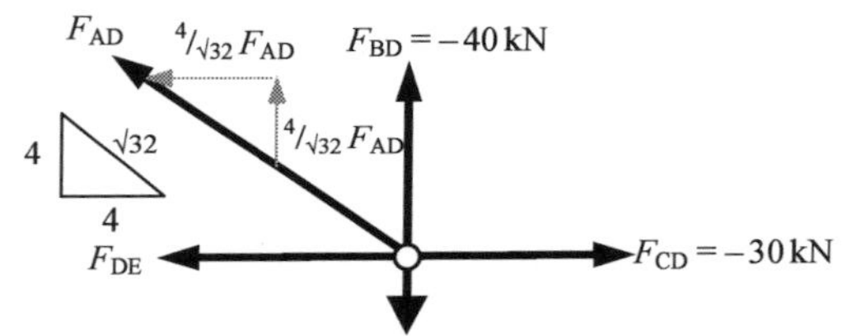

Sum vertical forces:

$F_{BD} - 20 + F_{AD(vert)} = 0$

$-40 - 20 + F_{AD(vert)} = 0$

$-60 + 0.7071F_{AD} = 0$

$F_{AD} = 84.853\,\text{kN(tens.)}$

Sum horizontal forces:

$F_{CD} - F_{DE} - F_{AD(horiz)} = 0$

$-30 - F_{DE} - 0.7071F_{AD} = 0$

$F_{DE} = -30 - 0.7071F_{AD}$

$F_{DE} = -90\,\text{kN(comp.)}$

4. Support A

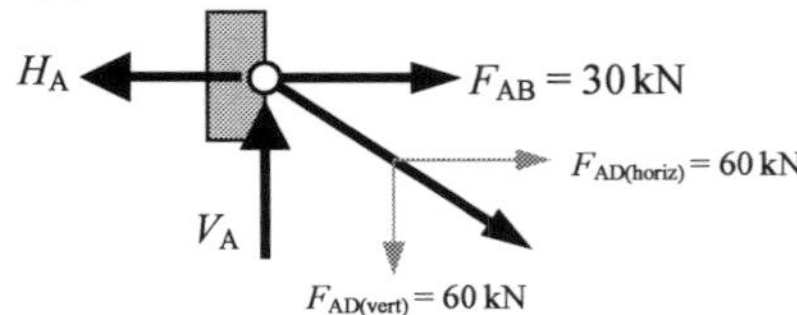

Sum vertical forces:

$V_A - F_{AD(vert)} = 0$

$V_A - 60\,\text{kN} = 0$

$V_A = 60\,\text{kN}$

Sum horizontal forces:

$-H_A + F_{AB} + F_{AD(horiz)} = 0$

$-H_A + 30 + 60 = 0$

$H_A = 90\,\text{kN}$

5. Support E

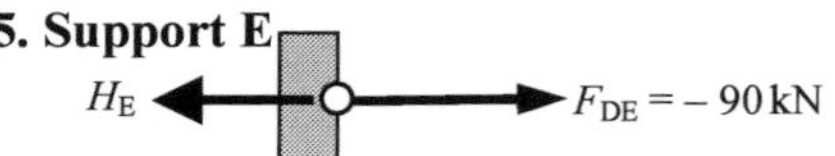

Sum horizontal forces:

$-H_E + F_{DE} = 0$

$-H_E + (-90) = 0$

$H_E = -90\,\text{kN(opp. direc.)}$

6. Check sum of external forces = 0

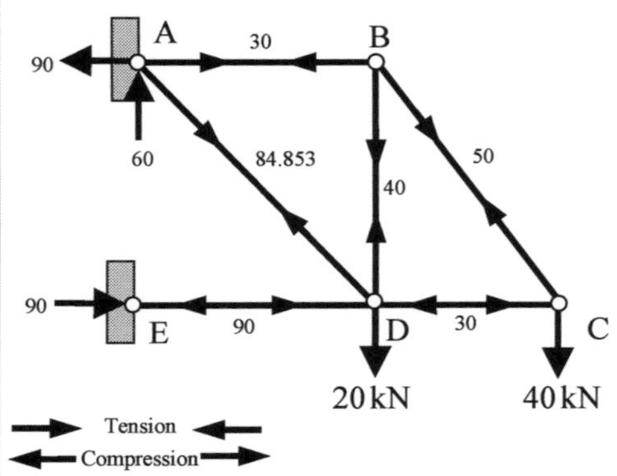

Final frame forces (kN)

Example 6.7: Analysis of a pin jointed frame supporting two loads

For the pin jointed frame of Example 6.7, determine the magnitude and type of forces in all members and the forces at the supports A and B.

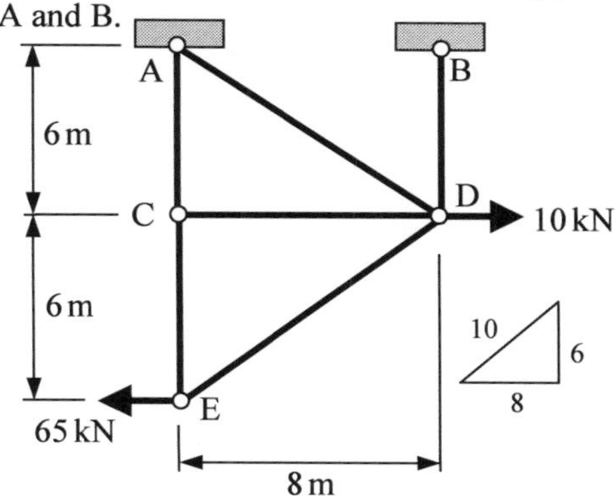

1. Joint E

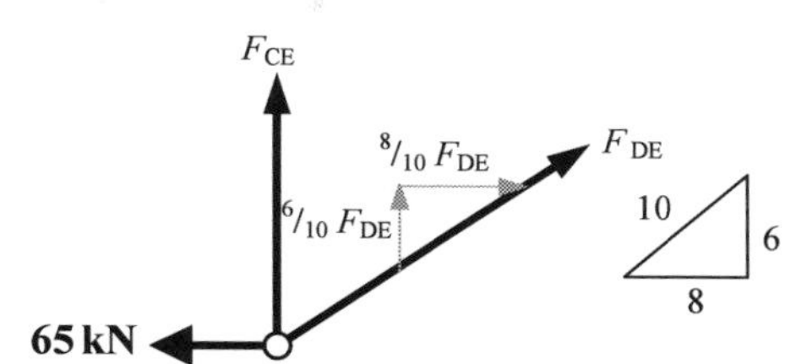

Sum vertical forces:

$F_{CE} + F_{DE(vert)} = 0$

$F_{CE} + 0.6F_{DE} = 0$

Sum horizontal forces:

$-65 + F_{DE(horiz)} = 0$

$-65 + 0.8F_{DE} = 0$

$F_{DE} = 81.25$ kN(tens.)

$F_{CE} + 0.6(81.25) = 0$

$F_{CE} = -48.75$ kN(comp.)

2. Joint C

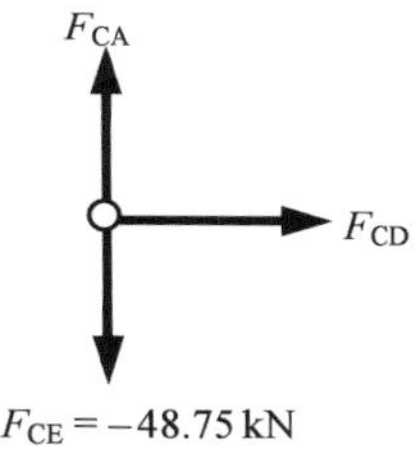

By observation:

$F_{CA} = -48.75$ kN (comp.) (equal & opposite to F_{CE})

$F_{CD} = 0$ kN (90° joint)

3. Joint D

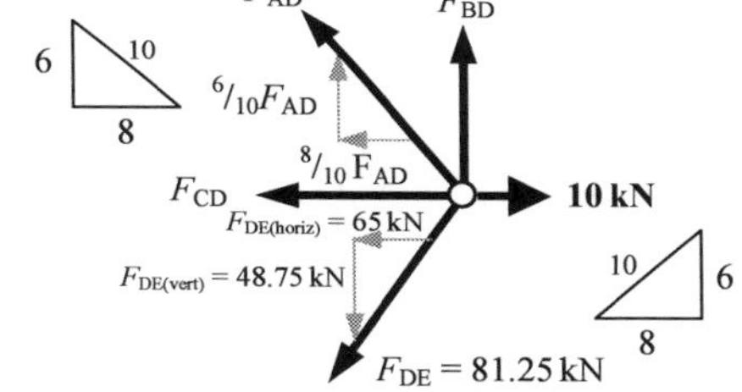

Sum vertical forces:

$-F_{DE(vert)} + F_{AD(vert)} + F_{BD} = 0$

$-48.75 + 0.6F_{AD} + F_{BD} = 0$

$0.6F_{AD} + F_{BD} = 48.75$

$F_{BD} = 90$ kN(tens.)

Sum horizontal forces:

$10 - F_{DE(horiz)} - F_{CD} - F_{AD(horiz)} = 0$

$10 - 65 - 0 - 0.8F_{AD} = 0$

$F_{AD} = -68.75$ kN(comp.)

4. Support A

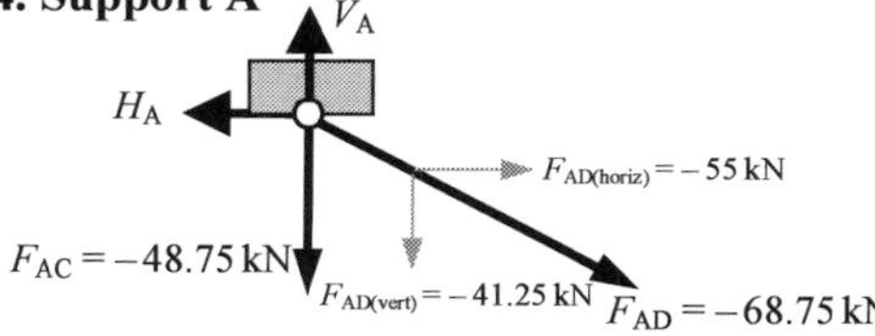

Sum vertical forces:

$V_A - F_{AD(vert)} - F_{AC} = 0$

$V_A - (-41.25)$
$\quad - (-48.75) = 0$

$V_A = -90$ kN(opp. direc.)

Sum horizontal forces:

$-H_A + F_{AD(horiz)} = 0$

$-H_A + (-55) = 0$

$H_A = -55$ kN(opp. direc.)

5. Support B

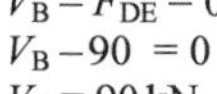

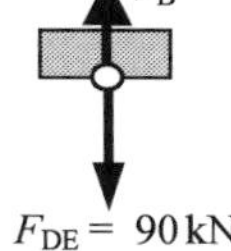

Sum vertical forces:

$V_B - F_{DE} = 0$

$V_B - 90 = 0$

$V_B = 90$ kN

6. Check sum of external forces = 0

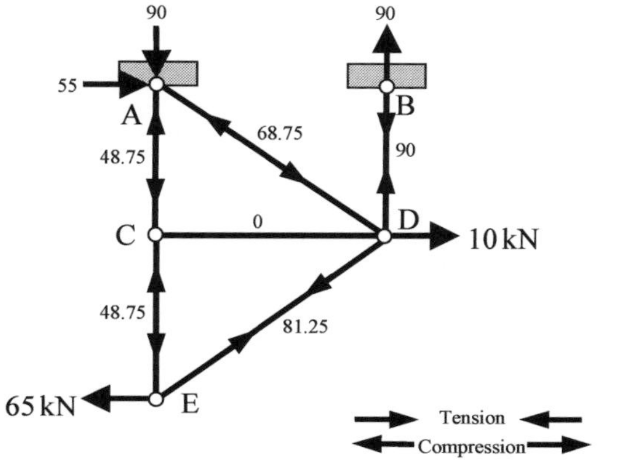

Final frame forces (kN)

Example 6.8: Analysis of a pin jointed frame supporting two loads

For the pin jointed frame of Example 6.8, determine the magnitude and type of forces in all members and the forces at the supports A and B.

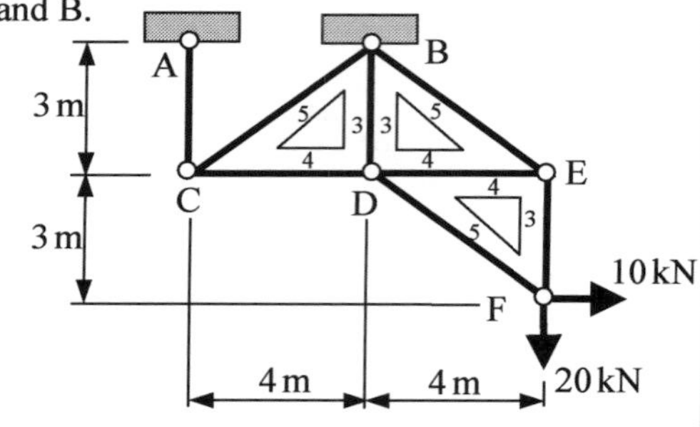

1. Joint F

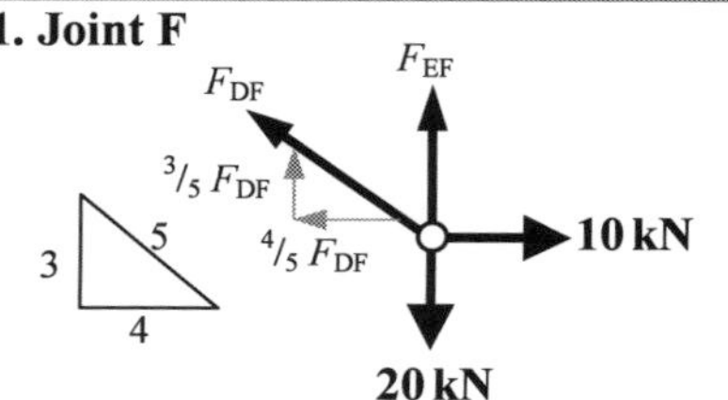

Sum vertical forces:

$-20 + F_{DF(vert)} + F_{EF} = 0$
$-20 + 0.6F_{DF} + F_{EF} = 0$
$0.6F_{DF} + F_{EF} = 20$
$0.6\,(12.5) + F_{EF} = 0$
$F_{EF} = 12.5$ kN(tens.)

Sum horizontal forces:

$10 - F_{DF(horiz)} = 0$
$10 - 0.8F_{DF} = 0$
$F_{DF} = 12.5$ kN(tens.)

2. Joint E

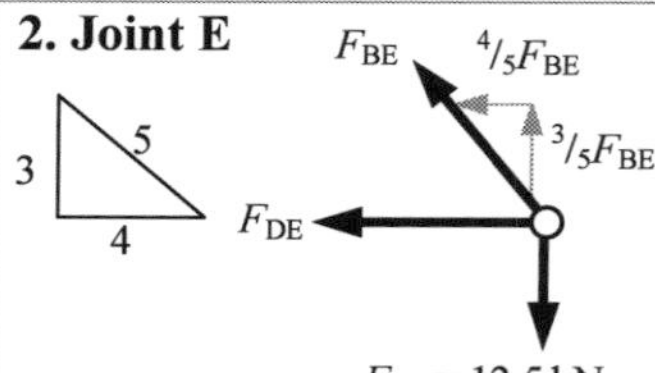

Sum vertical forces:

$-F_{EF} + F_{BE(vert)} = 0$
$-12.5 + 0.6F_{BE} = 0$
$F_{BE} = 20.83$ kN(tens.)

Sum horizontal forces:

$-F_{DE} - F_{BE(horiz)} = 0$
$-F_{DE} - 0.8F_{BE} = 0$

$-F_{DE} - 0.8(20.83) = 0$
$F_{DE} = -16.67$ kN(comp.)

3. Joint D

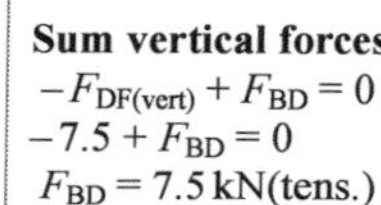
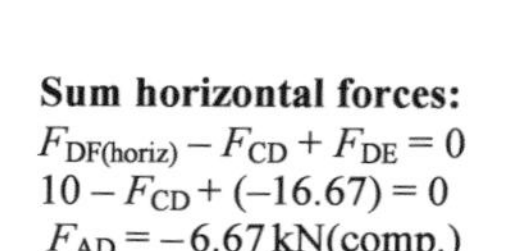

Sum vertical forces:

$-F_{DF(vert)} + F_{BD} = 0$
$-7.5 + F_{BD} = 0$
$F_{BD} = 7.5$ kN(tens.)

Sum horizontal forces:

$F_{DF(horiz)} - F_{CD} + F_{DE} = 0$
$10 - F_{CD} + (-16.67) = 0$
$F_{AD} = -6.67$ kN(comp.)

4. Joint C

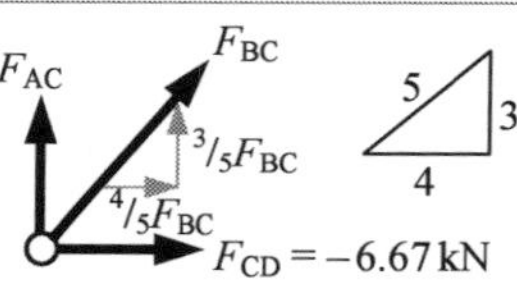

Sum vertical forces:

$F_{BC(vert)} + F_{AC} = 0$
$0.6F_{BC} + F_{AC} = 0$

$0.6(8.34) + F_{AC} = 0$
$F_{AC} = -5$ kN(comp.)

Sum horizontal forces:

$F_{BC(horiz)} + F_{CD} = 0$
$0.8F_{BC} + (-6.67) = 0$
$F_{BC} = 8.34$ kN(tens.)

5. Support B

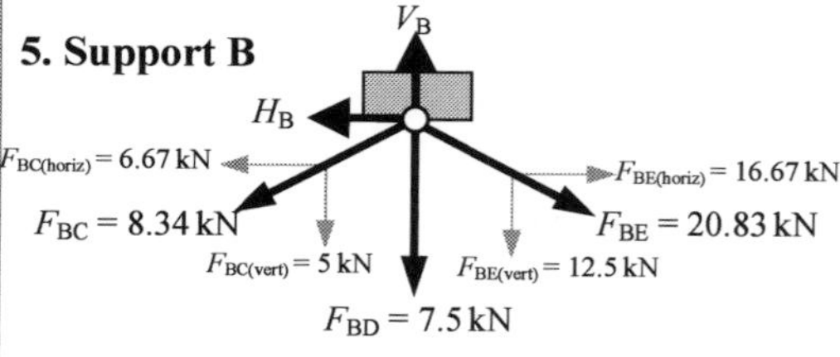

Sum vertical forces:

$V_B - F_{BE(vert)}$
$-F_{BD} - F_{BC(vert)} = 0$
$V_B - 5 - 7.5 - 12.5 = 0$
$V_B = 25$ kN

Sum horizontal forces:

$-F_{BC(horiz)} - H_B$
$+ F_{BE(horiz)} = 0$
$-6.67 - H_B + 16.67 = 0$
$H_B = 10$ kN

6. Support A

Sum vertical forces:

$V_A - F_{AC} = 0$
$V_A - 5 = 0$
$V_A = 5$ kN

7. Check sum of external forces = 0

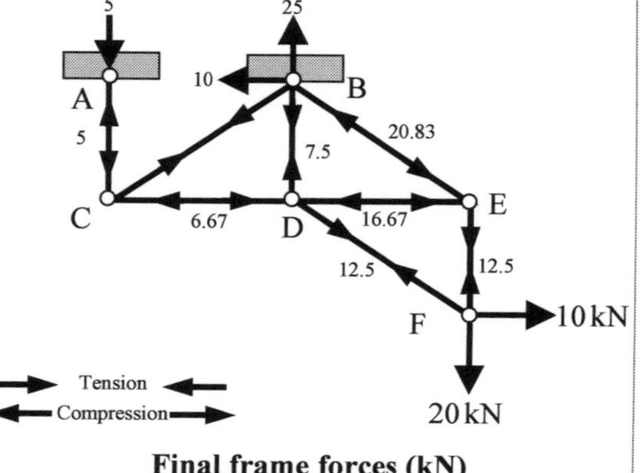

Final frame forces (kN)

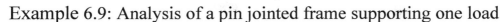

Example 6.9: Analysis of a pin jointed frame supporting one load

For the pin jointed frame of Example 6.9, determine the magnitude and type of forces in all members and the forces at the supports A and E.

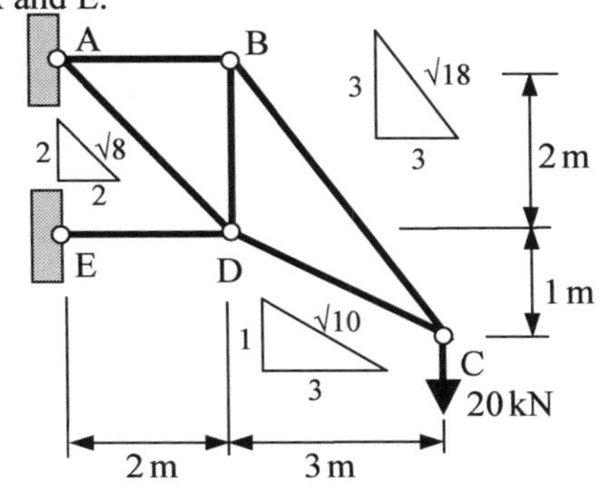

1. Joint C

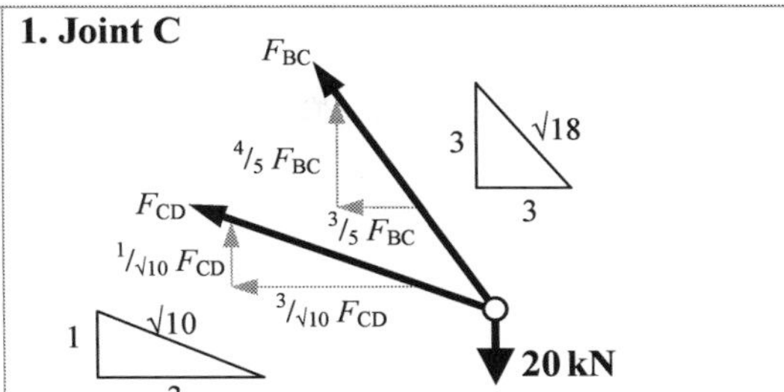

Sum vertical forces:

$-20 + F_{CD(vert)} + F_{BC(vert)} = 0$

$-20 + 0.316F_{CD} + 0.8F_{BC} = 0$

$0.316F_{CD} + 0.8F_{BC} = 20$ [1]

Sum horizontal forces:

$-F_{CD(horiz)} - F_{BC(horiz)} = 0$

$-0.949F_{CD} - 0.6F_{BC} = 0$ [2]

Solving equations [1] and [2] simultaneously will give:

$F_{BC} = 42.426\,\text{kN(tens.)}$ $F_{CD} = -31.6067\,\text{kN(comp.)}$

2. Joint B

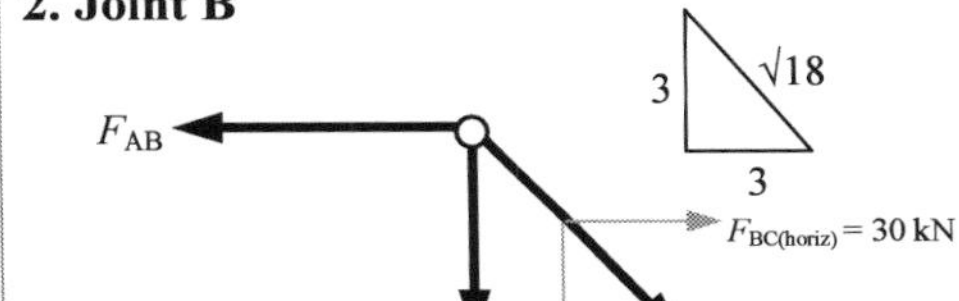

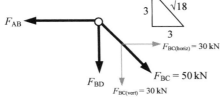

Sum vertical forces:

$-F_{BD} - F_{BC(vert)} = 0$

$-F_{BD} - 30 = 0$

$F_{BD} = -30\,\text{kN(comp.)}$

Sum horizontal forces:

$-F_{AB} + F_{BC(horiz)} = 0$

$-F_{AB} + 30 = 0$

$F_{AB} = 30\,\text{kN(tens.)}$

3. Joint D

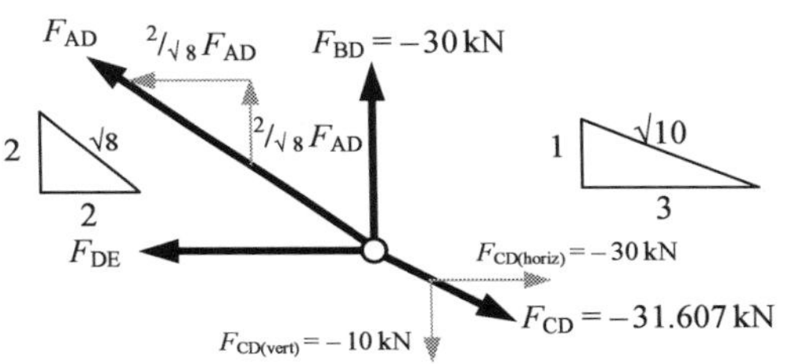

Sum vertical forces:

$F_{BD} - F_{CD(vert)} + F_{AD(vert)} = 0$

$-30 - (-10) + F_{AD(vert)} = 0$

$-20 + 0.7071F_{AD} = 0$

$F_{AD} = 28.285\,\text{kN(tens.)}$

Sum horizontal forces:

$F_{CD(horiz)} - F_{DE} - F_{AD(horiz)} = 0$

$-30 - F_{DE} - 0.7071F_{AD} = 0$

$F_{DE} = -30 - 0.7071F_{AD}$

$F_{DE} = -50\,\text{kN(comp.)}$

4. Support A

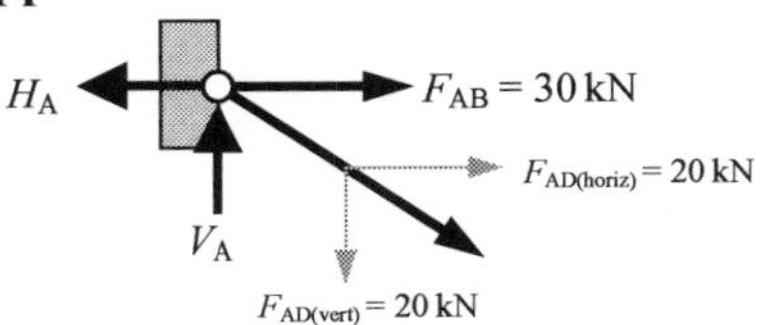

Sum vertical forces:

$V_A - F_{AD(vert)} = 0$

$V_A - 20\,\text{kN} = 0$

$V_A = 20\,\text{kN}$

Sum horizontal forces:

$-H_A + F_{AB} + F_{AD(horiz)} = 0$

$-H_A + 30 + 20 = 0$

$H_A = 50\,\text{kN}$

5. Support E

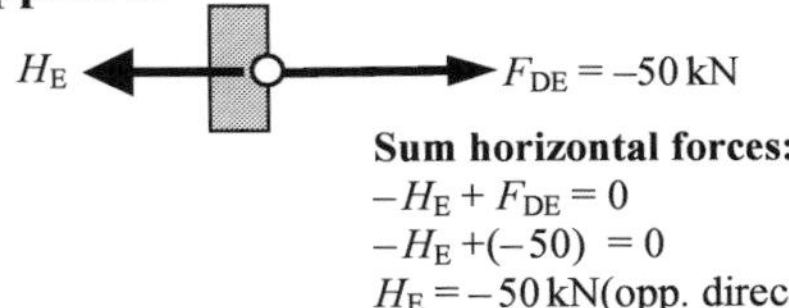

Sum horizontal forces:

$-H_E + F_{DE} = 0$

$-H_E + (-50) = 0$

$H_E = -50\,\text{kN(opp. direc.)}$

6. Check sum of external forces = 0

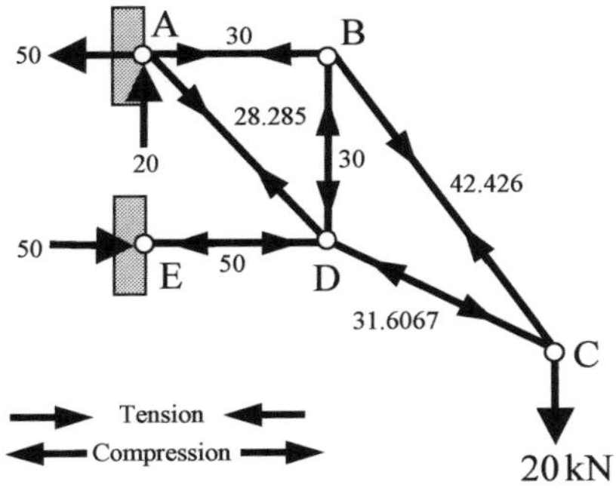

Final frame forces (kN)

Example 6.10: Analysis of a pin jointed frame supporting three loads

For the pin jointed frame of Example 6.10, determine the magnitude and type of forces in all members and the forces at the supports A and E.

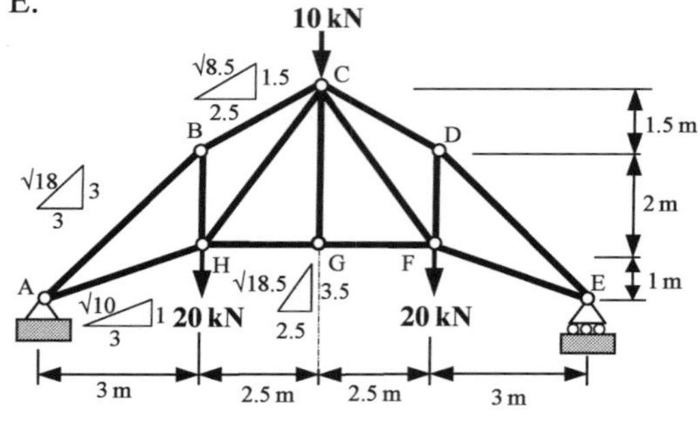

1. Calculate forces at supports

By observation $V_A = V_E = 50\,\text{kN}/2 = 25\,\text{kN}$

$H_A = 0\,\text{kN}$

OR:

Sum all horizontal forces:

$\Sigma F_x = H_A + 0\,\text{kN} = 0 \qquad \therefore H_A = 0\,\text{kN}$

Sum all vertical forces:

$\Sigma F_y = V_A + V_E - (10 + 20 + 20)\,\text{kN} = 0$

$V_A + V_E = 50\,\text{kN} \qquad [1]$

Moments about A:

$\Sigma M_z = 20\,\text{kN} \times 3\,\text{m} + 10\,\text{kN} \times 5.5\,\text{m} + 20\,\text{kN} \times 8\,\text{m} - V_E \times 11\,\text{m} = 0$

$\therefore V_E = 25\,\text{kN}$ and from [1] $V_A = 25\,\text{kN}$

Note: (i) Structure is symmetrical

(ii) Resolution of joint G gives $F_{CG} = 0\,\text{kN}$

$F_{CG} = 0\,\text{kN}$

F_{GH} F_{GF}

2. Joint A

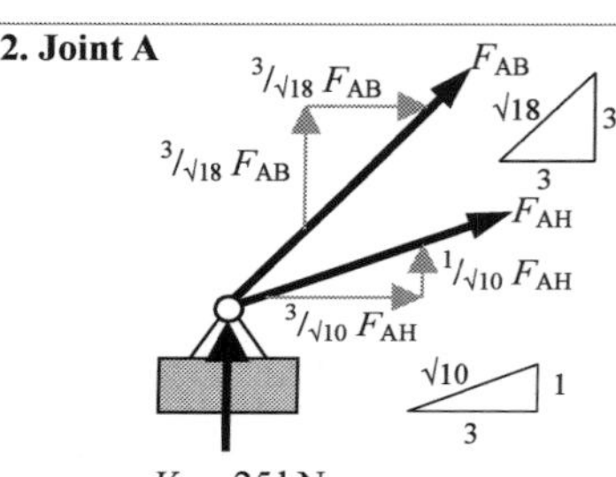

Sum all vertical forces:	Sum all horizontal forces:
$V_A + F_{AB(vertical)} + F_{AH(vertical)} = 0$	$F_{AB(horizontal)} + F_{AH(horizontal)} = 0$
$25 + 0.707F_{AB} + 0.316F_{AH} = 0$	$0.7071F_{AB} + 0.9488F_{AH} = 0$
$0.707F_{AB} + 0.316F_{AH} = -25$	$F_{AB} + 1.3417F_{AH} = 0 \quad [2]$
$F_{AB} + 0.447F_{AH} = -35.356\ [1]$	

Solving equations [1] and [2] simultaneously gives:

$F_{AH} = 39.53\,\text{kN}$ (tens.) $\qquad F_{AB} = -53.03\,\text{kN}$ (comp.)

3. Joint B

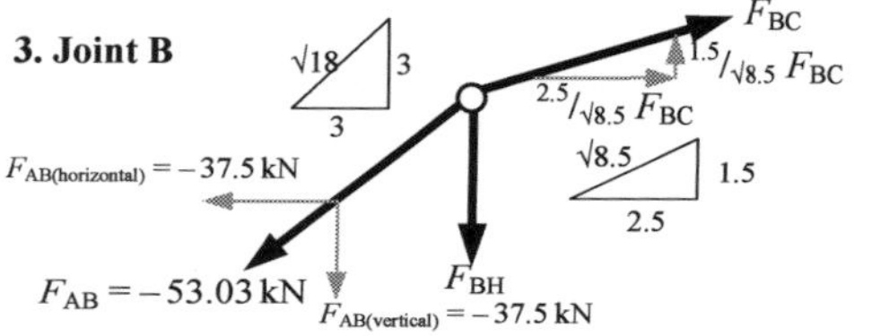

Sum vertical forces:	Sum horizontal forces:
$-F_{BH} - F_{AB(vertical)} + F_{BC(vertical)} = 0$	$-F_{AB(horiz)} + F_{BC(horiz)} = 0$
$-F_{BH} - (-37.5) + 0.515F_{BC} = 0$	$-(-37.5) + 0.858F_{BC} = 0$
$-F_{BH} + 37.5 + 0.515(-43.729) = 0$	$F_{BC} = -43.729\,\text{kN}$
$F_{BH} = 15\,\text{kN}$ (tens.)	(comp.)

4. Joint H

$F_{BH} = 15\,\text{kN}$; F_{CH}; $3.5/\sqrt{18.5}\,F_{CH}$; $2.5/\sqrt{18.5}\,F_{CH}$; F_{GH}; $\sqrt{10}$, 1, 3; $\sqrt{18.5}$, 3.5, 2.5; $F_{AH(horiz)} = 37.5\,\text{kN}$; $F_{AH} = 39.532\,\text{kN}$; 20 kN; $F_{AB(vert)} = 12.5\,\text{kN}$

Sum vertical forces:	Sum horizontal forces:
$F_{BH} + F_{CH(vert)} - 20 - F_{AH(vert)} = 0$	$F_{CH(horiz)} + F_{GH} - F_{AH(horiz)} = 0$
$15 + 0.814F_{CH} - 20 - 12.5 = 0$	$0.581F_{CH} + F_{GH} - 37.5 = 0$
$F_{CH} = 21.51\,\text{kN}$ (tens)	$F_{GH} = 25\,\text{kN}$ (tens.)

5. Joint C (check)

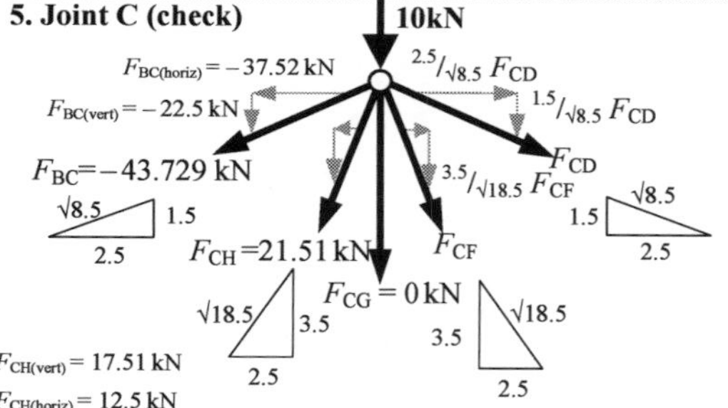

Sum vertical forces:

$-10 - 0.515F_{CD} - 0.814F_{CF} - 0 - 17.51 + 22.5 = 0$

$-0.515F_{CD} - 0.814F_{CF} = 5\,\text{kN} \qquad [1]$

Sum horizontal forces:

$0.858F_{CD} + 0.581F_{CF} - 12.5 + 37.52 = 0$

$0.858F_{CD} + 0.581F_{CF} = -25.02\,\text{kN} \qquad [2]$

Solving equations [1] and [2] simultaneously gives:

$F_{CD} = -43.729\,\text{kN}$ (comp.) $\qquad F_{CF} = 21.51\,\text{kN}$ (tens.)

6. Check sum of external forces = 0

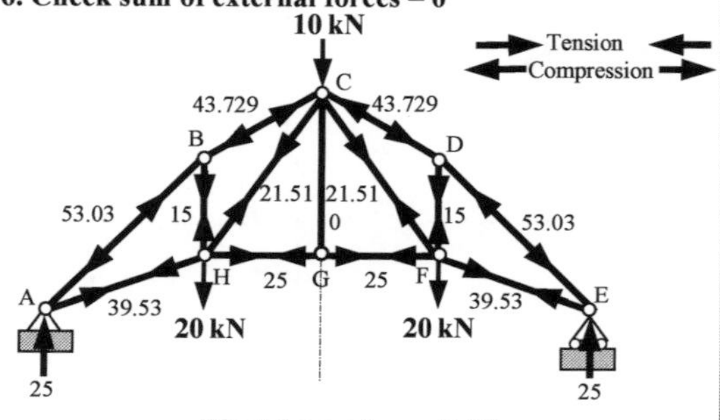

Final frame forces (kN)

Example 6.11: Analysis of a pin jointed frame supporting one load

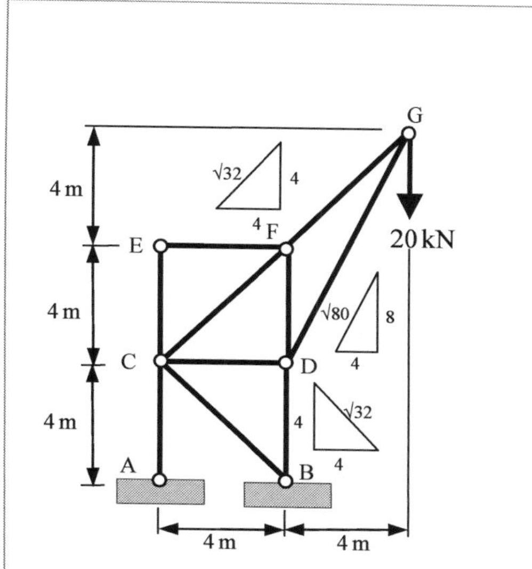

1. Joint G

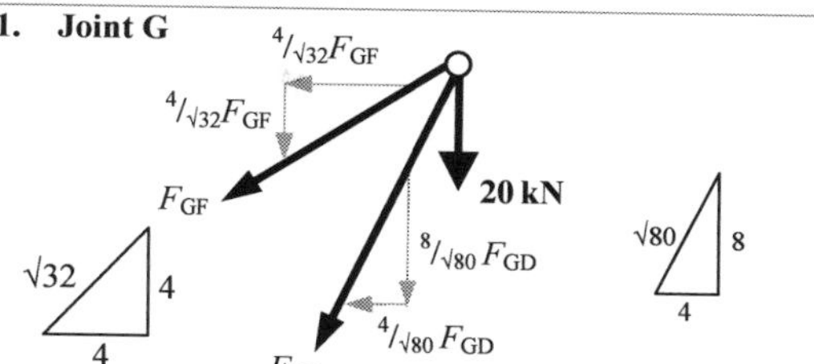

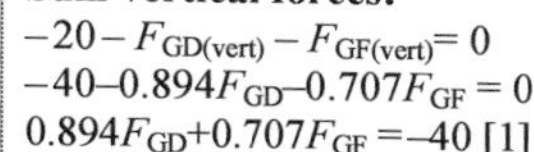

Sum vertical forces:

$-20 - F_{GD(vert)} - F_{GF(vert)} = 0$

$-40 - 0.894F_{GD} - 0.707F_{GF} = 0$

$0.894F_{GD} + 0.707F_{GF} = -40$ [1]

Sum horizontal forces:

$-F_{GD(horiz)} - F_{GF(horiz)} = 0$

$-0.447F_{GD} - 0.7071F_{GF} = 0$ [2]

Solving equations [1] and [2] simultaneously gives:

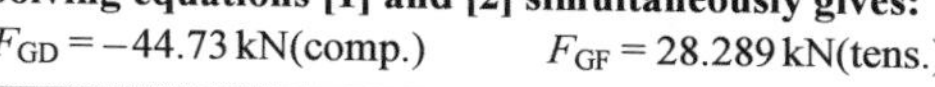

$F_{GD} = -44.73$ kN(comp.) $F_{GF} = 28.289$ kN(tens.)

2. Joints F and E

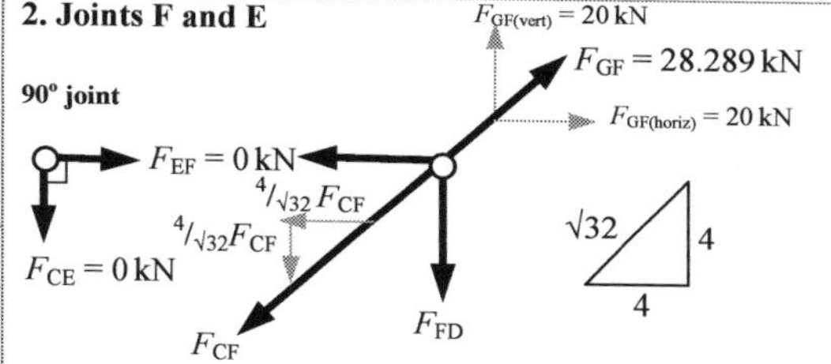

Sum vertical forces:

$-F_{FD} - F_{CF(vert)} + F_{GF(vert)} = 0$

$-F_{FD} - 0.7071F_{CF} + 20 = 0$

$F_{FD} = 0$ kN

Sum horizontal forces:

$-F_{CF(horiz)} - F_{EF} + F_{GF(horiz)} = 0$

$-0.7071F_{CF} - 0 + 20 = 0$

$F_{CF} = 28.289$ kN(tens.)

3. Joint D

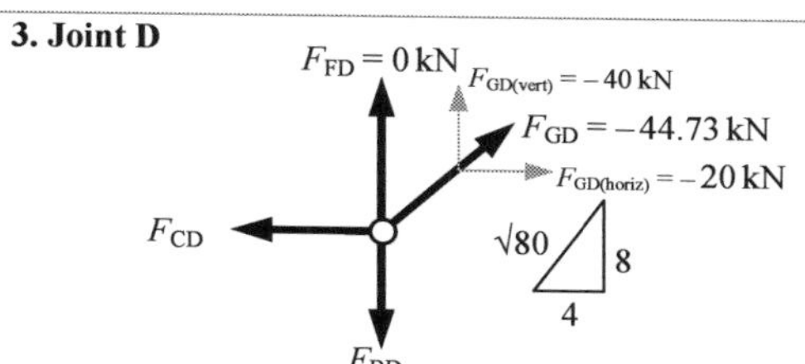

Sum vertical forces:

$F_{GD(vert)} - F_{BD} + F_{FD} = 0$

$-40 - F_{BD} + 0 = 0$

$F_{BD} = -40$ kN(comp.)

Sum horizontal forces:

$F_{GD(horiz)} - F_{CD} = 0$

$-20 - F_{CD} = 0$

$F_{CD} = -20$ kN(comp.)

4. Joint C

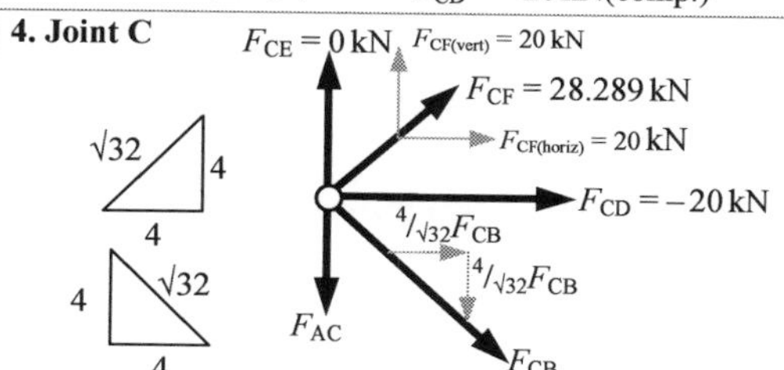

Sum vertical forces:

$F_{CE} + F_{CF(vert)}$

$- F_{CB(vert)} - F_{AC} = 0$

$0 - 20 - 0.7071F_{CB} + F_{AC} = 0$

$0.7071F_{CB} - F_{AC} = -20$

$F_{AC} = 20$ kN(tens.)

Sum horizontal forces:

$F_{CF(horiz)} + F_{CD} + F_{CB(horiz)} = 0$

$20 - 20 + 0.7071F_{CB} = 0$

$F_{CB} = 0$ kN

5. Support B

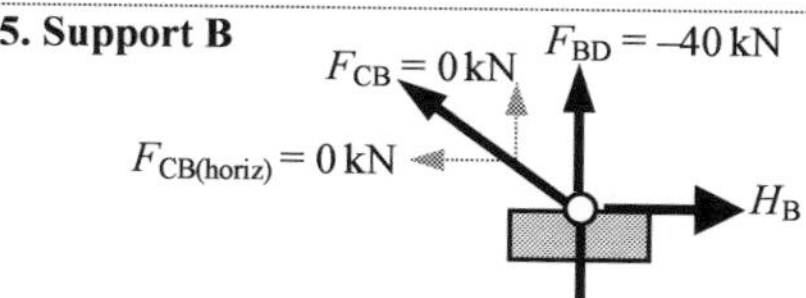

Sum vertical forces:

$-V_B + F_{CB(vert)} + F_{BD} = 0$

$-V_B + (-40) + 0 = 0$

$V_B = -40$ kN(opp. direc.)

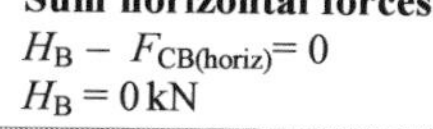

Sum horizontal forces:

$H_B - F_{CB(horiz)} = 0$

$H_B = 0$ kN

6. Support A

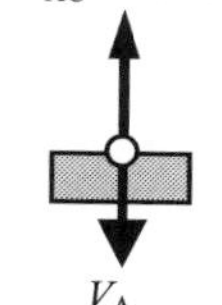

Sum vertical forces:

$-V_A + F_{AC} = 0$

$-V_A + 20 = 0$

$V_A = 20$ kN

7. Check sum of external forces = 0

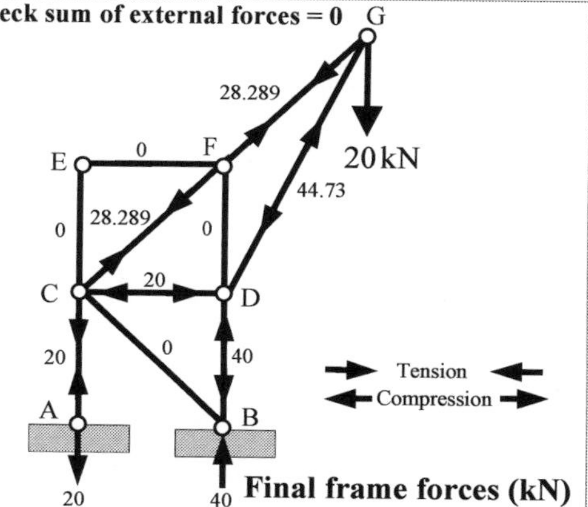

Final frame forces (kN)

Example 6.12: Analysis of a pin jointed frame supporting two loads

For the pin jointed frame of Example 6.12, determine the magnitude and type of forces in all members and the forces at the supports A and E.

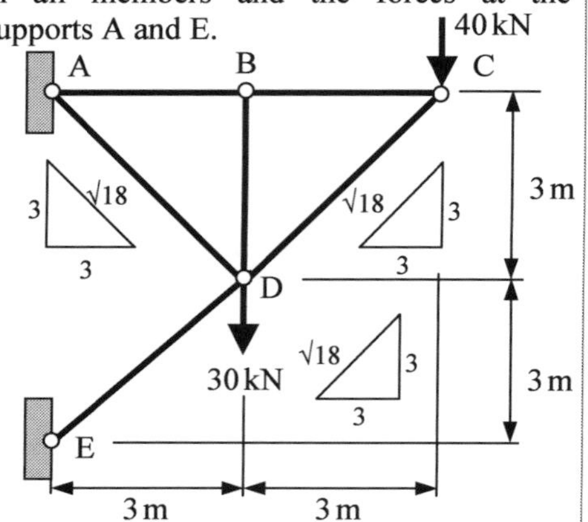

1. Joint C

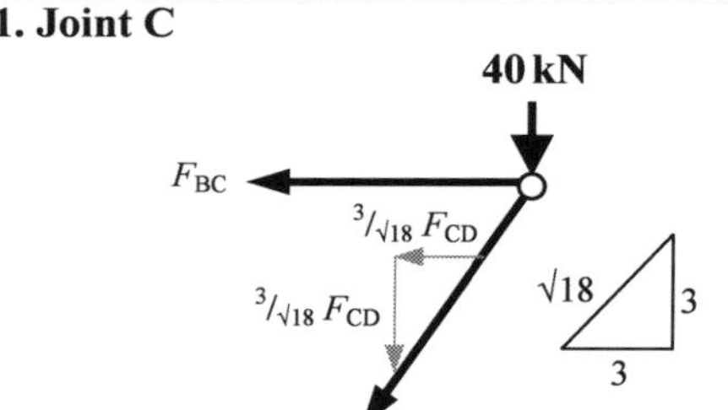

Sum vertical forces:

$-40 - F_{CD(vert)} = 0$

$-40 - 0.7071F_{CD} = 0$

F_{CD}=–56.569kN(comp.)

Sum horizontal forces:

$-F_{BC} - F_{CD(horiz)} = 0$

$-F_{BC} - 0.7071F_{CD} = 0$

$-F_{BC} - 0.7071(56.569) = 0$

$F_{BC} = 40$ kN(tens.)

2. Joint B

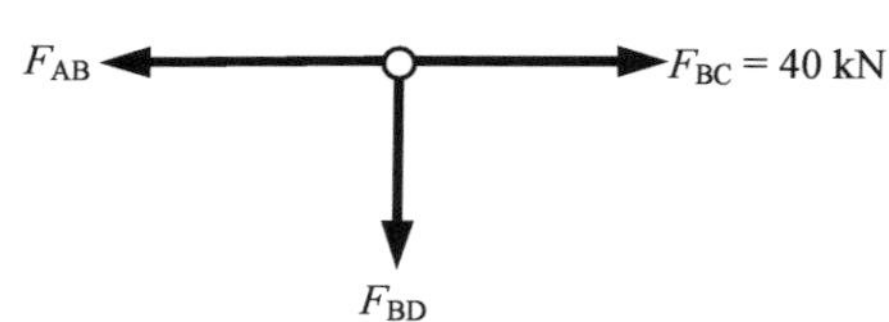

Sum vertical forces:

$-F_{BD} = 0$

Sum horizontal forces:

$-F_{AB} + F_{BC} = 0$

$-F_{AB} + 40 = 0$

$F_{AB} = 40$ kN(tens.)

3. Joint D

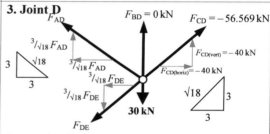

F_{AD} $F_{BD} = 0$ kN $F_{CD} = -56.569$ kN

$^3/\sqrt{18}\,F_{AD}$ $F_{CD(vert)} = -40$ kN

$^3/\sqrt{18}\,F_{AD}$ $F_{CD(horiz)} = -40$ kN

$^3/\sqrt{18}\,F_{DE}$ $^3/\sqrt{18}\,F_{DE}$

30 kN

F_{DE}

Sum vertical forces:

$F_{CD(vert)} - 30 - F_{DE(vert)}$
$+ F_{AD(vert)} + F_{BD} = 0$

$-40 - 30 - 0.7071F_{DE}$
$+ 0.7071F_{AD} + 0 = 0$

$0.7071F_{DE} - 0.7071F_{AD}$
$= -70$ [1]

Sum horizontal forces:

$F_{CD(horiz)} - F_{DE(horiz)}$
$- F_{AD(horiz)} = 0$

$-40 - 0.7071F_{DE}$
$- 0.7071F_{AD} = 0$

$0.7071F_{DE} + 0.7071F_{AD}$
$= -40$ [2]

Solving equations (1) and [2] simultaneously gives:

F_{AD} = 21.216kN(tens.) F_{DE}=–77.845 kN(comp.)

4. Support A

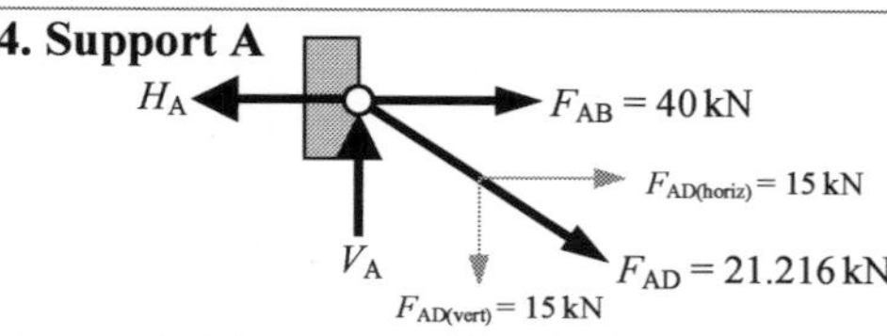

Sum vertical forces:

$V_A - F_{AD(vert)} = 0$

$V_A - 15$ kN $= 0$

$V_A = 15$ kN

Sum horizontal forces:

$-H_A + F_{AB} + F_{AD(horiz)} = 0$

$-H_A + 40 + 15 = 0$

$H_A = 55$ kN

5. Support E

$F_{DE(horiz)} = -55$ kN

$F_{DE} = -77.845$ kN

$F_{DE(vert)} = -55$ kN

H_E

V_E

Sum vertical forces:

$V_E + F_{DE(vert)} = 0$

$V_E - 55$ kN $= 0$

$V_E = 55$ kN

Sum horizontal forces:

$-H_E + F_{DE} = 0$

$-H_E + (-55) = 0$

$H_E = -55$ kN(opp. direc.)

6. Check sum of external forces = 0

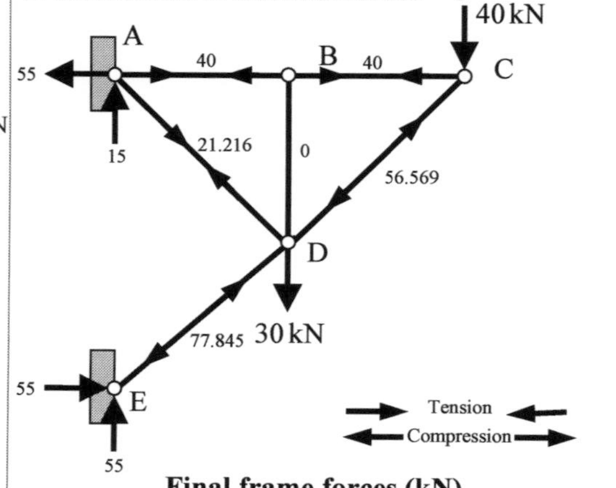

Final frame forces (kN)

Example 6.13: Analysis of a pin jointed frame supporting three loads

For the pin jointed frame of Example 6.13, determine the magnitude and type of forces in all members and the forces at the supports A and F.

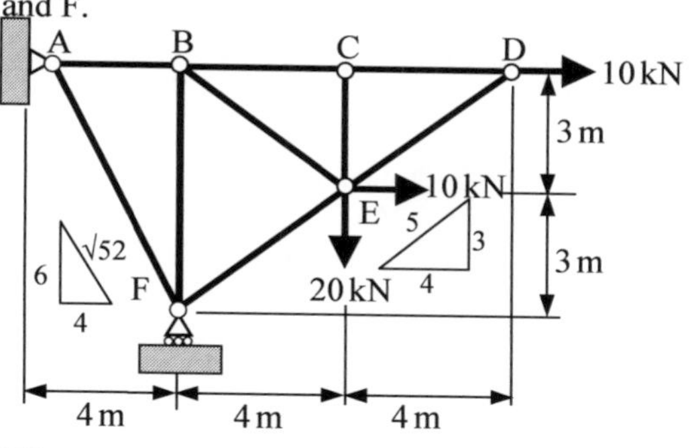

1. Joint D

F_{CD} 10 kN $^3/_5 F_{DE}$ F_{DE} $^3/_5 F_{DE}$ 5 3 4

Sum vertical forces:

$F_{DE(vert)} = 0\,\text{kN}$

Sum horizontal forces:

$10 - F_{DE(horiz)} - F_{CD} = 0\,\text{kN}$

$10 - 0.6(0) - F_{CD} = 0\,\text{kN}$

$F_{CD} = 10\,\text{kN}$ (tens.)

2. Joint C

F_{BC} $F_{CD} = 10\,\text{kN}$ F_{CE}

Sum vertical forces:

$F_{CE} = 0\,\text{kN}$

Sum horizontal forces:

$F_{CD} - F_{BC} = 0\,\text{kN}$

$10 - F_{BC} = 0\,\text{kN}$

$F_{BC} = 10\,\text{kN}$ (tens.)

3. Joint E

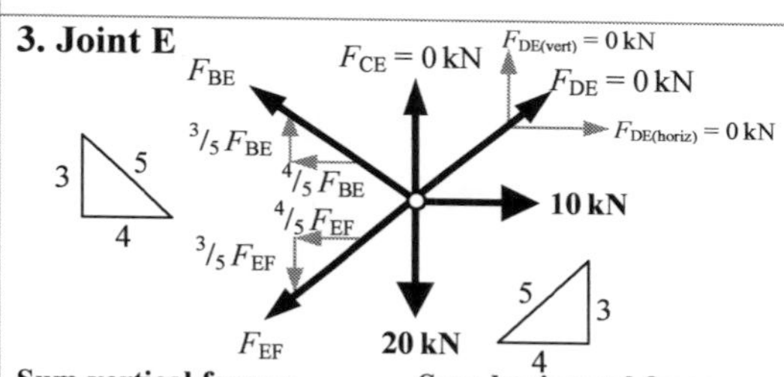

Sum vertical forces:

$-20 - F_{EF(vert)} + F_{BE(vert)} + F_{CE} = 0$

$-20 - 0.6F_{EF} + 0.6F_{BE} + 0 = 0$

$0.6F_{EF} - 0.6F_{BE} = -20\,\text{kN}$ [1]

Sum horizontal forces:

$10 - F_{EF(horiz)} - F_{BE(horiz)} + F_{DE(horiz)} = 0$

$10 - 0.8F_{EF} - 0.8F_{BE} + 0 = 0$

$0.8F_{EF} + 0.8F_{BE} = -10$ [2]

Solving equations [1] and [2] simultaneously gives:

$F_{BE} = 22.9\,\text{kN}$(tens.) $F_{EF} = -10.4\,\text{kN}$ (comp.)

4. Joint B

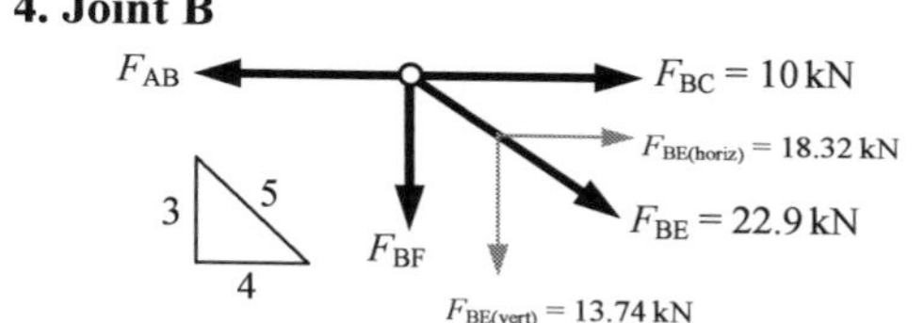

Sum vertical forces:

$-F_{BE(vert)} - F_{BF} = 0$

$-13.74 - F_{BF} = 0\,\text{kN}$

$F_{BE} = -13.74\,\text{kN}$ (comp.)

Sum horizontal forces:

$F_{BC} + F_{BE(horiz)} - F_{AB} = 0\,\text{kN}$

$10 + 18.32 - F_{AB} = 0\,\text{kN}$

$F_{AB} = 28.32\,\text{kN}$ (tens.)

5. Support F $F_{BF} = -13.74\,\text{kN}$

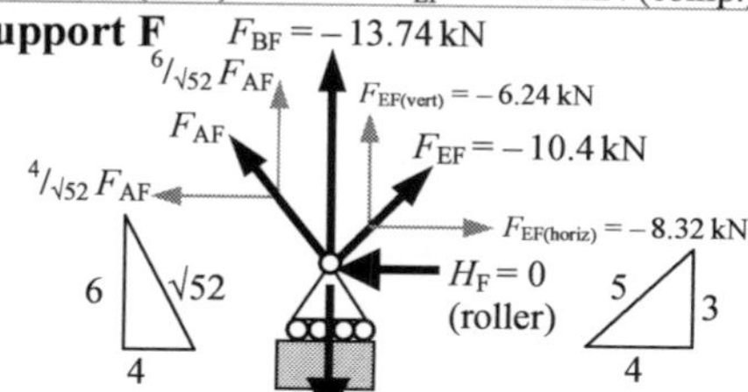

Sum vertical forces:

$-V_F + F_{AF(vert)} + F_{BF} + F_{EF(vert)} = 0$

$-V_F + 0.832F_{AF} - 13.74 - 6.24 = 0$

$V_F = -32.46\,\text{kN}$

Sum horizontal forces:

$-H_F - F_{AF(horiz)} + F_{EF(horiz)} = 0$

$0 - 0.555F_{AF} + (-8.32) = 0$

$F_{AF} = -15\,\text{kN}$ (comp.)

6. Support A

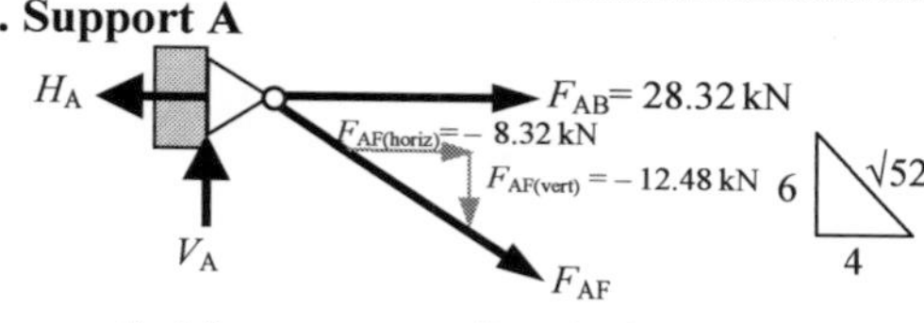

Sum vertical forces:

$V_A - F_{AF(vert)} = 0$

$V_A = -12.46\,\text{kN}$ (opp. direc.)

Sum horizontal forces:

$-H_A + F_{AF(horiz)} + F_{AB} = 0$

$-H_A - 8.32 + 28.32 = 0$

$H_A = 20\,\text{kN}$

7. Check the sum of external forces = 0

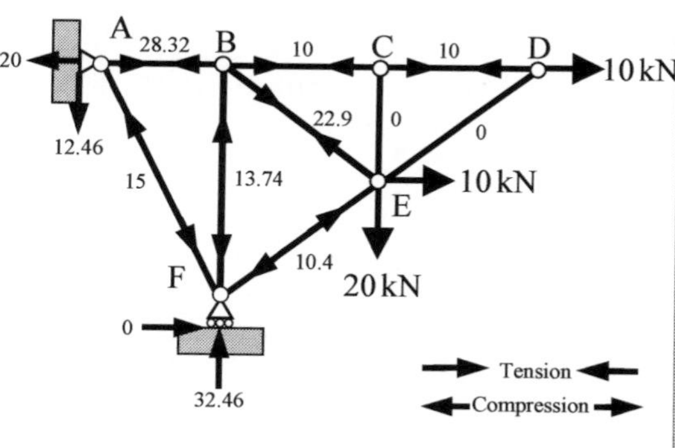

Final frame forces (kN)

Example 6.14: Analysis of a pin jointed frame supporting three loads

For the pin jointed frame of Example 6.14, determine the magnitude and type of forces in all members and the forces at the supports A and C.

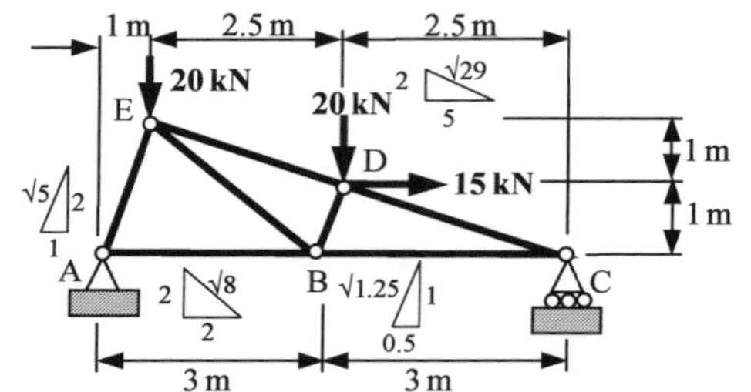

1. Calculate reactions

Sum all horizontal forces:

$\Sigma F_x = -H_A + 15\,\text{kN} = 0 \qquad \therefore\ H_A = 15\,\text{kN}$

Sum all vertical forces:

$\Sigma F_y = V_A + V_C - 20\,\text{kN} - 20\,\text{kN} = 0$

$V_A + V_C = 40\,\text{kN}$

Taking moments about support A

$\Sigma M_z = 20\,\text{kN}\times 1\,\text{m} + 20\,\text{kN}\times 3.5\,\text{m} + 15\,\text{kN}\times 1\,\text{m} - V_C \times 6\,\text{m} = 0$

$20 + 70 + 15 - 6\,V_C = 0$

$\therefore\ V_C = 17.5\,\text{kN}$ and $V_A = 22.5\,\text{kN}$

2. Support A

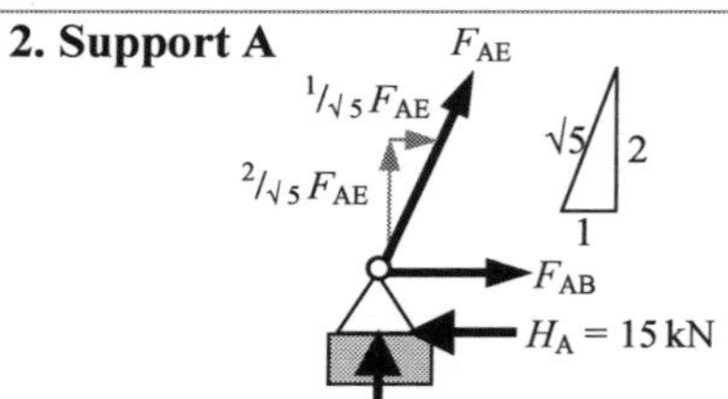

Sum vertical forces:

$V_A + F_{AE(vert)} = 0$

$22.5 + 0.894\,F_{AE} = 0$

$F_{AE} = -25.16\,\text{kN}$ (comp.)

Sum horizontal forces:

$-H_A + F_{AE(horiz)} + F_{AB} = 0$

$-15 + 0.447(-25.168) + F_{AB} = 0$

$F_{AB} = 26.25\,\text{kN}$ (tens.)

3. Support C

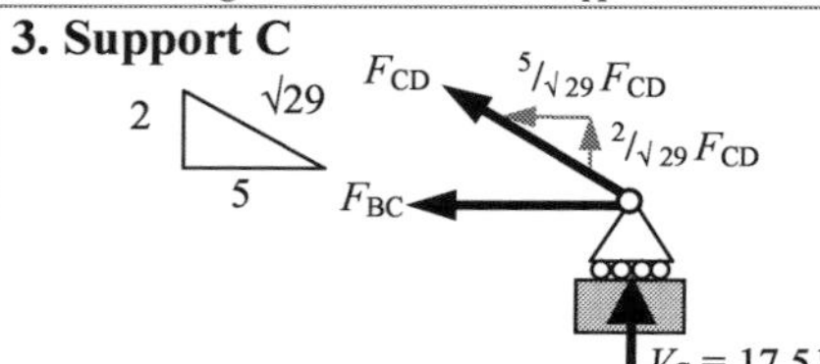

Sum vertical forces:

$V_C + F_{CD(vert)} = 0$

$17.5 + 0.371F_{CD} = 0$

$F_{CD} = -47.12\,\text{kN}$ (comp.)

Sum horizontal forces:

$-F_{BC} - F_{CD(horiz)} = 0$

$-F_{BC} - 0.928(-47.12) = 0$

$F_{BC} = 43.75\,\text{kN}$ (tens.)

4. Joint B

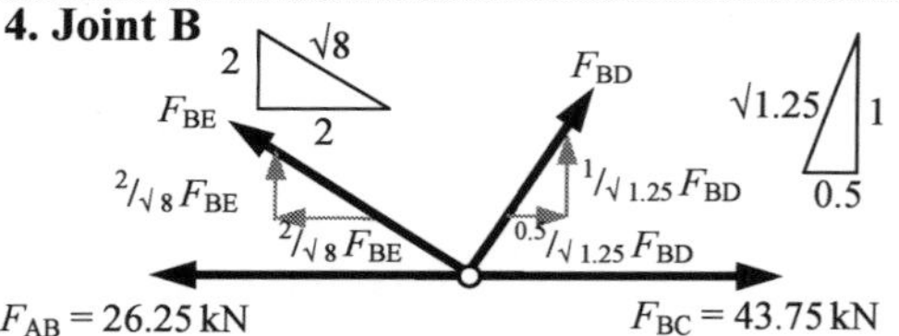

Sum vertical forces:

$F_{BE(vert)} + F_{BD(vert)} = 0$

$0.707F_{BE} + 0.894F_{BD} = 0$ [1]

Sum horizontal forces:

$F_{BC} - F_{AB} - F_{BE(horiz)} + F_{BD(horiz)} = 0$

$43.8 - 26.25 - 0.707F_{BE} + 0.447F_{BD} = 0$

$0.707F_{BE} - 0.447F_{BD} = 17.55$ [2]

Solving equations [1] and [2] simultaneously gives:

$F_{BD} = -13.05\,\text{kN}$ (comp.) $\quad F_{BE} = 16.5\,\text{kN}$ (tens.)

5. Joint E

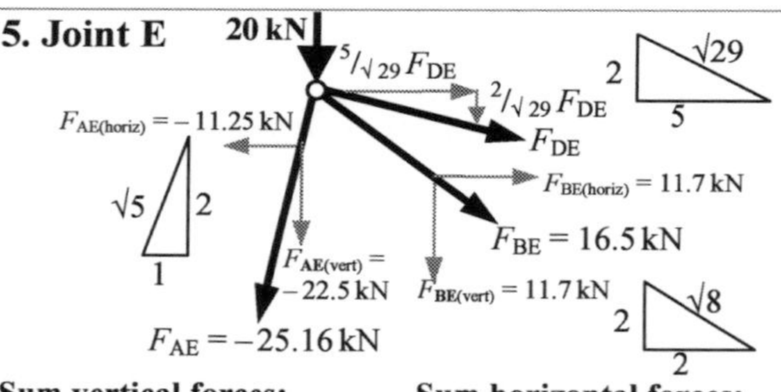

Sum vertical forces:

$-F_{BE(vert)} - F_{AE(vert)} - 20 - F_{DE(vert)} = 0$

$-11.7 + 22.5 - 20 - 0.371F_{DE} = 0$

$F_{DE} = -24.68\,\text{kN}$ (comp.)

Sum horizontal forces:

Will give:

$F_{DE} = -24.68\,\text{kN}$ (comp.)

6. Joint D (check)

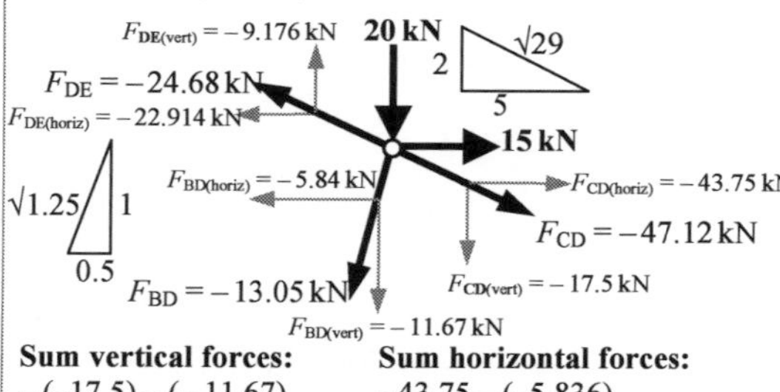

Sum vertical forces:

$-(-17.5) - (-11.67) + (-9.176) - 20 = 0$ (OK)

Sum horizontal forces:

$-43.75 - (-5.836) - (-22.914) + 15 = 0$ (OK)

7. Check sum of external forces = 0

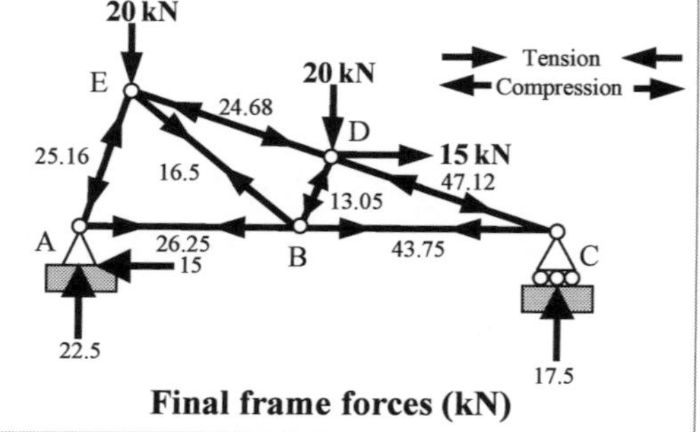

Final frame forces (kN)

Example 6.15: Analysis of a pin jointed frame supporting three loads

For the pin jointed frame of Example 6.15, determine the magnitude and type of forces in all members and the forces at the supports A and E.

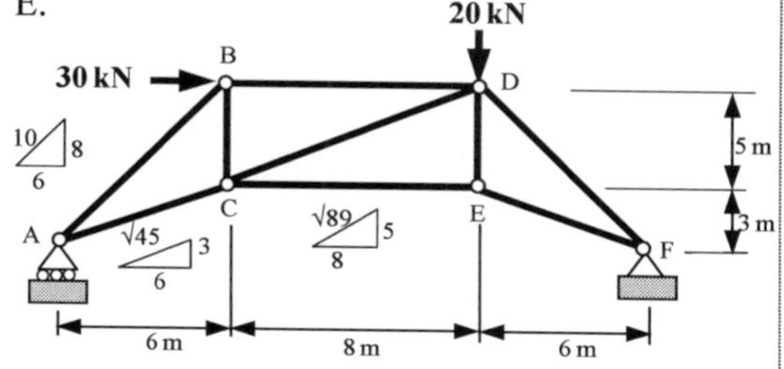

1. Calculate forces at supports

Sum all horizontal forces $\Sigma F_x = -H_F + 30\,\text{kN} = 0$

$\therefore H_F = 30\,\text{kN}$

Sum all vertical forces $\Sigma F_y = V_A + V_F - 20\,\text{kN} = 0$

$V_A + V_F = 20\,\text{kN}$ [1]

Moments about A

$\Sigma M_z = 30\,\text{kN} \times 8\,\text{m} + 20\,\text{kN} \times 14\,\text{m} - V_F \times 20\,\text{m} = 0$

$520 = 20 V_F$

$\therefore V_F = 26\,\text{kN}$ and from [1]

$V_A = -6\,\text{kN}$ (opp. direc.)

2. Support F

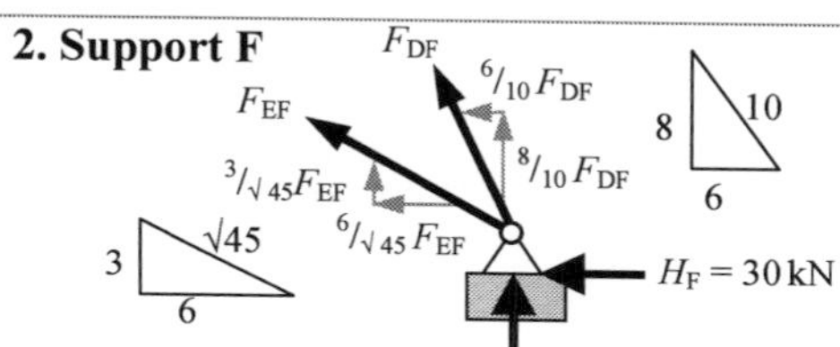

Sum vertical forces:

$F_{DF(vert)} + V_F + F_{EF(vert)} = 0$

$0.8F_{DF} + 26 + 0.447F_{EF} = 0$

$0.8F_{DF} + 0.447F_{EF} = -26$ [1]

Sum horizontal forces:

$-F_{DF(horiz)} - H_F - F_{EF(horiz)} = 0$

$-0.6F_{DF} - 30 - 0.894F_{EF} = 0$

$0.6F_{DF} + 0.894F_{EF} = -30$ [2]

Solving equations [1] and [2] simultaneously gives:

$F_{EF} = -18.8\,\text{kN}$(comp.) $F_{DF} = -22\,\text{kN}$(comp.)

3. Joint E

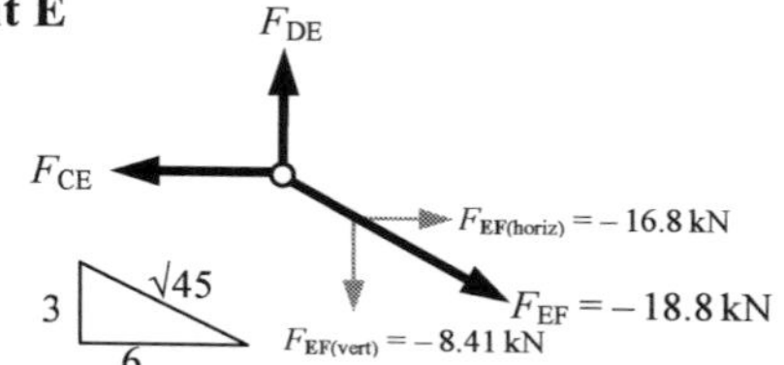

Sum vertical forces:

$-F_{EF(vert)} + F_{DE} = 0$

$-(-8.41) + F_{DE} = 0$

$F_{DE} = -8.41\,\text{kN}$ (comp.)

Sum horizontal forces:

$F_{EF(horiz)} - F_{CE} = 0$

$-16.8 - F_{CE} = 0$

$F_{CE} = -16.8\,\text{kN}$ (comp.)

4. Joint D

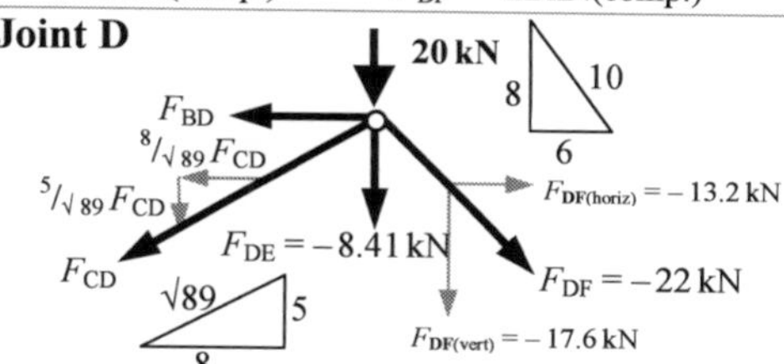

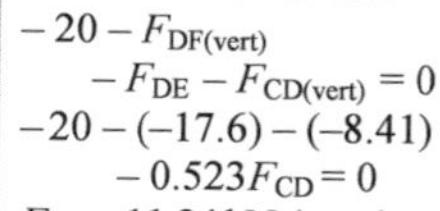

Sum vertical forces:

$-20 - F_{DF(vert)}$

$- F_{DE} - F_{CD(vert)} = 0$

$-20 - (-17.6) - (-8.41)$

$- 0.523F_{CD} = 0$

$F_{CD} = 11.34\,\text{kN}$ (tens.)

Sum horizontal forces:

$F_{DF(horiz)} - F_{CD(horiz)}$

$- F_{BD} = 0$

$-13.2 - 0.848(11.5)$

$- F_{BD} = 0$

$F_{BD} = -22.816\,\text{kN}$(comp.)

5. Joint B

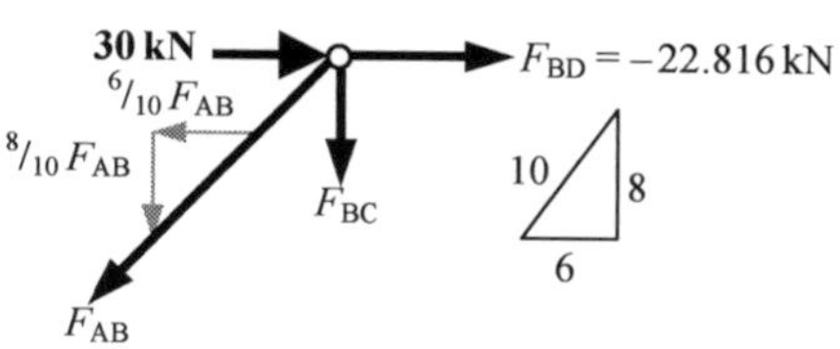

Sum vertical forces:

$-F_{BC} - F_{AB(vert)} = 0$

$-F_{BC} - 0.8F_{AB} = 0$

$F_{BC} = 9.58\,\text{kN}$ (comp.)

Sum horizontal forces:

$30 + F_{BD} - F_{AB(horiz)} = 0$

$30 - 22.816 - 0.6\,F_{AB} = 0$

$F_{AB} = 11.97\,\text{kN}$ (tens.)

6. Joint C

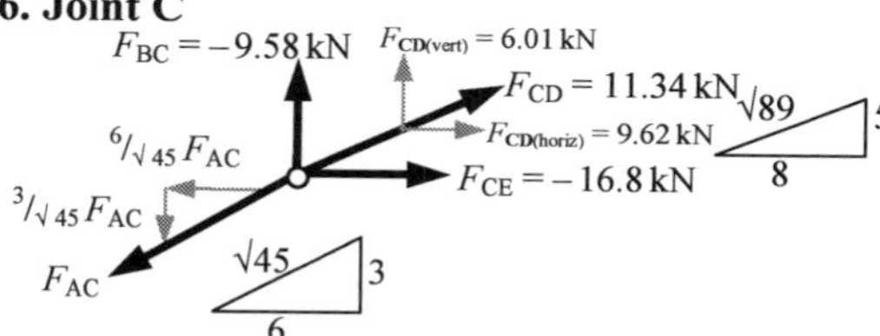

Sum vertical forces:

$F_{CD(vert)} - F_{AC(vert)} + F_{BC} = 0$

$6.01 - 0.4472F_{AC} - 9.58 = 0$

$F_{AC} = -7.99\,\text{kN}$ (comp.)

Sum horizontal forces:

$F_{CD(horiz)} + F_{CE} - F_{AC(horiz)} = 0$

$9.62 - 16.8 - 0.894\,F_{AC} = 0$

$F_{AC} = -8.02\,\text{kN}$(comp.)

(slight rounding error)

Check $V_A = 6\,\text{kN}$ and $H_A = 0\,\text{kN}$

7. Check sum of external forces = 0

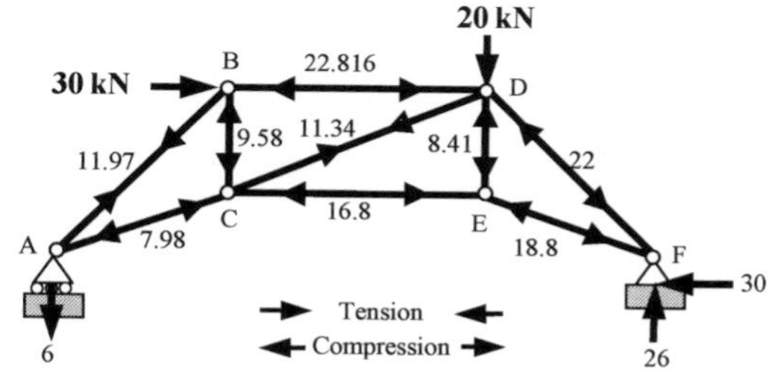

Final frame forces (kN)

Having completed this chapter you should now be able to:

- ***Determine the external and internal forces on pin jointed frames using:***
  - ***A graphical technique***
  - ***The method of sections***
  - ***Joint resolution***
- ***Plot frame force diagrams***

Further Problems 6

1. For the pin jointed frame of Figure Question 1, determine the magnitude and type of forces in the members FG, FH and EG.

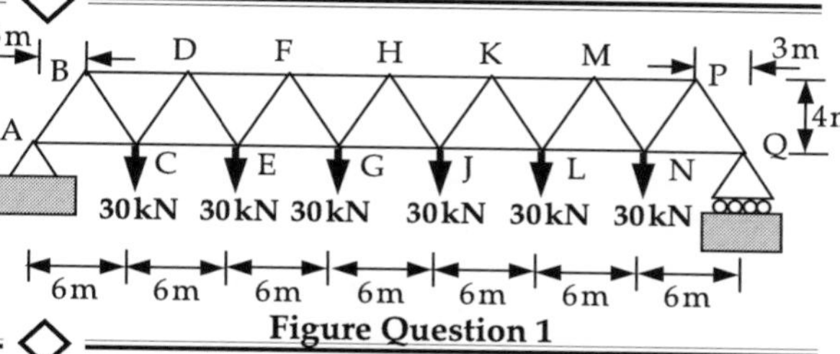

Figure Question 1

2. For the pin jointed frame of Figure Question 2, determine the magnitude and type of forces in all members and the forces at the supports A and E.

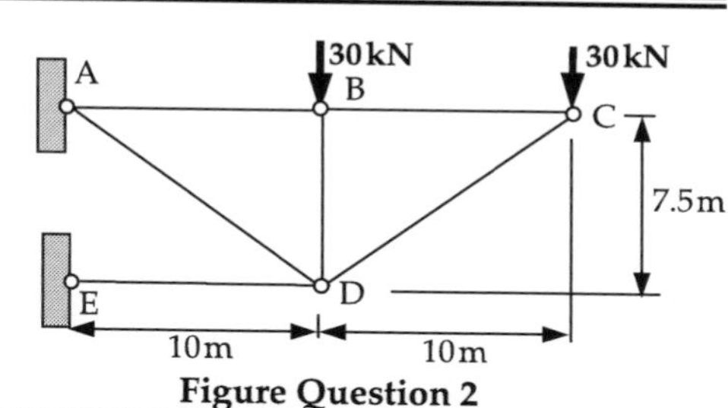

Figure Question 2

3. For the pin jointed frame of Figure Question 3, determine the magnitude and type of forces in all members and the forces at the supports A and B.

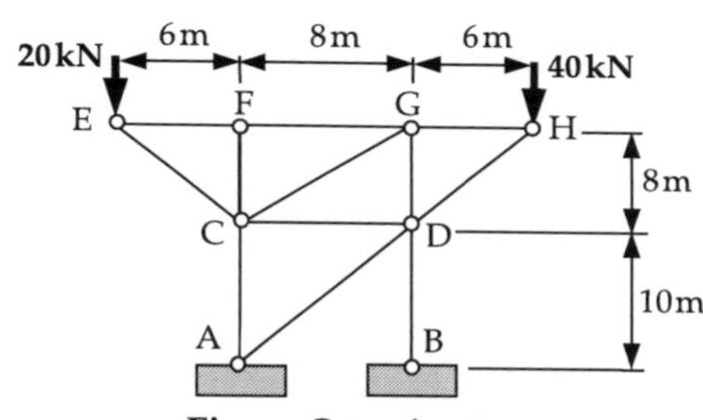

Figure Question 3

4. For the pin jointed frame of Figure Question 4, determine the magnitude and type of forces in all members and the forces at the supports A and B.

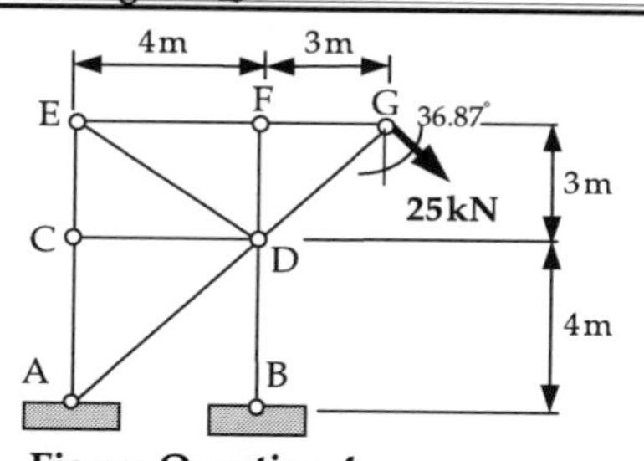

Figure Question 4

5. For the pin jointed frame of Figure Question 5, determine the magnitude and type of forces in all members and the forces at the supports A and D.

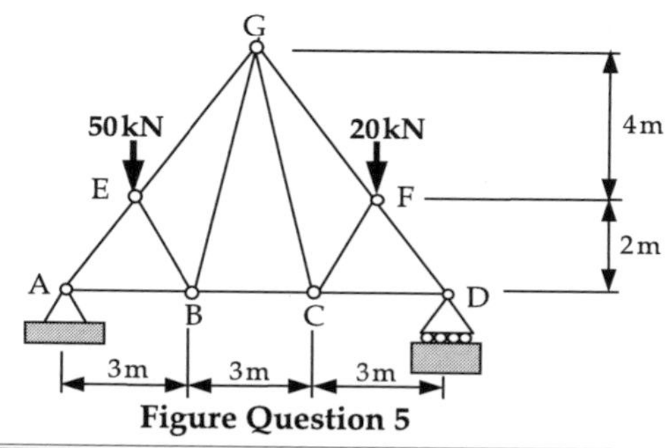

Figure Question 5

Solutions to Further Problems 6

1.

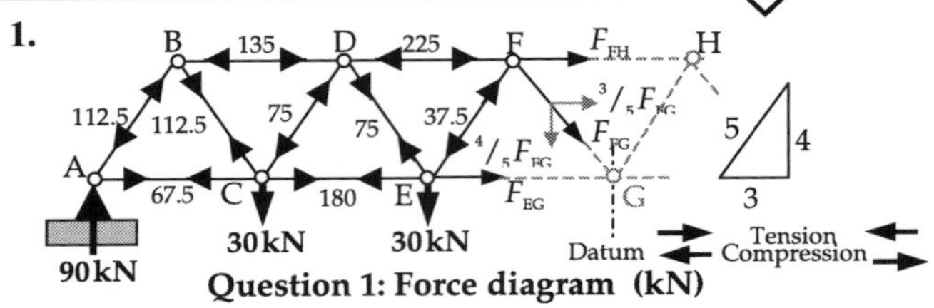

Question 1: Force diagram (kN)

Method of sections:

Sum horizontal forces:

$\sum F_x = F_{FH} + 0.6\,F_{FG} + F_{EG} = 0$

Sum vertical forces:

$\sum F_y = 90 - 30 - 30 - 0.8\,F_{FG} = 0$

$F_{FG} = 187.5\,\text{kN}$

Moments about datum $\sum M_z$

$F_{FH} \times 4 + 90 \times 18 - 30 \times 12 - 30 \times 6 = 0$

$F_{FH} = -270\,\text{kN}$

and $F_{EG} = 247.5\,\text{kN}$

2.

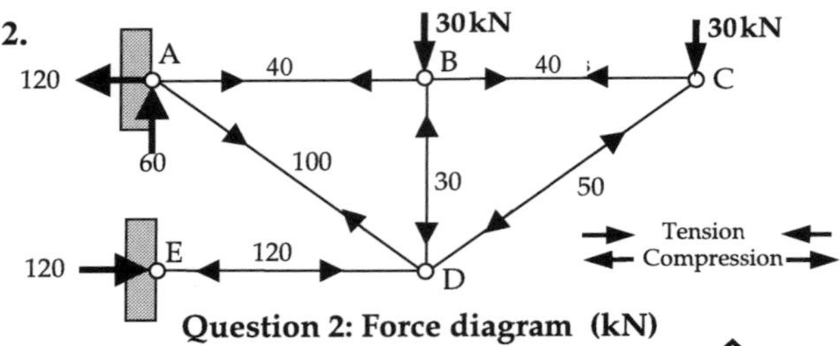

Question 2: Force diagram (kN)

Sum external forces:

Sum of horizontal forces:

$120\,\text{kN} - 120\,\text{kN} = 0$

Sum of vertical forces:

$60\,\text{kN} - 30\,\text{kN} - 30\,\text{kN} = 0$

3.

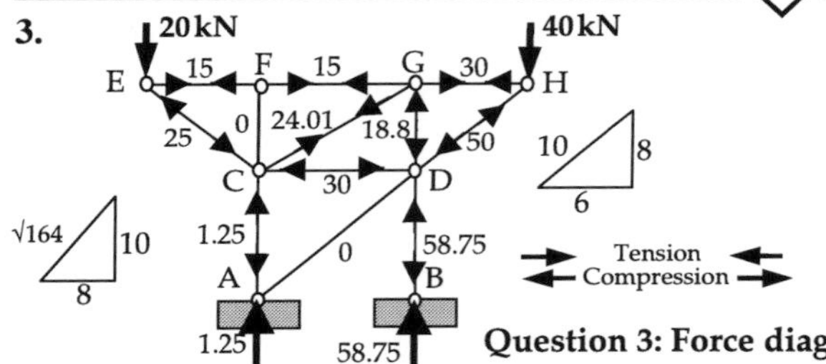

Question 3: Force diagram (kN)

Sum external forces:

Sum of horizontal forces:

$0\,\text{kN} - 0\,\text{kN} = 0$

Sum of vertical forces:

$1.25\,\text{kN} + 58.75\,\text{kN} - 20\,\text{kN} - 40\,\text{kN} = 0$

4.

Question 4: Force diagram (kN)

Sum external forces:

Sum of horizontal forces:

$15\,\text{kN} - 15\,\text{kN} = 0$

Sum of vertical forces:

$61.25\,\text{kN} - 41.25\,\text{kN} - 20\,\text{kN} - 40\,\text{kN} = 0$

5.

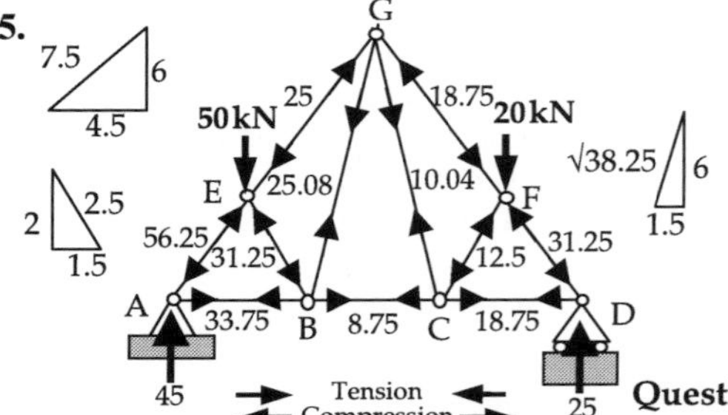

Question 5: Force diagram (kN)

Sum external forces:

Sum of horizontal forces:

$0\,\text{kN} - 0\,\text{kN} = 0$

Sum of vertical forces:

$45\,\text{kN} + 25\,\text{kN} - 50\,\text{kN} - 20\,\text{kN} = 0$

7. Torsion and Torque

In this chapter we will learn about:

- ***Torsion and torque***
- ***Torque diagrams***
- ***The analysis of***
  - ***Circular sections***
  - ***Thin walled closed sections***
  - ***Open sections***

7.1 Introduction

In previous chapters we have studied two-dimensional structures and calculated the axial thrust shear force and bending moments to which they are subjected. We will now extend our studies to consider moments which act along the line of the member. This type of moment is known as a *torque* and its effect will be to twist the member along its length. Consider the examples of Figure 7.1.

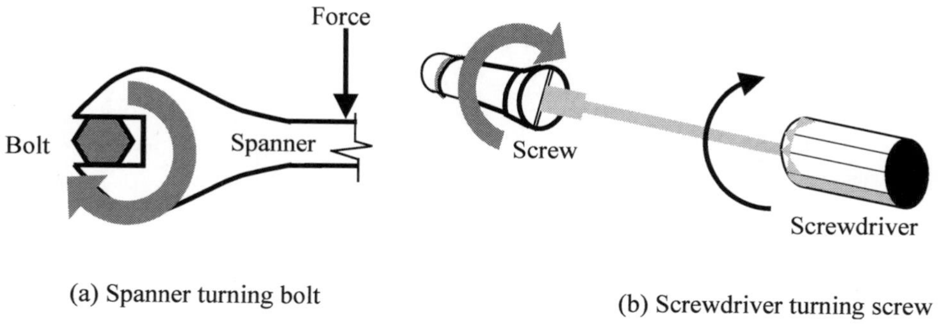

(a) Spanner turning bolt

(b) Screwdriver turning screw

Figure 7.1: Some common examples of torque

In Figure 7.1a, a force applied to the spanner will produce a moment at the head of the bolt. The resultant twisting force along the length of the bolt is a torque. In the example of Figure 7.1b, the twisting force applied to the screwdriver will produce a torque along the length of its shaft which will be transferred as a twisting moment to the head of the screw, causing it to rotate.

In three-dimensional space, using Cartesian co-ordinates, we can describe the forces applied to a structure as shown in Figure 7.2.

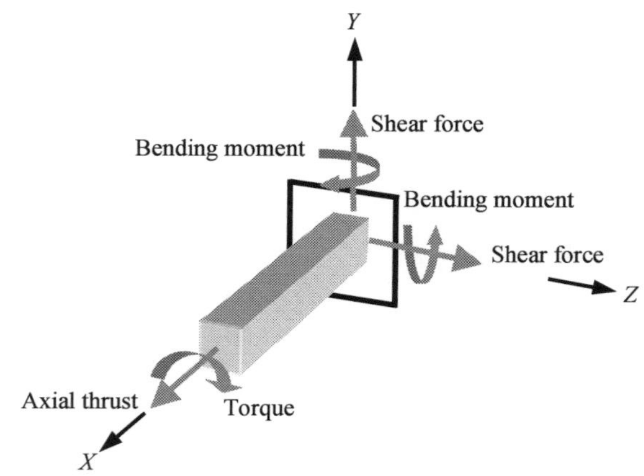

Figure 7.2: Three-dimensional system of moments and forces

Torque can therefore be considered to be a rotational moment acting along the axis of the member. It is induced by a force applied at a distance from the member, just like a bending moment. The effect is to cause twisting of the member, as illustrated in Figure 7.3.

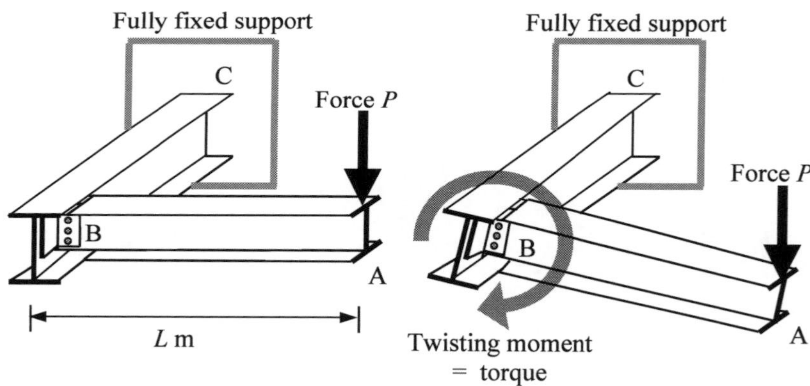

Figure 7.3: Torsion in a cantilever beam member

From Figure 7.3, we note that the beam BC supports another cantilevered beam AB. The beam AB has a load of P kN applied at a distance of L m from B. The load will cause member BC to twist as shown. The amount of twist, or torsion, in member BC will be directly related to the magnitude and direction of the applied torque,which as a moment will be:

$$\text{Torque} = \text{Force} \times \text{Distance}$$
$$T = P \times L = PL \text{ kN m}$$

Torsion stresses can be developed in various ways and it is important that, in the design of structures, torsional resistance is considered. In the example of Figure 7.3, torsional stresses are induced by the applied bending moment in member AB. The connection at B is considered to be fully fixed and, therefore, the moment is transferred in its entirety to member BC as a torque. A similar situation can develop in the connection of beams and columns in the frame of a building. Further examples of torsion in structures, as shown in Figure 7.4, might include the torsion induced in the deck of a bridge or that induced in a framed building due to differential horizontal loading.

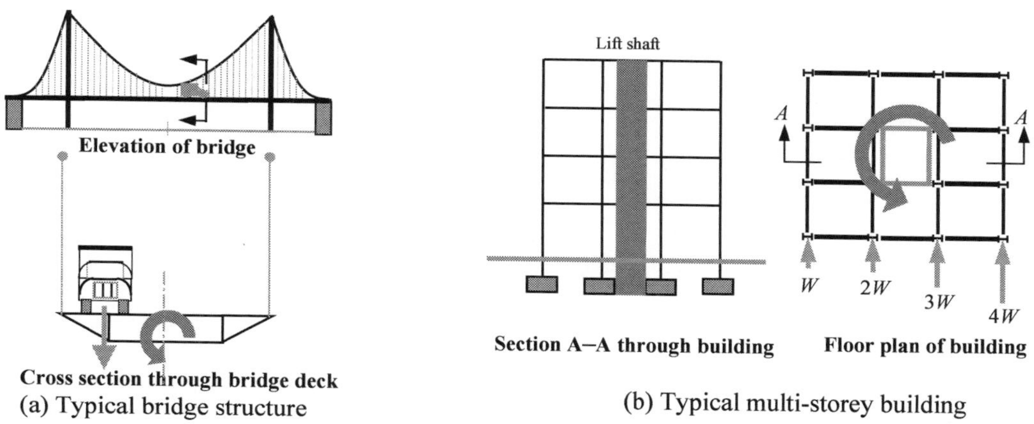

(a) Typical bridge structure

(b) Typical multi-storey building

Figure 7.4: Examples of torsion in structures

In the example of Figure 7.4a, the torque might be induced by differential loading either side of the bridge deck. In the frame of the multi-storey building of Figure 7.4b, differential horizontal load on the structure will induce torsion in the building. It is common in the design of such buildings to provide torsional resistance in the form of a stiff core, often a lift shaft or stair well which is designed to resist the torque induced by the loading. Other common examples of torque are the forces induced in a shaft which drives a pump or motor, such as the propeller shaft on a ship.

The analysis of structures subject to torque and the stresses induced is very complex and is commonly completed using powerful computer programs. At this stage of our studies we will only concern ourselves with an elementary analysis of torsion and torque problems.

7.2 Torsion and Torque Diagrams

We have already studied the plotting of bending moment diagrams for various structural members and frames and, since torque is another form of bending moment, we can plot a similar diagram to show the distribution of torque along a structural member. This diagram is known as a *torque diagram*.

As with bending moment diagrams, the analysis is completed by taking sections through the member and calculating the sum of all torques about a datum through the cut. The analysis is simplified because the torsion in a member will only change at the point at which a torque is applied. Consider the cantilever beam of Figure 7.5.

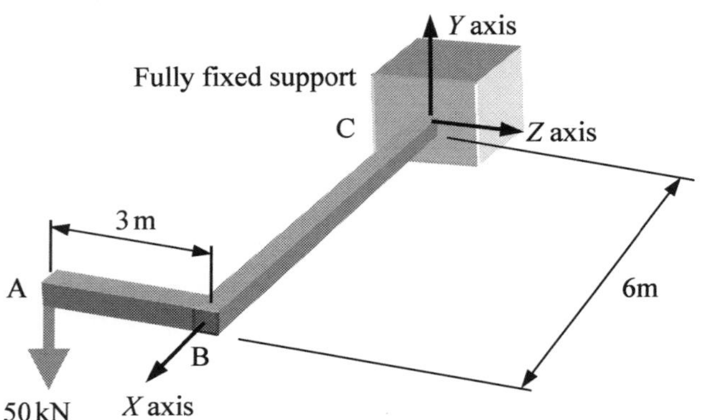

Figure 7.5: Cantilever beam system

We can analyse the cantilever beam system of Figure 7.5 as a simple frame and plot the shear force and bending moment diagrams. We know, from our previous studies, that it is only necessary to calculate the reactions and the principal values hence:

Sum of all horizontal forces: $\Sigma F_x = 0\,\text{kN}$
Sum of all vertical forces: $\Sigma F_y = -50\,\text{kN} + V_C$ (vertical reaction at support)
$\therefore\ V_C = 50\,\text{kN}$
Sum of moments about support: $\Sigma M_z = -50\,\text{kN} \times 6\,\text{m} = -300\,\text{kN m}$

Sum of moments about support: ΣM_x (torque) $= -50\,\text{kN} \times 3\,\text{m} = -150\,\text{kN m}$

We can also calculate the principal values for the members such that:

Shear force:

At A = 0 kN (start point)
At A = –50 kN (point load)
At B = –50 kN
At C = –50 kN
At C = –50 kN + 50 kN = 0 kN (reaction)

Bending moment:

At A = 0 kN m
At B (in member AB) = –50 kN × 3 m = –150 kN m
At B (in member BC) = 0 kN m
At C = –50 kN × 6 m = –300 kN m

Note: the moment at point B varies in each member. This is because the moment is calculated from the load's relative distance in the longitudinal direction of the member. We therefore use the z distance for member AB and the x distance for member BC.

For torque, we note that member AB resists bending moment only. There is no rotational moment along the line of the member. Along BC the torque is equal to the bending moment in AB = 150 kN m. For the structure to be in equilibrium this torque must be resisted at the support. The shear force, bending moment and torque diagrams are shown in Figure 7.6.

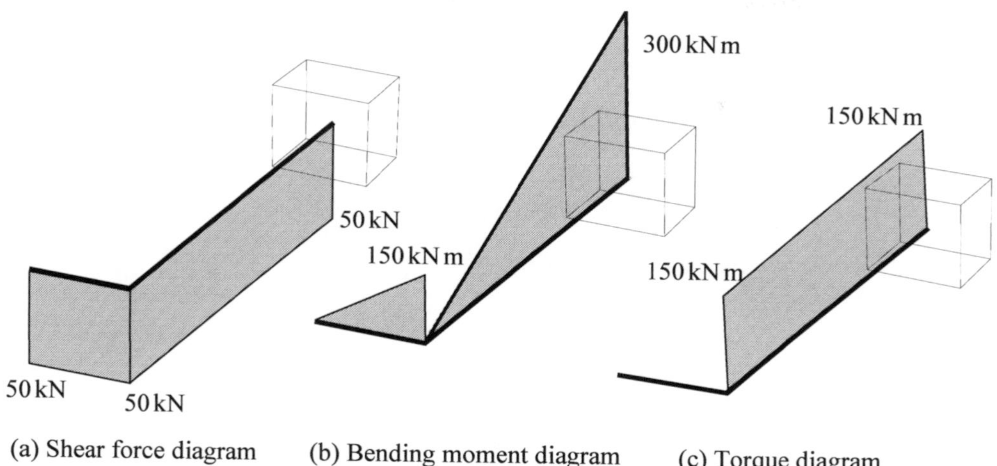

(a) Shear force diagram (b) Bending moment diagram (c) Torque diagram

Figure 7.6: Analysis of a cantilever beam

The plotting of the diagrams in three-dimensional space can be difficult and, as will be demonstrated in the following example, it is often better to plot the diagrams for each member separately in a two-dimensional form.

Example 7.1: Torque diagram for a cantilever beam supporting four other beams

In Example 7.1, shown in Figure 7.7, we have a cantilever beam AD supporting four secondary beams which are fully fixed to it. The secondary beams support loads which will induce torsion in the main beam AD. Calculate the torque induced by each secondary beam and plot the torque diagram for the member AD.

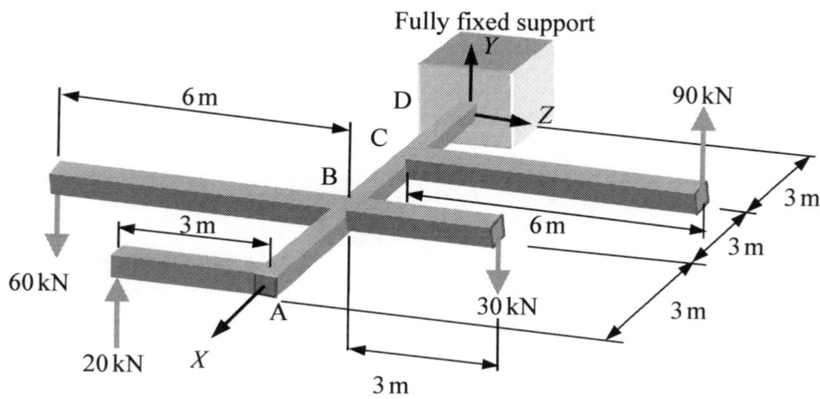

Figure 7.7: Cantilever beam – Example 7.1

We can calculate the torque at each point along the beam as the bending moment for each of the attached members, remembering to adopt the convention of defining clockwise moments as positive, anti-clockwise moments as negative.

Torque:

At A $= 20\,\text{kN} \times 3\,\text{m} = 60\,\text{kN m}$

At B $= 60\,\text{kN m (from A)} - 60\,\text{kN} \times 6\,\text{m} + 30\,\text{kN} \times 3\,\text{m}$
$= 60\,\text{kN m} - 360\,\text{kN m} + 90\,\text{kN m} = -210\,\text{kN m}$

At C $= -210\,\text{kN m (from B)} - 90\,\text{kN} \times 6\,\text{m} = -750\,\text{kN m}$

For equilibrium of the system, the torque must be resisted at the support. The torque diagram for member AD is shown in Figure 7.8.

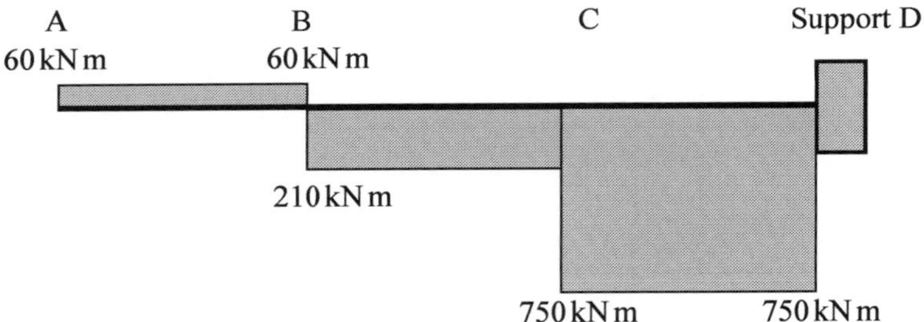

Figure 7.8: Torque diagram – Example 7.1

We note that the torque diagram is similar in every respect to the bending moment diagrams plotted earlier. Using the beam as a zero datum, positive (clockwise) torques are plotted above the line and negative (anticlockwise) are plotted below the line. For equilibrium, the torque at the support must be equal and opposite to the applied torque. We can therefore plot a torque diagram using the same methodology developed for bending moment diagrams.

Example 7.2: Torque diagram for a cantilever beam supporting three other beams

For the cantilever beam of Example 7.2, plot the torque diagram for member AD.

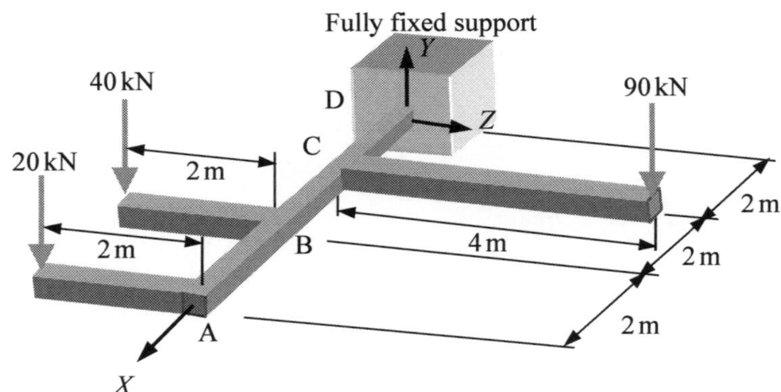

Figure 7.9: Cantilever beam – Example 7.2

We again calculate the torque at each point along the beam as the bending moment for each of the attached members, remembering the convention of clockwise moments positive.

Torque: At A $= -20\,\text{kN} \times 2\,\text{m} = -40\,\text{kN m}$

At B $= -40\,\text{kN m}$ (from A) $- 40\,\text{kN} \times 2\,\text{m} = -40\,\text{kN m} - 80\,\text{kN m} = -140\,\text{kN m}$

At C $= -140\,\text{kN m}$ (from B) $+ 90\,\text{kN} \times 4\,\text{m} = 220\,\text{kN m}$

The torque diagram for member AD is shown in Figure 7.10.

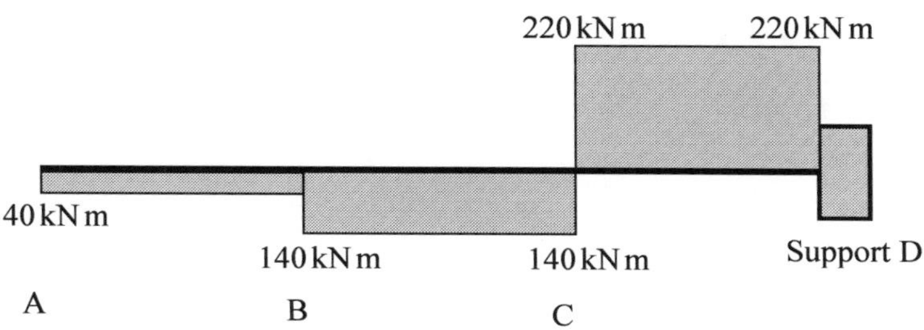

Figure 7.10: Torque diagram – Example 7.2

◇

The plotting of torque diagrams is therefore a fairly straightforward process, however the analysis of the torsion in members can be a little more complex. We will now proceed to look at the analysis of such members in more detail.

7.3 Simple Torsion Problems

In order to enable us to solve simple torsion problems we must first introduce a relationship similar to the Engineers' bending equation presented in Chapter 2 which can be used to calculate factors such as the shear stress distribution across circular members. This relationship for circular cross-sections may be defined as:

$$\frac{T}{J} = \frac{\tau}{r} = \frac{G\theta}{L} \qquad [7.1]$$

This again is a very important equation, which will prove to be useful in the analysis of torsion problems. You should therefore endeavour to remember it. It is extremely important when using this equation that a consistent set of units be adopted. The most common units for shear stress are N/mm^2 and, it may therefore be necessary to convert values to ensure consistency.

We will now consider each part of the equation in turn, noting its relationship to the Engineer's bending equation which was given as:

$$\frac{M}{I} = \frac{\sigma}{y} = \frac{E}{R}$$

T = Torque (N mm)

This is the bending moment along the axis of the member, normally taken as the maximum torque applied, since this will induce the maximum extreme fibre shear stress.

J = Polar Second Moment of Area (for circular sections) or Torsion Constant for thin walled open and closed sections (mm^4)

This relates to a geometric property for the section and is determined by its shape and dimensions. It is a property which is very similar to the Second Moment of Area discussed in Chapter 2. Some typical values for the Polar Second Moment of Area and Torsion Constant are given in the table of Figure 7.11.

Section	Polar Second Moment of Area (J)
D	$\dfrac{\pi D^4}{32}$
d, D	$\dfrac{\pi (D^4 - d^4)}{32}$
	Torsion Constant (J)
t_f, t_w, b, d, D, B	$\dfrac{2B^2D^2t_wt_f}{Bt_w + Dt_f}$

Figure 7.11: Polar Second Moment of Area and Torsion Constant (J) for some common sections

τ = Shear stress (N/mm^2)

This is the stress caused by the shear forces and is normally calculated at the outermost fibre of the section at some distance from its centroid. Consider the rubber block shown in Figure 7.12.

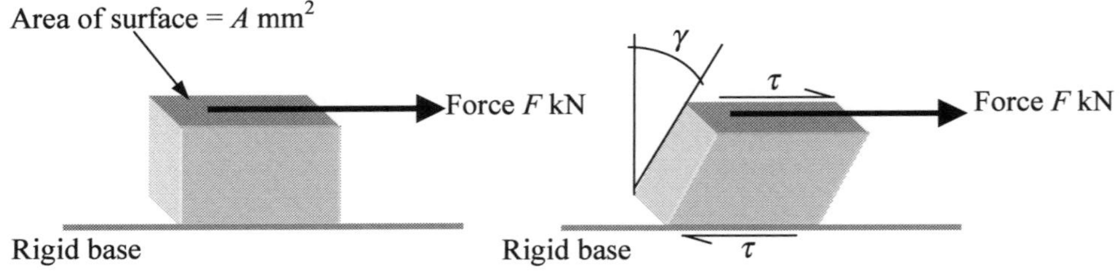

Figure 7.12: Force applied to rubber block

The rubber block is rigidly fixed to the base and a force of F kN is applied over the whole of the top surface of area A mm^2. The force will cause the block to distort and will induce a shearing stress in the block at the top and bottom surfaces which, for equilibrium, must be equal and opposite. These shear stresses will be equal to:

$$\tau = \frac{F}{A} \qquad [7.2]$$

Also note that the angle to which the block is distorted by the force is the shear strain γ measured in radians.

When a torque is applied to a thin walled closed section, such as the rectangular hollow box shown in Figure 7.11, it will cause shear stresses at the surface. The shear will 'flow' through the section and the amount of shear force which can be carried by any part of a member will depend on its thickness. The shear stress can also be related to this shear flow q of the section by the equation:

$$\tau = \frac{q}{t} \qquad [7.3]$$

where q = shear flow and t = thickness of the part.

The shear flow can also be calculated from the relationship:

$$q = \frac{T}{2A} \qquad [7.4]$$

where T = applied torque and A = area of section.

Hence we can derive a further relationship for shear stress in thin walled closed sections such that:

$$q = \tau\ t = \frac{T}{2A} \qquad [7.5]$$

For thin walled open sections, such as Universal Beams and Columns, the shear stress can be found from the relationship:

$$\tau = \frac{Tt_{max}}{J} \qquad [7.6]$$

It should be noted that torque is resisted by shear stresses alone, the distribution of which will be the same for any part of a member as long as both the member properties and torque remain constant.

r = radius (mm)

This is the distance from the centroid of the circular section to the point at which we want to calculate the shear stress. Maximum shear stress will normally occur in the extreme fibres of the member so that r is normally the distance from the centroid to the furthest point on the section. However, we can calculate the stress at any point through the section by varying the distance r.

G = Shear modulus OR Rigidity modulus (N/mm²)

This is a property of the material used to form the member and relates its shear stress to its shear strain characteristics. With reference to Figure 7.12, we will find that the shear modulus is:

$$G = \frac{\text{Shear stress}}{\text{Shear strain}} = \frac{\tau}{\gamma} \qquad [7.7]$$

As with the elastic modulus, this property has been determined by experimentation for various commonly used materials.

θ = Angle of twist (radians)

As discussed previously, when a torque is applied to a member it will cause it to rotate and twist. The angle of twist is calculated as θ and is specified in radians. This angle is commonly specified per metre length of the member and in such cases would be defined as the rate of twist as shown in Figure 7.13.

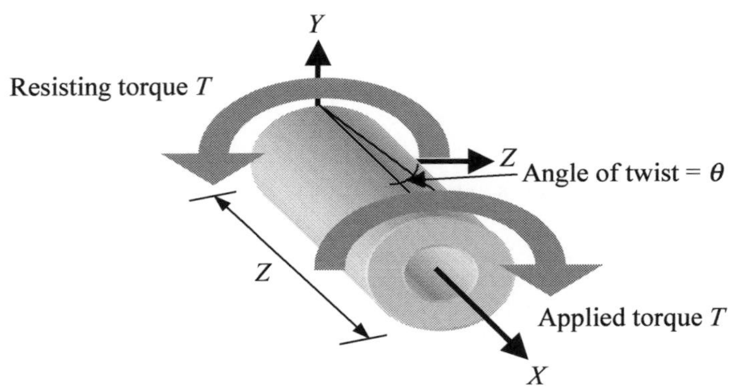

Figure 7.13: Torque applied to a hollow shaft

From Figure 7.13 we note that for equilibrium of the system the applied torque must be resisted by an equal and opposite resisting torque. The applied torque will cause the shaft to twist and the angle of twist is given as θ. The rate of twist for the whole shaft will be given by:

$$\frac{\mathrm{d}\theta}{\mathrm{d}z} = \frac{\theta}{Z} \qquad [7.8]$$

and from the torsion equation we find that, in the limit:

$$\frac{\theta}{L} = \frac{T}{GJ} \qquad [7.9]$$

L = Length of member (mm)

This is the distance of the whole or part of the member to be analysed.

Having discussed each of the elements of the torsion equation [7.1], we will now apply it to various examples. However, before proceeding further, a relationship exists between power and torque which can prove useful and is given for information in equation [7.10] below:

$$P = \frac{2\pi nT}{60} = \frac{\pi nT}{30} \qquad [7.10]$$

where P = power (in Watts), T = torque (N m) and n = number of revolutions per minute.

Example 7.3: Torsion of a solid circular section shaft

The circular solid shaft of Figure 7.14, is 1.5 m long and has a diameter of 100 mm. It transmits a torque of 3500 N m and has a shear modulus (G) = 80 kN/mm^2. Calculate:

(a) the maximum shear stress in the shaft;
(b) the total angle of twist.

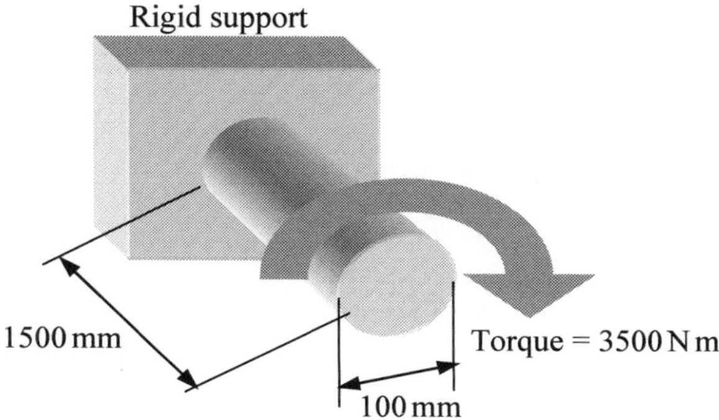

Figure 7.14: Circular solid shaft – Example 7.3

Solution based on the torsion equation: $\dfrac{T}{J}=\dfrac{\tau}{r}=\dfrac{G\theta}{L}$

(a) Shear stress given by $\dfrac{T}{J}=\dfrac{\tau}{r}$ from which $\tau=\dfrac{Tr}{J}$

For a circular shaft, $J=\dfrac{\pi D^4}{32}=\dfrac{\pi(100)^4}{32}=9.82\times10^6\,\text{mm}^4$ and radius $r = 50$ mm

and shear stress $\tau=\dfrac{3500\times10^3\times50}{9.82\times10^6}=17.821\ \text{N/mm}^2$

(b) Angle of twist is given by $\dfrac{T}{J}=\dfrac{G\theta}{L}$ from which:

$$\theta=\frac{TL}{GJ}=\frac{3500\times10^3\times1500}{80\times10^3\times9.82\times10^6}=6.7\times10^{-3}\,\text{radians}$$

Example 7.4: Torsion of a part hollow and part solid circular section shaft

The circular shaft of Figure 7.15 is 3 m long and has an outside diameter of 100 mm. It transmits a torque of 5600 N m and has a shear modulus (G) = 80 kN/mm^2. For 1 m of its length the shaft is hollow with an internal diameter of 50 mm whilst for the other 2 m the shaft is solid. Calculate:

(a) the maximum shear stress in the solid part of the shaft;
(b) the maximum shear stress in the hollow part of the shaft;
(c) the total angle of twist in degrees.

Sketch the shear stress distribution across the solid and hollow parts.

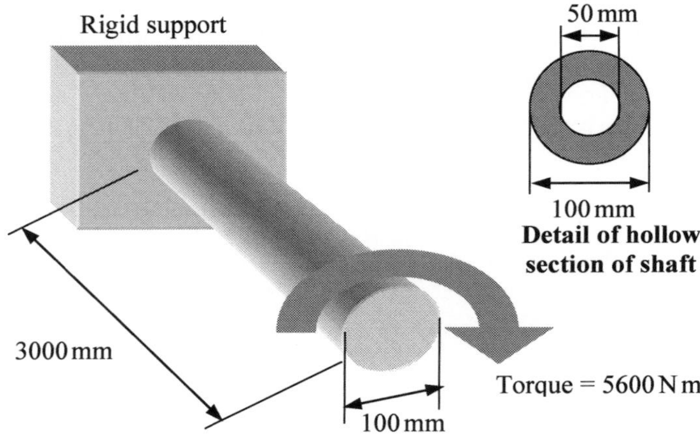

Figure 7.15: Circular shaft – Example 7.4

Solution based on the torsion equation: $\frac{T}{J}=\frac{\tau}{r}=\frac{G\theta}{L}$

(a) For the solid section:

Shear stress is given by $\frac{T}{J}=\frac{\tau}{r}$ from which $\tau=\frac{Tr}{J}$

For a circular shaft, $J=\frac{\pi D^4}{32}=\frac{\pi(100)^4}{32}=9.82\times10^6\,\text{mm}^4$ and radius r = 50 mm

and shear stress $\tau=\frac{5600\times10^3\times50}{9.82\times10^6}=28.51\ \text{N/mm}^2$

(b) For the hollow section:

Shear stress is given by $\tau=\frac{Tr}{J}$

For a circular shaft, $J=\frac{\pi(D^4-d^4)}{32}=\frac{\pi(100^4-50^4)}{32}=920.39\times10^3\,\text{mm}^4$

For maximum stress, radius r_{external} = 50 mm

and shear stress $\tau=\frac{5600\times10^3\times50}{920.39\times10^3}=304.22\ \text{N/mm}^2$

To calculate internal stress, radius r_{internal} = 25 mm

and shear stress $\tau=\frac{5600\times10^3\times25}{920.39\times10^3}=152.11\ \text{N/mm}^2$

(c) Angle of twist is given by $\frac{T}{J}=\frac{G\theta}{L}$

In order to calculate the total angle of twist we will first calculate the angle of twist for the solid section followed by that for the hollow section.

For solid section $\theta=\frac{TL}{GJ}=\frac{5600\times10^3\times2000}{80\times10^3\times9.82\times10^6}=0.0143$ radians

For hollow section $\theta=\frac{TL}{GJ}=\frac{5600\times10^3\times1000}{80\times10^3\times920.39\times10^3}=0.0761$ radians

Total angle of twist = 0.0143 + 0.0761 = 0.0904 radians.

We note that the total angle of twist is in radians and we must convert this to degrees to complete this example. We know that a circle contains 2π radians or 360°. A relationship therefore exists to convert from radians to degrees based on:

$$1 \text{ radian} = \frac{360}{2\pi} = 57.296°$$

The angle of twist, θ, for Example 7.4 = $0.0904 \times 57.296° = 5.177°$

The stress distributions for the hollow and solid parts are shown in Figure 7.16.

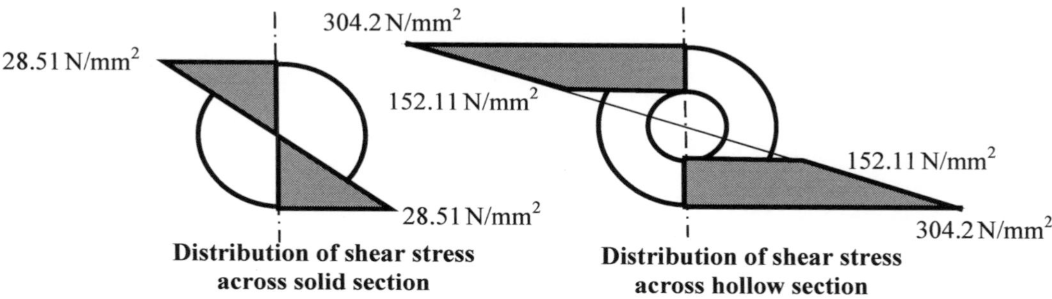

Figure 7.16: Stress distributions across solid and hollow sections of Example 7.4

From Figure 7.16, we note that in the solid part the shear stress is distributed across the whole section whilst for the hollow section the shear stress is distributed across the walls of the shaft.

◇

Example 7.5: Torsion of a hollow circular section shaft

A circular hollow shaft has an external diameter to internal diameter ratio of 1:0.6 and is to be used to drive a hydraulic pump. The length of the shaft is 12 m and it is required to transmit a torque of 150×10^6 N mm. The shaft will be formed in steel with a shear modulus (G) = 80 kN/mm². If the maximum allowable shear stress is to be limited to 60 N/mm², and the maximum angle of twist must not exceed 1°, calculate the minimum internal and external diameters for the shaft.

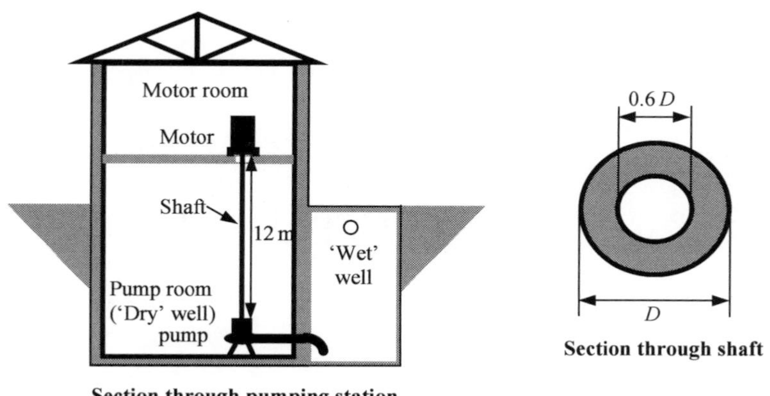

Figure 7.17: Details of pumping station and shaft – Example 7.5

Solution based on the torsion equation: $\frac{T}{J}=\frac{\tau}{r}=\frac{G\theta}{L}$

(a) Shear stress not to exceed 60 N/mm^2:

$\frac{T}{J}=\frac{\tau}{r}$ from which $r=\frac{\tau J}{T}$

We know that: τ (maximum) = 60 N/mm^2

and T = torque = 150 × 10^6 N mm

For a circular hollow shaft, $J=\frac{\pi(D^4-d^4)}{32}$ and for maximum shear stress $r=\frac{D}{2}$ mm

In this case $d = 0.6D$ and

$$J=\frac{\pi(D^4-(0.6D)^4)}{32}=\frac{\pi(D^4-0.6^4D^4)}{32}=0.0855D^4$$

We can now substitute all our values into the equation for r such that:

$$\frac{D}{2}=\frac{60\times0.0855D^4}{150\times10^6}$$

Solving for D we find D (external diameter) = 244.5 mm

and $0.6D$ (internal diameter) = 146.7 mm

(b) Angle of twist must not exceed 1°:

First we need to convert 1° to radians such that: θ = (1/57.296) = 0.0175 radians

Angle of twist is given by $\frac{T}{J}=\frac{G\theta}{L}$

Substituting our previously calculated values we find:

$$\frac{150\times10^6}{0.0855D^4}=\frac{80\times10^3\times0.0175}{12\ 000}$$

Solving for D we find D (external diameter) = 350.2 mm

and $0.6D$ (internal diameter) = 210.1 mm

Therefore, in order to satisfy both criteria, the shaft must have an external diameter of 350.2 mm and an internal diameter of 210.1 mm (note that shear stress = 20.42 N/mm^2).

◇

Example 7.6: Torsion of a thin walled closed rectangular section

The square hollow box section of Figure 7.18 is subjected to a clockwise torsional force of 75 kN applied at a distance of 0.9 m above its centroid. Assume the shear modulus (G) for the material is 80 kN/mm^2 and the torsion constant (J) is given by:

$$J=\frac{2B^2D^2t_wt_f}{Bt_w+Dt_f}$$

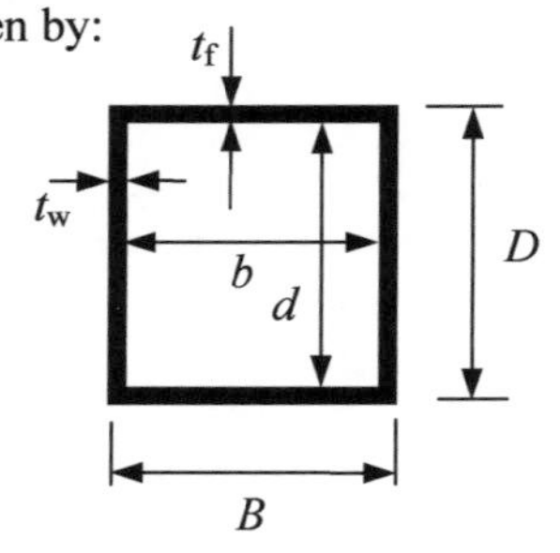

t_f t_w b d D B

Determine:

(a) the maximum shear stress developed in the steel box section due to the applied torque;
(b) the angle of twist per unit length for the section.

Solution

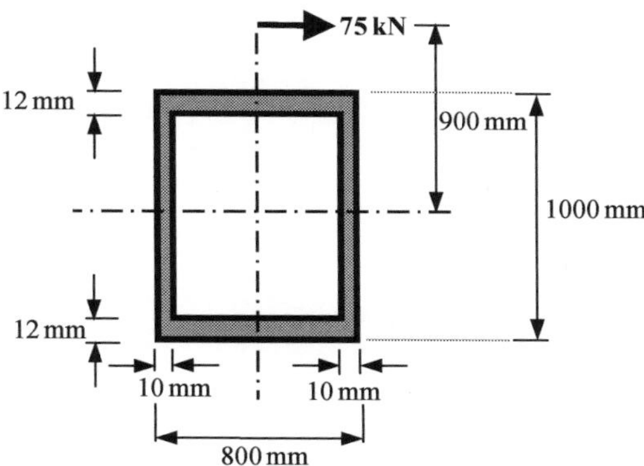

Figure 7.18: Details of hollow box section – Example 7.6

Torque T = Force × Distance = 75 kN × 0.9 m = 67.5 kN m

Enclosed area of section along mid-line of the wall $= (1000-(2\times 12/2))\times(800-(2\times 10/2))$
$= 780\ 520\,\text{mm}^2$

Shear flow is given by equation [7.4] as $q = \dfrac{T}{2A} = \dfrac{67.5\times 10^6}{2\times 780\ 520} = 43.24\ \text{N/mm}$

Now, maximum shear stress will occur in the thinner walls of the section, so from equation [7.5]:

$$q = \tau\ t$$

(a) Maximum shear stress $= \tau = \dfrac{q}{t} = \dfrac{43.24}{10} = 4.324\ \text{N/mm}^2$

The torsion constant is given by $J = \dfrac{2B^2D^2t_w t_f}{Bt_w + Dt_f}$

and for the box section of Example 7.6 $\quad J = \dfrac{2(800^2)(1000^2)\times 10\times 12}{800(10)+1000(12)} = 7.68\times 10^9\,\text{mm}^4$

(b) Angle of twist is given by $\dfrac{T}{J} = \dfrac{G\theta}{L}$

In this case we are required to calculate the rate of twist, i.e. θ/m, therefore the length of box section required for this calculation = 1 m = 1000 mm

$$\theta = \frac{TL}{GJ} = \frac{67.5\times 10^6\times 1000}{80\times 10^3\times 7.68\times 10^9} = 1.098\times 10^{-4}\ \text{radians/metre}$$

Example 7.7: Torsion of a circular hollow section shaft

A circular hollow shaft has an external diameter of 100 mm and is 10 mm thick. It is to be used to drive a hydraulic pump similar to the arrangement shown in Figure 7.17. The maximum power developed by the motor will be 750 kW at 500 revolutions per minute. The shaft will be formed in steel with a shear modulus (G) = 80 kN/mm^2. Determine the shear stress in the shaft and the twist per metre length.

Solution:

From equation [7.9]: $P = \dfrac{\pi n T}{30}$

$$750 \times 10^3 = \frac{\pi \times 500 \times T}{30}$$ gives $T = 14\ 324$ Nm

Solution based on torsion equation: $\dfrac{T}{J} = \dfrac{\tau}{r} = \dfrac{G\theta}{L}$

Shear stress:

$\dfrac{T}{J} = \dfrac{\tau}{r}$ from which $\tau = \dfrac{T\,r}{J}$

For a circular hollow shaft $J = \dfrac{\pi(D^4 - d^4)}{32}$ and for maximum shear stress $r = \dfrac{D}{2}$ mm

$$J = \frac{\pi\left(100 - \left(100^4 - (100 - 2(10))^4\right)\right)}{32} = \frac{\pi(100^4 - 80^4)}{32} = 5.796 \times 10^6\,\text{mm}^4$$

$r = 100/2 = 50$ mm

$$\tau = \frac{Tr}{J} = \frac{14\ 324 \times 10^3 \times 50}{5.796 \times 10^6} = 123.57\ \text{N/mm}^2$$

Angle of twist per metre:

Angle of twist is given by $\dfrac{T}{J} = \dfrac{G\theta}{L}$ and $\theta = \dfrac{TL}{GJ}$

Substituting our previously calculated values, we find:

$$\theta = \frac{14\ 324 \times 10^3 \times 1000}{80 \times 10^3 \times 5.796 \times 10^6} = 0.031\ \text{radians/metre}$$

Example 7.8: Torsion of a solid circular section shaft with varying diameter

The solid circular shaft of Figure 7.19 is 3 m long and has an outside diameter of 100 mm for 1.5 m and an outside diameter of 50 mm for 1.5 m. It transmits a torque of 90 kN m applied at its free end and has a shear modulus (G) = 80 kN/mm^2. Calculate:

(a) the maximum shear stress in the each part of the shaft;
(b) the total angle of twist.

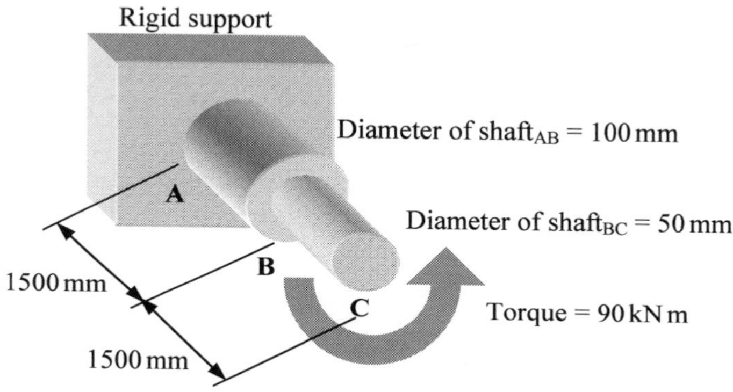

Figure 7.19: Circular shaft – Example 7.8

Solution based on torsion equation: $\frac{T}{J} = \frac{\tau}{r} = \frac{G\theta}{L}$

(a) Shear stress:

Shear stress is given by $\frac{T}{J} = \frac{\tau}{r}$ from which $\tau = \frac{Tr}{J}$

For part AB:

For a circular shaft, $J = \frac{\pi D^4}{32} = \frac{\pi (100)^4}{32} = 9.82 \times 10^6 \text{ mm}^4$ and radius $r = 50$ mm

and shear stress $\tau_{AB} = \frac{-90 \times 10^6 \times 50}{9.82 \times 10^6} = -0.46 \text{ N/mm}^2$

For part BC:

For a circular shaft, $J = \frac{\pi D^4}{32} = \frac{\pi (50)^4}{32} = 613.6 \times 10^3 \text{ mm}^4$ and radius $r = 25$ mm

and shear stress $\tau_{BC} = \frac{-90 \times 10^6 \times 25}{613.6 \times 10^3} = -3.67 \text{ N/mm}^2$

(b) Angle of twist is given by $\frac{T}{J} = \frac{G\theta}{L}$

In order to calculate the total angle of twist we will first calculate the angle of twist for the section AB followed by that for the section BC.

For section AB, $\theta_{AB} = \frac{TL}{GJ} = \frac{-90\times10^6 \times 1500}{80\times10^3 \times 9.82\times10^6} = -171.84\times10^{-6}$ radians

For section BC, $\theta_{BC} = \frac{TL}{GJ} = \frac{-90\times10^6 \times 1500}{80\times10^3 \times 613.6\times10^3} = -2.75\times10^{-3}$ radians

Total angle of twist for AC $= -171.84 \times 10^{-6} - 2.75 \times 10^{-3} = -2.922 \times 10^{-3}$ radians.

◇

Example 7.9: Torsion of a thin walled closed rectangular section shaft

The torsional resistance of the framed building of Figure 7.20 is provided by a concrete lift shaft located at its centre. The total torque to which the shaft will be subjected is 6500 kN m, and in this example we will consider it to be applied at roof level. The shaft is fully fixed at its foundation. All openings in the shaft can be ignored and the shaft treated as a rectangular hollow box section. Assume the shear modulus (G) for the material is 40 kN/mm^2 and torsion constant (J) is given by:

$$J = \frac{2B^2D^2t_w t_f}{Bt_w + Dt_f}$$

Determine the thickness of the concrete walls required, assuming $t_w = t_f$ and:
(a) the maximum allowable shear stress must not exceed 1.6 N/mm^2;
(b) the maximum angle of twist must not exceed 0.21°.

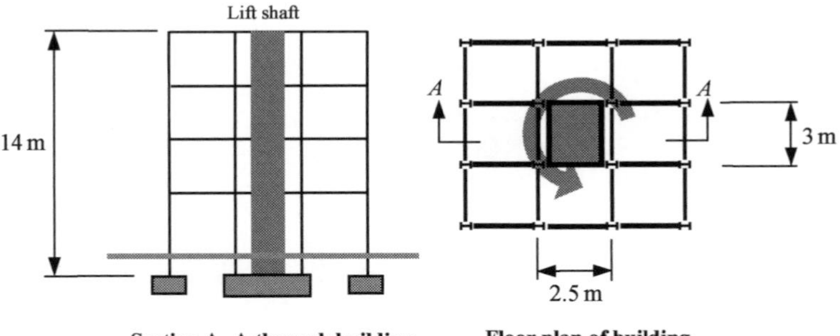

Figure 7.20: Details of building and shaft – Example 7.9

Calculate thickness of shaft walls = t.

Torque = $T = -6500$ kN m

The torsion constant is given by $J = \frac{2B^2D^2t_w t_f}{Bt_w + Dt_f}$

and for the box section of Example 7.6, $J = \dfrac{2(2500^2)(3000^2) \times t \times t}{2500(t) + 3000(t)} = 20.455 \times 10^9 t \text{ mm}^4$

(a) Shear stress must not exceed 1.6 N/mm^2

Shear stress is given by $q = \dfrac{T}{2A} = \dfrac{-6500 \times 10^6}{2 \times 2500 \times 3000} = 433.33$ N/mm

and

$$\tau = \frac{q}{t} = \frac{-433.33}{t}$$

which gives $t = 270.83$ mm (negative sign can be ignored)

(b) Angle of twist must not exceed 0.21° (Note angle of twist = 0.21/57.296 radians)

Angle of twist is given by $\dfrac{T}{J} = \dfrac{G\theta}{L}$

$$\frac{-6500 \times 10^6}{20.455 \times 10^9\ T} = \frac{80 \times 10^3 \times \left(0.21/57.296\right)}{14\ 000}$$ which gives $t = 303$ mm

Therefore the concrete shaft thickness must be a minimum of 303 mm to meet both conditions.

Example 7.10: Torsion of a open section 'I' beam

The Universal Beam of Figure 7.21 is 3 m long and has the properties listed in the table of Figure 7.22. It transmits an anticlockwise torque of 500 N m applied at its free end. The shear modulus (G) for steel may be taken as 80 kN/mm^2. Determine the maximum shear stress in the beam and the total angle of twist in degrees.

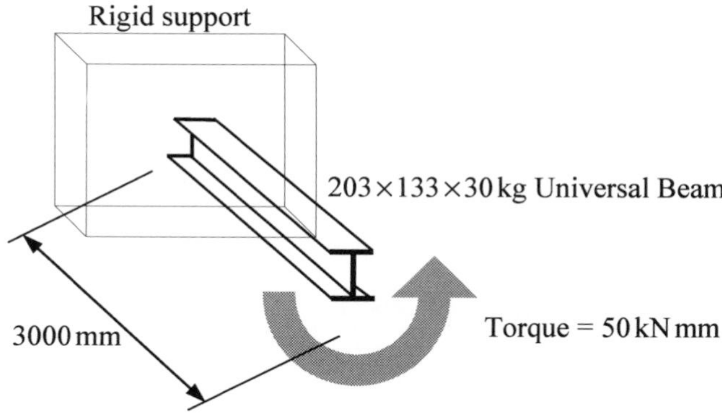

Figure 7.21: Beam – Example 7.10

Serial size	Weight per metre (kg)	Depth (mm)	Thickness (mm)		Area (cm^2)	Torsion constant (J) (cm^4)
			Web	Flange		
203×133	30	206.8	6.3	9.6	38	10.2

Figure 7.22: Section properties of beam – Example 7.10

In this case we have a thin walled open section, and in order to calculate the shear stress we must use equation [7.6].

(a) Shear stress:

Shear stress is given by $\tau = \dfrac{Tt_{\max}}{j}$

and shear stress $\tau = \dfrac{-500\times10^{3}\times9.6}{10.2\times10^{4}} = -47.06\ \text{N/mm}^{2}$

(b) Angle of twist is given by $\dfrac{T}{J} = \dfrac{G\theta}{L}$

$$\theta = \frac{TL}{GJ} = \frac{-500\times10^{3}\times3000}{80\times10^{3}\times10.2\times10^{4}}\times57.296 = -10.53°$$

◇

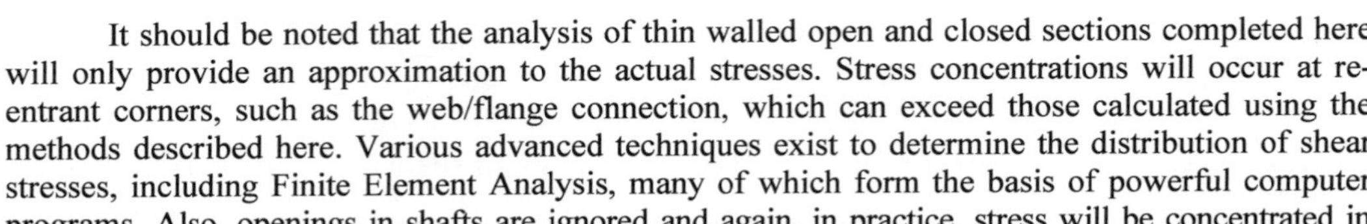

It should be noted that the analysis of thin walled open and closed sections completed here will only provide an approximation to the actual stresses. Stress concentrations will occur at re-entrant corners, such as the web/flange connection, which can exceed those calculated using the methods described here. Various advanced techniques exist to determine the distribution of shear stresses, including Finite Element Analysis, many of which form the basis of powerful computer programs. Also, openings in shafts are ignored and again, in practice, stress will be concentrated in these areas.

It must be remembered that the application of the torsion equations detailed in this chapter are based on a number of assumptions and simplifications. However, despite these and other simplifications in the methods proposed, it does provide us with a preliminary analysis of members subject to torsional forces.

Having completed this chapter you should now be able to:

- ***Draw simple torque diagrams***
- ***Calculate the effect of a torque applied to simple structures including***
  - ***circular shafts***
  - ***thin walled closed sections***
  - ***thin walled open sections***

Further Problems 7

1. The cantilever beam AD shown in Figure Question 1 supports four secondary beams which are fully fixed to it. The secondary beams support loads that will induce torsion in the main beam AD. Calculate the torque induced by each secondary beam and plot the torque diagram for the member AD.

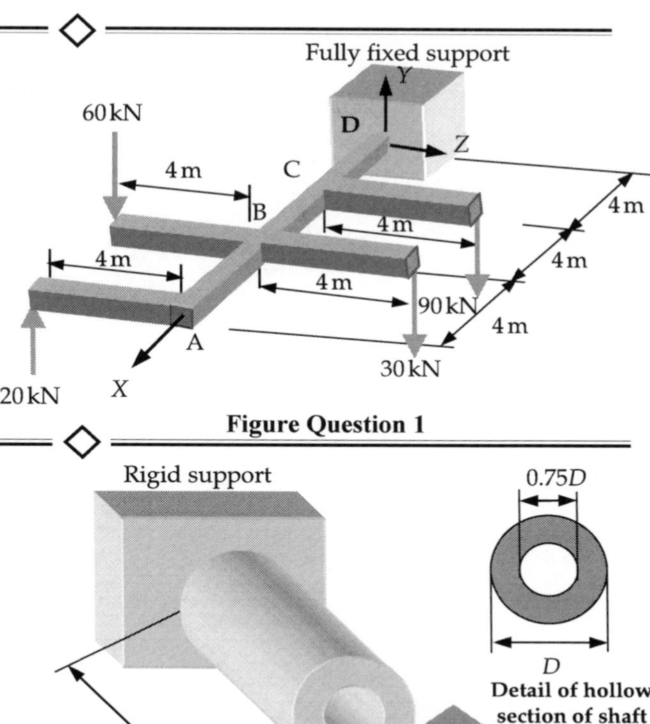

Figure Question 1

2. The hollow circular shaft of Figure Question 2 is 3 m long and has an external diameter to internal diameter ratio of 1 : 0.75. It transmits a torque of 6600 N m and has a shear modulus (G) = 80 kN/mm^2. The shear stress in the shaft must not exceed 50 N/mm^2 and the total angle of twist must not be greater than 0.085 radians.

   Determine the minimum external and internal diameters for the shaft.

Figure Question 2

3. The square hollow box section of Figure Question 3 is subjected to a clockwise torsional force of 100 kN applied at a distance of 2.0 m above its centroid. Assume the shear modulus (G) for the material is 80 kN/mm^2 and the torsion constant (J) is given by (as defined in Figure 7.11):

$$J = \frac{2B^2D^2t_w t_f}{Bt_w + Dt_f}$$

Determine:

(a) the maximum shear stress developed in the steel box section due to the applied torque;

(b) the angle of twist per unit length for the section.

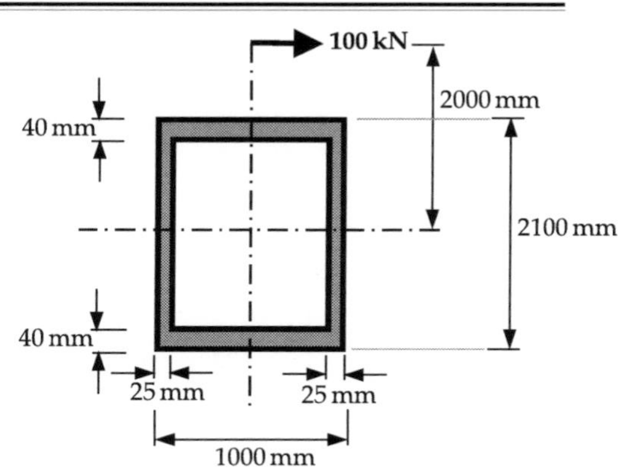

Figure Question 3

4. A circular hollow shaft of external diameter = 90 mm and internal diameter = 60 mm is to be used to drive a hydraulic pump similar to the arrangement shown in Figure Question 4. The maximum power developed by the motor will be 600 kW at 3000 rpm (revolutions per minute). The shaft will be formed in steel with a shear modulus (G) = 80 kN/mm^2. Determine the shear stress in the shaft and the twist per metre length.

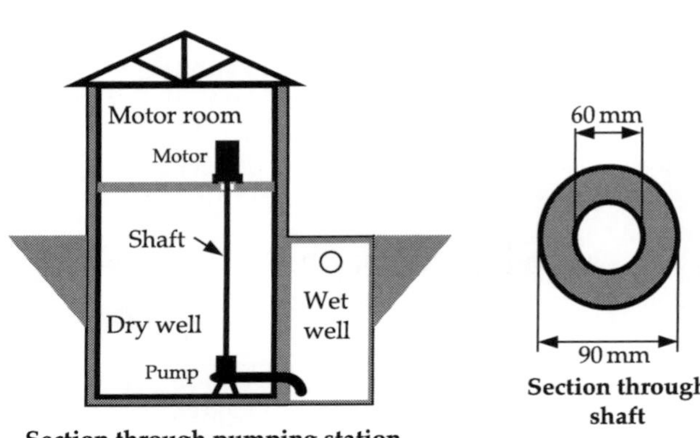

Figure Question 4

5. The Universal Beam of Figure Question 5 is 5 m long and has the properties listed in the table. It transmits an anticlockwise torque of 90 kN mm applied to its free end. The shear modulus (G) for steel may be taken as 80 kN/mm^2. Determine the maximum shear stress in the beam and the total angle of twist in degrees.

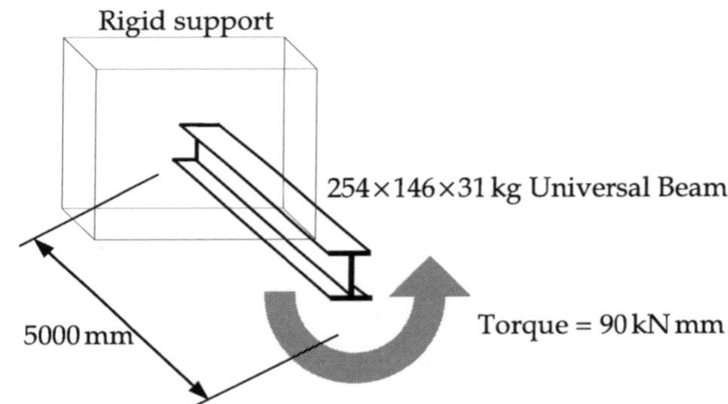

Figure Question 5

Serial size	Weight per metre (kg)	Depth (mm)	Thickness (mm)		Area (cm^2)	Torsion constant (J) (cm^4)
			Web	Flange		
254 × 146	31	251.5	6.1	8.6	40	8.73

Section properties of beam – Question 5

Solutions to Further Problems 7

Torsion equation: $\frac{T}{J}=\frac{\tau}{r}=\frac{G\theta}{L}$

1.

Bending moments:
At A = $20\,\text{kN}\times 4\,\text{m} = 80\,\text{kN m}$
At B $= 20\,\text{kN}\times 4\,\text{m} - 60\,\text{kN}\times 4\,\text{m} + 30\,\text{kN}\times 4\,\text{m} = -40\,\text{kN m}$
At C $= 20\,\text{kN}\times 4\,\text{m} - 60\,\text{kN}\times 4\,\text{m} + 30\,\text{kN}\times 4\,\text{m} + 90\,\text{kN}\times 4\,\text{m} = 320\,\text{kN m}$

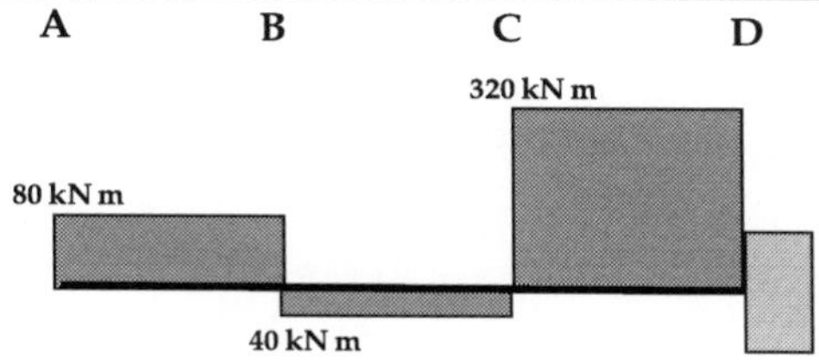

A B C D
320 kN m
80 kN m
40 kN m

Torque diagram for Question 1

2.

$$J=\frac{\pi(D^4-d^4)}{32}=\frac{\pi(D^4-(0.75d)^4)}{32}=0.0671D^4\,\text{mm}^4$$

From: $\frac{T}{J}=\frac{\tau}{r}$

(a) Shear stress not to exceed $50\,\text{N/mm}^2$ gives:

$$\frac{-6600\times10^3}{0.0671D^4}=\frac{50}{\left(\frac{D}{2}\right)}$$

from which: $D_{\text{external}} = 99.45$ mm
$D_{\text{internal}} = 74.6$ mm

From: $\frac{T}{J}=\frac{G\theta}{L}$

(b) Angle of twist not to exceed 0.085 radians gives:

$$\frac{-6600\times10^3}{0.0671D^4}=\frac{80\times10^3\times0.085}{3000}$$

from which: $D_{\text{external}} = 81.16$ mm
$D_{\text{internal}} = 60.87$ mm

$\therefore$ **$D_{\text{external}} = 99.45$ mm and $D_{\text{internal}} = 74.6$ mm**

3.

Torque = force × distance
= $90\,\text{kN}\times 2\,\text{m}$
= $180\,\text{kN m}$

$$J=\frac{2B^2D^2t_wt_f}{Bt_w+Dt_f}$$

$q = 85.714\,\text{N/mm}$
$\tau = 2.14\,\text{N/mm}^2$

$$J=\frac{2(1000^2)(2100^2)\times25\times40}{1000\times25+2100\times40}$$
$=8.09174\times10^{10}\,\text{mm}^4$

and $\theta = 30.896\times10^{-6}$ radians

4.

$P = \dfrac{\pi n T}{30}$, hence

$$600 \times 10^3 = \frac{\pi \times 3000 \times T}{30}$$

gives $T = 1909.86$ N m

J for circular hollow section = 5.1689×10^6 mm^4

and $\tau = 16.63\,\text{N/mm}^2$ and $\theta = 4.6186 \times 10^{-3}$ radians

5.

$T = -90\,\text{kN mm}$

$J = 8.73 \times 10^4\,\text{mm}^4$
gives:

$\tau = -8.8659\,\text{N/mm}^2$
$\theta = -0.0644$ radians
$= 3.69°$

8. Combined Stress

> ***In this chapter we will learn about:***
> - ***Direct stress***
> - ***Bending stress***
> - ***Combined stress***

8.1 Introduction

The calculation of combined stress, that is stresses caused by two or more types of loading, can be very complex. We have already investigated various forms of stress in structures. In Chapter 2, we considered direct and bending stress. In Chapter 3 we calculated the shear forces that will cause shear stress and in Chapter 7 we studied torsional stresses. Many 'real' structures will be subjected to various types of load combinations and hence forces. All such forces will induce stress which will be cumulative, and their combined effect may promote failure of a structural member. However, they often act in different planes and therefore can only be combined using complex analysis techniques such as the Finite Element Method. Stresses developed due to direct loading in the vertical or horizontal plane (x or y axis) and bending about the z axis will however have a cumulative effect which can be calculated using equations developed earlier in this text.

We will now consider methods for analysing the stress developed in structures subjected to both direct force and bending moment. These structures include short columns, retaining and free-standing walls in which combined stresses are of major importance in design.

8.2 Direct and Bending Stresses

Direct forces acting at a point or over a small area of the cross-section of a structure, as shown in Figure 8.1a, cause direct stress. The stress caused by the load can be uniform compression or tension, and will be considered as distributed evenly across the entire cross-section. At a point directly under the load, the stresses developed will have a maximum intensity which will dissipate through the structure as we move further away from the point of application. However, in this analysis we will consider that the analysis is undertaken at a level where a uniform stress has been developed. Bending stress is caused by a bending moment or 'couple' and will produce a stress distribution which varies across the cross-section, as shown in Figure 8.1b. The stress will vary from tension to compression as shown.

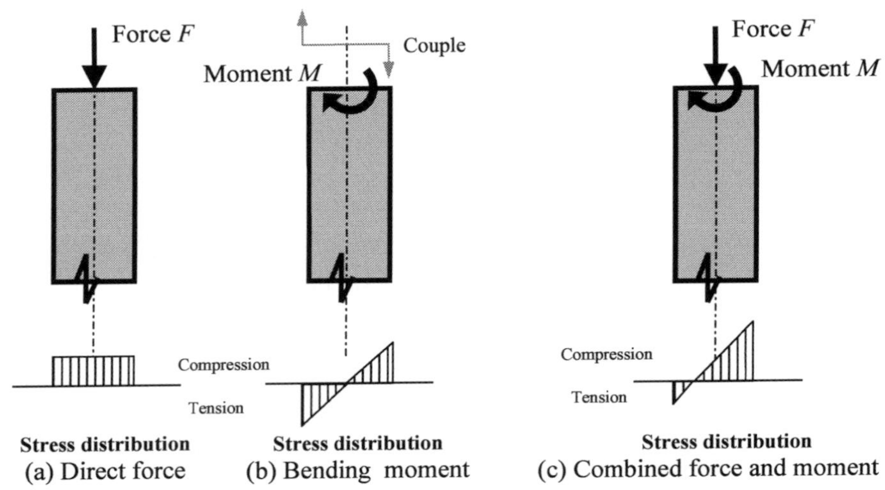

Figure 8.1: Direct, bending and combined stress

It can be seen that the stresses developed by the direct force and those induced by the bending moment, in this case a couple, are of the same type. The vertical direct load will cause a compressive stress in the structure, whilst the application of the bending moment in the form of a couple at the point indicated in Figure 8.1b will cause a compressive stress in the right side of the structure and a tensile stress in the left. As the stress is related directly to the moment, and the moment is determined by 'Force × Distance from a datum', the stress will vary linearly as the distance from the force varies. Also, as these stresses are of the same type we can combine them in a combined stress diagram as shown in Figure 8.1c. From this combined stress diagram we note that the combined effect of the direct force and bending moment is to increase the compressive stress in the right side of the structure and to reduce the tensile stress in the left side. The combined effect can be determined by calculating sufficient ordinates to accurately plot both the direct and bending stress graphs separately and then adding these to form a combined stress diagram, noting here that compressive stresses are plotted as positive whilst tensile stresses are negative. Alternatively, as the stresses vary linearly, we can plot the stress distributions by adding the maximum bending stress values located at the left and right of the structure to the value calculated for the direct stress, remembering again that compressive stress values are taken as positive, tensile values as negative.

Depending on the ratio of the magnitude of direct forces and bending moments, various forms of the combined stress diagram can be postulated. Consider the distribution of stresses in the three diagrams of Figure 8.2.

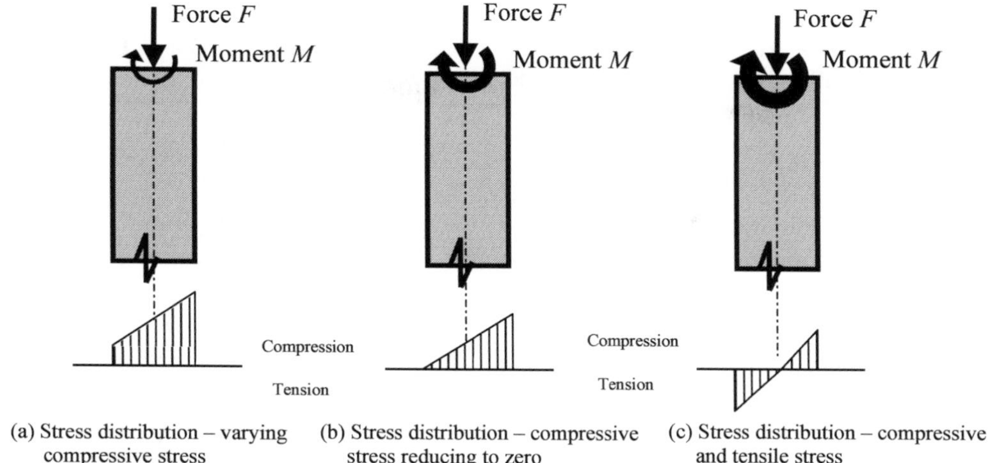

(a) Stress distribution – varying compressive stress

(b) Stress distribution – compressive stress reducing to zero

(c) Stress distribution – compressive and tensile stress

Figure 8.2: Stress distribution resulting from various load and moment combinations

In Figure 8.2a, the compressive stress induced by the direct force is very much greater than the tensile stress developed by the moment, such that it will cancel out any tension in the structure and the stress distribution across the structure will vary from a maximum at the point where the compressive stress caused by the direct force and that caused by the moment combine, to a minimum where the tensile stress from the moment causes a reduction in the compressive stress from the direct force. Figure 8.2b shows the stress distribution across a structure where the tensile stress developed by the moment is equal to the compressive stress developed by the direct force and, therefore, when combined, they will equate to zero. The stress distribution developed by this combination of direct force and moment is very important in the design of many structures in which the development of tensile stress can cause failure. Walls constructed of clay bricks with mortar joints (generally known as 'masonry' walls) have little or no resistance to tensile forces and will fail under very low tensile loads. Unreinforced concrete is also very weak in tension and cracks can develop under moderate tensile loads. This type of stress distribution is also important in the design of many structural members that are simply supported on a base or directly on the soil below. The ground beneath the structure's foundation will, in general, not provide resistance to tensile stress, however it will support some compressive load. We will find that one of the limiting criteria, which can be adopted for the design of such structures, will be to ensure that tensile forces do not develop at the interface between the base of the structure and the ground that supports it. Figure 8.2c shows the stress distribution for a structure in which the tensile force developed by the applied moment is greater than the compressive force induced by the direct force. In such cases, part of the structure will be required to resist the tensile stresses developed.

It should be noted that we have only looked here at direct compressive forces, however the effect of direct tensile force can be considered in a similar manner.

Depending on the materials and type of structure to be designed, we can adopt limits to ensure that the building remains serviceable. For instance, if walls are constructed of masonry we would normally want to ensure that only compressive stresses are developed. The inclusion of some reinforcement may be required if tensile stresses are to be resisted. In the design of some types of retaining walls, that is vertical walls required to support soil and/or fluids, or the design of mass concrete dams, we must ensure that, at the interface between the structure and the soil, no tensile stress is developed and that the maximum compressive stress does not exceed the allowable compressive stress for the soil. For structures designed in steel and reinforced concrete, the material can resist tensile stresses and therefore it is possible to design elements in which both tensile and compressive stresses are developed.

In order to analyse structures for the effects of combined stress we must first calculate the stresses caused by direct forces and also those induced by bending moments.

In Chapter 2 we found that:

$$\text{Direct stress} = \sigma_{\text{DIRECT}} = \frac{\text{Force}}{\text{Area}} = \frac{F}{A} \qquad [8.1]$$

Also from Chapter 2, we have the Engineers' bending equation, which was given as:

$$\frac{M}{I} = \frac{\sigma}{y} = \frac{E}{R}$$

from which we found that bending stress can be determined from:

$$\text{Bending stress} = \sigma_{\text{BENDING}} = \frac{My}{I} \qquad [8.2]$$

Therefore the combined stress due to both direct forces and bending moments can be found by combining equations [8.1] and [8.2] such that:

$$\text{Combined stress} = \sigma_{\text{COMBINED}} = \sigma_{\text{DIRECT}} \pm \sigma_{\text{BENDING}} = \frac{F}{A} \pm \frac{My}{I} \qquad [8.3]$$

The '+' and '–' between the two expressions for direct and bending stress is determined by the type of stresses developed due to bending. A compressive bending stress will produce a positive stress whilst a tensile stress will be considered as a negative value. The calculation of bending stress may also be simplified with the introduction of a further section property known as the section modulus.

8.2.1 Section Modulus (*Z*)

The section modulus for a particular section is given by:

$$\text{Section modulus } Z = \frac{I}{y} \text{mm}^3 \qquad [8.4]$$

where I = Second Moment of Area
and y = the distance to the extreme fibre of the section.

Since the section modulus for a particular section is related to its Second Moment of Area, and since both the Second Moment of Area and the distance to the extreme fibre are related to the axis about which they are calculated (see Chapter 2), then the Z value must also be related to the axis about which it is calculated, such that:

$$Z_{xx} = \frac{I_{xx}}{y_{(\max)}} \text{mm}^3 \quad \text{and} \quad Z_{yy} = \frac{I_{yy}}{x_{(\max)}} \text{mm}^3 \qquad [8.5a \,\&\, b]$$

as shown in Figure 8.3.

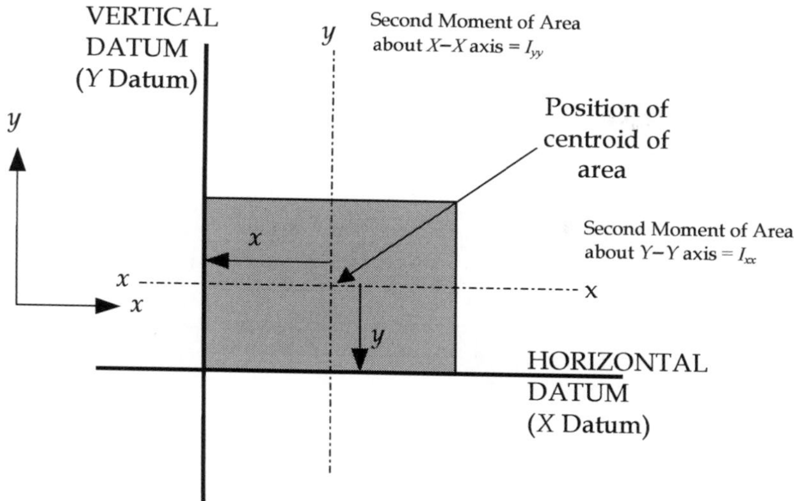

Figure 8.3: Co-ordinate system for calculation of section modulus (*Z*)

For a rectangular section we know from Chapter 2 that the Second Moment of Area about the *X–X* and *Y–Y* axis are given by:

$$I_{xx} = \frac{bd^3}{12} \quad \text{and} \quad I_{yy} = \frac{db^3}{12}$$

where b = breadth of section
and d = depth of section.

We also know that the distance from the centroid to the extreme fibre in the vertical and horizontal directions are given by:

$$y_{\max} = \frac{d}{2} \quad \text{and} \quad x_{\max} = \frac{b}{2} \qquad [8.6]$$

Substituting the equations for Second Moment of Area and those for y and x given in [8.6] into the equations of [8.5] will give (for a rectangular section):

$$Z_{xx} = \frac{\left(\dfrac{bd^3}{12}\right)}{\left(\dfrac{d}{2}\right)} = \frac{bd^2}{6}\,\text{mm}^3 \quad \text{and} \quad Z_{yy} = \frac{\left(\dfrac{db^3}{12}\right)}{\left(\dfrac{b}{2}\right)} = \frac{db^2}{6}\,\text{mm}^3 \qquad [8.7a \,\&\, b]$$

The section modulus can therefore be calculated if both the Second Moment of Area and the distance to the extreme fibre are known. It should be noted that, as was the case when calculating the moment of resistance for various non-symmetric shapes, the distance to the extreme fibre and hence the section modulus can vary about the axis, e.g. $Z_{xx(\text{top})}$ for a particular section can be different to $Z_{xx(\text{bottom})}$ of the same section.

Having developed this further section property, we can include it in equation [8.3] for combined stress such that:

$$\text{Combined stress} = \sigma = \frac{F}{A} \pm \frac{M}{Z} \quad \text{since } Z = \frac{I}{y} \quad \text{and} \therefore \frac{1}{Z} = \frac{y}{I} \qquad [8.8]$$

Again, it should be noted that this general equation can be used to calculate the combined stress about any axis by inserting the appropriate value for the section modulus (Z) as required.

Having developed the general combined stress equation [8.8], we should also note that the stress distribution of Figure 8.2b is of great importance in the design of many structures, particularly walls, in which tension in the structural member must be avoided. However, the development of tension in such structures will not necessarily cause them to fail. Whilst tension cracks might be propagated within the material, so long as the structure continues to act monolithically (as a single structure) failure will not occur until it 'topples' or 'overturns'. We will now consider the failure of walls due to overturning.

8.2.2 Overturning Moments

Consider the wall shown in Figure 8.4 supported directly on the soil at its base. A force F is applied horizontally which varies in magnitude from zero to F_3.

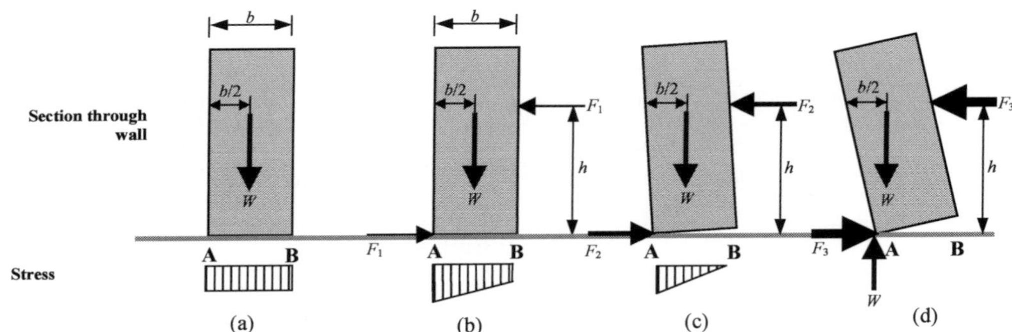

Figure 8.4: Variable horizontal force applied to vertical wall

Assuming the structure remains monolithic throughout, and with reference to Figure 8.4, we find:

(a) The wall is supported by the soil at its base. For equilibrium, the soil must exert an upward reaction equal and opposite to the force due to the weight of the wall, W. The force W is distributed over the base such that the compressive stress induced is evenly distributed across it.

(b) A force F_1 is applied to the wall at a height h above its base. In order to maintain equilibrium, and to prevent the wall from sliding, an equal and opposite horizontal force must be developed at point A. The horizontal applied force on the wall will produce an anticlockwise (overturning) moment about point A = $\{F_1 \times h\}$. The anticlockwise moment will develop a tensile stress at point B which will induce a reduction in compressive stress at that point and an increase in compressive stress at A (wall rotating about point A). If the wall is not to overturn, this anticlockwise moment will be resisted by a clockwise (restoring) moment about A, equal to the weight of the wall acting through the centroid of its base, which equals $\{W \times (b/2)\}$.

(c) As the horizontal force increases to F_2, so the reduction in compressive stress at point B will continue until it is reduced to zero. At this point the stress developed will be distributed linearly across the base. A horizontal force of F_2 must also be developed at point A to prevent the wall from sliding.

(d) As the horizontal force increases further, to F_3, the anticlockwise (overturning) moments cause the structure to rotate further about point A as the compressive stress approaches a maximum value. For equilibrium at this stage we must have:

(1)	Downward force	=	Vertical reaction
	Weight of wall W	=	Reaction $W_{upwards}$
(2)	Horizontal applied force	=	Horizontal reaction
	Horizontal force F_3	=	Horizontal reaction F_3
(3)	Overturning moment	=	Restoring moment
	$F_3 \times h$	=	$W \times (b/2)$

If these three conditions are met the structure will remain in equilibrium and will not overturn, however if F_3 increases further, such that the overturning moment exceeds the restoring moment, the structure will not be in equilibrium and the wall will 'topple' over.

In the design of such structures we can therefore check for two conditions:

Condition 1: Zero tension at the interface of the wall and its base or supporting material.
Condition 2: Overturning moment should not exceed restoring moment.

Compliance with these two conditions will ensure that the structure will not overturn and allows us to complete a preliminary design of the structure. However, it must be noted that this analysis assumes that the structure is monolithic, that sliding will not take place and that the structure can be designed to its limit, i.e. without a factor of safety.

It is clear that bending moments can be induced in columns and walls due to horizontal applied loads. Moments can also be induced if the vertical applied load does not act through the centroid of the section. This type of load, known as *eccentric loading*, will now be discussed.

8.2.3 Eccentric Loading

Eccentric load will occur in a structure if the applied load does not act through the centroid of the section. Some common examples of eccentric loading are shown in Figure 8.5 and include the vertical load applied by a beam connected to a column, and a point load applied to a non-symmetric section.

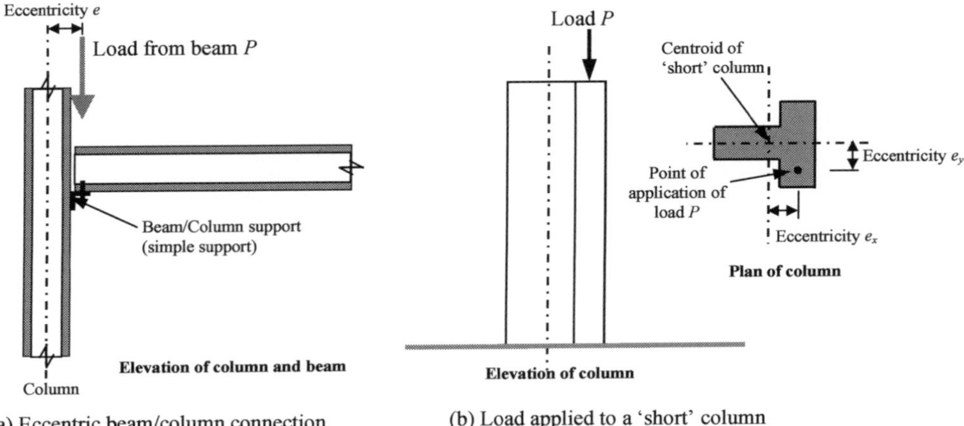

(a) Eccentric beam/column connection

(b) Load applied to a 'short' column

Figure 8.5: Examples of eccentric loading

From Figure 8.5a it can be seen that the beam bears onto a bracket that provides a simple (pin) support. Therefore no moment is induced at the support, however the load from the beam will induce a vertical load which must be supported by the column. There will also be a moment induced in the column because the reaction from the beam does not act through the column's centroid. The distance from the point of application of the load to the centroid of the supporting column is known as its eccentricity. The moment induced by the load from the beam will be equal to:

$$\text{Moment} = \text{force} \times \text{distance} = P \times e = Pe \text{ kN m}$$

In Figure 8.5b we first note the use of the term 'short' column. This term is used to define a column in which bending, or 'buckling', does not predominate. That is, the column is very 'stiff' and, in terms of its height to width ratio, it is not considered to be 'slender'. The analysis of slender columns is complex and is detailed in many other texts, so it will therefore not be considered here. In Figure 8.4b the point load P is applied at a point with eccentricity e_x and e_y from the centroid of the section. The point load will therefore induce moments about the X–X and Y–Y axis such that:

$$\text{Moment about } X\text{–}X \text{ axis} = M_{xx} = Pe_y$$
$$\text{Moment about } Y\text{–}Y \text{ axis} = M_{yy} = Pe_x$$

We must analyse such a structure about both axes in order to calculate the effects of these 'bi-axial' bending moments induced by the point load.

In analysing structures which support eccentric loads, we must ensure that the direct load from the vertical forces, in this case force P, is included in the direct stress calculations and the moment developed due to its eccentricity is included in the calculation of the bending stress.

Having developed the combined stress equation and investigated the principles of both overturning moments and eccentric loading, we will now apply these concepts to a number of different structural problems.

Example 8.1: Analysis of a free-standing wall

The 4 m high, 300 mm thick free-standing concrete wall of Figure 8.6 is fixed at its base. It is required to resist a maximum wind load of $1.2\,kN/m^2$ evenly distributed over one face. Given that the concrete mass is $2400\,kg/m^3$, determine the maximum tensile and compressive stress at the base and plot the stress distribution across the cross-section.

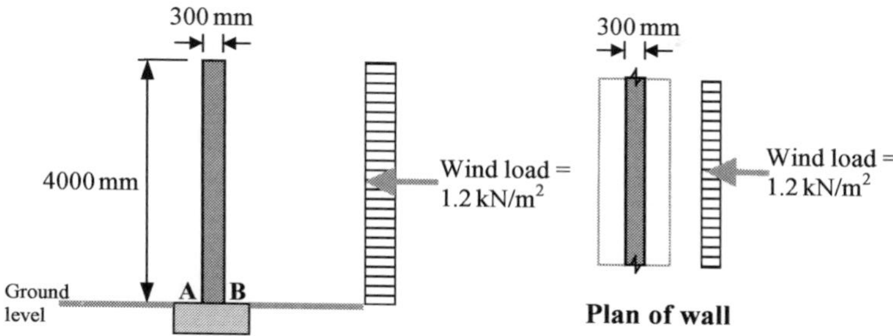

Figure 8.6: Free-standing concrete wall – Example 8.1

We first note that the wall is free-standing, that is unsupported along its height, and that it is fixed at its base and hence can resist both tensile and compressive stresses. (Consider the wall as a vertical cantilever.) We also note that no length is given for the wall and so, in order to simplify our calculations we will use a unit length of 1 m.

The combined stress equation [8.8] is given as:

$$\sigma = \frac{F}{A} \pm \frac{M}{Z}$$

and we must now solve this equation to determine both tensile and compressive stresses at the base.

We will first calculate the section properties for the structure, remembering that we are using a 1 m unit length.

$$\text{Area } A = 0.3\,\text{m} \times 1\,\text{m} = 0.3\,\text{m}^2$$

To calculate the section modulus Z we note that rotation will take place along an axis parallel to the longitudinal axis of the wall, as shown in Figure 8.7a.

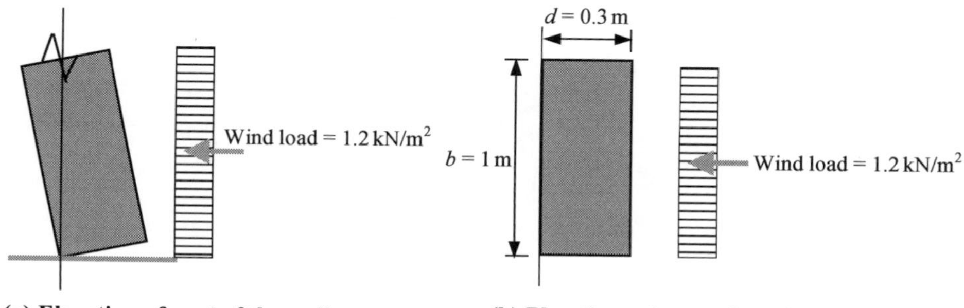

Figure 8.7: Details of a section of the wall – Example 8.1

The section modulus must therefore be calculated parallel to this axis (Figure 8.7b) such that:

For a rectangular section $Z = \dfrac{bd^2}{6} = \dfrac{1 \times 0.3^2}{6} = 0.015 \text{ m}^3$

We must now determine the direct force from the weight of the wall, and the bending moment induced by the wind load.

Direct force F = Force from mass of structure $= 2400\,\text{kg/m}^3 \times 0.3\,\text{m} \times 4\,\text{m} \times 9.81\,\text{m/s}^2$
$= 28\,252.8\,\text{N/m run} = 28.2528\,\text{kN/m run}$

Note here that we need to convert the concrete mass to a force. This is achieved by including the gravitational constant $g = 9.81\,\text{m/s}^2$. Note also that our calculations will be completed per metre run (or length) of the wall. This is a consequence of using a 1 m unit length.

We can now calculate the moment induced by the wind, noting that this is a uniformly distributed load (UDL) evenly spread over one face.

Moment from wind force $M = 1.2\,\text{kN/m}^2 \times 4\,\text{m} \times (4\,\text{m}/2) = 9.6\,\text{kN m/m run}$

Again the moment is per metre run of the wall.

We now have all the elements necessary to complete our calculations for both the tensile and compressive stresses, per metre run, at the base of the wall using the combined stress equation such that:

$$\sigma_{\text{COMPRESSION}} = \frac{F}{A} + \frac{M}{Z} = \frac{28.2528}{0.3} + \frac{9.6}{0.015} = 734.18 \text{ kN/m}^2 = 0.734 \text{ N/mm}^2$$

$$\sigma_{\text{TENSION}} = \frac{F}{A} - \frac{M}{Z} = \frac{28.2528}{0.3} - \frac{9.6}{0.015} = -545.824 \text{ kN/m}^2 = -0.545 \text{ N/mm}^2$$

A plot of the stress distribution diagram is shown in Figure 8.8.

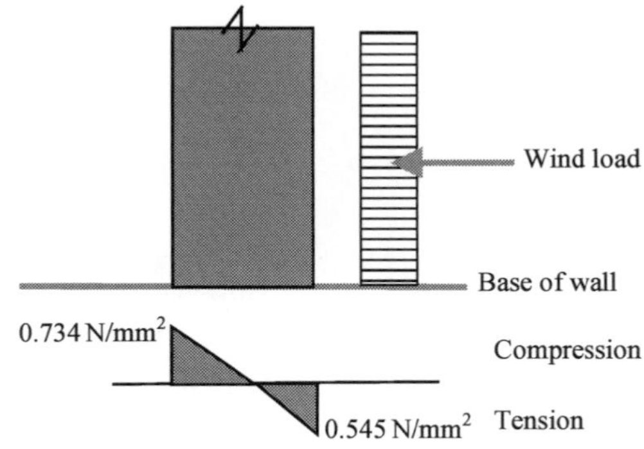

Figure 8.8: Stress distribution, per metre run, across base of wall – Example 8.1

◇

In all cases when the length of the structure is not given, we will use a unit length in order to complete the calculations for combined stress. The results can then be converted to calculate the total stress for any length of the structure.

Example 8.2: Analysis of a free-standing chimney

A square section plain (unreinforced) concrete chimney 12 m high is shown in Figure 8.9. The chimney is subjected to an evenly distributed wind load on the windward face of P kN/m, whilst on the leeward side a suction load equivalent to half that on the windward face acts at the top of the chimney, which reduces to zero at the base. Given that the concrete mass is 2500 kg/m^3 and that the maximum wind pressure anticipated for the site is 1.65 kN/m^2, determine whether the chimney will be safe in this situation.

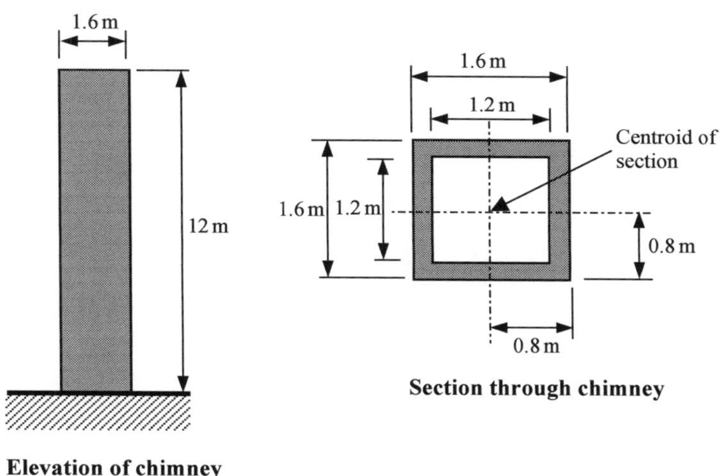

Figure 8.9: Details of concrete chimney – Example 8.2

Again we must solve the combined stress equation [8.8]. However, we know that unreinforced concrete is weak in tension and, therefore, in order to remain serviceable and prevent cracking in the chimney, we will want to ensure that the concrete remains in compression. The limiting stress distribution will be that where maximum compression is achieved in one face whilst the stress in the other face is zero. We also note that we are required to determine whether the structure will be safe for the anticipated wind load. We will need to check that the structure can comply with two criteria:

Criteria 1: **Serviceability:** What is the maximum wind load that the structure can carry and remain serviceable (i.e. without the structure cracking)?

Criteria 2.: **Overturning:** What is the maximum wind load that the structure can carry before it overturns and collapses (assuming the structure remains monolithic)?

We will therefore use the combined stress equation to determine the maximum wind load before tension will be induced in the base of the structure, and we will then calculate the maximum wind pressure before the structure overturns.

From Figure 8.9, we can first determine the section properties and direct force from the chimney:

$$\text{Area of chimney } A = [(1.6\,\text{m} \times 1.6\,\text{m}) - (1.2\,\text{m} \times 1.2\,\text{m})] = 1.12\,\text{m}^2$$

In order to determine the section modulus for the hollow chimney section, we must revert to first principles. Equation [8.4] gives:

$$\text{Section modulus} \quad Z = \frac{I}{y}$$

For the square hollow section chimney $I_{xx} = I_{yy} = \dfrac{bd^3}{12} = \dfrac{[(1.6 \times 1.6^3) - (1.2 \times 1.2^3)]}{12} = 0.373 \text{ m}^4$

From Figure 8.9 we note that, in this case, $y_{max} = x_{max} = 0.8$ m and therefore:

$$\text{Section modulus} \quad Z = \frac{I}{y} = \frac{0.373}{0.8} = 0.467 \text{ m}^3$$

Note: both the area and section modulus are calculated for the hollow box section of the chimney.

The direct force will be equal to the force exerted by the mass of the chimney, such that:

$$\text{Direct force} \quad F = 2500\,\text{kg/m}^3 \times [(1.6\,\text{m} \times 1.6\,\text{m}) - (1.2\,\text{m} \times 1.2\,\text{m})] \times 12\,\text{m} \times 9.81\,\text{m/s}^2$$
$$= 329\,616\,\text{N} = 329.616\,\text{kN}$$

We must now determine the moments due to the wind load on the two faces of the chimney. The distribution of the loads will be as detailed in the question and shown diagrammatically in Figure 8.10.

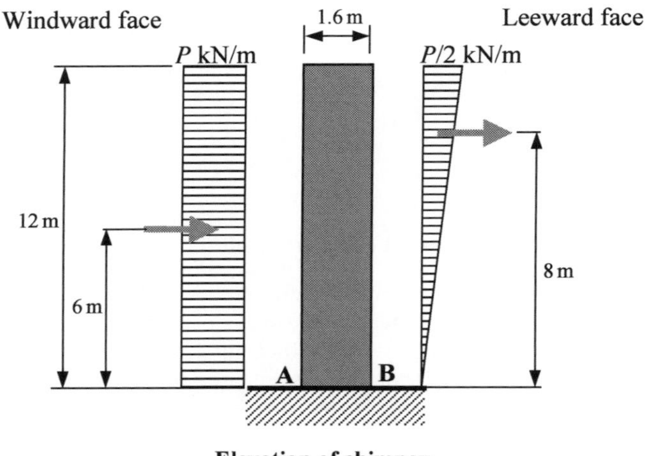

Figure 8.10: Details of wind loading on chimney – Example 8.2

Rotation will occur about point B. Taking moments at the base of the chimney, at B, we have:

$$\begin{aligned} \text{Moments due to wind load} \quad M &= (P\,\text{kN/m} \times 12\,\text{m} \times 1.6\,\text{m} \times 6\,\text{m})(\text{windward face}) \\ &\quad + (P/2\,\text{kN/m} \times 0.5 \times 12\,\text{m} \times 1.6\,\text{m} \times 8\,\text{m})\ (\text{leeward face}) \\ &= 115.2P + 38.4P\,\text{kNm} = 153.6P\,\text{kNm} \end{aligned}$$

Note here that the wind load is three-dimensional since the chimney has a finite width. We must therefore calculate the load on the windward face as a rectangular block and that on the leeward face as a wedge-shaped block 1.6 m wide.

We can now use the combined stress equation to calculate the value of P for zero tensile stress, such that:

$$\sigma_{\text{TENSION}} = 0 = \frac{F}{A} - \frac{M}{Z} = \frac{329.616}{1.12} - \frac{153.6P}{0.467}$$

Solving for P we find: $P = 0.895\,\text{kN/m} < 1.65\,\text{kN/m}$

From this result we note that, at the base of the chimney structure, tensile stresses will be induced if the wind pressure exceeds 0.895 kN/m. This is less than the 1.65 kN/m anticipated for the site. The structure might therefore develop cracks at the maximum loading and may well become unserviceable.

We must now determine the wind pressure at which the structure will overturn. We have already calculated the overturning moment due to wind load as:

Overturning moment $OM = 153.6P$ kN m

The restoring moment will be equal to the force from the chimney multiplied by the distance to the centroid of the section from the point at which overturning will take place. From Figure 8.10 we note that the chimney will rotate about point B, hence the restoring moment (*RM*) is:

Restoring moment $RM = 329.616\,\text{kN} \times 0.8\,\text{m} = 263.693\,\text{kN m}$

At the limit, that is the point just before overturning is about to take place, the overturning moment will equal the restoring moment and hence:

$$\text{Overturning moment} = \text{Restoring moment}$$
$$153.6P = 263.693\,\text{kN m}$$

and hence

$$P = 1.72\,\text{kN/m} > 1.65\,\text{kN/m}$$

We note that the structure is able to support a maximum wind load of 1.72 kN/m which is greater than the maximum of 1.65 kN/m anticipated for the site.

The chimney structure, so long as it remains monolithic, is therefore safe for the site conditions anticipated, however tension cracks may form in the structure which could render it unserviceable.

◇

Example 8.3: Analysis of a column carrying an eccentric load from a beam

The short steel column of Figure 8.11 supports a beam on a steel angle bracket connected to its flange. Assuming the column is fully fixed to a steel base plate, calculate the maximum tensile and compressive stresses induced by the loads indicated. The section properties for the column are given in Figure 8.12. (Assume the self-weight of the beam and column are included in the specified loads.)

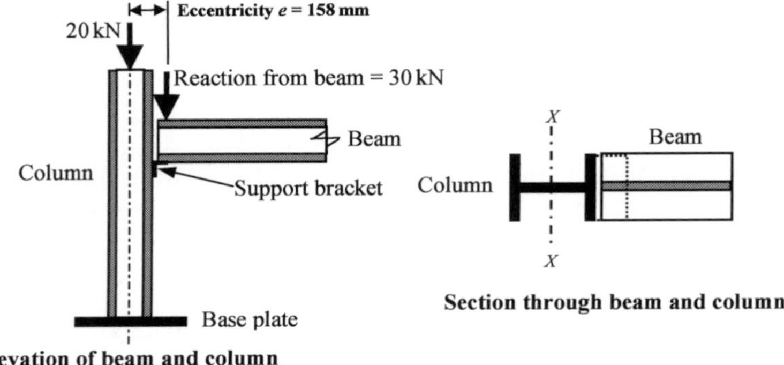

Figure 8.11: Short steel column and beam details – Example 8.3

Serial size (mm)	Weight per metre (kg)	Depth (mm)	Area A (cm^2)	Section modulus Z_{xx} (cm^3)
152 × 152	23	152.4	29.8	184

Figure 8.12: Section properties for column – Example 8.3

In this example we have an eccentric load from a beam which will induce a moment in the column. We must also remember that the reaction from the beam will increase the stresses due to the direct loads. The section properties for the column are given as:

Area $A = 29.8\,\text{cm}^2 = 2980\ \text{mm}^2$
Section modulus $Z_{xx} = 184\,\text{cm}^3 = 184\ 000\,\text{mm}^3$

The forces and moments are:

Direct force $F = 20\,\text{kN} + 30\,\text{kN} = 50\,\text{kN} = 50\ 000\,\text{N}$
Moment due to eccentric load $= 30\ 000\,\text{N} \times 158\,\text{mm} = 4\ 740\ 000\,\text{N mm}$

From the combined stress equation [8.8], we find:

$$\sigma_{\text{COMPRESSION}} = \frac{F}{A} + \frac{M}{Z} = \frac{50\ 000}{2980} + \frac{4\ 740\ 000}{184\ 000} = 42.54\ \text{N/mm}^2$$

$$\sigma_{\text{TENSION}} = \frac{F}{A} - \frac{M}{Z} = \frac{50\ 000}{2980} - \frac{4\ 740\ 000}{184\ 000} = -8.98\ \text{N/mm}^2$$

◇

Example 8.4: Analysis of a short column carrying an eccentric point load

Figure 8.13 shows a short unreinforced concrete column that is subjected to a 100 kN vertical downward point load in the position indicated. Determine:

(a) the load eccentricity and hence the applied moment about the Y–Y axis;

(b) the maximum tensile and compressive stresses to the left and to the right of the Y–Y axis in the column.

Assume that the self-weight of the column can be ignored.

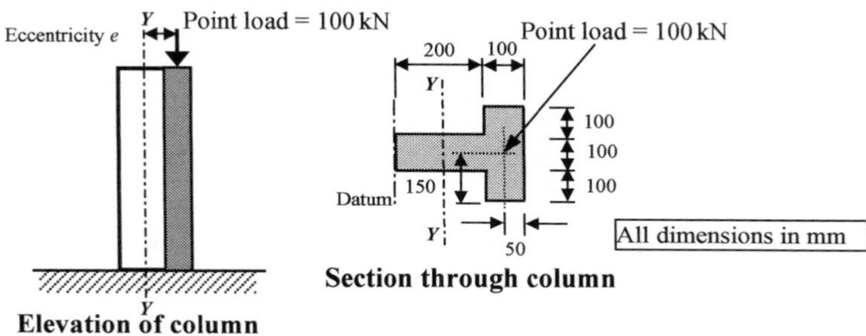

Figure 8.13: Details of short concrete column – Example 8.4

In order to complete this example, we will first need to calculate the section properties of the column. We have already studied section properties in Chapter 2 and the calculations for the column are contained in the table of Figure 8.14. The datum for these calculations is located to the left of the section, as shown in Figure 8.13.

Part No.	Area A (mm^2)	Distance x (mm)	$A \times x$ (mm^3)	I_{yy} (mm^4)	$(\bar{x} - x)$ (mm)	$A(\bar{x} - x)^2$ (mm^4)
PART 1 (200 × 100)	20 000	100	2 000 000	66 666 666.7	90	1 800 000
PART 2 (100 × 300)	30 000	200 +100/2 = 250	7 500 000	25 000 000	60	108 000 000
	ΣA = 50 000		ΣAx = 9 500 000	91 666 666.7		109 800 000

Figure 8.14: Table to calculate position of centroid and I_{yy} for column section – Example 8.4

From Figure 8.14, position of centroid $= \bar{x} = (9\,500\,000/50\,000) = 190\,\text{mm}$

Second Moment of Area for section = I_{yy} = 91 666 666.7 + 109 800 000 = 201.647×10^6 mm^4

Distance from centroid to extreme fibre = x $\quad x_{\text{Left}} = 190\,\text{mm}$

$x_{\text{Right}} = (300\,\text{mm} - 190\,\text{mm}) = 110\,\text{mm}$

Section modulus to left of centroid $\quad Z_{yy\,\text{Right}} = \dfrac{I_{yy}}{x_{\text{Right}}} = \dfrac{201.467\times10^6}{110} = 1\ 831\ 518.18\text{mm}^3$

Section modulus to left of centroid $\quad Z_{yy\,\text{Left}} = \dfrac{I_{yy}}{x_{\text{Left}}} = \dfrac{201.467\times10^6}{190} = 1\ 060\ 350.9\text{mm}^3$

The direct load F is given in Figure 8.13 as 100 kN, so all that remains is to calculate the moment induced by the eccentric load. The eccentricity will be the distance from the centroid of the section to the point of application of the load:

$$e = 250\,\text{mm} - 190\,\text{mm} = 60\,\text{mm}$$

Therefore: $\quad$ Moment $\quad M = 100\,\text{kN} \times 0.06\,\text{m} = 6\,\text{kN m}$

We can now calculate the compressive and tensile stresses to the left and right of the datum Y–Y such that:

$$\sigma_{\text{COMPRESSION(Right)}} = \frac{F}{A} + \frac{M}{Z} = \frac{100\times10^3}{50\ 000} + \frac{6\times10^6}{1\ 831\ 518.18} = 2.328\ \text{N/mm}^2$$

$$\sigma_{\text{TENSION(Left)}} = \frac{F}{A} - \frac{M}{Z} = \frac{100\times10^3}{50\ 000} - \frac{6\times10^6}{1\ 060\ 350.9} = -3.658\ \text{N/mm}^2$$

The tension properties of the concrete section could be improved by placing steel reinforcement in the left stem of the section to resist the tensile stresses.

◇

Example 8.5: Analysis of a trapezoidal gravity dam structure

Figure 8.15 shows a section through a mass concrete gravity dam which is to be used to form one end of a freshwater reservoir. Assuming that the water level in the reservoir is at the top of the dam, determine:

(a) the maximum height H if the dam is not to overturn;
(b) the maximum height H if there is to be no tension in the concrete;
(c) the maximum compressive stress for the dam for the height found in (b).

Assume: Density of the concrete = 2400 kg/m^3
Density of water = 1000 kg/m^3

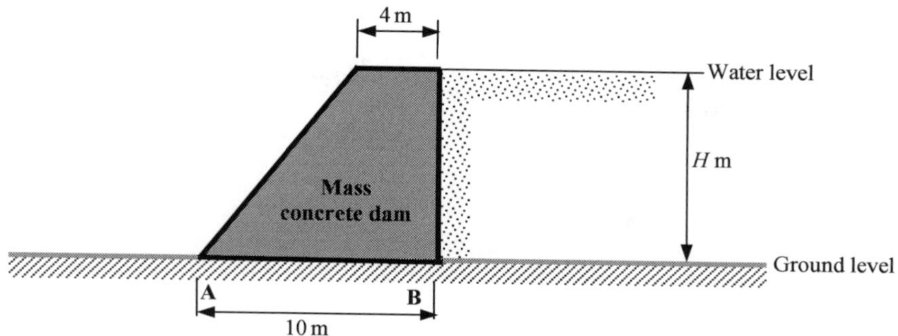

Figure 8.15: Section through mass concrete dam structure – Example 8.5

In our solution we will consider a 1 m length of the dam. We also note that the dam is formed from a trapezoidal section. In order to calculate the restoring moment for the structure, we will need to determine the position of its centroid. This can be difficult for complex shapes, however in order to simplify the problem we can consider the section as being made up of a triangular and a rectangular part. Finally we will need to determine the forces and moments exerted by the water pressure on the structure. The load from the water will be a triangular distribution which will be zero at the water surface and a maximum at the base. The maximum force at the base will be given by:

$$\text{Maximum force from water} = \gamma g h \quad \text{where } \gamma = \text{denstity of water}$$
$$\text{Maximum force from water} = 1000\,\text{kg/m}^3 \times 9.81\,\text{m/s}^2 \times H$$
$$= 9.81\ H\,\text{kN/m}$$

Noting that rotation will take place about point A (at the toe of the dam), the forces and lever arms required for our calculations are shown in Figure 8.16.

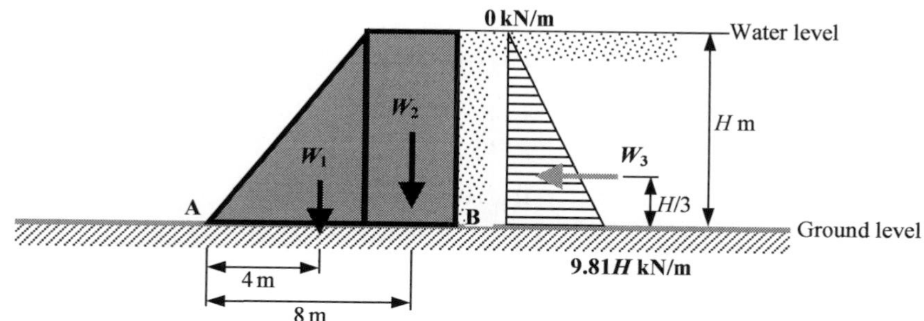

Figure 8.16: Detail of sections and lever arms for mass concrete dam – Example 8.5

From Figure 8.16 we note:

W_1 = Force from section 1 (triangular part of structure)
W_2 = Force from section 2 (rectangular part of structure)
W_3 = Force from water

To solve for part (a) of this example, we must calculate the overturning and restoring moments about point A, per metre length, of the dam.

Overturning moment: $OM = W_3 \times H/3\ \text{m} = (0.5 \times 9.81H\ \text{kN/m} \times H\ \text{m}) \times H/3 = 1.635\ H^3\ \text{kNm}$

Restoring moments: RM Section 1 = $W_1 \times 4$ m = (2400 kg/m^3 × 9.81 m/s^2 × (0.5×6 m×Hm))×4 m
= 282.528H kN m
RM Section 2 = $W_2 \times 8$ m = (2400 kg/m^3 × 9.81 m/s^2 × (4 m × Hm)) × 8 m
= 753.408H kN m

For equilibrium:

Overturning moment = Restoring moment
1.635 H^3 kN m = 282.528H kN m + 753.408 H kN m
1.635 H^2 kN m = 1035.936 kN m
from which: H = 25.17 m

In order to solve for part (b) we must first consider a section along the base of the dam structure as shown in Figure 8.17.

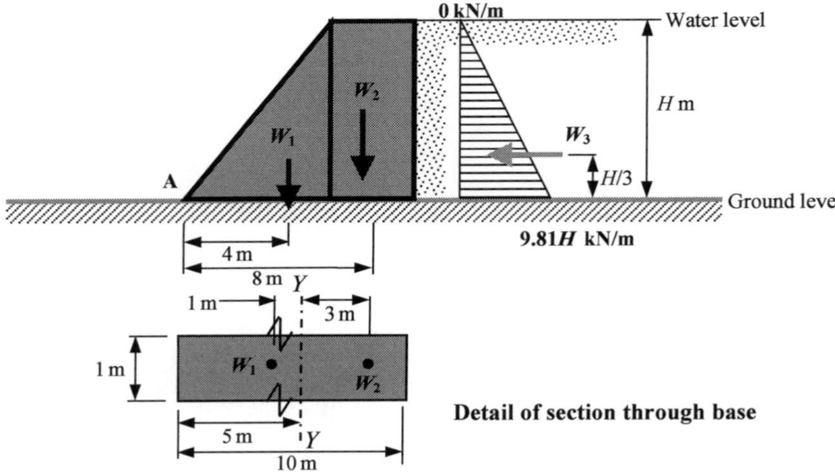

Figure 8.17: Detail of base of mass concrete dam structure – Example 8.5

To calculate the compressive stresses in the base, assuming zero stress at point B, we must calculate the properties for the section at the base. These are:

$$\text{Area}\quad A = 1\,\text{m} \times 10\,\text{m} = 10\,\text{m}^2$$

$$\text{Section modulus}\quad Z = \frac{bd^2}{6} = \frac{1\ \text{m}\times 10\ \text{m}^2}{6} = 16.67\ \text{m}^2$$

We next calculate the bending moment M. In this case the two parts of the trapezoidal section must be included in the moment calculation, since the centroid of the parts do not coincide with the axis Y–Y which passes through the centroid of the rectangular base. We therefore take moments about the axis Y–Y indicated in Figure 8.17, remembering that we are working in three-dimensional space and must include the moment due to the water in our calculations such that:

Moments about Y–Y

$$\begin{aligned} M &= -W_1 \times 1\,\text{m} + W_2 \times 3\,\text{m} - W_3 \times H/3\,\text{m} \\ &= -(2400\,\text{kg/m}^3 \times 9.81\,\text{m/s}^2 \times (0.5 \times 6\,\text{m} \times H\text{m})) \times 1\,\text{m} \\ &\quad + (2400\,\text{kg/m}^3 \times 9.81\,\text{m/s}^2 \times (4\,\text{m} \times H\text{m})) \times 3\,\text{m} \\ &\quad -(0.5 \times 9.81H\,\text{kN/m} \times H\text{m}) \times H/3 \\ &= -70.632H + 282.528H - 1.635H^3\ \text{kN m} \\ &= 211.896H - 1.635H^3\ \text{kN m} \end{aligned}$$

The direct force exerted by the structure will be:

$$\text{Direct force} \quad F = W_1 + W_2 = (2400\,\text{kg/m}^3 \times 9.81\,\text{m/s}^2 \times (0.5 \times 6\,\text{m} \times H\text{m})) + (2400\,\text{kg/m}^3 \times 9.81\,\text{m/s}^2 \times (4\,\text{m} \times H\text{m})) = 164.808H\,\text{kN m}$$

We can now substitute these values into the combined stress equation, remembering that the tensile stress must equal zero:

$$\sigma_{\text{TENSION}} = 0 = \frac{F}{A} - \frac{M}{Z} = \frac{164.808H}{10} - \frac{(211.896H - 1.635H^3)}{16.67}$$

We can simplify this equation to:

$$16.67\,(-164.808H) = 10\,(211.896H - 1.635H^3)$$
$$-2747.35\,H = 2118.96H - 16.35H^3$$
$$-4866.31 = -16.35H^2$$

from which:

$$H = 17.25\,\text{m}$$

To solve for part (c) we can substitute the value for $H = 17.25$ m into the compressive stress equation such that:

$$\sigma_{\text{COMPRESSION}} = \frac{F}{A} + \frac{M}{Z} = \frac{164.808(17.25)}{10} + \frac{(211.896(17.25) - 1.635(17.25)^3)}{16.67} = 474.38\ \text{kN/m}^2$$

Example 8.6: Analysis of a trapezoidal gravity dam structure

Figure 8.18 shows a section through a mass concrete gravity dam which is to be used to form one end of a freshwater reservoir. Determine the required width for the base, assuming there will be zero stress at point B and that:

Density of the concrete $= 22.7\,\text{kN/m}^3$
Density of water $= 9.81\,\text{kN/m}^3$

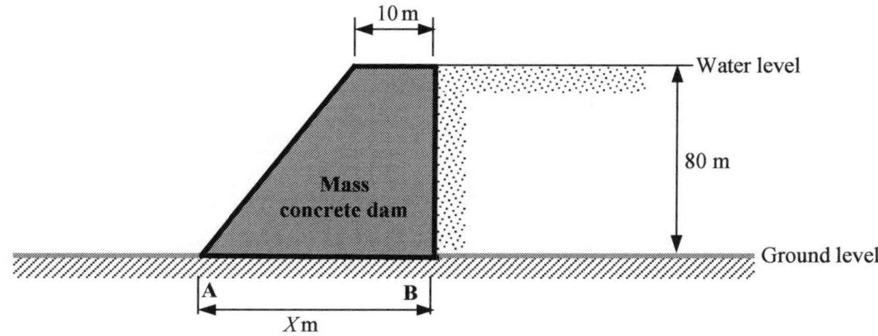

Figure 8.18: Section through mass concrete dam structure – Example 8.6

We will consider a 1 m length of the dam. In order to simplify the problem, we will consider the section as being made up of a triangular and a rectangular part.

In this case we are only required to determine the width of the base for zero stress at point B. We must first consider a section along the base of the dam structure as shown in Figure 8.19.

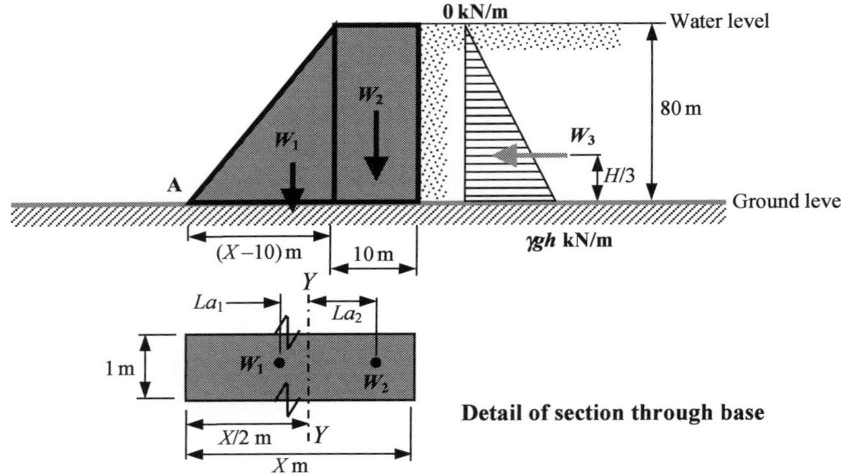

Figure 8.19: Detail of base of mass concrete dam structure – Example 8.6

We can calculate the section properties for the dam at the base as:

Area $A = 1\,\text{m} \times X\text{m} = X\text{m}^2$

Section modulus $Z = \dfrac{bd^2}{6} = \dfrac{1 \times X^2}{6} = 0.167X^2\ \text{m}^3$

We next calculate the bending moment M and again the two parts of the trapezoidal section must be included in the moment calculation, since the centroid of the parts does not coincide with the axis through the centroid of the rectangular base. Before taking moments about the Y–Y axis indicated, we must first calculate the lever arms for each part. With reference to Figure 8.19 we find:

$$La_1 = (X/2) - (^2/_3(X - 10))$$
$$La_2 = (X/2) - 5\,\text{m}$$

Taking moments about Y–Y

$$M = -(W_1 \times ((X/2) - (^2/_3(X - 10)))) + (W_2 \times ((X/2) - 5\,\text{m})) - W_3 \times 80/3\ \text{m}$$
$$M = -((22.7\,\text{kN/m}^3 \times (0.5 \times (X - 10) \times 80\,\text{m})) \times ((X/2) - (^2/_3(X-10))))$$
$$+ ((22.7\,\text{kN/m}^3 \times (10\,\text{m} \times 80\,\text{m})) \times ((X/2) - 5\,\text{m})) - (0.5 \times 9.81\,\text{kN/m}^3 \times (80\,\text{m})^2 \times 80/3)$$
$$M = -(-154.36X^2 - 4512.8X + 60\,563.6) + (9080X - 90\,800) - 837\,120\ \text{kNm}$$
$$M = 154.36X^2 + 13\,592.76X - 988\,483.6\ \text{kNm}$$

Having found an equation for the moments we must now determine the direct forces:

$$F = W_1 + W_2 = (22.7\,\text{kN/m}^3 \times (0.5 \times (X-10) \times 80\,\text{m})) + (22.7\,\text{kN/m}^3 \times (10\,\text{m} \times 80\,\text{m}))$$
$$F = (908X - 9080) + 18\,160\ \text{kN} = 908X + 9080\ \text{kN}$$

We can now substitute these values into the combined stress equation, remembering that the tensile stress must equal zero:

$$\sigma_{\text{TENSION}} = 0 = \frac{F}{A} - \frac{M}{Z} = \frac{(908X + 9080)}{X} - \frac{(154.36X^2 + 13\,592.76X - 988\,483.6)}{0.167X^2}$$

We can simplify this equation to:

$$0.167X^2(908X + 9080) = X(151.64X^2 + 13\,620X - 988\,483.6)$$

$$151.64X^3 + 1516.36X^2 = 151.64X^3 + 13\,620X^2 - 988\,483.6X$$

from which $988\,483.6 = 12\,103.64X$

and $X = 81.67\,\text{m}.$

◇

Example 8.7: Analysis of a simple retaining wall structure

The reinforced concrete retaining wall of Figure 8.20 is required to support a soil embankment along the side of a motorway. Determine the required width for the base in order to ensure the structure will not overturn given:

Density of the reinforced concrete $= 24\,\text{kN/m}^3$
Density of retained soil $= 18\,\text{kN/m}^3$

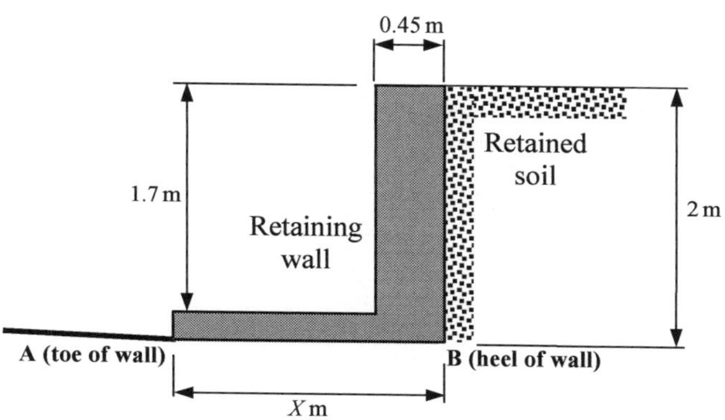

Figure 8.20: Section through concrete retaining wall – Example 8.7

It should be noted that, in general, the distribution of soil pressure on a wall is influenced by a number of factors related to its properties. In this case we are concerned with the structural rather than the geotechnical aspects of the design, and to simplify the problem we will assume the pressure exerted by the soil is in the form of a triangular distribution with a minimum value of 0 kN/m per metre run at the soil surface, rising to a maximum of $18\,\text{kN/m}^3 \times 2\,\text{m} = 36\,\text{kN/m}$ per metre run at the base. We first split the wall into parts and, noting that rotation of the wall will take place about point A, we determine the lever arms required to calculate the moments as detailed in Figure 8.21.

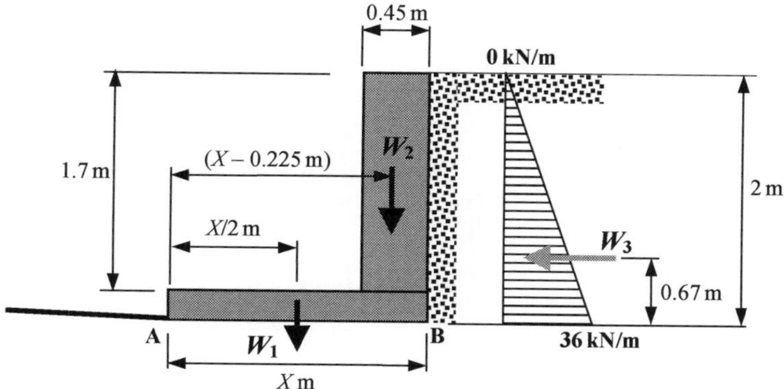

Figure 8.21: Details of parts and lever arms for concrete retaining wall – Example 8.7

With reference to Figure 8.21, the moments per metre run of wall will be:

Restoring moment from section 1 $RM_1 = W_1 \times (X/2)$
$= (24\,\text{kN/m}^3 \times 0.3\,\text{m} \times X\text{m}) \times (X/2) = 3.6\,X^2\,\text{kNm}$

Restoring moment from section 2 $RM_2 = W_2 \times (X - 0.225\,\text{m})$
$= (24\,\text{kN/m}^3 \times 0.45\,\text{m} \times 1.7\,\text{m}) \times (X - 0.225\,\text{m})$
$= 12.24X - 2.754\,\text{kNm}$

Overturning moment from soil $OM = W_3 \times (2\,\text{m}/3)$
$= 0.5 \times 36\,\text{kN/m} \times 2\,\text{m} \times (2\,\text{m}/3) = 24\,\text{kNm}$

For equilibrium: Overturning moment = Restoring moment
$24\,\text{kNm} = 3.6X^2 + 12.24X - 2.754\,\text{kNm}$

and simplifying we have: $3.6X^2 + 12.24X - 26.754 = 0$

We must now solve this equation to find X using:

$$x = \frac{-b \pm \sqrt{b^2 - 4ac}}{2a}$$

Given: $a = 3.6$, $b = 12.24$ and $c = -26.754$ then:

$$x = \frac{-12.24 \pm \sqrt{12.24^2 - 4(3.6)(-26.754)}}{2(3.6)}$$

From which we find $x = 1.513\,\text{m}$.

Therefore, to prevent overturning of the wall, the base must be at least 1.513 m wide.

◇

Example 8.8: Analysis of a simple retaining wall structure

The mass concrete retaining wall of Figure 8.22 is required to support a soil embankment. It has a rectangular cross-section 1.5 m wide and 6 m high. The lateral load from the soil behind is, in this case, to be considered as a triangular distributed load varying from $0\,\text{kN/m}^2$ at the top to $60\,\text{kN/m}^2$ at the base of the wall. A horizontal tie exerts a force of P kN/m at a point 1.5 m below the top of the wall, along its length, in a direction opposing the applied load from the soil. Determine:

(a) the value of P/unit length of the wall required to ensure zero stress at point B;
(b) the resulting compressive stress at point A.

Density of the concrete = $24\,\text{kN/m}^3$

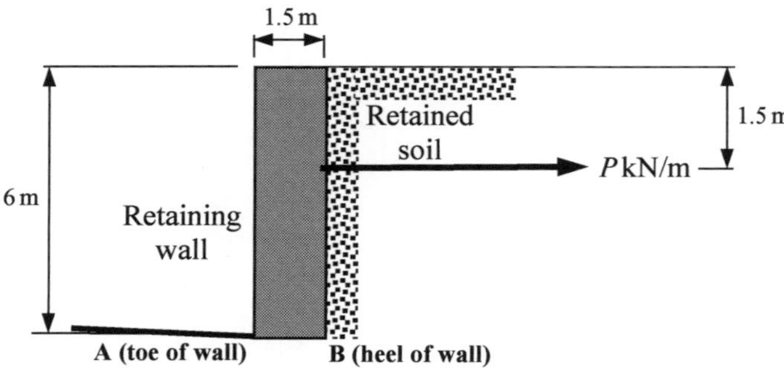

Figure 8.22: Section through concrete retaining wall – Example 8.8

We will analyse a 1 m section of the wall as before. We first determine the lever arm about point A for each of the forces, as shown in Figure 8.23.

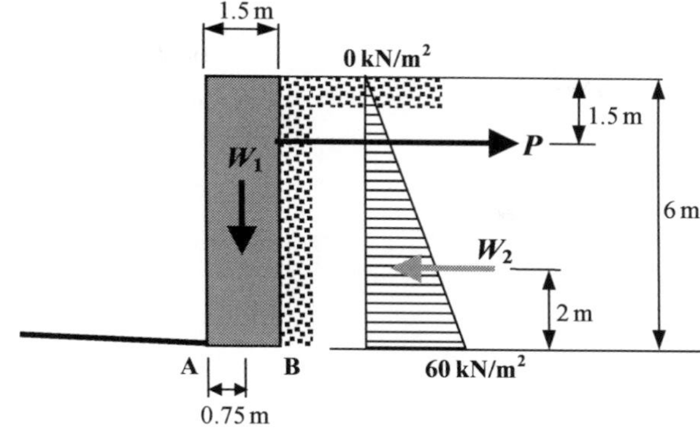

Figure 8.23: Details of sections and lever arms for mass concrete retaining wall – Example 8.8

Weight of mass concrete retaining wall $W_1 = 24\,\text{kN/m}^3 \times 6\,\text{m} \times 1.5\,\text{m} = 216\,\text{kN/m run}$

$$\text{Area} \quad A = 1\,\text{m} \times 1.5\,\text{m} = 1.5\,\text{m}^2$$

$$\text{Section modulus} \quad Z = \frac{bd^2}{6} = \frac{1 \times 1.5^2}{6} = 0.375 \ \text{m}^3$$

Bending moment due to lateral pressure from soil and force from tie:

$$\begin{aligned} M &= W_1 \times 0.75\,\text{m} - W_2 \times (^1/_3 \times 6) + P \times (6\,\text{m} - 1.5\,\text{m}) \\ &= 216\,\text{kN/m} \times 0.75\,\text{m} - ((0.5 \times 60\,\text{kN/m}^2 \times 6\,\text{m}) \times 2\,\text{m}) + 4.5P \\ &= 215.25 - 360 + 4.5P \\ &= -144.75 + 4.5P \ \text{kN m/m run} \end{aligned}$$

For zero pressure at point B:

$$\sigma_{\text{TENSION}} = 0 = \frac{F}{A} - \frac{M}{Z} = \frac{216}{1.5} - \frac{(-144.75 + 4.5P)}{0.375}$$

from which $\quad 144 = -386 + 12P$

and $\quad P = 44.167\,\text{kN/m/m run}$

For part (b) we must find the resulting pressure at point A:

$$\sigma_{\text{COMPRESSION}} = 0 = \frac{F}{A} + \frac{M}{Z} = \frac{216}{1.5} + \frac{(-144.75 + 4.5(44.167))}{0.375} = 288 \ \text{kN/m}^2$$

◇

In the design of many structures, including retaining walls, we are normally required to determine or apply some factor of safety to ensure stability of the structure. This can easily be included in our calculations, for instance, if a factor of safety of 2.0 is required against overturning, this means that our restoring moment must be twice the overturning moment. Hence:

$$\text{Factor of safety against overturning} = \frac{\text{restoring moments}}{\text{overturning moments}}$$

It is essential to include such factors of safety to account for unknown circumstances and situations which can affect the stability of our structure, such as changes in geology, water pressure and increased loading on the surface behind the wall. Factors of safety may also be applied when the quality of workmanship and materials cannot be strictly controlled.

Having completed this chapter you should now be able to determine the combine stress effects on:

- ***Free-standing walls***
- ***Eccentrically loaded columns***
- ***Simple retaining walls***
- ***Trapezoidal section dams***

Further Problems 8

1. The 250 mm thick free-standing concrete wall of Figure Question 1 is 3 m high and fixed at its base. It is required to resist a maximum wind load of 1.5 kN/m^2 which is to be considered as a triangular distributed load over one face. Given that the concrete mass is 2400 kg/m^3, determine the maximum tensile and compressive stress at the base.

2. Figure Question 2 shows a section through a mass concrete gravity dam which is to be used to form one end of a freshwater reservoir.

   Determine the required width for the base, assuming there will be zero stress at point B and that:
   Density of the concrete = 24 kN/m^3
   Density of water = 9.81 kN/m^3

3. The reinforced concrete retaining wall of Figure Question 3 is required to support the soil to the side of an excavation.

   Determine the factor of safety against overturning for the structure.

   Density of the concrete = 24 kN/m^3
   Density of soil = 30 kN/m^3

   Note: the distribution of the soil pressure may, in this case, be assumed to be triangular.

4. The mass concrete retaining wall of Figure Question 4 is required to support a lateral load from the soil behind which may be considered as a triangular distributed load varying from 0 kN/m^2 at the top to 50 kN/m^2 at the base of the wall. A horizontal tie exerts a force of 40 kN/m along its length, in a direction opposing the applied load from the soil. Determine the thickness of wall required for zero stress at point B.

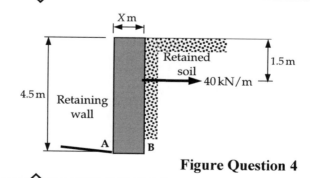

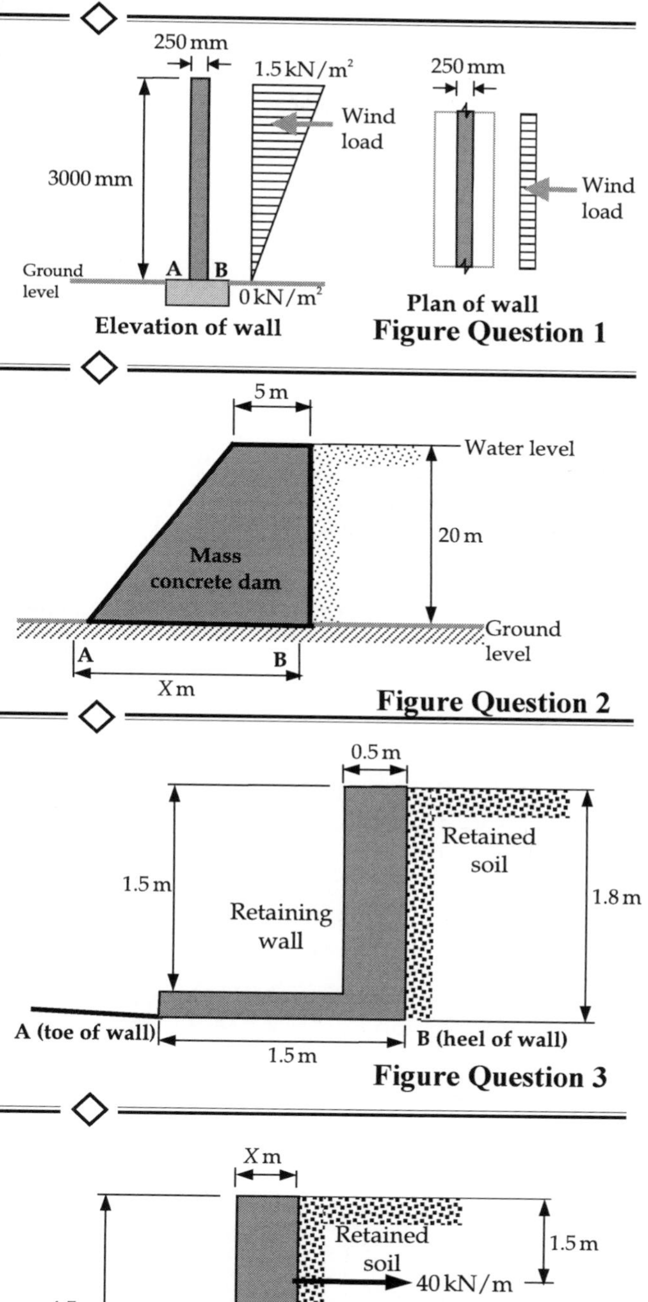

Figure Question 1

Figure Question 2

Figure Question 3

Figure Question 4

Solutions to Further Problems 8

1.

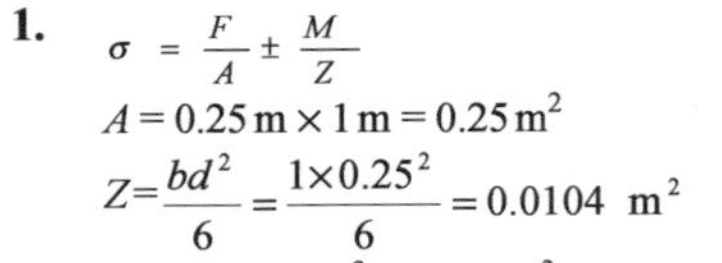

$$\sigma = \frac{F}{A} \pm \frac{M}{Z}$$

$A = 0.25\,\text{m} \times 1\,\text{m} = 0.25\,\text{m}^2$

$$Z = \frac{bd^2}{6} = \frac{1 \times 0.25^2}{6} = 0.0104\ \text{m}^2$$

$F = 2400\,\text{kg/m}^3 \times 9.81\,\text{m/s}^2 \times 3\,\text{m} \times 0.25\,\text{m}$
$= 17.658\,\text{kN/m run}$

$M = 0.5 \times 1.5\,\text{kN/m}^2 \times 3\,\text{m} \times (^2/_3(3\,\text{m}))$
$= 4.5\,\text{kNm/m run}$

$$\sigma_{\text{TENSION}} = \frac{17.658}{0.25} - \frac{4.5}{0.0104} = -0.361\ \text{N/mm}^2$$

$$\sigma_{\text{COMPRESSION}} = \frac{17.658}{0.25} + \frac{4.5}{0.0104} = 0.502\ \text{N/mm}^2$$

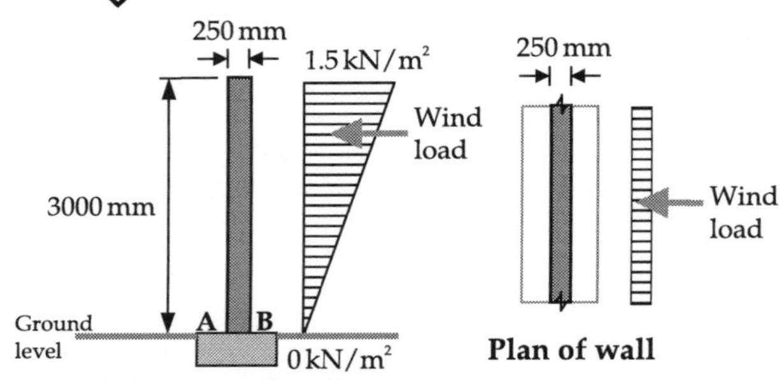

Question 1

2. $A = 1\,\text{m} \times X = X\,\text{m}^2$

$$Z = \frac{bd^2}{6} = \frac{1 \times X^2}{6} = 0.167X^2\,\text{m}^2$$

$F = W_1 + W_2 = 240X + 1200$

M = overturning moment – restoring moment

(moments about Y–Y axis at base)

$= -W_1 \times (^X/_2 - (^2/_3(X-5))) + W_2 \times (^X/_2 - 2.5)$
$- W_3 \times 6.67$

$= 40X^2 + 1800X - 23\,080$

$$\sigma_{\text{TENSION}} = 0 = \frac{240X + 1200}{X} - \frac{(40X^2 + 1800X - 23\,080)}{0.167X^2}$$

From which $X = 14.43$ m

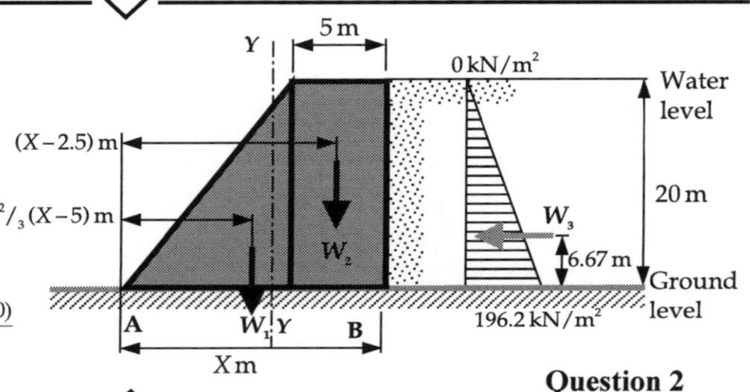

Question 2

3.

$$\text{Factor of safety} = \frac{\text{restoring moment}}{\text{overturning moment}}$$

Moments about A:

Restoring moment:

$RM = W_1 \times 0.75\,\text{m} + W_2 \times 1.25\,\text{m}$
$= 30.6\,\text{kN m/m run}$

Overturning moment:

$OM = W_3 \times 0.6\,\text{m} = 29.16\,\text{kN m/m run}$

$$\textbf{Factor of safety} = \frac{30.6}{29.16} = 1.05$$

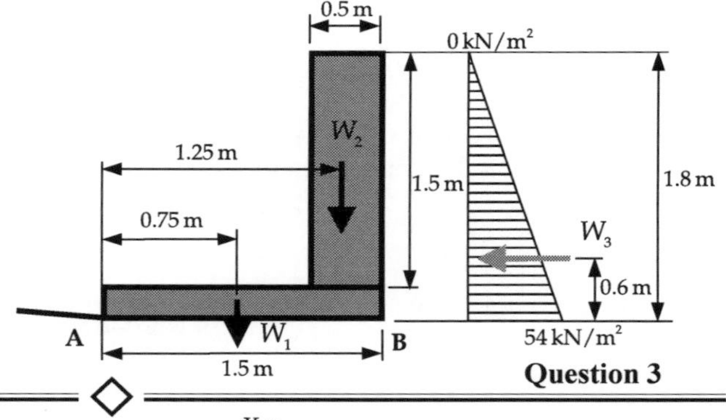

Question 3

4. $A = 1\,\text{m} \times X = X\ \text{m}^2$

$$Z = \frac{bd^2}{6} = \frac{1 \times X^2}{6} = 0.167X^2\ \text{m}^2$$

$F = W_1 = 24\,\text{kN/m}^3 \times 4.5\,\text{m} \times X\,\text{m}$
$= 108X$

Moments about A:

$M = (W_1 \times {}^X/_2) + (40\,\text{kN/m} \times 3\,\text{m})$
$- (W_2 \times 1.5\,\text{m}) = 54X^2 - 48.75$

$$\sigma = 0 = \frac{108X}{X} - \frac{(54X^2 - 48.75)}{0.167X^2}$$

From which $X = 0.754$ m

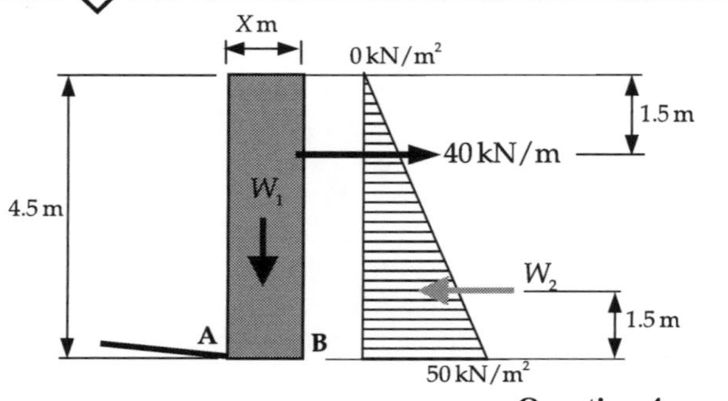

Question 4

Index